Methods in Enzymology

Volume 414
MEASURING BIOLOGICAL RESPONSES WITH AUTOMATED MICROSCOPY

METHODS IN ENZYMOLOGY

EDITORS-IN-CHIEF

John N. Abelson   Melvin I. Simon

DIVISION OF BIOLOGY
CALIFORNIA INSTITUTE OF TECHNOLOGY
PASADENA, CALIFORNIA

FOUNDING EDITORS

Sidney P. Colowick and Nathan O. Kaplan

Methods in Enzymology

Volume 414

Measuring Biological Responses with Automated Microscopy

EDITED BY

James Inglese

NIH CHEMICAL GENOMICS CENTER
NATIONAL INSTITUTES OF HEALTH
BETHESDA, MARYLAND

AMSTERDAM • BOSTON • HEIDELBERG • LONDON
NEW YORK • OXFORD • PARIS • SAN DIEGO
SAN FRANCISCO • SINGAPORE • SYDNEY • TOKYO
Academic Press is an imprint of Elsevier

ELSEVIER

Academic Press is an imprint of Elsevier
525 B Street, Suite 1900, San Diego, California 92101-4495, USA
84 Theobald's Road, London WC1X 8RR, UK

This book is printed on acid-free paper. ∞

For information on all Elsevier Academic Press publications
visit our Web site at www.books.elsevier.com

ISBN-13: 978-0-12-182819-6
ISBN-10: 0-12-182819-0

PRINTED IN THE UNITED STATES OF AMERICA
06 07 08 09 10 9 8 7 6 5 4 3 2 1

Table of Contents

Contributors to Volume 414

Article numbers are in parentheses following the names of contributors.
Affiliations listed are current.

CYNTHIA L. ADAMS (24), *Cytokinetics, Inc., South San Francisco, California*

LARBI AMAZIT (11), *Integrated Microscopy Core, Department of Molecular and Cellular Biology, Baylor College of Medicine, Houston, Texas*

DOUGLAS S. AULD (29), *NIH Chemical Genomics Center, National Institutes of Health, Bethesda, Maryland*

NATHALIE AULNER (15), *The Judith P. Sulzberger, MD Columbia Genome Center, Department of Physiology and Cellular Biophysics, Columbia University, New York, New York*

CHRISTOPHER P. AUSTIN (29), *NIH Chemical Genomics Center, National Institutes of Health, Bethesda, Maryland*

AUDREY BAKER (23), *Scynexis, Inc., Research Triangle Park, North Carolina*

MATTHEW BEARD (15), *The Judith P. Sulzberger, MD Columbia Genome Center, Department of Physiology and Cellular Biophysics, Columbia University, New York, New York*

GEOFFREY BARGER (15), *The Judith P. Sulzberger, MD Columbia Genome Center, Department of Physiology and Cellular Biophysics, Columbia University, New York, New York*

VALERIA BERNO (11), *Integrated Microscopy Core, Department of Molecular and Cellular Biology, Baylor College of Medicine, Houston, Texas*

MALENE BERTELSEN (20), *Department of Biological Sciences, Astra Zeneca R&D, Lund, Sweden*

JOHN A. BEUTLER (2), *Molecular Targets Development Program, National Cancer Institute at Frederick, National Institutes of Health, Frederick, Maryland*

STÉPHANE BONNEAU (12), *Laboratoire Kastler Brossel, Ecole normale supérieure and Université Pierre et Marie Curie, Paris, France*

KRISTEN M. BORCHERT (9), *Becton-Dickinson Technologies, Research Triangle Park, North Carolina*

CÉDRIC BOUZIGUES (12), *Laboratoire Kastler Brossel, Ecole normale supérieure and Université Pierre et Marie Curie, Paris, France*

LARS BRANDÉN (15), *The Judith P. Sulzberger, MD Columbia Genome Center, Department of Physiology and Cellular Biophysics, Columbia University, New York, New York*

GARY BROOKER (18), *Department of Biology, Integrated Imaging Center, Johns Hopkins University, Montgomery County Campus, Rockville, Maryland*

PAUL J. BUSHWAY (17), *Burnham Institute for Medical Research, La Jolla, California*

SCOTT CALLAWAY (10), *Vala Sciences Inc., La Jolla, California*

CAROLINA CEBALLOS (10), *The Whitney Laboratory for Marine Bioscience, University of Florida, Department of Neuroscience, College of Medicine, St. Augustine, Florida*

SUMIT K. CHANDA (28), *Genomics Institute of the Novartis Research Foundation, San Diego, California*

WILLIAM CIESLIK (33), *Hamamatsu Photonic Systems, Bridgewater, New Jersey*

DANIEL A. COLEMAN (24), *Cytokinetics, Inc., South San Francisco, California*

GE CONG (24), *Cytokinetics, Inc., South San Francisco, California*

SÉBASTIEN COURTY (12), *Laboratoire Kastler Brossel, Ecole normale supérieure and Université Pierre et Marie Curie, Paris, France*

ANNE MOON CROMPTON (24), *Cytokinetics, Inc., South San Francisco, California*

MAXIME DAHAN (12), *Laboratoire Kastler Brossel, Ecole normale supérieure and Université Pierre et Marie Curie, Paris, France*

MARIA A. DEBERNARDI (18), *Department of Biology, Integrated Imaging Center, Johns Hopkins University, Montgomery County Campus, Rockville, Maryland*

ANGIE B. DULL (2), *Basic Research Program, SAIC-Frederick, Inc., Molecular Targets Development Program, National Cancer Institute at Frederick, National Institutes of Health, Frederick, Maryland*

MARIE-VIRGINIE EHRENSPERGER (12), *Laboratoire Kastler Brossel, Ecole normale supérieure and Université Pierre et Marie Curie, Paris, France*

KATHLEEN A. ELIAS (24), *Cytokinetics, Inc., South San Francisco, California*

JEFFREY T. FINER (26), *Cytokinetics, Inc., South San Francisco, California*

MASANOBU FUJIWARA (33), *Hamamatsu Photonic Systems, Bridgewater, New Jersey*

FRANK ULRICH GANDENBERGER (30), *Drug Discovery Technologies, Evotec Technologies GmbH, Hamburg, Germany*

RALPH J. GARIPPA (7), *Roche Inc., Roche Discovery Technologies, Nutley, New Jersey*

BENJAMIN GEIGER (13), *Department of Molecular Cell Biology, The Weizmann Institute of Science, Rehovot, Israel*

KENNETH A. GIULIANO (31), *Cellumen, Inc., Pittsburgh, Pennsylvania*

ALBERT GOUGH (31), *Cellumen, Inc., Pittsburgh, Pennsylvania*

GABRIELE GRADL (7), *Evotec Technologies GmbH, Cell Handling and Analysis Department, Berlin, Germany*

DOROTHEA HAASEN (8), *Boehringer Ingelheim Pharma GmbH & Co. KG, Biberach, Germany*

GORDON L. HAGER (2, 3), *Laboratory of Receptor Biology and Gene Expression, Hormone Action and Oncogenesis Section, National Cancer Institute, National Institutes of Health, Bethesda, Maryland*

KLAUS HAHN (32), *Department of Pharmacology, Lineberger Cancer Center, University of North Carolina, Chapel Hill, North Carolina*

LAURA V. HALE (9), *Lilly Research Laboratories, Eli Lilly & Co, Indianapolis, Indiana*

DAVID L. HALLADAY (9), *Lilly Research Laboratories, Eli Lilly & Co, Indianapolis, Indiana*

RICHARD L. HANSEN (26), *Cytokinetics, Inc., South San Francisco, California*

RALF HEILKER (8), *Boehringer Ingelheim Pharma GmbH & Co. KG, Biberach, Germany*

ARNE HEYDORN (27), *Fisher BioImage ApS, Søborg, Denmark*

CRUZ A. HINOJOS (11), *Integrated Microscopy Core, Department of Molecular and Cellular Biology, Baylor College of Medicine, Houston, Texas*

LOUIS HODGSON (32), *Department of Pharmacology, Lineberger Cancer Center, University of North Carolina, Chapel Hill, North Carolina*

ANN F. HOFFMAN (7), *Roche Inc., Roche Discovery Technologies, Nutley, New Jersey*

KEITH A. HOUCK (9), *National Center for Computational Taxicology (D343-03), Office of Research and Development, U.S. Environmental Protection Agency, Research Triangle Park, North Carolina*

BONNIE J. HOWELL (6, 16, 25), *Department of Cancer Research, Merck Research Laboratories, West Point, Pennsylvania*

KUO-SEN HUANG (30), *Research Discovery Technologies, Hoffmann-La Roche, Inc., Nutley, New Jersy*

CHRISTINE C. HUDSON (4, 5), *Xsira Pharmaceuticals, North Carolina*

MITCHELL V. HULL (28), *Genomics Institute of the Novartis Research Foundation, San Diego, California*

EDWARD HUNTER (10), *Vala Sciences Inc., La Jolla, California*

JAMES INGLESE (29), *NIH Chemical Genomics Center, National Institutes of Health, Bethesda, Maryland*

RAY ISMAIL (14), *GE Healthcare, The Maynard Centre, Forest Farm, Whitchurch, Cardiff, United Kingdom*

ANDREW D. JACK (19), *Independent Statistical Consultant, Berkshire, United Kingdom*

AJIT JADHAV (29), *NIH Chemical Genomics Center, National Institutes of Health, Bethesda, Maryland*

BERND JAGLA (15), *The Judith P. Sulzberger, MD Columbia Genome Center, Department of Physiology and Cellular Biophysics, Columbia University, New York, New York*

RONALD L. JOHNSON (29), *NIH Chemical Genomics Center, National Institutes of Health, Bethesda, Maryland*

PATRICIA A. JOHNSTON (21, 22, 23, 31), *Sphinx RTP Laboratories, Eli Lilly and Company, Research Triangle Park, North Carolina*

PAUL A. JOHNSTON (21, 22, 23), *Sphinx RTP Laboratories, Eli Lilly and Company, Research Triangle Park, North Carolina*

ZVI KAM (13), *Department of Molecular Cell Biology, The Weizmann Institute of Science, Rehovot, Israel*

RAMANI KANDASAMY (21, 22, 23), *Senior Research Scientist, BASF Corporation, RTP, North Carolina*

PETER D. KELLY (15), *The Judith P. Sulzberger, MD Columbia Genome Center, Department of Physiology and Cellular Biophysics, Columbia University, New York, New York*

JONATHAN M. KENDALL (14), *GE Healthcare, The Maynard Centre, Forest Farm, Whitchurch Cardiff, United Kingdom*

BRIGITTE H. KEON (26), *Cytokinetics, Inc., South San Francisco, California*

ACHIM KIRSCH (7), *Evotec Technologies GmbH, Cell Handling and Analysis Department, Berlin, Germany*

PRIYA KUNAPULI (6), *Merck & Co., Inc., West Point, Pennsylvania*

VADIM KUTSYY (24), *Cytokinetics, Inc., South San Francisco, California*

CARMEN L. LAETHEM (21, 22, 23), *Sphinx RTP Laboratories, Eli Lilly and Company, Research Triangle Park, North Carolina*

IRENA LAVELIN (13), *Department of Molecular Cell Biology, The Weizmann Institute of Science, Rehovot, Israel*

SEUNGTAEK LEE (6, 16, 25), *Department of Automated Biotechnology, North Wales, Pennsylvania*

DAVID L. LENZI (26), *Cytokinetics, Inc., South San Francisco, California*

YUVALAL LIRON (13), *Department of Molecular Cell Biology, The Weizmann Institute of Science, Rehovot, Israel*

CARSON R. LOOMIS (4, 5), *Xsira Pharmaceuticals, Research Triangle Park, North Carolina*

CAMILLA LUCCARDINI (12), *Laboratoire Kastler Brossel, Ecole normale supérieure and Université Pierre et Marie Curie, Paris, France*

BETINA K. LUNDHOLT (27), *Fisher BioImage ApS, Søborg, Denmark*

MICHAEL A. MANCINI (11), *Integrated Microscopy Core, Department of Molecular and Cellular Biology, Baylor College of Medicine, Houston, Texas*

ELISABETH D. MARTINEZ (3), *Laboratory of Receptor Biology and Gene Expression, Hormone Action and Oncogenesis Section, National Cancer Institute, National Institutes, of Health, Bethesda, Maryland*

DAVID MARK (30), *Research Discovery Technologies, Hoffmann-La Roche, Inc., Nutley, New Jersy*

ELISABETH D. MARTINEZ* (2, 3, 29), *Laboratory of Receptor Biology and Gene Expression, Hormone Action and Oncogenesis Section, National Cancer Institute, National Institutes of Health, Bethesda, Maryland*

THOMAS MAYER (15), *The Judith P. Sulzberger, MD Columbia Genome Center, Department of Physiology and Cellular Biophysics, Columbia University, New York, New York*

PATRICK M. McDONOUGH (10), *Vala Sciences Inc., La Jolla, California*

MARK MERCOLA (17), *Burnham Institute for Medical Research, La Jolla, California*

IVANA MIKIC (10), *Vala Sciences Inc., La Jolla, California*

LOREN MIRAGLIA (28), *Genomics Institute of the Novartis Research Foundation, San Diego, California*

SUHA NAFFAR-ABU-AMARA (13), *Department of Molecular Cell Biology, The Weizmann Institute of Science, Rehovot, Israel*

DEBRA R. NICKISCHER (9, 21, 22, 23), *Sphinx RTP Laboratories, Eli Lilly and Company, Research Triangle Park, North Carolina*

ROBERT H. OAKLEY (4, 5), *Xsira Pharmaceuticals, North Carolina*

DONALD R. OESTREICHER (24), *Cytokinetics, Inc., South San Francisco, California*

ANTHONY P. ORTH (28), *Genomics Institute of the Novartis Research Foundation, San Diego, California*

LEN PAGLIARO (27), *Fisher BioImage ApS, Søborg, Denmark*

YAEL PARAN (13), *Department of Molecular Cell Biology, The Weizmann Institute of Science, Rehovot, Israel*

SONIA PLANEY (10), *The Whitney Laboratory for Marine Bioscience, University of Florida, Department of Neuroscience, College of Medicine, St. Augustine, Florida*

AMY PLATTS (19), *The Neuroscience Research Centre, Merck Sharp and Dohme Research Laboratories, Essex, United Kingdom*

MORTEN PRAESTEGAARD (27), *Fisher BioImage ApS, Søborg, Denmark*

JEFFREY H. PRICE (10), *Vala Sciences Inc., Burnham Institute for Medical Research, La Jolla, California*

GILLIAN R. RICHARDS (19), *The Neuroscience Research Centre, Merck Sharp and Dohme Research Laboratories, Essex, United Kingdom*

DANIEL A. RINES (28), *Genomics Institute of the Novartis Research Foundation, San Diego, California*

*Corresponding address: Department of Pharmacology, University of Texas, Southwestern Medical Center, Dallas, Texas.

JAMES E. ROTHMAN (15), *The Judith P. Sulzberger, MD Columbia Genome Center, Department of Physiology and Cellular Biophysics, Columbia University, New York, New York*

ANDREAS SCHNAPP (8), *Department of Pulmonary Research, Biberach, Germany*

RACHELLE J. SELLS GALVIN (9), *Lilly Research Laboratories, Eli Lilly & Co, Indianapolis, Indiana*

LAURA V. SEPP-LORENZINO (16), *Department of Cancer Research, Merck Research Laboratories, West Point, Pennsylvania*

TERRI SERON (10), *The Whitney Laboratory for Marine Bioscience, University of Florida, Department of Neuroscience, College of Medicine, St. Augustine, Florida*

FEIMO SHEN (32), *Department of Pharmacology, Lineberger Cancer Center, University of North Carolina, Chapel Hill, North Carolina*

ANTON SIMEONOV (29), *NIH Chemical Genomics Center, National Institutes of Health, Bethesda, Maryland*

PETER B. SIMPSON (19), *Astrazeneca, Alderley Park, Macclesfield, Cheshire, United Kingdom*

MICHAEL D. SJAASTAD (4, 5), *Molecular Devices Corporation, California; National Human Genome Research Institute, NIH, Bethesda, Maryland*

DEBORAH H. SMITH (15), *The Judith P. Sulzberger, MD Columbia Genome Center, Department of Physiology and Cellular Biophysics, Columbia University, New York, New York*

SIMON STUBBS (1), *GE Healthcare, The Maynard Centre, Forest Farm, Whitchurch, Cardiff, United Kingdom*

ADAM T. SZAFRAN (11), *Integrated Microscopy Core, Department of Molecular and Cellular Biology, Baylor College of Medicine, Houston, Texas*

D. LANSING TAYLOR (31), *Cellumen, Inc., Pittsburgh, Pennsylvania*

NICK THOMAS (1, 14), *GE Healthcare, The Maynard Centre, Forest Farm, Whitchurch, Cardiff, United Kingdom*

UDO TÖBBEN (15), *The Judith P. Sulzberger, MD Columbia Genome Center, Department of Physiology and Cellular Biophysics, Columbia University, New York, New York*

OSCAR J. TRASK, JR. (9, 21, 22, 23), *Sphinx RTP Laboratories, Eli Lilly and Company, Research Triangle Park, North Carolina*

JAY K. TRAUTMAN (24), *Cytokinetics, Inc., South San Francisco, California*

BUU TU (28), *Genomics Institute of the Novartis Research Foundation, San Diego, California*

EUGENI A. VAISBERG (24, 26), *Cytokinetics, Inc., South San Francisco, California*

MARTIN J. VALLER (8), *Boehringer Ingelheim Pharma GmbH & Co. KG, Biberach, Germany*

HENRIKE VEITH (29), *NIH Chemical Genomics Center, National Institutes of Health, Bethesda, Maryland*

BENEDIKT VON MASSENBACH (10), *The Whitney Laboratory for Marine Bioscience, University of Florida, Department of Neuroscience, College of Medicine, St. Augustine, Florida*

RACHAEL WATSON (10), *The Whitney Laboratory for Marine Bioscience, University of Florida, Department of Neuroscience, College of Medicine, St. Augustine, Florida*

GENEVIEVE L. WELCH (28), *Genomics Institute of the Novartis Research Foundation, San Diego, California*

JOHN K. WESTWICK (29), *Odyssey Thera, Inc., San Ramon, California*

RHONDA GATES WILLIAMS (21, 22, 23), *Sphinx RTP Laboratories, Eli Lilly and Company, Research Triangle Park, North Carolina*

SABINA WINOGRAD-KATZ (13), *Department of Molecular Cell Biology, Weizmann Institute of Science, Rehovot, Israel*

MICHAEL R. WYLER (15), *The Judith P. Sulzberger, MD Columbia Genome Center, Department of Physiology and Cellular Biophysics, Columbia University, New York, New York*

ADAM YASGAR (29), *NIH Chemical Genomics Center, National Institutes of Health, Bethesda, Maryland*

DAVID ZACHARIAS (10), *The Whitney Laboratory for Marine Bioscience, University of Florida, Department of Neuroscience, College of Medicine, St. Augustine, Florida*

JUN ZHANG (10), *The Whitney Laboratory for Marine Bioscience, University of Florida, Departments of Neuroscience, College of Medicine, St. Augustine, Florida*

JIA ZHANG (28), *Genomics Institute of the Novartis Research Foundation, San Diego, California*

YA-QIN ZHANG (29), *NIH Chemical Genomics Center, National Institutes of Health, Bethesda, Maryland*

WEI ZHENG (29), *NIH Chemical Genomics Center, National Institutes of Health, Bethesda, Maryland*

Preface

Phenotypic cell-based assays provide a means to quantify specific biological processes, such as protein trafficking and cytokinesis. The assays, used for years in basic research, are becoming more sophisticated and are increasingly applied in drug discovery. Fluorescence microscopy is the primary mode of phenotypic analysis. For example, microscopy-based "imaging" of biological responses measured through the localization or expression of cellular proteins is most often achieved with immunofluorescent probes, genetically encoded fluorescent protein fusions, and fluorescent dyes.

Here we describe the use of automated fluoresence microscopy in the analysis of a wide cross-section of cellular biology, including: (chapter), the cell cycle (1, 16, 24), enzymes controlling epigenetic events (2, 29), translocation of ligand-dependent transcription factors exemplified with the steroid hormone nuclear receptors (3, 11, 14, 29), translocation of ligand-independent transcription factors, β-catenin in primary human preosteoblasts (9), NFAT measured through adenovirus delivery of an NFAT-nitroreductase reporter gene (14), NFκB p65 in TNF-α-induced HUVEC (15), NFκB p65 and c-Jun in IL-1α stimulated HeLa cells (20) and FKHR in response to PI3 kinase signaling (27), cell motility (13), lipid modification of proteins by palmitic acid (10), GABA A receptor trafficking in cultured neurons visualized with quantum dots (12), focal adhesion complex remodeling via labeled paxillin (13), E-selectin and VCAM-1 expression in TNF-α-induced HUVEC (15), apoptosis (16), embryonic stem cell cardiogenesis (17), intracellular calcium dynamics in drug-sensitive and -insensitive human breast cancer MCF-7 cell lines (18) and heterogeneous primary neuronal cultures (19), mitogen-activated protein kinase signaling (20, 21–23), cellular morphology (24), the p53:hdm2 protein-protein interaction (27), ATPases (30), and G protein-coupled receptor signaling (4–8, 25, 29), a major therapeutic target class. Additionally, chapter 13 describes the use of a cDNA expression library to facilitate the identification of structural or organelle-specific proteins based on localization of the expressed proteins as fusions to YFP.

Development of automated light microscopes capable of rapidly imaging cell populations in 96-, 384- and 1536-well plates enables the aforementioned breadth of biological assays for screening of libraries of small molecules (13, 23, 29) and nucleic (13–15, 27, 28) acids. An example of a chemical library screen is described in Chapter 23. Accompanied by Chapters 21 and 22, this segment outlines the development of a high throughput screening (HTS) cellular assay to identify p38 MAPK inhibitors, deployment of the assay in a 32,000 compound screen, and the selectivity profiling of candidate actives identified from the screen. Defining a compound's specificity through classification based on cellular phenotype is becoming an increasingly important application for automated microscopy. In Chapter 24, a multivariate approach using a cell-line panel categorizes compounds having different mechanisms of action, illustrating the power of cellular phenotype to distinguish the potential "off-target" effects of new substances.

Recently available genome-scale cDNA and siRNA libraries (see Chapter 28, Table I) are complemented by current advances in automated high-throughput microscopy and quantitative image analysis. As with small molecule library handling, storage, and delivery to cell cultures, genomic libraries have their own unique requirements. Nucleic acid-based libraries are made to enter mammalian cells by transfection or viral delivery, and the advantages and limitations of each are discussed in several chapters of this volume (13–15, 27, 28). Methods describing the preparation, handling, and arraying in microtiter plate format of plasmid cDNA or shRNA libraries and high-throughput transfection protocols (28), the use of pooled RNA duplexes (13), and strategies for the assessment of transfection efficiency variation in the context of HTS (27, 28) are also explored.

The microscope has undergone a remarkable evolution since Hans and Zacharias Janssen made their first compound microscope in Holland *ca.* 1590. Likewise, developments in image acquisition and analysis have made equally great strides from the first sketches of "cells" within slices of cork published by Robert Hooke in 1665. The systems described throughout this volume represent the current state-of-the-art custom and commercial automated microscope-based imagers and software. Differences between platforms include the nature of sample illumination, light collection, field of view, autofocus mechanism (the subject of Chapter 32), throughput, and analysis software, to name a few; the merits of each depend on the application (25). For example, in the high-throughput screening of large chemical libraries, the need for sample throughput may outweigh the wide range of excitation and emission wavelength options required in the detailed kinetic investigation of multiple subcellular fluorescent markers for a more limited number of samples.

The instruments described in this volume include the INCell 1000 and 3000 (1, 4–6, 8, 17, 27), Discovery-1 (2, 3, 26), Cell Lab IC100 Image Cytometer (3, 10, 11), Opera (7), ArrayScan HCS (9, 20–23), Pathway BioImager (18, 19), several custom platforms based on the Olympus Model IX71 (12, 13, 32) and Model IX81 inverted microscopes (13), and the inverted Zeiss Axiovert 100M epifluorescent microscope (24). In addition, a cytometry-based cellular imaging instrument, the Explorer, containing no microscope optics and designed for cellular assays where moderate resolution of cellular features is acceptable, but high sample throughput is required, is described in Chapter 29. Another detector, the Plate::Vision multimodal reader, illustrates how microscopy can be used to enhance the screening of non-cellular assays. Here an array of 96-minilenses creates a 'quasiconfocal detection zone' that samples a ~10 nL volume within the well of a microtiter plate and can enable very high screening throughput (30). The refinement of components for automated microscopy is the subject of the final chapters. A "systems cell biology" approach aims to correlate multiple cellular biomarkers from relevant cell types to develop predictive tools for drug discovery in Chapter 31. On the hardware side, a digital autofocus solution to accompany high numerical aperture oil immersion lenses is described in Chapter 32, and a review of the fluorescence lifetime imaging microscope, enabled by the streak camera (33), illustrate technologies that will enable the next generation automated microscopes.

Regardless of the specific microscopy platform, investigators operating with large sample populations will benefit from an integration of sample preparation, imaging, data collection, processing, and storage. Chapters 2, 5, 6, 7, 13, 26, and 28 describe in varying detail the approaches taken by several laboratories to prepare a scaleable screening infrastructure.

Currently the applications of automated light microscopy are bounded only by our imagination and the fundamental physical properties of light. The latter limitation is surmountable by techniques that allow microscopic inspection of individual molecules and atoms, such as atomic force and electron microscopy. Applying these technologies in a manner described here for light microscopy will be a topic for future volumes in this series.

I would like to thank the authors of this volume for their contributions, and the publisher for providing the excellent logistic and project management assistance of Cindy Minor and Jamey Stegmaier.

JAMES INGLESE

METHODS IN ENZYMOLOGY

[1] Dynamic Green Fluorescent Protein Sensors for High-Content Analysis of the Cell Cycle

By SIMON STUBBS and NICK THOMAS

Abstract

We have developed two dynamic sensors that report cell cycle position in living mammalian cells. The sensors use well-characterized components from proteins that are spatially and temporally regulated through the cell cycle. Coupling of these components to Enhanced Green Fluorescent Protein (EGFP) has been used to engineer fusion proteins that report G1/S and G2/M transitions during the cell cycle without perturbing cell cycle progression. Expression of these sensors in stable cell lines allows high content analysis of the effects of drugs and gene knockdown on the cell cycle using automated image analysis to determine cell cycle position and to abstract correlative data from multiplexed sensors and morphological analysis.

Introduction

The cell cycle is one of the most fundamental and complex processes in biology and influences the development, life, and, in cases where it goes wrong, the death of all eukaryotes. Progression of a cell through four phases, G1, S, G2, and M, is exquisitely regulated by a series of checks and balances to ensure that DNA is correctly maintained, replicated, and segregated into daughter cells at division.

The cell cycle and its control mechanisms have been studied extensively for over a century (Nurse, 2000), and significant advances have been made in identifying and describing the role of components and their regulatory mechanisms. Understanding of the cell cycle has progressed to a point where the basic circuitry can be mapped (Kohn, 1999), and ongoing research continues to refine our understanding of the interplay between the players in this game of life (Aleem *et al.*, 2005; Cobrinik, 2005; Fu *et al.*, 2004; Sherr and Roberts, 2004). Much work remains, and study of the basic molecular mechanisms followed by integration of these mechanisms into a full understanding of cell cycle control and regulation will engage researchers for many years to come (Nurse, 2000a).

The aim of these studies has been twofold: to understand cell cycle control as a key biological process and to gain a greater understanding of the process as an aid in the selection of targets and development of effective drugs for use against the aberrant cell cycle in oncology

METHODS IN ENZYMOLOGY, VOL. 414   0076-6879/06 $35.00
          DOI: 10.1016/S0076-6879(06)14001-X

(Carnero, 2002; Gillessen *et al.*, 2002; Hamel and Covell, 2002; Sampath and Plunkett, 2001; Stewart *et al.*, 2003; Vermeulen *et al.*, 2003) and other fields, including cardiovascular (Bicknell *et al.*, 2003), neurological (Arendt, 2002) and hepatic (Horie *et al.*, 2003) disease, stroke (O'Hare *et al.*, 2002), and HIV infection (Galati *et al.*, 2002).

Cancer is characterized by deregulated cell cycle control. In contrast to normal cells proliferating only in response to developmental or other mitogenic signals, tumor cell proliferation proceeds automatously. The cell cycle in a cancer cell is not necessarily different from that of a normal cycling cell, but in the cancer cell the accelerator and braking mechanisms that normally control cell cycle progression to give a cell time to repair damaged DNA and to respond to mitogenic stimuli or differentiative inhibition have become decoupled from the cell cycle engine.

In oncology, opportunities for target identification and drug development exist at both G1/S and G2/M transitions as intervention points to target the cell cycle in tumor cells. The G2/M transition has traditionally been the focus of attention, resulting in the development of drugs such as taxol (Hadfield *et al.*, 2003), which interfere with the mechanics of cell division. In more recent programs, new G2/M cell cycle control targets, such as histone deacetylase (Wong *et al.*, 2005), aurora kinases (Mortlock *et al.*, 2005), Polo-like kinases (PLK) (Blagden and de Bono, 2005), and the G2 checkpoint proteins CHK1 (Li and Zhu, 2002) and CHK2 (Kawabe, 2004), continue to be evaluated and targeted. At the G1/S transition, targets involved in DNA repair and replication, including DNA helicases (Sharma *et al.*, 2005), poly(ADP-ribose) polymerase (Jagtap and Szabo, 2005), topoisomerases (Pommier *et al.*, 2003), and the MCM complex (Lei, 2005), together with targets controlling G1–S progression, including CDKs (Owa *et al.*, 2001), mTOR (Dutcher, 2004), and pRB/E2F (Seville *et al.*, 2005), offer alternative approaches to selectively targeting the aberrant cell cycle in cancer cells.

Despite advances in knowledge of the mechanics of the process, the techniques routinely used to study the cell cycle and related events have remained essentially unchanged for many years. Proliferation assays that measure cell numbers (Denizot and Lang, 1986) or radiolabeled thymidine (Graves *et al.*, 1997) incorporation can be used to obtain a relatively crude population averaged response to an experimental condition, that is, whether cell growth is stimulated or inhibited, but in asynchronous cells these methods cannot give any indication of which part of the cell cycle is being affected. Higher-resolution methods, including flow cytometry (Smith *et al.*, 2000), bromodeoxyuridine incorporation (Humbert *et al.*, 1990), and other immunofluorescence techniques (Yuan *et al.*, 2002), allow analysis of cell cycle status at the individual cell level and can be used to determine the distribution of a population of cells around the cell cycle. Despite their widespread use, all of the aforementioned methods lack the ability to provide a fully dynamic description of the cell cycle in individual cells.

Recent developments in understanding of the control mechanisms of the cell cycle and methods for engineering genetically encoded fluorescent sensors (Tsien, 1998; Zhang *et al.*, 2002) have made it possible to design novel cell cycle sensors based on key cell cycle control molecules. When coupled with advances in high-throughput cellular analysis instruments (Lundholt *et al.*, 2005; Ramm and Thomas, 2003), these sensors enable high-content analysis of the cell cycle and a high-definition view of the effects of candidate drugs.

Design and Construction of Dynamic Cell Cycle Sensors

Application of fluorescent proteins to cell cycle analysis has enabled significant advances to be made in understanding the timing of molecular events that control the cell cycle. While green fluorescent protein (GFP) fusions with key cell cycle control proteins (Arnaud *et al.*, 1998; Huang and Raff, 1999; Raff *et al.*, 2002; Weingartner *et al.*, 2001; Zeng *et al.*, 2000) and other proteins (Kanda *et al.*, 1998; Reits *et al.*, 1997; Tatebe *et al.*, 2001) have provided very significant insights into the molecular mechanics of the cell cycle, expression of cell cycle protein fusions that retain enzyme or structural activity have the potential to perturb the cell cycle and are therefore not suitable as cell cycle sensors (Clute and Pines, 1999).

To provide nonperturbing stealth cell cycle sensors we have developed constructs (Fig. 1) based on the fusion of EGFP to domains isolated from well-characterized cell cycle control and response proteins. The first of these sensors reports on the G1/S transition and the second reports on the G2/M transition.

The G1/S cell cycle phase marker (Fig. 1A and B) is derived from the human homologue of helicase B (HELB), a protein that is essential for G1/S transition (Taneja *et al.*, 2002). HELB has been demonstrated to be localized at nuclear foci induced by DNA damage (Gu *et al.*, 2004), where the protein operates during G1 to process endogenous DNA damage prior to the G1/S transition. Consistent with the proposed action of HELB, the protein resides in the nucleus during G1 but is predominantly cytoplasmic in S and G2 phase cells. Cell cycle-coordinated nuclear and cytoplasmic residence is controlled by a 131 amino acid C-terminal phosphorylation-dependent subcellular localization control domain (PSLD) containing a nuclear localization sequence that retains the protein in the nucleus in G1 . In late G1 phase, serine residues within the PSLD become phosphorylated by increasing levels of active cyclin E/ Cdk2, resulting in unmasking of a nuclear export sequence leading to protein export to the cytoplasm.

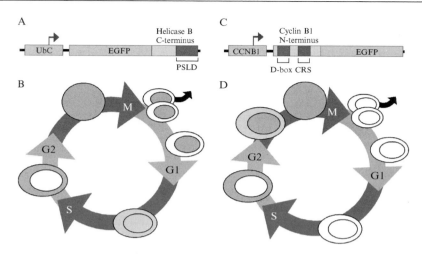

FIG. 1. EGFP cell cycle phase markers. Constitutive expression of the G1/S sensor (A) is achieved via a ubiquitin C (UbC) promoter driving production of a fusion protein between EGFP and the C-terminal region of human DNA helicase B containing a phosphorylation and subcellular localization domain (PSLD). The fluorescent fusion protein localizes to the nucleus in G1 cells (B) and undergoes translocation to the cytoplasm as cells progress through S phase and into G2. Expression of the G2/M sensor (C) is controlled by the cyclin B1 (CCNB1) promoter, which initiates production of the cyclin B1–EGFP fusion protein in late S phase. As cells progress through late S phase into G2, fluorescence increases in the cytoplasm (D) until phosphorylation of the cytoplasmic retention sequence (CRS) causes translocation of the sensor to the nucleus at prophase. At anaphase the sensor is degraded rapidly mediated by the cyclin B1 destruction box (D-box) producing two nonfluorescent daughter cells following mitosis.

The G1/S cell cycle phase marker is a fusion of the PSLD domain of HELB to EGFP, with expression under the control of the human ubiquitin C promoter. The sensor exhibits subcellular localization changes that mimic those of HELB (Figs. 1B and 2B), but does not interfere with cell cycle progression, as the fusion protein lacks the enzymatic and structural domains of the parent protein.

The G2/M cell cycle phase marker (Fig. 1C and D) utilizes functional elements from cyclin B1. Cyclin B1 expression and destruction (Pines, 1999) are tightly regulated and act as a major cell cycle control switch that can be applied to engineering a sensor suitable for following the transition from S phase through G2 into mitosis. The sensor (Thomas, 2003; Thomas and Goodyer, 2003) comprises a fusion of amino acids 1–170 from the amino terminus of cyclin B1 coupled to EGFP, with expression under the control of the cyclin B1 promoter (Hwang et al., 1995). The sensor is switched on in late S phase (Fig. 1D), switched off during mitosis by the destruction box (D-box) (Clute and Pines, 1999), and, in the intervening period, translocates from the cytoplasm to the nucleus at prophase, regulated by the cytoplasmic retention signal (Hagting *et al.*, 1999). The fusion protein is consequently expressed and degraded in concert with endogenous cyclin B1. Because the fusion protein lacks the C-terminal sequences used in the cyclin B1–CDK interaction, it does not interfere with cell cycle progression as reported previously for a full cyclin B1–GFP fusion protein (Takizawa and Morgan, 2000).

To ensure minimal perturbation of the cell cycle, stable U-2 OS cell lines expressing the G1/S and G2/M sensors were derived through rigorous screening of a large number of clones and selection of single clones that demonstrate minimal levels of EGFP expression compatible with determination of cell cycle status by microscopy and image analysis.

Time-lapse imaging (Fig. 2A) reveals the dynamic behavior of the EGFP sensors through the cell cycle. Both sensors exhibit multiphasic characteristics for EGFP intensity and/or localization (Fig. 2B). The G2/M sensor shows a dramatic drop in intensity following mitosis (0–0.5 h), maintains low fluorescence through G1 (0.5–3 h), and a slow increase in intensity through S phase (3–15 h) followed by a stepped increase in sensor expression through G2 (15–21 h). As cells complete the cell cycle the sensor increases rapidly in intensity as cells pass through prophase (21–22 h) with intensity reaching a maximum at mitosis (23.5 h). The G1/S sensor distribution varies in an opposed fashion. Following mitosis (1.5 h) the sensor is almost entirely restricted to the nucleus and during G1 (1.5–5.5 h) undergoes export to the cytoplasm, resulting in equal sensor distribution between the nucleus and the cytoplasm. Nuclear export continues at a slower rate through to completion of S phase (5.5–14 h), at which time the sensor is predominantly cytoplasmic. Progression through G2 (14–23 h) is accompanied by a slow increase in cytoplasmic intensity as the residual sensor is cleared from the nucleus, followed by a rapid increase in the nuclear/cytoplasmic distribution ratio at mitosis (24 h).

Validation of Sensors

Both G1/S and G2/M sensors have been validated extensively by testing with known cell cycle inhibitors. Analysis of colchicine-induced mitotic blockage in G2/M sensor expressing cells (Fig. 3) confirmed that the sensor reported cell cycle transition and blockage accurately and that quantitative data could be extracted from images acquired by high-throughput microscopy. Treatment of U-2 OS cells expressing the G1/S sensor with a range of cell cycle inhibitors confirmed the cell cycle-related subcellular distribution of the G1/S sensor (Fig. 4) with strong nuclear localization of the sensor in

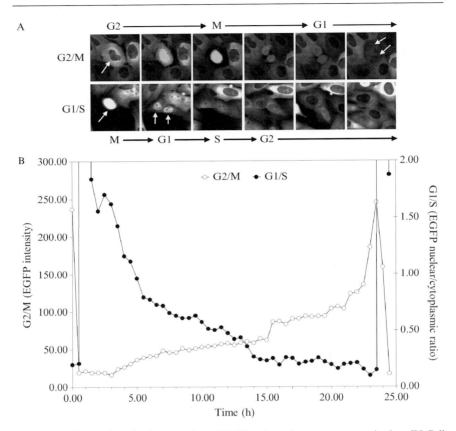

FIG. 2. EGFP cell cycle phase markers. (A) Time-lapse images were acquired on IN Cell Analyzer 3000 of U-2 OS cells stably expressing G2/M (top) and G1/S (bottom) EGFP sensors, and typical cells transitioning the key reporting periods for each cell line are shown. In the G2/M series the central cell in the first frame (arrowed) is in G2 and transitions via prophase (EGFP in nucleus) through mitosis and cytokinesis with destruction of the sensor producing two daughter cells (arrowed) in the final frame with minimal EGFP fluorescence. In the G1/S series the central cell in the first frame (arrowed) is in M and divides to produce two daughter cells with brightly fluorescent nuclei (second frame arrowed), which both transition through S and into G2 with an associated movement of the EGFP sensor from the nucleus to the cytoplasm. (B) Cells from time-lapse images were analyzed for EGFP intensity and subcellular distribution. Typical traces for G1/S (EGFP nuclear/cytoplasmic ratio) and a G2/M (EGFP intensity) sensor expressing cells are shown, each trace starts at mitosis, tracks the sensor output through the cell cycle in a single daughter cell, and finishes at mitosis.

the presence of the G1 blockers roscovitin and olomoucine and distinct cytoplasmic localization in cells exposed to the G2 blockers nocodazole, colcemid, and paclitaxel.

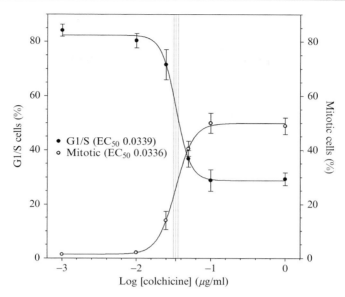

FIG. 3. Cell cycle analysis. G2/M CCPM expressing cells were incubated with increasing concentrations of colchicine and imaged on IN Cell Analyzer 3000. Cells were analyzed for EGFP intensity and distribution using automated image analysis and designated as G1/S, G2, prophase, or mitotic. Dose–response curves are shown for G1/S and M populations with associated EC_{50} derived from the two curves. ($EC_{50} \pm 1$ SD values are shown on the curves by vertical dotted lines).

Coanalysis of the G1/S sensor with BrdU incorporation (Fig. 5) confirmed the indication from time-lapse images (Fig. 2A) that cytoplasmic relocation of this sensor occurs prior to initiation of DNA replication, with BrdU incorporation confined to cells with EGFP nuclear/cytoplasmic ratios indicative of equivalent localization of the sensor in nucleus and cytoplasm following nuclear export of the sensor. Specific changes in the subcellular distribution of the sensor during the cell cycle were further corroborated through comparison with cellular DNA content measured by Hoechst staining. Scatter plot analysis (Fig. 6) of these data shows a U-shaped distribution with cells with a $2n$ genomic DNA complement having a high nuclear to cytoplasmic ratio for the G1S sensor. As cells progress into S phase and DNA content increases, the EGFP nuclear/cytoplasmic ratio decreases, reaching a minimum in cells in G2 ($4n$ DNA). Post-G2, the EGFP ratio increases as cells enter mitosis and have a high nuclear to cytoplasmic ratio for the G1/S sensor due to cell rounding.

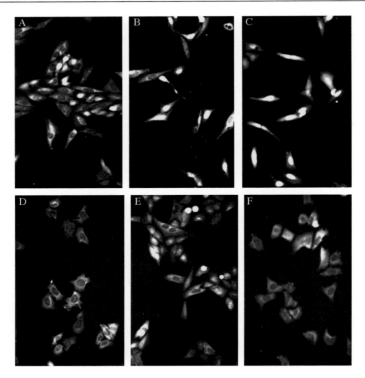

FIG. 4. Cell cycle phase-specific chemical arrest in U-2 OS cells expressing the G1S sensor. Cells were untreated (A) or treated for 24 h with roscovitin (B), olomoucine (C), nocodazole (D), colcemid (E), and paclitaxel (F). Fluorescent, fixed-cell images were acquired on IN Cell Analyzer 1000 using EGFP excitation and emission filters. Cell cycle blocks were confirmed using propidium iodide staining and flow cytometry (data not shown).

In engineering any dynamic sensor it is a critical design requirement that the sensor does not perturb the process it is designed to measure. To evaluate the effect of sensor expression on the cell cycle we compared cell cycle duration and phase distribution by cell proliferation assays and flow cytometry in sensor expressing and parental cells. Cells expressing the G2/M sensor at levels equivalent to endogenous cyclin B1 (7000 copies/cell in G2) (Thomas *et al.*, 2005) showed identical doubling times and cell cycle distributions. Microarray analysis showed no significant differences in expression of cyclins, CDKs, or CDK inhibitors between the two cell types. Similarly, analysis of G1/S sensor expressing cells showed that cell cycle duration and distribution were not affected by sensor expression (data not shown).

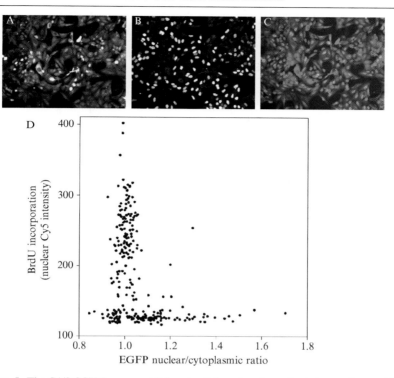

FIG. 5. The G1/S CCPM sensor exhibits subcellular relocation prior to bromodeoxyuridine (BrdU) incorporation. U-2 OS cells stably expressing the G1/S sensor were incubated with BrdU for 1 h and fixed in 4% formaldehyde and imaged on IN Cell Analyzer 1000. (A) EGFP distribution, (B) nuclear BrdU incorporation detected with mouse anti-BrdU/DNAase and Cy5 antimouse antibodies (cell proliferation fluorescence assay, GE Healthcare), and (C) combined false-color image. Images were analyzed with the IN Cell 1000 Analyzer Morphology Analysis Module (GE Healthcare) and to quantify nuclear and cytoplasmic EGFP intensity and BrdU incorporation (D) in each cell. Cells with a predominantly nuclear distribution of the G1S sensor (high Nuc/Cyt ratio) do not exhibit BrdU incorporation, confirming that sensor export from the nucleus occurs before S phase.

Cell Cycle Analysis by Automated High-Throughput Imaging for Drug Profiling and Target Validation

One of the most powerful aspects of high content analysis is the ability to measure dependencies between cellular processes, an area of investigation precluded by population-averaged analyses. The ability to use dynamic sensors for multiplexed analysis of cell cycle status with other probes or with morphological measurements in live or fixed cells now allows study of the interrelationships between cellular events and the cell cycle.

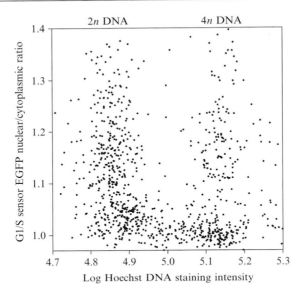

Fig. 6. Sensor localization correlates with cell cycle changes in cellular DNA content. U-2 OS cells expressing the G1/S sensor were fixed and DNA quantitatively stained with Hoechst 33342. Cells were imaged on IN Cell Analyzer 1000 and analyzed for EGFP distribution and DNA content.

Applying these sensors in multiplexed assays using coanalysis of the G2/M or G1/S sensors with measurement of BrdU incorporation and DNA content provide methods to characterize complex drug-induced effects. Figure 7 shows single cell three-parameter data analysis for cells expressing the G1/S sensor incubated in the presence and absence of a putative cell cycle inhibitor (compound A). Data from control cells showed the classical inverted horseshoe pattern typical of flow cytometry analysis of BrdU incorporation versus DNA content (Fig. 7A), with EGFP distribution data providing confirmatory information on cell cycle status. Cells exposed to compound A showed cytokinetic failure with significant numbers of cells exhibiting DNA endoreduplication from $4n$ to $8n$ DNA content after exposure for 24 h (Fig. 7B). G1/S sensor analysis indicated significant numbers of cells in G1 following endoreduplication (large circles on scatter plot). After exposure for 48 h (Fig. 7D) very little BrdU incorporation was evident, with the majority of cells exhibiting polyploid G1 phase arrest. These findings from a fixed cell assay were confirmed by time-lapse fluorescent imaging of cells expressing the G1/S sensor (Fig. 8). Cells treated with compound A showed aberrant cytokinesis with subsequent polyploid arrest in G1.

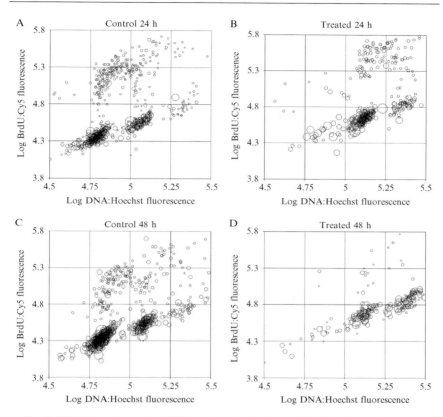

FIG. 7. Multiplexed analysis of DNA endoreduplication in U-2 OS cells stably expressing the G1/S sensor. Cells were grown for 24 h (a and b) or 48 h (c and d) in the presence (b and d) or absence (a and c) of compound A, pulse labeled with BrdU for 1 h, and fixed. Cells were stained with Hoechst 33342 (DNA content) and BrdU detected as described for Fig. 5. Scatter plots show three parameter data for individual cells imaged on the IN Cell Analyzer 1000 and analyzed with the IN Cell 1000 Analyzer Morphology Analysis Module (GE Healthcare). Cellular DNA content is displayed on the x axis (log total nuclear blue channel fluorescence from Hoechst:DNA binding), DNA replication as measured by BrdU incorporation on the y axis (log total nuclear red fluorescence from anti-BrdU/Cy5 antimouse) and G1/S sensor EGFP distribution (nuclear/cytoplasmic ratio) is proportional to the size of the circles (large circles are G1 phase cells, smaller circles are S phase and G2 phase cells).

We have used the G2/M sensor to multiplex with other fluorescence probes, including cholera toxin (CTX). Analysis of CTX binding to live G2/M CCPM expressing cells (Fig. 9) showed that binding of CTX was restricted to G1 cells confirming previous data obtained from immunofluorescence staining of fixed cells (Majoul *et al.*, 2002). CTX binding occurs through

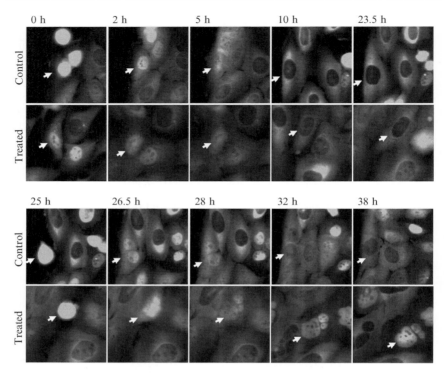

Fig. 8. Inhibition of cytokinesis. U-2 OS cells stably expressing the G1/S sensor were imaged over a 38-h period on IN Cell Analyzer 3000. Arrows highlight representative cells for control populations and populations treated with compound A. Control cells demonstrated a temporally normal cell cycle and entered mitosis around 25 h. Cells in the treated population fail to demonstrate cytokinesis, resulting in large polyploid daughter cells; these cells demonstrate a prolonged period of intense nuclear distribution of the G1/S sensor indicative of polyploid psuedo G1 arrest.

interaction with the plasma membrane ganglioside GM1, which is known to interact with a range of ligands and receptors, including FGF (Rusnati *et al.*, 2002), neuropeptides (Valdes-Gonzalez *et al.*, 2001), the EGF receptor (Miljan *et al.*, 2002), and opiate receptors (Crain and Shen, 1998). The influence of cell cycle-specific expression of a single molecular species over a wide range of biological processes highlights the significant potential for the cell cycle to affect responses in many cell-based assays. By enabling the capture of context-linked events, high content analysis provides a powerful tool for detecting and controlling for cell cycle dependencies in assay data.

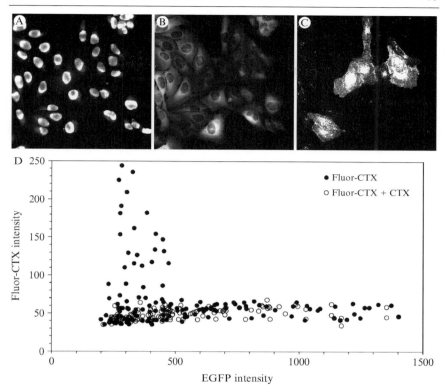

FIG. 9. Multiplexed analysis of ligand binding and cell cycle status. Binding of fluo-rescently labeled cholera toxin subunit B (CTX-B) to live G2/M CCPM cells was imaged on IN Cell Analyzer3000; (A) Hoechst-stained nuclei (B) EGFP, (C) Alexa-594 labeled CTX-B. (D) EGFP expression and CTX-B binding were determined by image analysis of duplicate wells in the presence and absence of a 100-fold molar excess of unlabeled CTX-B. Specific binding of fluorescent CTX-B is restricted to cells with low EGFP expression, that is, cells in G1.

Recent developments in small interfering RNA (siRNA) techniques for specifically modulating gene expression in a diverse range of cells and organisms have revolutionized the functional analysis of genes and proteins by providing synthetic and virally encoded siRNA methodologies to allow large-scale RNAi screens to be performed in mammalian cells. Combina-tion of siRNA techniques with high content analysis enables study of complex systems by allowing the combination of data from fluorescent cellular sensors with morphological parameters to provide a detailed description of the phenotypic effects of siRNAs in cellular screens.

We have used the G1/S and G2/M sensors in phenotypic screens using two siRNA libraries (Dharmacon) of 120 and 79 siRNA pools directed

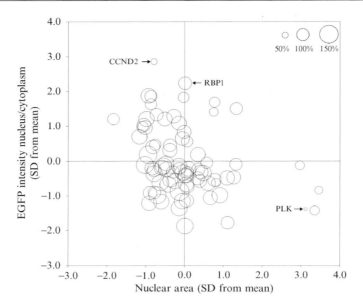

FIG. 10. Multiparameter analysis of siRNA screen in G1/S CCPM. Cells were reverse transfected with a panel of 79 siRNAs directed against cell cycle control and related genes. Cells were imaged on IN Cell Analyzer 3000 and image analysis performed using IN Cell Developer Toolbox to determine cell number, EGFP distribution, and a variety of nuclear morphology parameters. Data for nuclear area, EGFP nuclear/cytoplasmic ratio (high = G1, low = G2), and cell number (circle diameter proportional to cell number as percentage control siRNA) are shown. CCND2, cyclin D2 siRNA; RBP1, retinol-binding protein 1 siRNA; PLK, Polo-like kinase siRNA.

against key cell cycle control genes and other proteins. In screens using the G1/S sensor, a number of siRNAs (cyclin A2, cyclin B1, cyclin B2, cyclin D2, cyclin E1, CDK1, and PLK) produced a significant decrease in cell proliferation (Fig. 10) relative to transfection and siRNA controls. These siRNAs had very similar effects on cell proliferation in parallel screens carried out using the G2/M EGFP sensor (data not shown).

Analysis of the nuclear/cytoplasmic distribution of the EGFP G1/S sensor (Figs. 10 and 11) revealed a number of siRNAs, including cyclin D1, cyclin D2, cyclin E1, cyclin E2, CDK6, CDK7, MDM2, and retinol-binding protein 1 (RBP1), which caused a significant increase in cells in G1. These findings correlate well with the known roles of D- and E-type cyclins and associated CDKs in regulating G1/S transition and with p53-mediated arrest at the G1/S checkpoint via increased p53 activity and consequent inhibition of retinoblastoma protein (RB) phosphorylation.

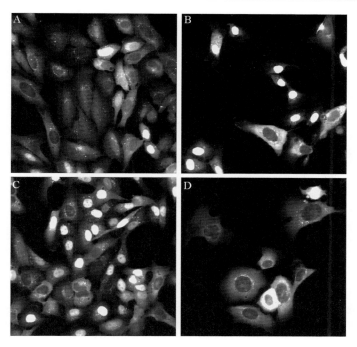

FIG. 11. Effects of siRNAs on cell cycle in G1/S CCPM. Cells were reverse transfected with control (A), cyclin D2 (B), retinal-binding protein (C), and Polo-like kinase (D) siRNAs and EGFP distribution imaged on IN Cell Analyzer 3000.

An interesting and previously unreported observation is the very significant blockage of cells in G1 induced by RBP1 knockdown (Fig. 11C). We postulate that this effect is mediated through intracellular release of retinol and subsequent inhibition of RB phosphorylation mediated through cyclin D and cyclin E and the CDK inhibitors p21 and p27 (Zhang et al., 2001; Yu et al., 2005).

To further define the effects of siRNAs, images were analyzed using a custom analysis protocol written using IN Cell Developer Toolbox to extract quantitative descriptors of nuclear morphology, including nuclear area (Fig. 10). Of those siRNAs having antiproliferative activity, cyclins A2 and B1, CDK1, and PLK showed significant changes in nuclear morphology. siRNAs producing significant arrest of cells in G1, for example, cyclin D2, MDM2, and RBP1, did not induce significant changes in nuclear morphology. Consequently, abstraction of multiparameter sensor and nonsensor data can be used to characterize siRNA activity into G1 blockers (nuclear EGFP, nuclear morphology unchanged), early G2 blockers (cytoplasmic

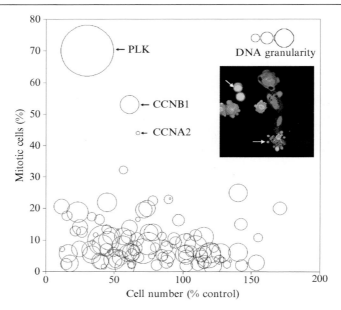

FIG. 12. Multiparameter analysis of siRNA screen in G2/M CCPM. Cells were transfected with a panel of 120 siRNAs directed against cell cycle control and related genes. Cells were imaged on IN Cell Analyzer 1000, and images were converted to IN Cell Analyzer 3000 format and analyzed using cell cycle and nuclear granularity analysis software. Data for cell number, mitotic cells, and DNA granularity (circle diameter proportional to DNA granularity) are shown. PLK, Polo-like kinase siRNA; CCNB1, cyclin B1 siRNA; CCNA2, cyclin A2 siRNA. (Inset) Mitotic and apoptotic cells following treatment with PLK siRNA.

EGFP, nuclear morphology unchanged), and late G2 blockers (cytoplasmic EGFP, significant changes in nuclear morphology).

Similarly, for the screens carried out using the G2/M sensor, applying morphological analysis in addition to cell cycle classification allowed higher resolution characterization of the effects of siRNA knockdown (Fig. 12). Treatment of cells with cyclin A2 siRNA (CCNA2) resulted in a significant accumulation of cells in prophase and mitosis to a similar degree as that observed for cyclin B1 (CCNB1), corresponding with the requirement for cyclin A for G1/S and G2/M transitions. In confirmation of the specificity of siRNA knockdown, cell cycle perturbation was not observed for the germ line functional homologue cyclin A1 (CCNA1), which is not expressed in differentiated U-2OS cells. Morphological analysis of images derived from cells treated with cyclin A2 siRNA revealed a significant increase in nuclear area (395.3 $\pm$ 173.7 μm^2) compared to control cells (219.1 $\pm$ 95.7 μm^2) and cyclin A1 siRNA-treated cells (229.8 $\pm$ 98.5 μm^2).

Knockdown of PLK with siRNA has been shown previously to inhibit cell proliferation, arrest cells in mitosis, and induce apoptosis (Liu and Erikson, 2003). Cell cycle analysis of G2/M sensor cells treated with siRNA directed against PLK showed a dramatic increase in mitotic cells (Fig. 12). Secondary analysis of images for DNA granularity as a measure of apoptotic DNA fragmentation showed a high incidence of apoptosis in PLK siRNA-treated cells. As for the screen described earlier using the G1/S sensor, combining high-content data derived from the same image stack allows subclassification of siRNA knockdown effects, for example, cyclin B1 (G2/M transition) G2/M block, no change in nuclear granularity, cyclin A2 (G1/S and G2/M transition) G2/M block and reduction in nuclear granularity; PLK (G2/M transition) G2/M block and increased nuclear granularity.

Conclusions and Future Directions

Accurate, noninvasive dynamic monitoring of the cell cycle position of individual live cells is of enormous value in the exploration of novel and potentially multiple actions of genes and drug candidates. We believe that cell cycle surveillance sensors of the type described here will enable increasingly sophisticated studies of the mode of action of drugs and chemotherapies in cancer and other diseases by allowing use of a common assay reporter across a range of complementary analysis platforms, including high-throughput imaging and flow cytometry. The use of genetically encoded cell cycle sensors as a measurable phenotype also has significant potential for application in functional genomics. Cellular phenotype screening using forward genetics (Stark and Gudkov, 1999) and large-scale RNAi libraries (Willingham et al., 2004) offer complementary approaches to discovery of the role of genes involved in cell cycle regulation and in signaling pathways that are cell cycle dependent.

We are currently developing a range of adenovirally encoded sensors using fluorescent fusion proteins and other sensors (Kendall et al., 2006) for use in target validation and lead profiling. Use of adenoviral vectors to deliver encoded cell cycle sensors will enable the development of high content dynamic cell cycle analysis in a wider range of cell types, including primary cells, which offer a more physiologically relevant background for the analysis of gene and drug function than standard transformed laboratory cell lines.

Acknowledgments

The authors thank Hayley Tinkler, Suzanne Hancock, and Mike Kenrick for providing much of the data described here.

References

Aleem, E., Kiyokawa, H., and Kaldis, P. (2005). Cdc2-cyclin E complexes regulate the G1/S phase transition. *Nature Cell Biol.* **7**(8), 831–836.

Arendt, T. (2002). Dysregulation of neuronal differentiation and cell cycle control in Alzheimer's disease. *J. Neural Transm. Suppl.* **62**, 77–85.

Arnaud, L., Pines, J., and Nigg, E. A. (1998). GFP tagging reveals human Polo-like kinase 1 at the kinetochore/centromere region of mitotic chromosomes. *Chromosoma* **107**(6–7), 424–429.

Bicknell, K. A., Surry, E. L., and Brooks, G. (2003). Targeting the cell cycle machinery for the treatment of cardiovascular disease. *J. Pharm. Pharmacol.* **55**(5), 571–591.

Blagden, S., and de Bono, J. (2005). Drugging cell cycle kinases in cancer therapy. *Curr. Drug Targets* **6**(3), 325–335.

Carnero, A. (2002). Targeting the cell cycle for cancer therapy. *Br. J. Cancer* **87**, 129–133.

Clute, P., and Pines, J. (1999). Temporal and spatial control of cyclin B1 destruction in metaphase. *Nature Cell Biol.* **1**, 82–87.

Cobrinik, D. (2005). Pocket proteins and cell cycle control. *Oncogene* **24**(17), 2796–2809.

Crain, S. M., and Shen, K. F. (1998). Modulation of opioid analgesia, tolerance and dependence by Gs-coupled, GM1 ganglioside-regulated opioid receptor functions. *Trends Pharmacol. Sci.* **19**(9), 358–365.

Denizot, F., and Lang, R. (1986). Rapid colorimetric assay for cell growth and survival. Modifications to the tetrazolium dye procedure giving improved sensitivity and reliability. *J. Immunol. Methods* **89**(2), 271–277.

Dutcher, J. P. (2004). Mammalian target of rapamycin (mTOR) inhibitors. *Curr. Oncol. Rep.* **6**(2), 111–115.

Fu, M., Wang, C., Li, Z., Sakamaki, T., and Pestell, R. G. (2004). Cyclin D1: Normal and abnormal functions. *Endocrinology* **145**(12), 5439–5447.

Galati, D., Bocchino, M., Paiardini, M., Cervasi, B., Silvestri, G., and Piedimonte, G. (2002). Cell cycle dysregulation during HIV infection: Perspectives of a target based therapy. *Curr. Drug Targets Immune Endocr. Metabol. Disord.* **2**(1), 53–61.

Gillessen, S., Groettup, M., and Cerny, T. (2002). The proteasome, a new target for cancer therapy. *Onkologie* **25**(6), 534–539.

Graves, R., Davies, R., Brophy, G., O'Beirne, G., and Cook, N. (1997). Noninvasive, real-time method for the examination of thymidine uptake events—application of the method to V-79 cell synchrony studies. *Anal. Biochem.* **248**(2), 251–257.

Gu, J., Xia, X., Yan, P., Liu, H., Podust, V. N., Reynolds, A. B., and Fanning, E. (2004). Cell cycle-dependent regulation of a human DNA helicase that localizes in DNA damage foci. *Mol. Biol. Cell.* **15**(7), 3320–3332.

Hadfield, J. A., Ducki, S., Hirst, N., and McGown, A. T. (2003). Tubulin and microtubules as targets for anticancer drugs. *Prog. Cell Cycle Res.* **5**, 309–325.

Hagting, A., Jackman, M., Simpson, K., and Pines, J. (1999). Translocation of cyclin B1 to the nucleus at prophase requires a phosphorylation-dependent nuclear import signal. *Curr. Biol.* **9**(13), 680–689.

Hamel, E., and Covell, D. G. (2002). Antimitotic peptides and depsipeptides. *Curr. Med. Chem. Anti-Cancer Agents* **2**(1), 19–53.

Horie, T., Sakaida, I., Yokoya, F., Nakajo, M., Sonaka, I., and Okita, K. (2003). L-cysteine administration prevents liver fibrosis by suppressing hepatic stellate cell proliferation and activation. *Biochem. Biophys. Res. Commun.* **305**(1), 94–100.

Huang, J., and Raff, J. W. (1999). The disappearance of cyclin B at the end of mitosis is regulated spatially in Drosophila cells. *EMBO J.* **18**(8), 2184–2195.

Humbert, C., Giroud, F., and Brugal, G. (1990). Detection of S cells and evaluation of DNA denaturation protocols by image cytometry of fluorescent BrdUrd labelling. *Cytometry* **11**(4), 481–489.

Hwang, A., Maity, A., McKenna, W. G., and Muschel, R. J. (1995). Cell cycle-dependent regulation of the cyclin B1 promoter. *J. Biol. Chem.* **270**(47), 28419–28424.

Jagtap, P., and Szabo, C. (2005). Poly(ADP-ribose) polymerase and the therapeutic effects of its inhibitors. *Nature Rev. Drug Discov.* **4**(5), 421–440.

Kanda, T., Sullivan, K. F., and Wahl, G. M. (1998). Histone-GFP fusion protein enables sensitive analysis of chromosome dynamics in living mammalian cells. *Curr. Biol.* **8**(7), 377–385.

Kawabe, T. (2004). G2 checkpoint abrogators as anticancer drugs. *Mol. Cancer Ther.* **3**(4), 513–519.

Kendall, J. M., Ismail, R., and Thomas, N. (2006). Adenoviral sensors for high-content cellular analysis. *Methods Enzymol.* **414** (this volume).

Kohn, K. W. (1999). Molecular interaction map of the mammalian cell cycle control and DNA repair systems. *Mol. Biol. Cell.* **10**(8), 2703–2734.

Lei, M. (2005). The MCM complex: Its role in DNA replication and implications for cancer therapy. *Curr. Cancer Drug Targets* **5**(5), 365–380.

Li, Q., and Zhu, G. D. (2002). Targeting serine/threonine protein kinase B/Akt and cell-cycle checkpoint kinases for treating cancer. *Curr. Top. Med. Chem.* **2**(9), 939–971.

Liu, X., and Erikson, R. L. (2003). Polo-like kinase (Plk)1 depletion induces apoptosis in cancer cells. *Proc. Natl. Acad. Sci. USA* **100**(10), 5789–5794.

Lundholt, B. K., Linde, V., Loechel, F., Pedersen, H. C., Moller, S., Praestegaard, M., Mikkelsen, K., Scudder, K., Bjorn, S. P., Heide, M., Arkhammar, P. O., Terry, R., and Nielsen, S. J. (2005). Identification of Akt pathway inhibitors using redistribution screening on the FLIPR and the IN Cell 3000 Analyzer. *J. Biomol. Screen.* **10**(1), 20–29.

Majoul, I., Schmidt, T., Pomasanova, M., Boutkevich, E., Kozlov, Y., and Soling, H. D. (2002). Differential expression of receptors for Shiga and Cholera toxin is regulated by the cell cycle. *J. Cell Sci.* **115**(Pt. 4), 817–826.

Miljan, E. A., Meuillet, E. J., Mania-Farnell, B., George, D., Yamamoto, H., Simon, H. G., and Bremer, E. G. (2002). Interaction of the extracellular domain of the epidermal growth factor receptor with gangliosides. *J. Biol. Chem.* **277**(12), 10108–10113.

Mortlock, A., Keen, N. J., Jung, F. H., Heron, N. M., Foote, K. M., Wilkinson, R., and Green, S. (2005). Progress in the development of selective inhibitors of Aurora kinases. *Curr. Top Med. Chem.* **5**(2), 199–213.

Nurse, P. (2000). The incredible life and times of biological cells. *Science* **289,** 1711–1716.

Nurse, P. (2000a). A long twentieth century of the cell cycle and beyond. *Cell* **100,** 71–78.

O'Hare, M., Wang, F., and Park, D. S. (2002). Cyclin-dependent kinases as potential targets to improve stroke outcome. *Pharmacol. Ther.* **93**(2–3), 135–143.

Owa, T., Yoshino, H., Yoshimatsu, K., and Nagasu, T. (2001). Cell cycle regulation in the G1 phase: A promising target for the development of new chemotherapeutic anticancer agents. *Curr. Med. Chem.* **12,** 1487–1503.

Pines, J. (1999). Four-dimensional control of the cell cycle. *Nature Cell Biol.* **1**(3), 73–79.

Pommier, Y., Redon, C., Rao, V. A., Seiler, J. A., Sordet, O., Takemura, H., Antony, S., Meng, Z., Liao, Z., Kohlhagen, G., Zhang, H., and Kohn, K. W. (2003). Repair of and checkpoint response to topoisomerase I-mediated DNA damage. *Mutat. Res.* **532**(1–2), 173–203.

Raff, J. W., Jeffers, K., and Huang, J. Y. (2002). The roles of Fzy/Cdc20 and Fzr/Cdh1 in regulating the destruction of cyclin B in space and time. *J. Cell. Biol.* **157**(7), 1139–1149.

Ramm, P., and Thomas, N. (2003). Image-based screening of signal transduction assays. *Sci. STKE* **177,** PE14.

Reits, E. A., Benham, A. M., Plougastel, B., Neefjes, J., and Trowsdale, J. (1997). Dynamics of proteasome distribution in living cells. *EMBO J.* **16**(20), 6087–6094.

Rusnati, M., Urbinati, C., Tanghetti, E., Dell'Era, P., Lortat-Jacob, H., and Presta, M.. (2002). Cell membrane GM1 ganglioside is a functional coreceptor for fibroblast growth factor 2. *Proc. Natl. Acad. Sci. USA* **99**(7), 4367–4372.

Sampath, D., and Plunkett, W. (2001). Design of new anticancer therapies targeting cell cycle checkpoint pathways. *Curr. Opin. Oncol.* **13,** 484–490.

Seville, L. L., Shah, N., Westwell, A. D., and Chan, W. C. (2005). Modulation of pRB/E2F functions in the regulation of cell cycle and in cancer. *Curr. Cancer Drug Targets* **5**(3), 159–170.

Sharma, S., Doherty, K. M., and Brosh, R. M., Jr. (2005). DNA helicases as targets for anti-cancer drugs. *Curr. Med. Chem. Anti-Cancer Agents* **5**(3), 183–199.

Sherr, C. J., and Roberts, J. M. (2004). Living with or without cyclins and cyclin-dependent kinases. *Genes Dev.* **18**(22), 2699–2711.

Smith, P. J., Blunt, N., Wiltshire, M., Hoy, T., Teesdale-Spittle, P., Craven, M. R., Watson, W. B., Amos, W. B., Errington, R. J., and Patterson, L. H. (2000). Characteristics of a novel deep red/infrared fluorescent cell-permeant DNA probe, DRAQ5, in intact human cells analyzed by flow cytometry, confocal and multiphoton microscopy. *Cytometry* **40**(4), 280–291.

Stark, G. R., and Gudkov, A. V. (1999). Forward genetics in mammalian cells: Functional approaches to gene discovery. *Hum. Mol. Genet.* **8**(10), 1925–1938.

Stewart, Z. A., Westfall, M. D., and Pietenpol, J. A. (2003). Cell-cycle dysregulation and anticancer therapy. *Trends Pharmacol. Sci.* **24**(3), 139–145.

Takizawa, C. G., and Morgan, D. O. (2000). Control of mitosis by changes in the subcellular location of cyclin-B1-Cdk1 and Cdc25C. *Curr. Opin. Cell Biol.* **12**(6), 658–665.

Taneja, P., Gu, J., Peng, R., Carrick, R., Uchiumi, F., Ott, R. D., Gustafson, E., Podust, V. N., and Fanning, E. (2002). A dominant-negative mutant of human DNA helicase B blocks the onset of chromosomal DNA replication. *J. Biol. Chem.* **277**(43), 40853–40861.

Tatebe, H., Goshima, G., Takeda, K., Nakagawa, T., Kinoshita, K., and Yanagida, M. (2001). Fission yeast living mitosis visualized by GFP-tagged gene products. *Micron* **32** (1), 67–74.

Thomas, N. (2003). Lighting the circle of life: Fluorescent sensors for covert surveillance of the cell cycle. *Cell Cycle* **2**(6), 545–549.

Thomas, N., and Goodyer, I. (2003). Stealth sensors: Real time monitoring of the cell cycle. *Drug Disc. Today Targets* **2**(1), 26–33.

Thomas, N., Kenrick, M., Giesler, T., Kiser, G., Tinkler, H., and Stubbs, S. (2005). Characterization and gene expression profiling of a stable cell line expressing a cell cycle GFP sensor. *Cell Cycle* **4**(1), 191–195.

Tsien, R. Y. (1998). The green fluorescent protein. *Annu. Rev. Biochem.* **67,** 509–544.

Valdes-Gonzalez, T., Inagawa, J., and Ido, T. (2001). Neuropeptides interact with glycolipid receptors: A surface plasmon resonance study. *Peptides.* **22**(7), 1099–1106.

Vermeulen, K., Van Bockstaele, D. R., and Berneman, Z. N. (2003). The cell cycle: A review of regulation, deregulation and therapeutic targets in cancer. *Cell Prolif.* **36**(3), 131–149.

Weingartner, M., Binarova, P., Drykova, D., Schweighofer, A., David, J. P., Heberle-Bors, E., Doonan, J., and Bogre, L. (2001). Dynamic recruitment of Cdc2 to specific microtubule structures during mitosis. *Plant Cell* **13**(8), 1929–1943.

Willingham, A. T., Deveraux, Q. L., Hampton, G. M., and Aza-Blanc, P. (2004). RNAi and HTS: Exploring cancer by systematic loss-of-function. *Oncogene* **23**(51), 8392–8400.

Wong, C. F., Guminski, A., Saunders, N. A., and Burgess, A. J. (2005). Exploiting novel cell cycle targets in the development of anticancer agents. *Curr. Cancer Drug Targets* **5**(2), 85–102.

Yuan, J., Eckerdt, F., Bereiter-Hahn, J., Kurunci-Csacsko, E., Kaufmann, M., and Strebhardt, K. (2002). Cooperative phosphorylation including the activity of polo-like kinase 1 regulates the subcellular localization of cyclin B1. *Oncogene* **21**(54), 8282–8292.

Yu, Z., Lin, J., Xiao, Y., Han, J., Zhang, X., Jia, H., Tang, Y., and Li, Y. (2005). Induction of cell-cycle arrest by all-trans retinoic acid in mouse embryonic palatal mesenchymal (MEPM) cells. *Toxicol. Sci.* **83**(2), 349–354.

Zeng, Y., Hirano, K., Hirano, M., Nishimura, J., and Kanaide, H. (2000). Minimal requirements for the nuclear localization of p27(Kip1), a cyclin-dependent kinase inhibitor. *Biochem. Biophys. Res. Commun.* **274**(1), 37–42.

Zhang, D., Vuocolo, S., Masciullo, V., Sava, T., Giordano, A., Soprano, D. R., and Soprano, K. J. (2001). Cell cycle genes as targets of retinoid induced ovarian tumor cell growth suppression. *Oncogene* **20**(55), 7935–7944.

Zhang, J., Campbell, R. E., Ting, A. Y., and Tsien, R. Y. (2002). Creating new fluorescent probes for cell biology. *Nature Rev. Mol. Cell. Biol.* **3**(12), 906–918.

[2] High-Content Fluorescence-Based Screening for Epigenetic Modulators

By ELISABETH D. MARTINEZ, ANGIE B. DULL, JOHN A. BEUTLER, and GORDON L. HAGER

Abstract

Epigenetic processes have gained a great amount of attention in recent years, particularly due to the influence they exert on gene transcription. Several human diseases, including cancer, have been linked to aberrant epigenetic pathways. Consequently, the cellular enzymes that mediate epigenetic events, including histone deacetylases and DNA methyltransferases, have become prime molecular targets for therapeutic intervention. The effective and specific chemical inhibition of these activities is a top priority in cancer research and appears to have therapeutic potential. This chapter describes the development of mammalian cell-based fluorescent assays to screen for epigenetic modulators using an innovative combination of approaches. Detailed protocols for the use of the assays in drug screens,

METHODS IN ENZYMOLOGY, VOL. 414   0076-6879/06 $35.00

as well as for the initial characterization of hits, are provided. Furthermore, options for evaluating the mechanism of action of these compounds are presented and principles to govern the choice of hit compounds for the development of leads are discussed.

Introduction

Proper physiological function of organs requires the correct temporal and spatial expression of genes and the regulation of their expression levels. Chromosomal aberrations, mutations of transcription factors, and changes in chromatin structure are among the events that can give rise to abnormal silencing or abnormal activation of gene expression. Chromatin architecture is in large part determined by chemical modifications of DNA sequences and of histones, the proteins on which DNA is wound. The methylation status of DNA, as well as the acetylation, methylation, and phosphorylation status of histone proteins, determines the level of chromatin compaction in the cell and, thus, the accessibility of transcription factors to their target genes. In general, unmethylated DNA and acetylated histones at promoter sites are permissive to transcriptional activity, whereas methylated DNA and hypo-acetylated histones prevent transcription. The expression and/or function of the enzymes that mediate histone acetylation, histone deacetylation (HDACs), and DNA methylation (DNMTs) is often altered in human cancers (Dhordain et al., 1998; Kitabayashi et al., 2001; Melki et al., 1998; Mizuno et al., 2001; Shigeno et al., 2004; Warrell et al., 1998), consequently affecting the regulation of transcriptional processes and thus cellular growth and function.

Several compounds have been identified that inhibit the methylation or deacetylation pathways mediated by these enzymes. These compounds have had immediate application in the treatment of cancers because of their ability to reactivate aberrantly silenced tumor suppressor genes (Cote et al., 2002; Sandor et al., 2002; Suzuki et al., 2002; Vigushin and Coombes, 2002). A number of these inhibitors are being evaluated in current clinical trials (Kelly et al., 2005; McLaughlin and La Thangue, 2004; Ryan et al., 2005; Villar-Garea and Esteller, 2004), but the search for structurally diverse inhibitors with improved pharmacological proper-ties and selectivity continues (Haggarty et al., 2003; Hennessy et al., 2003). Most efforts to date to screen for such drugs have used in vitro approaches or have looked for compounds capable of overinducing already tran-scriptionally active genes. This chapter discusses the development of a fluorescence-based cellular assay to identify small molecules based on their ability to reactivate the expression of a silenced reporter gene. The princi-ples that should guide the development of similar cell-based assays, the actual techniques that can be used to develop and use these systems in drug screens, and detailed protocols for hit analysis are provided.

Epigenetic Regulators of Gene Expression as Drug Targets

The mechanism by which HDACs and DNMTs generally repress transcription has been partly elucidated over the last few years. Methylation of DNA at CpG islands (genomic regions of statistically high CG dinucleotide density, found near or on gene promoters) is an uncommon event in active euchromatin. Rather, cellular regulatory cascades or pathological signaling pathways specifically target particular promoter sequences, leading to their methylation. The methylated DNA then becomes a docking site for methyl-binding proteins (Magdinier and Wolffe, 2001). These, in turn, recruit transcriptional repressor complexes, containing histone deacetylases. HDACs deacetylate local histones, causing the compaction of chromatin, thus hindering the access of transcription factors to promoter sequences. HDACs can also exert their transcriptional inhibitory activity in a DNA methylation-independent manner. In either case, the resulting chromatin compaction, if aberrant, deregulates transcriptional control and can lead to unchecked growth or proliferation due to the silencing of tumor suppressor genes or to functionally equivalent events (Konishi *et al.*, 2002; Nakamura *et al.*, 2001; Roman-Gomez *et al.*, 2002). It follows that chemical inhibition of HDACs and/or of DNMTs should allow for the reactivation of silenced genes, potentially restoring transcriptional balance to the cell.

Several compounds have been shown to inhibit the activity of HDACs, such as the hydroxamic acid trichostatin A, the short chain fatty acid butyrate, the epoxide apicidin, and the depsipeptide FR901228. Compounds that inhibit DNMTs have also been developed based on the structure of cytidine, the DNA base methylated by this enzyme family. These include 5-aza-2′-deoxycytidine (also known as decitabine), azacytidine, and zebularine. These agents reactivate the expression of aberrantly silenced tumor suppressor or differentiation genes and/or downregulate (presumably by indirect mechanisms) the expression of cell proliferation or oncogenes in human cancer cells (Cote *et al.*, 2002; Sandor *et al.*, 2002; Suzuki *et al.*, 2002; Vigushin and Coombes, 2002). In general, these inhibitors do not produce global transcriptional changes, and normal cells can overcome the transient G2/M arrest that is sometimes induced by treatment (Johnstone, 2002). Early clinical trials have shown some promise in cancer treatment (Kelly *et al.*, 2005; McLaughlin and La Thangue, 2004; Ryan *et al.*, 2005; Villar-Garea and Esteller, 2004), yet a wider structural variety of inhibitors is necessary to address drug absorption, distribution, stability, efficacy, and toxicity, as well as variations in patient responses.

Rationale for the Development of Cell-Based Assays to Screen for Epigenetic Modulators

Because of their potential usefulness as therapeutics and as research tools, we sought to develop a system to identify novel inhibitors of HDACs and DNMTs that would have unique advantages over existing methods and to establish a standard protocol for developing such assays. Our optimal design includes the following requirements: (1) the system should be mammalian cell based for maximal biological relevance, (2) it should allow simultaneous screening for small molecule inhibitors of both HDACs and DNMTs, (3) it should measure a biological event closely related to the events one seeks to target in cancer cells (i.e., gene silencing), and (4) it should produce a signal that is easy to measure and amenable to automation. This design is represented in Fig. 1.

Our design rationale is based on several points. Because their functions overlap, a system that involves the reversal of transcriptional repression could be used to interfere with either DNA methylation or histone deacetylation pathways. To achieve this, we could stably introduce a reporter gene into mammalian cells together with a selection marker and screen for cells that, while retaining selection, showed no reporter expression. We decided to use the CMV promoter to drive our reporter gene because it is known to be a strong and constitutive promoter. In this way, the lack of

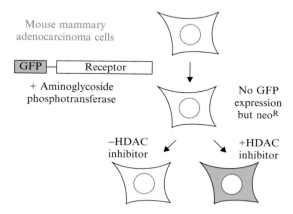

FIG. 1. Schematic of the strategy used in the development of the mammalian cell-based assay. Briefly, C127-derived cells were cotransfected with an expression vector for a GFP-tagged protein driven by a constitutively active CMV promoter and with a gene for neomycin resistance. After selection, clones that survived in the presence of the antibiotic but did not express GFP were expanded. GFP expression was obtained by treatment with inhibitors of HDACs or DNMTs.

promoter activity would likely be the result of a repressive chromatin structure on or near the promoter rather than the lack of a stimulatory signal (such as a hormone or a particular transcription factor or a signal arising during a stage of the cell cycle). Similarly, the reversal of repression by candidate compounds would likely be the consequence of changes in chromatin architecture rather than the induction of unrelated stimulatory signals. We then chose a green fluorescent protein (GFP)-fusion protein as our reporter to facilitate its detection by automated fluorescence microscopy. The methods that follow describe the principles and techniques applied in the development and use of this cell-based system, which can serve as a template for developing similar cell-based assays for use in drug screen applications.

Methodological Considerations

Choice of Cell Line, Reporter System, and Selection Marker for High Content Transcriptional Assays

The choice of cell line should be based on the following principles: (1) the cell line should express various HDACs and/or DNMTs, (2) it should be responsive to chemical epigenetic modulators, that is, exhibiting histone hyperacetylation when treated with known HDAC inhibitors, (3) it should not express high levels of multidrug resistance pumps or drug efflux transporters, (4) it should exhibit a morphology that is amenable to imaging, that is, flatter cells that spread well on a surface and are relatively homogeneous and, (5) it should grow under standard conditions that can facilitate miniaturization (for high-throughput applications). All other things being equal, the user may also want to consider the advantages of using a human cell line for greater relevance in drug screening, although our experience has been very positive using C127 mouse mammary adenocarcinoma cells.

While there are currently many reporter systems available for cellular transcriptional studies, we chose GFP over other reporters, such as CAT or luciferase, because of the following advantages. GFP expression can be measured in live cells, and its monitoring does not require cell lysis or the addition of reagents to the cells (Hager, 1999; Martinez *et al.*, 2005). Fluorescence detection is simple, friendly to automation, and requires only an initial investment in equipment without the ongoing cost of substrates. Furthermore, the use of GFP as a reporter allows the user to query the experimental results at any time before the intended end point of the study without irreversibly interfering with biological processes. This permits added flexibility in experimental design. In terms of promoter choice, we strongly recommend use of the CMV promoter to drive the reporter gene. The assay presented here is based on silencing of the integrated promoter and thus this promoter must be constitutive and strong to be useful in this application. Any selection marker can be used; we recommend neomycin because of its availability and the low cost of G418 compared to other selection antibiotics.

Development and Basic Characterization of Cell-Based Fluorescent Assays to Screen for Epigenetic Modulators

Once the mammalian cell line, promoter–reporter gene construct, and antibiotic resistance gene are chosen, most cells can be stably transfected using the following procedure.

1. Exponentially growing mammalian cells are transfected with expression vectors for a GFP-tagged construct (or other reporter) driven by a strong promoter (we recommend the CMV promoter) and for aminoglycoside phosphotransferase, which confers neomycin resistance, in a 20:1 molar ratio using Lipofectamine 2000. Fugene 4 is an effective alternate reagent for transfection. The transfected cells are then incubated under standard conditions.

2. Forty-eight to 72 h posttransfection, the cells are split 1:5 or 1:10 into separate dishes and allowed to reattach.

3. Once attached, the cells are exposed to selection media consisting of growth media supplemented with 1 mg/ml G418. The amount of G418 may have to be optimized depending on the cell line used. Most manufacturers of G418 have sample killing curves on their web sites that can serve as reference.

4. The selection is continued for 2 to 3 weeks with selection media being replaced every other day.

5. Once the majority of the cells are dead, colony formation is monitored by inspection and the corresponding areas are marked on the bottom of the dish.

6. When colonies reach approximately 1 to 3 mm in diameter, they are detached with 1 to 2 μl of trypsin and placed directly into wells of a 24-well dish already containing fresh selection media.

7. Antibiotic-resistant colonies are progressively expanded and monitored for fluorescence under a fluorescent microscope or plate reader. Colonies that are consistently negative for GFP expression are further expanded and some of these are chosen for characterization. Note: depending on the type of fluorescent microscope and objective or plate reader being used, plates where the colonies are growing may need to have glass or optic plastic bottom surfaces for optimal imaging or monitoring of GFP expression.

Once neomycin-resistant clones with a stable GFP-negative phenotype are obtained, they should be characterized for their response to known epigenetic modulators such as known inhibitors of HDACs or DNMTs. This can be done as follows.

1. Logarithmically growing stable cell lines are plated on glass bottom chamber slides (e.g., Labtek II from Nalge-Nunc) at about 30 to 40% confluency and are allowed to attach overnight.

2. Cells are then treated with known epigenetic modulators for 6–18 h and the induction of GFP expression is monitored by fluorescence microscopy on a fluorescent plate reader or by FACS analysis. Although the concentrations of HDAC and DNMT inhibitors may have to be optimized, we have found that 50 ng/ml of TSA, 25 mM sodium butyrate or 1 μM 5-aza-deoxycytidine give excellent GFP expression. Note: 5-aza-deoxycytidine is very unstable. Only use stocks freshly made the day of the induction. In our hands, frozen stocks of TSA and sodium butyrate are stable for 1 to 3 months at −20°.

3. Based on the GFP expression profile of the cell lines, one can determine which cell lines show the more robust and homogeneous response, which can then be characterized further. An example of a robust and homogeneous response is shown in Fig. 2. It is important to freeze down a number of stocks of early passages of the chosen cell lines. Note: stable cell lines should be monitored regularly to ensure the stability of the negative-GFP phenotype. After prolonged culture, some cells may begin to

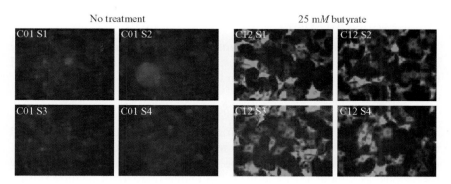

FIG. 2. Robust and homogeneous cellular response to inhibitors. Cells were plated on glass-bottom 96-well plates and the next day they were treated with 25 mM sodium butyrate or with 0.5% DMSO vehicle (labeled no treatment). After overnight incubation, cells were fixed and imaged using a Discovery-1 automated system (Molecular Devices). Four areas per well were imaged showing clear fluorescent signal only in wells that were treated with HDAC inhibitor. Induction of GFP expression by sodium butyrate is the positive control for the assay and is the basis for identification of hit compounds.

spontaneously express GFP. These cells will not be useful for screening purposes and thus should be discarded.

Basic molecular characterization of the stable cell line(s) that shows a clear and constant negative GFP phenotype and a robust response to known epigenetic modulators can then proceed. This characterization can include the following.

1. ChIP analysis of histone acetylation status at the integrated CMV promoter.

2. Southern blot analysis to determine the approximate copy number of inserts integrated in the genome.

3. Metaphase DNA spreads combined with fluorescence *in situ* hybridization (FISH) to determine the chromosomal location of insertions.

4. GFP induction dose–response curves and EC_{50} calculations to determine the potency and efficacy of the various chemical epigenetic modulators.

5. FACS analysis of cells induced with maximal doses of chemical epigenetic modulators to determine the total percentage of cells that respond by expressing GFP. For automated screening purposes the cell line of choice should have greater than a 60% response rate.

6. Before proceeding on to primary screening, the cell line of choice should exhibit, in addition to the qualities described already, reproducibility of induction by the various drugs. Several frozen stocks of varying passages should be tested to ensure reproducibility, robustness, and at least moderate homogeneity in the response. Subcloning of cells may be necessary if these qualities are not maintained over time.

Note: while some of the aforementioned characterization is not strictly necessary for drug screening purposes, we strongly recommend that the cell line be studied as noted in 1, 4, 5, and 6 *before* any screening is done.

Setup and Optimization of Primary Screens

Depending on the imaging platform or reader to be used and on the cell type, the first step is to examine how well the cells grow on various kinds of multiwell plates, including doubling time, and how they tolerate assay conditions, as well as the quality of the image or readout acquired using the different plates. The goal is to establish the best conditions for both cell growth and imaging/readout quality. Using the Discovery-1 imaging system from Molecular Devices, we found that optimal cell adhesion and image quality were obtained with Nunc glass-bottom 96-well plates. These plates were selected for use in the screen due to the clarity of acquired images and the ability of the cells to grow on the glass and remain attached after repeated washings. The cells were cultured in Dulbecco's modified Eagle's medium (DMEM) containing 4 mM L-glutamine, 1 mM sodium pyruvate, 1% penicillin/streptomycin, 1 mg/ml G418, and 10% fetal bovine serum for the entirety of the assay. Optimal growth conditions for the assay may need to be established for each cell type.

Before proceeding further, cells should be tested for their tolerance to the standard vehicle in which the compounds in the library are dissolved. This is typically dimethyl sulfoxide (DMSO). DMSO tolerance can be evaluated in a 96-well plate format using a sulforhodamine staining protocol for cell viability. The DMSO tolerance threshold is critical in determining compound dilution protocols and in deciding on the final concentrations to be tested in the screen. The cells are treated with increasing doses of DMSO for 48 h. Then the cells are fixed, stained, and quantitated with sulforhodamine B staining as detailed next.

1. Plate cells into 96-well Nunc plastic plates at 60,000 cells/well in complete media at 100 μl/well. Add increasing doses of DMSO diluted in growth media in a 50-μl total volume to each well for final concentrations of 0, 0.05, 0.1, 0.25, 0.5, 1, 2, 5, and 10% (repeat each concentration in at least 3–4 wells). Allow the cells to grow for 48 h in the presence of DMSO under standard conditions. Each well contains a 150-μl volume at this point in the procedure.

2. Fix the cells by adding 50 μl/well of 40% cold trichloroacetic acid (TCA) for 30 min at room temperature (for final concentration of 10% TCA).

3. Decant the media containing TCA and rinse the plates five times with water. Air dry the plates.

4. Add 100 μl SRB (0.4% sulforhodamine B [SRB], 1% acetic acid) per well and stain for 1 h at room temperature.

5. Rinse the plates five times with 1% acetic acid. Air dry the plates (must be completely dry).

6. Extract SRB with 100 μl of 10 mM Tris base per well.

7. Shake and read at 520 nm using a SpectraMax 250 plate reader from Molecular Devices or similar equipment.

As can be seen in Fig. 3, our cells were able to survive for 48 h in up to 1% DMSO. Consequently, the assay was designed so that compounds were added for a 0.5% final DMSO concentration.

Among known inhibitors of epigenetic pathways active in our system, sodium butyrate was selected as a positive control compound for the screen due to its stability in cell culture, low cytotoxicity, and availability. Various doses in the millimolar range were tested and examined for GFP expression on the imaging system. A dose of 25 mM sodium butyrate for 24 h was

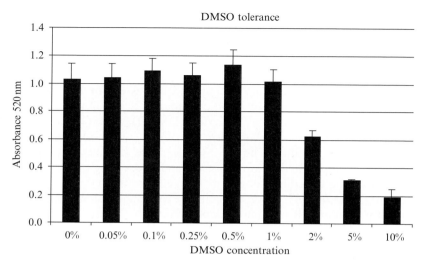

FIG. 3. DMSO tolerance assay. C127-derived cells were treated with increasing amounts of DMSO for 48 h and cell viability was measured by staining with SRB. These cells are able to tolerate 1% DMSO for 48 h.

routinely utilized in the assay (see Fig. 2). In our case, because the GFP signal was not bright enough for reliable measurement on a fluorescence plate reader, the images were acquired on the Discovery-1 imaging system, which has greater sensitivity. Each system should be optimized to obtain strong positive control signals and a large dynamic range of fluorescence detection. To facilitate location of the focal plane on the imaging system, nuclei are stained using Hoechst 33342 (2 μg/ml). Hits can be identified by visual inspection of the images for GFP expression, or by automatic intensity measurements, depending on the robustness of the assay, and the degree of automation used.

We suggest starting with small validation libraries such as the Library of Pharmacologically Active Compounds (Sigma) before other chemical and natural products libraries are screened in the assay. Libraries can be selected based on screening rationale. For example, the presence of chemical diversity is conducive for the discovery of novel compounds, whereas libraries of structurally related compounds may aid in finding analogs of existing drugs with improved pharmacokinetics. Compounds are added to the plates containing the cells using liquid-handling systems such as the Biomek FX (Beckman Coulter). The FX program may need to be modified for attached cells to ensure that the monolayer is not disrupted during the addition of compounds. Pipette tip height above the bottom of the well, speed of addition, mixing volume, and speed of mixing may need to be optimized to maintain the integrity of the cell monolayer during compound addition.

Screening for epigenetic modulators is as follows.

1. On day one, seed cells at 70,000 cells/well into Nunc glass-bottom 96-well plates (100 μl/well) in complete media. Grow the cells overnight at 37°, 5% CO_2, 95% air in a humidified incubator. Note: the cell number may need to be adjusted according to cell type and doubling time.

2. On the following morning, treat the 96-well plate as follows for 24 h.
   a. Column 1 (negative control): add 50 μl of 1.5% DMSO solution in complete media for 0.5% final DMSO concentration.
   b. Column 12 (positive control): add 50 μl of 75 mM butyrate for 25 mM final concentration.
   c. Columns 2–11: add 30 μl complete media + 20 μl compounds from chemical or natural products library for 0.5% final DMSO concentration and 5 μM concentration of compound. Compounds are added to the plates using the Biomek FX liquid handling system from Beckman Coulter or similar device.

3. On the third day, after 24 h of drug treatment, fix the cells in media containing compounds and stain nuclei by adding 4% final formaldehyde and 2 μg/ml final Hoechst 33342 for 1 h. Wash the plates three times with 200 μl of phosphate-buffered saline and add a final volume of 100 μl/well for imaging. Acquire images or read on a plate reader. For the Discovery-1 imaging system, we recommend using a 20× Nikon Plan Fluor objective and exposing for 600 ms in the fluorescein isothiocyanate (FITC) channel and 30 ms in the DAPI channel. Four sites per well can be acquired for each GFP and Hoechst signal for thorough querying of each well. It is best to autofocus on the DAPI channel first, acquire these images, and then image in the FITC channel without refocusing. Note: fixing the cells and staining nuclei are optional. Depending on the imaging system used, these steps can enhance image quality greatly.

A number of compounds may exhibit autofluorescence, which can produce false-positive hits in the screen. In order to circumvent this issue, compounds should be screened in parallel against the untransfected parental cell line. Hits identified in the assay can be compared with the parental cell line to determine if the compound was a hit or a false-positive fluorescent compound.

Cherry picking is used to confirm hit compounds by selecting hits from the original compound plate and transferring to a plate containing only hits from the primary screen for retesting. Hits from the primary screen can be cherry picked using the Multiprobe robotic handling system from Packard and reevaluated in the assay to confirm activity. After hit compounds are confirmed with cherry-picking experiments, the compounds are tested in secondary and tertiary assays. We recommend running retesting experiments at least twice for validation of the primary hits.

Evaluation of Hit Compounds

A screening assay represents only the very first step in discovery. It is essential following any screen to counterscreen the confirmed hits in assays that complement the initial screening assay in order to establish specificity. In the case of the current assay, the primary goal is to determine if the hit compound actually hits target enzymes in epigenetic pathways (e.g., HDACs or DNMTs). This is established by conducting cell-free enzymatic assays (see later).

A second consideration when prioritizing hits is to consider the chemical and pharmaceutical properties of the hit. In a perfect world, all library compounds would have desirable "drug-like" properties; however, most libraries contain compounds that are less than ideal as molecular probes or drug leads. The most common type of undesirable compound are those with reactive functional groups capable of binding covalently to target molecules, at sulfhydryl groups of cysteine residues, β-amino group of lysines, or other sites within biological macromolecules. Because such sites occur on nearly all proteins, the activity of compounds with reactive functional groups observed in a single assay is likely to be nonspecific in nature, as the compound may react with many proteins. With a cell-based assay, reactive compounds pose less of a problem than in cell-free biochemical assays; however, they do turn up. One example of this found in our screening campaign was a platinum complex. A second class of undesirable compound is those that are broadly toxic, such as perhalogenated organics derived from pesticides, a group that we also encountered as active with our assay.

Drug likeness has been defined by Lipinski's well-known rule of five, which is aimed at identifying synthetic molecules that will have a good chance of being orally bioavailable drug candidates. These rules require that a compound have less than 5 hydrogen bond donors, less than 10 hydrogen bond acceptors, a log P of <5, and a molecular mass of <500 Da (Lipinski *et al.*, 2001). However, the rules explicitly exempt natural products due to the built-in evolutionary pressures that demand a measure of bioavailability to be useful toxins and modulators or to serve other ecological roles. We chose to ignore the rule of five in evaluation of hits from this assay, as we were willing to accept compounds as molecular probes, for which the physicochemical requirements are less stringent than for drugs. Nonetheless, many of the hits fell within the limits of the rule of five.

Once a hit has been confirmed, it is desirable to explore the chemical space surrounding its structure to determine if a lead series of analogs can be developed. This can be done in the case of NCI structural diversity set compounds by substructure searches of the full NCI repository to identify a number of related compounds, which can then be tested in the primary assay. Commercially available analogs can also be identified using Chem-Navigator software (http://www.chemnavigator.com), which facilitates procurement of compounds from a wide variety of commercial sources. If a limited number of samples are screened, it is possible to analyze the structural relationships of hit compounds by inspection. If a larger number of hits require analysis, a software package such as Leadscope (http://www.leadscope.com) may be used to cluster hits by structure.

Available Secondary Screens for the Evaluation of Candidate Epigenetic Modulators

Once the candidate compounds have been confirmed in cherry-picking experiments and/or characterized as described earlier, before going on to perform mechanistic secondary assays, we recommend that the hits be tested on a different clone than the one used in the primary screens, that is, one of the original cell clones that was characterized as GFP negative, antibiotic resistant, that responded to known HDAC or DNMT inhibitors but which was not selected for use in the screen. This will ensure that the compounds have a transcriptional effect independent of the insertion site of the GFP construct, of copy number, etc.

Secondary mechanistic assays can then proceed. Among the available secondary assays, the ones listed here are particularly relevant. We limit ourselves to brief descriptions of each assay as detailed protocols can be found elsewhere and fall outside the scope of this chapter.

1. Western analysis of global histone acetylation. Candidate epigenetic modulators can be assayed for their ability to cause global increases in histone acetylation by treating cells overnight with the individual compounds and appropriate positive and negative controls and then performing Western analysis on the total cell lysates with antibodies specific to the various modifications of histones 3 and 4. Although many antibodies against modified histone 3 and 4 are available commercially, our experience with Upstate antibodies has been very good.

2. ChIP to measure histone acetylation at the integrated silenced promoter. As has been described by others (El Osta and Wolffe, 2001; Lambert and Nordeen, 2001), chromatin immunoprecipitation assays can be useful in identifying the modification status of particular histones at a given genomic loci. Cells can be treated with classical epigenetic

modulators as described earlier in this chapter and with candidate compounds in parallel for various lengths of time, as determined by GFP induction time courses, taking into consideration that histone modifications presumably are a relatively early event in the induction process. Standard ChIP assays can then be performed using primers specific for the promoter of interest, for example, CMV, or the promoter of a silenced tumor suppressor gene for the polymerase chain reaction (PCR).

3. *In vitro* HDAC inhibition assays. Several active, purified recombinant human HDACS, including HDAC 8, Sirt 1, 2, and 3, are currently available from Biomol, Upstate, and other companies. In addition, several investigators have developed protocols for the purification of active enzymes (Li *et al.*, 2004; Schultz *et al.*, 2004). Active, purified recombinant proteins can be combined with labeled hyperacetylated histone substrates or commercial fluorophore conjugated substrates, and the activity of the enzymes can be monitored in the presence or absence of candidate inhibitors or control drugs.

4. *In vitro* DNMT inhibition assays. These are performed using purified recombinant mammalian DNMTs or bacterial methylases using appropriate oligonucleotide substrates in the presence of candidate or control inhibitors (Brueckner *et al.*, 2005; Yokochi and Robertson, 2004). Presumably, this assay will only detect inhibitors that do not require incorporation into DNA, that is, those that can inhibit soluble DNMTs.

5. Reexpression of silenced tumor suppressor genes in human cancer cell lines. The RNA expression levels of a silent tumor suppressor gene (by RT-PCR) or its protein expression levels (by Western blotting) can be measured after treatment with candidate or control compounds.

6. Demethylation of hypermethylated promoters in human cancer cell lines. The methylation status of aberrantly hypermethylated promoters can be measured by bisulfite sequencing or by methylation-specific PCR after treatment of cells with candidate or control compounds.

Hits that are demonstrated to act via acetylation or methylation mechanisms should be then submitted for chemical optimization and developed further as drug leads. Evaluation of how the compounds may interact with detoxification pathways should be a component of lead development.

General Conclusions and Perspectives

The methods described in this chapter provide a simple, relatively economic way of developing a fluorescence-based mammalian cell assay for the identification of novel epigenetic modulators. The detailed protocols provided for the basic characterization of hits can be useful not only in analyzing epigenetic modulators, but also in dealing with hits from other assay systems. Furthermore, the principles upon which the assay is built can be directly applied as described in this chapter or can be used to develop assay variations such as to screen for inducers or silencers of a particular promoter. Combining the power of GFP technology with the biology of the cell can give rise to creative avenues to search for solutions to cellular conundrums and therapeutic needs.

References

Brueckner, B., Boy, R. G., Siedlecki, P., Musch, T., Kliem, H. C., Zielenkiewicz, P., Suhai, S., Wiessler, M., and Lyko, F. (2005). Epigenetic reactivation of tumor suppressor genes by a novel small-molecule inhibitor of human DNA methyltransferases. *Cancer Res.* **65,** 6305–6311.

Cote, S., Rosenauer, A., Bianchini, A., Seiter, K., Vandewiele, J., Nervi, C., and Miller, W. H., Jr. (2002). Response to histone deacetylase inhibition of novel PML/RARalpha mutants detected in retinoic acid-resistant APL cells. *Blood* **100,** 2586–2596.

Dhordain, P., Lin, R. J., Quief, S., Lantoine, D., Kerckaert, J. P., Evans, R. M., and Albagli, O. (1998). The LAZ3(BCL-6) oncoprotein recruits a SMRT/mSIN3A/histone deacetylase containing complex to mediate transcriptional repression. *Nucleic Acids Res.* **26,** 4645–4651.

El Osta, A., and Wolffe, A. P. (2001). Analysis of chromatin-immunopurified MeCP2-associated fragments. *Biochem. Biophys. Res. Commun.* **289,** 733–737.

Hager, G. L. (1999). Studying nuclear receptors with green fluorescent protein fusions. *Methods Enzymol.* **302,** 73–84.

Haggarty, S. J., Koeller, K. M., Wong, J. C., Butcher, R. A., and Schreiber, S. L. (2003). Multidimensional chemical genetic analysis of diversity-oriented synthesis-derived deacetylase inhibitors using cell-based assays. *Chem. Biol.* **10,** 383–396.

Hennessy, B. T., Garcia-Manero, G., Kantarjian, H. M., and Giles, F. J. (2003). DNA methylation in haematological malignancies: The role of decitabine. *Expert. Opin. Investig. Drugs* **12,** 1985–1993.

Johnstone, R. W. (2002). Histone-deacetylase inhibitors: Novel drugs for the treatment of cancer. *Nature Rev. Drug Discov.* **1,** 287–299.

Kelly, W. K., O'Connor, O. A., Krug, L. M., Chiao, J. H., Heaney, M., Curley, T., MacGregore-Cortelli, W., Tong, W., Secrist, J. P., Schwartz, L., Richardson, S., Chu, E., Olgac, S., Marks, H., Scher, H., and Richon, V. M. (2005). Phase I study of an oral histone deacetylase inhibitor, suberoylanilide hydroxamic acid, in patients with advanced cancer. *J. Clin. Oncol.* **23,** 3923–3931.

Kitabayashi, I., Aikawa, Y., Yokoyama, A., Hosoda, F., Nagai, M., Kakazu, N., Abe, T., and Ohki, M. (2001). Fusion of MOZ and p300 histone acetyltransferases in acute monocytic leukemia with a t(8;22)(p11;q13) chromosome translocation. *Leukemia* **15,** 89–94.

Konishi, N., Nakamura, M., Kishi, M., Nishimine, M., Ishida, E., and Shimada, K. (2002). DNA hypermethylation status of multiple genes in prostate adenocarcinomas. *Jpn. J. Cancer Res.* **93,** 767–773.

Lambert, J. R., and Nordeen, S. K. (2001). Analysis of steroid hormone-induced histone acetylation by chromatin immunoprecipitation assay. *Methods Mol. Biol.* **176,** 273–281.

Li, J., Staver, M. J., Curtin, M. L., Holms, J. H., Frey, R. R., Edalji, R., Smith, R., Michaelides, S. K., Davidsen, S. K., and Glaser, K. B. (2004). Expression and functional characterization of recombinant human HDAC1 and HDAC3. *Life Sci.* **74,** 2693–2705.

Lipinski, C. A., Lombardo, F., Dominy, B. W., and Feeney, P. J. (2001). Experimental and computational approaches to estimate solubility and permeability in drug discovery and development settings. *Adv. Drug Deliv. Rev.* **46,** 3–26.

Magdinier, F., and Wolffe, A. P. (2001). Selective association of the methyl-CpG binding protein MBD2 with the silent p14/p16 locus in human neoplasia. *Proc. Natl. Acad. Sci. USA* **98,** 4990–4995.

Martinez, E. D., Rayasam, G. V., Dull, A. B., Walker, D. A., and Hager, G. L. (2005). An estrogen receptor chimera senses ligands by nuclear translocation. *J. Steroid Biochem. Mol. Biol.* **97**(4), 307–321.

McLaughlin, F., and La Thangue, N. B. (2004). Histone deacetylase inhibitors open new doors in cancer therapy. *Biochem. Pharmacol.* **68,** 1139–1144.

Melki, J. R., Warnecke, P., Vincent, P. C., and Clark, S. J. (1998). Increased DNA methyltransferase expression in leukaemia. *Leukemia* **12,** 311–316.

Mizuno, S., Chijiwa, T., Okamura, T., Akashi, K., Fukumaki, Y., Niho, Y., and Sasaki, Y. (2001). Expression of DNA methyltransferases DNMT1, 3A, and 3B in normal hematopoiesis and in acute and chronic myelogenous leukemia. *Blood* **97,** 1172–1179.

Nakamura, M., Sakaki, T., Hashimoto, H., Nakase, H., Ishida, E., Shimada, K., and Konishi, N. (2001). Frequent alterations of the p14(ARF) and p16(INK4a) genes in primary central nervous system lymphomas. *Cancer Res.* **61,** 6335–6339.

Roman-Gomez, J., Castillejo, J. A., Jimenez, A., Gonzalez, M. G., Moreno, F., Rodriguez, M. C., Barrios, J., Maldonado, J., and Torres, A. (2002). 5' CpG island hypermethylation is associated with transcriptional silencing of the p21(CIP1/WAF1/SDI1) gene and confers poor prognosis in acute lymphoblastic leukemia. *Blood* **99,** 2291–2296.

Ryan, Q. C., Headlee, D., Acharya, M., Sparreboom, A., Trepel, J. B., Ye, J., Figg, W. D., Hwang, Chung, E. J., Murgo, A., Melillo, G., Elsayed, Y., Monga, M., Kalnitskiy, M., Zwiebel, J., and Sausville, E. A. (2005). Phase I and pharmacokinetic study of MS-275, a histone deacetylase inhibitor, in patients with advanced and refractory solid tumors or lymphoma. *J. Clin. Oncol.* **23,** 3912–3922.

Sandor, V., Bakke, S., Robey, R. W., Kang, M. H., Blagosklonny, M. V., Bender, J., Brooks, R., Piekarz, R., Tucker, E., Figg, W. D., Chan, K. K., Goldspiel, B., Fojo, A. T., Balcerzak, S. P., and Bates, S. E. (2002). Phase I trial of the histone deacetylase inhibitor, depsipeptide (FR901228, NSC 630176), in patients with refractory neoplasms. *Clin. Cancer Res.* **8,** 718–728.

Schultz, B. E., Misialek, S., Wu, J., Tang, J., Conn, M. T., Tahilramani, R., and Wong, L. (2004). Kinetics and comparative reactivity of human class I and class IIb histone deacetylases. *Biochemistry* **43,** 11083–11091.

Shigeno, K., Yoshida, H., Pan, L., Luo, J. M., Fujisawa, S., Naito, K., Nakamura, S., Shinjo, K., Takeshita, R., Ohno, R., and Ohnishi, K. (2004). Disease-related potential of mutations in transcriptional cofactors CREB-binding protein and p300 in leukemias. *Cancer Lett.* **213,** 11–20.

Suzuki, H., Gabrielson, E., Chen, W., Anbazhagan, R., van, E. M., Weijenberg, M. P., Herman, J. G., and Baylin, S. B. (2002). A genomic screen for genes upregulated by demethylation and histone deacetylase inhibition in human colorectal cancer. *Nature Genet.* **31,** 141–149.

Vigushin, D. M., and Coombes, R. C. (2002). Histone deacetylase inhibitors in cancer treatment. *Anticancer Drugs* **13,** 1–13.

Villar-Garea, A., and Esteller, M. (2004). Histone deacetylase inhibitors: Understanding a new wave of anticancer agents. *Int. J. Cancer* **112,** 171–178.

Warrell, R. P., Jr., He, L. Z., Richon, V., Calleja, E., and Pandolfi, P. P. (1998). Therapeutic targeting of transcription in acute promyelocytic leukemia by use of an inhibitor of histone deacetylase. *J. Natl. Cancer Inst.* **90,** 1621–1625.

Yokochi, T., and Robertson, K. D. (2004). DMB (DNMT-magnetic beads) assay: measuring DNA methyltransferase activity *in vitro*. *Methods Mol. Biol.* **287,** 285–296.

[3] Development of Assays for Nuclear Receptor Modulators Using Fluorescently Tagged Proteins

By Elisabeth D. Martinez and Gordon L. Hager

Abstract

This chapter describes a method for designing cell-based assays to screen for nuclear receptor modulators. The basic strategy consists in following the movement of the receptors from the cytoplasm into the nucleus in response to ligand binding or analogous activating events. The receptors are tagged with green fluorescent protein for automated, fluorescent detection. In the case of constitutively nuclear receptors, they are engineered for cytoplasmic retention in the absence of an activating signal by fusing them to specific regions of the glucocorticoid receptor, which is found predominantly in the cytoplasm of cultured cells. The resulting chimeras respond to ligands or receptor modulators by translocating into the nucleus. This movement is monitored easily by automated fluorescent microscopy and serves as the basis for screening libraries. Finally, secondary assays built into the cell system can differentiate between modulators that stimulate, inhibit, or do not affect the transcriptional activity of the receptor under study. This approach has been validated for both the estrogen receptor and the retinoic acid receptor and should be applicable to any member of the superfamily, facilitating the identification of new ligands and selective receptor modulators.

Introduction

Nuclear Receptor Biology

Nuclear receptors are ligand-activated transcription factors that mediate a variety of important biological functions, including ion transport and salt balance, glucose and lipid metabolism, and the development and maintenance of reproductive organs. At the molecular level, a large number of human diseases result from the malfunctioning of these receptors. Examples include various forms of cancer caused, in part, by mutations in receptors such as the androgen, estrogen, or thyroid hormone receptors (Crescenzi *et al.*, 2003; Kato *et al.*, 2004; Shi *et al.*, 2002; Yamamoto *et al.*, 2001), congenital adrenal hypoplasia caused by DAX-1 mutations (Muscatelli *et al.*, 1994), mild early onset obesity caused by SHP mutations (Nishigori *et al.*, 2001), vitamin D-resistant rickets caused by mutations in the vitamin D receptor

METHODS IN ENZYMOLOGY, VOL. 414
0076-6879/06 $35.00
DOI: 10.1016/S0076-6879(06)14003-3

(Van Maldergem *et al.*, 1996), familial glucocorticoid resistance caused by glucocorticoid receptor mutations (de Lange *et al.*, 1997), and type I pseudohypoaldosteronism caused by mutations in the mineralocorticoid receptor (Tajima *et al.*, 2000). Agonist or antagonist ligands can be utilized to modulate the transcriptional activity of these receptors and some have been used successfully in the clinic. Two challenges exist, however. First, finding selective ligands that have the desired effect in one target tissue (e. g., in the breast in the case of antiestrogens) without deregulating receptor signaling in other tissues (e.g., in the ovary or bone) has been generally difficult. Second, for a number of receptors, the "orphan nuclear receptors," the endogenous ligands are still unknown. These orphan receptors have been the subject of much study and although ligands have been found relatively recently for several of them, the ligands of other orphans whose structures contain a ligand-binding pocket remain elusive. Given the clinical significance of the nuclear receptor superfamily and the therapeutic potential that modulators of these receptors do and can offer, we decided to develop cell-based assays to facilitate the identification of novel ligands and receptor modulators.

Rationale for the Development of Translocation Assays for Nuclear Receptor Ligand Discovery

Our goal was to develop assays for the discovery of nuclear receptor modulators that would offer maximal biological relevance and would detect a broad range of ligands (not just full agonists) through a signal that could be measured easily by automation and thus would be amenable to high-throughput applications. Conventional screens for receptor ligands usually use transcriptional activity as a readout. Transcriptional activity in response to ligand activation, however, can be cell line dependent, due to the coregulators that a cell type may express, and thus is not the best system for a broad screen. In the case of growth-promoting steroids, proliferation assays have also been used to screen for novel ligands. These assays, however, require several days to complete and are restricted in scope, as they may miss ligands that interact with the receptor and affect the transcription of a subset of genes without having an effect on proliferation (as could be the case for some selective modulators). Taking the aforementioned into consideration, we decided to exploit an early event in receptor activation as our assay readout: the cytoplasmic to nuclear translocation of the receptor that occurs upon ligand binding. This strategy can be used without the need for receptor manipulation in the case of receptors that are predominantly cytoplasmic in cell culture in the absence of hormone and undergo nuclear translocation only when bound by ligand, such as the glucocorticoid receptor and the aryl hydrocarbon receptor. Receptors that are predominantly nuclear even when unliganded, such as the estrogen receptor, the vitamin D receptor, retinoic acid receptor (RAR), the peroxisome proliferator-activated receptors (PPARs), and others, first need to be engineered for cytoplasmic retention. A strategy for this engineering is presented later. Advantages of using the movement of the receptor from the cytoplasm to the nucleus as a readout in ligand screens are several. First, this event can be triggered by a wide variety of ligands (pure agonists, pure antagonists, selective modulators), giving the screen breath in its scope. Second, the movement of receptors into the nucleus can be monitored easily by fusing the receptor to green fluorescent protein (GFP) or another similar fluorescent protein. In addition, this biological process is less dependent on the cellular context and takes place in the order of minutes, giving a real-time live cell readout if desired.

GFP-Based Technology as a Powerful Tool for Automated Drug Discovery of Nuclear Receptor Modulators

With recent advances in fluorescent microscopy and the development of various forms of the GFP for labeling of fusion proteins *in vivo*, technologies for drug discovery have emerged that combine the strengths of automation with the power of *in vivo* or *in vitro* fluorescence detection. GFP is a particularly attractive reporter system for drug discovery because it can be measured in live cells, it does not require a cumbersome or lengthy assay for its detection but simply light excitation, it can be fused to most protein targets without altering their basic biological activity, its cellular location can be tracked, and its detection is automated easily. For these reasons, the cell-based assays described in this chapter are based on GFP technology. We and others have found that fusing GFP to members of the nuclear receptor superfamily or to chimeric receptors does not alter their function or their native cellular localization (Hager, 1999; Wan *et al.*, 2001). Because the nucleus can be stained with DNA dyes for easy detection and used as a reference, increases in GFP nuclear localization can be monitored effectively by automated microscopy.

Methods

Design of Translocating Nuclear Receptor Chimeras

Although all receptors shuttle within the cell, as mentioned earlier, some nuclear receptors are predominantly cytoplasmic in the absence of hormonal stimulation. Such receptors can be fused to GFP and used directly to establish stable cell lines for high-content screening of nuclear receptor ligands/modulators. Many of the members of the superfamily, however, are found in the nucleus constitutively and require some engineering before they can be used in translocation-based drug screens. The general strategy for the design of translocating nuclear receptors consists in making chimeras between a constitutively nuclear receptor and one that translocates in response to hormone. The main idea is to combine domains such that the chimera remains in the cytoplasm when uninduced and shuttles into the nucleus when induced by ligands of the nuclear receptor of interest. Because its cellular location is very clearly cytoplasmic in uninduced culture cells, the glucocorticoid receptor can be used as a building block to construct chimeras between it and any constitutively nuclear receptor. Because it has been proven to work effectively, the principle that should be applied is to express the glucocorticoid receptor N terminus, DNA-binding domain, and partial ligand-binding domain sequences upstream of partial ligand-binding domain sequences from the nuclear receptor under study (illustrated in Fig. 1). A detailed protocol for the design of these chimeras is given next.

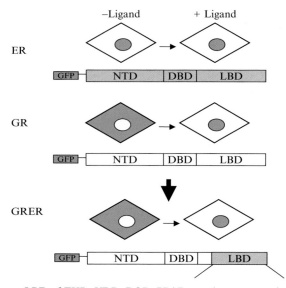

or LBD of THR, VDR, ROR, PPAR or other receptors including orphans

Fig. 1. Graphical representation of the general strategy used for creating nuclear receptors that shuttle between the cytoplasm and the nucleus in response to ligands/receptor modulators. The ligand-binding domain of any constitutively nuclear receptor could be used in place of the ER, which is the example shown here.

Protocol for the Design of Translocating Chimeras

1. Obtain (or construct using standard subcloning procedures) a glucocorticoid receptor (GR) mammalian expression vector tagged with GFP at the N terminus and containing an antibiotic marker under a bacterial promoter. The rat GR works well, yet other species such as the mouse or human GR should give similar results. Note: there should be a short polypeptide between GFP and GR to isolate the GR structure from the GFP structure, which is quite rigid (Hager, 1999).

2. Analyze the restriction enzyme sites in the construct to identify unique sites (or double cutters if necessary) that cut the GR in the sequence corresponding to the loop between helices 1 and 3 of the ligand-binding domain (LBD) (by convention GR does not have a helix 2). Additionally, identify a restriction enzyme that cuts at the very end of the GR expression sequence or just into the vector sequence. The idea is to remove a fragment corresponding to the C terminus of the LBD of the GR, maintaining only sequences that encode helix 1 and a region of the loop between helices 1 and 3. It may be necessary to insert a polylinker with unique restriction enzyme sites or to create a unique site by silent point mutagenesis using polymerase chain reaction in the desired location (Mackem *et al.*, 2001). Although addition of a polylinker will introduce extra sequence, it may facilitate future chimera construction. We have found that the net addition of three amino acids (from polylinker sequences) to the loop between helices 1 and 3 does not disturb the overall receptor structure or its function (see Fig. 2 for an example). Note: when choosing the final digestion strategy, keep in mind the subsequent steps in chimera construction to ensure subcloning compatibility.

3. Digest 1–2 μg of the expression vector with the single cutter restriction enzyme (or do a partial digestion if not using a single cutter) to obtain a linear fragment lacking the desired region of the GR LBD as defined in step 2. Gel purify the fragment. This is the "vector fragment."

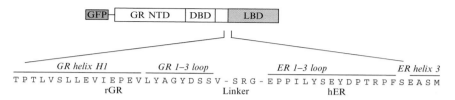

Fig. 2. Detailed view of the amino acid sequence at the fusion site between glucocorticoid and estrogen receptor chimeras showing the addition of three residues introduced by the presence of a linker used for cloning purposes. A small duplication of the loop between helices 1 and 3 results from this chimera construction based on the homology comparison of the receptor sequences (see Fig. 3).

```
rGR    438  KLCLVCSDEASGCHYGVLTCGSCKVFFKRAVEGQHNYLCAGRNDCIIDKIRRKNCPACRY  497
hER α  183  RYCAVCNDYASGYHYGVWSCEGCKAFFKRSIQGHNDYMCPATNQCTIDKNRRKSCQACRL  242

       498  RKCLQAGMN----LEARKTKKKIKGIQQ---------ATAGVSQDTSENPNKTIV----  539
       243  RKCYEVGMMKGGIRKDRRGGRMLKHKRQRDDGEGRGEVGSAGDMRAANLWPSPLMIKRSK  302

       540  --PAALPQLTPTLVSLLEVIEPEVLYAGYDSSVPDSAWRIMTTLNMLGGRQVIAAVKWAK  597
       303  KNSLALSLTADQMVSALLDAEPPILYSEYDPTRPFSEASMMGLLTNLADRELVHMINWAK  362

       598  AILGLRNLHLDDQMTLLQYSWMFLMAFALGWRSYRQSSGNLLCFAPDLIIN-EQRMSLPC  656
       363  RVPGFVDLTLHDQVHLLECAWLEILMIGLVWRS-MEHPGKLL-FAPNLLLDRNQGKCVEG  420

       657  MYDQCKHMLFVSSELQRLQVSYEEYLCMKTLLLLSS----VPKEGLKSQE-------LFD  705
       421  MVEIFDMLLATSSRFRMMNLQGEEFVCLKSIILLNSGVYTFLSSTLKSLEEKDHIHRVLD  480

       706  EIRMTYIKELGKAIVKREGNSSQNWQRFYQLTKLLDSMHEV  746
       481  KITDTLIHLMAKAGLTLQ----QQHQRLAQLLLILSHIRHM  517
```

FIG. 3. Amino acid sequence homology between rGR and hER α proteins shows how the receptors align in their secondary structure. Residues corresponding to helix 1 are underlined for reference.

4. Obtain an expression vector or cDNA for the nuclear receptor of interest. Run a pairwise comparison of the amino acid sequences for the GR and this receptor, marking the various helices to get an idea of how the two LBDs align (an example of the comparison between rGR and hER is shown in Fig. 3). This can be done using Blast or other public software. Analyze the DNA sequence of the receptor under study for restriction enzyme sites to obtain the region of the LBD equivalent to the deleted GR. Design the final construct to have as close to a full LBD as may be feasible, minimizing areas of deletion or duplication. We have found that duplicating a small region of the loop between helices 1 and 3 is preferable to having a net deletion in this area (Martinez *et al.*, 2005).

5. Once the best digestion strategy has been found, digest 3 to 4 μg of the receptor of interest and gel isolate the main LBD fragment. This is the "insert."

6. Analyze the compatibility of the vector and insert fragments for ligation. Blunt end fragments if necessary using standard procedures to fill in or digest overhangs. Ensure that no coding sequences are deleted in this process.

7. Perform conventional ligation reactions and use the ligation products to transform competent bacteria. Plate the bacteria on agar plates with the selection antibiotic and grow colonies overnight. Pick colonies and prepare plasmid DNA. Digest plasmid DNA with restriction enzymes that can predict the presence and orientation of the insert. Sequence candidate constructs to ensure the correct fusion point and the presence of wild-type sequences. Prepare batches of plasmid DNA for storage and for use in subsequent steps.

8. Use the plasmid DNA obtained in step 7 to transiently transfect adherent mammalian cells. For example, transfect NIH 3T3 cells using the Lipofectamine 2000 reagent following the manufacturer's protocol. Briefly, plate cells on cover glass chambers (such as on the Nunc Lab Tek II series) and allow them to attach to the plate at least overnight. Transfect the cells (at about 60 to 70% confluency) the next day and incubate them for 24 h. View the cells growing on cover glass chambers in an inverted fluorescent microscope to detect the expression and localization of the GFP-tagged chimera. The chimera should be found mainly in the cytoplasm. Next, test its ability to translocate into the nucleus by adding a cognate ligand for the receptor whose LBD is present in the construct. Treat a separate well with glucocorticoids (i.e., dexamethasone) as a negative control (dexamethasone should not bind to or translocate the chimera unless the LBD is from a receptor that cross-reacts with glucocorticoids such as potentially the mineralocorticoid receptor). Incubate for 1 to 3 h and monitor translocation by fluorescence microscopy. It is important to evaluate the phenotype of the chimeric receptor in more than one cell type to establish if cytoplasmic retention is a generalized phenomenon.

Note: for specific examples of how this complete design strategy can be applied, please see the published studies by Mackem *et al.* (2001) and Martinez *et al.* (2005).

This protocol should yield cytoplasmic chimeric receptors that respond to ligands by translocating into the nucleus. If the chimera is not efficiently retained in the cytoplasm or if it does not translocate in response to ligand, the fusion site may need to be optimized. The studies described in the references mentioned earlier should serve as guides for specific design strategies. From our experience, however, the optimal fusion site for the chimeras is between helix 1 and helix 3 of the LBD.

Construction of Mammalian Cell Lines Stably Expressing GFP-Tagged Chimeric Receptors

It has been reported that nuclear receptors can behave differently when they are stably expressed versus transiently expressed in cells (Smith *et al.*, 1993). This is thought to be in part due to processing differences and to differential association with heat shock proteins and immunophilins (Botos *et al.*, 2004). Because these associations can influence cytoplasmic/nuclear retention of the receptors, it is advisable to establish stable expression of the chimeras to ensure their cytoplasmic localization and their response to ligands by nuclear translocation. Regardless, stable cell lines are necessary for drug screening, as they avoid the need for expensive and repeated transient transfections and they offer a more homogeneous expression.

Before establishing stable cell lines, the investigator should consider whether to use a constitutive or an inducible system for expression of the chimeras. While constitutive expression is more economic and simple and does not require activation steps, some cells may not tolerate high receptor levels and may silence receptor expression or lose the inserted construct after a few passages. Constitutive systems are less problematic in this regard, yet they require extra steps during the establishment of the cell line and during experimental procedures to induce receptor expression. Our experience has been that fibroblasts and nonendocrine cells usually tolerate receptor expression well, while some cells of endocrine origin (such as C127) tend to be less tolerant of exogenous receptors.

Note: see section entitled "Automated Measurements of Nuclear Translocation" on secondary assays before establishing stable cell lines.

If the end user decides to establish stable cell lines with constitutive expression of the chimera, standard protocols can be used, such as the one outlined by Martinez *et al.* (2006). If inducible systems are preferred, we recommend following the procedures outlined by Walker *et al.* (1999). In either case, established cell clones should be analyzed for GFP-chimera expression (see later), and clones that are homogeneous and show good expression over several passages should be chosen for further studies.

Preliminary Evaluation of a Cell-Based Assay for Nuclear Receptor Ligands Using Fluorescence Microscopy

Before proceeding onto drug screens, the stably expressed chimeric receptor should be evaluated for its translocation efficiency in response to known ligands. This is usually done manually before miniaturization for automated procedures. It is important to establish the response of the chimera to a broad set of ligands and to measure the dose responsiveness of the translocation event, as both of these parameters constitute the foundation of successful high-content drug screens. Although the endogenous ligands of a number of orphan receptors have not been identified, synthetic ligands exist for several of them. These are available commercially through Calbiochem, Cayman, Tocris, and other providers.

Simple Protocol for Evaluation of Known Ligands

1. Plate 50,000 to 200,000 cells per well of a two-well cover glass chamber (Nunc Lab Tek II series) and allow cells to attach and grow for a day.

2. Treat cells with vehicle or with individual known ligands (full and partial agonists and antagonists) for 1 to 5 h.

3. Visualize cells as described earlier for transiently transfected cells. If the GFP expression is not very strong or if the medium gives off

autofluorescence, it may be necessary to visualize the cells in phosphate-buffered saline (PBS) or Hank's balanced salt solution (HBSS). When imaging is prolonged for longer than 10 min, we recommend adding serum to the buffers. The cells can be observed and imaged live or can be observed first, then fixed, and imaged after fixation. Fixation protocols (we use 3.5% paraformaldehyde in PBS for 10 min at room temperature) should not significantly alter the distribution of the chimeric receptor. It is good practice, however, to always visualize the cells before fixing them to ensure this. Although the fixation state is optional, it can give added flexibility, as these cells can be imaged up to 48 h after fixation if stored properly. Store the fixed cells at 4° in PBS containing 0.1% paraformaldehyde and protected from the light. Some signal loss may be experienced after 48 h of storage.

Note: It is expected that all cognate ligands of the receptor under study will cause the receptor to translocate into the nucleus, although some antagonists may require high concentrations, that is, micromolar, and/or may not result in complete translocation of the chimera. If any known ligand does not induce translocation, it would be advantageous to evaluate whether it can inhibit translocation in response to other ligands. If this is the case, other such translocation inhibitors could be screened for by cotreating the cells with an agonist that induces translocation together with drugs or molecules from compound libraries and monitoring for cytoplasmic retention.

Protocol for Manual Evaluation of Dose Responsiveness and EC$_{50}$ Calculations

To determine if the assay is feasible for measuring the potency and efficacy of novel ligands/modulators, the dose responsiveness of the translocation process can be measured and EC$_{50}$ values calculated. Although several imaging systems are equipped with cytoplasmic to nuclear translocation algorithms available in their software packages (discussed later), it is advisable to go through these measurements manually at least once to ensure proper understanding of the calculations and correct setup of the analysis journals.

1. Plate and treat cells as described in the previous section. Fix the cells as described earlier and stain nuclei with DAPI. Image on both the GFP and the DAPI channels at least 20 cells per concentration of ligand whose contours are clearly delineated (we recommend using a 40× objective and imaging areas with multiple cells in them). Superimpose or combine the GFP and the DAPI images and save these dual-channel files without overriding the original separate files.

2. Using any standard imaging software (such as MetaVue), open the combined files and define the area of the nucleus by the DAPI signal. Use

the contour of the cell to define the full area of each individual cell (shown in Fig. 4). Choose an area outside the cell to measure the intensity of the background in the GFP channel. Measure and record the average background intensity per pixel, the integrated intensity of the GFP signal over the full cell area, and the integrated intensity of the GFP signal over the nucleus. Multiply the background value per pixel times the area of the cell or the nucleus. Subtract this value from the corresponding integrated intensities of the whole cell or the nucleus. Repeat these calculations for about 20 cells per ligand concentration. Express the fluorescence in the nucleus as a percentage of the total cell fluorescence and average the percentage nuclear fluorescence for each concentration.

3. Plot out the average percentage nuclear fluorescence against drug concentration. Using graphing software, define a semilog scale by choosing

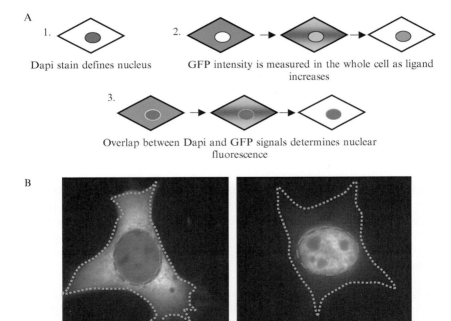

FIG. 4. Measurements of receptor nuclear translocation. (A) Schematic of the procedure for defining the nucleus and the area of the whole cell showing increased overlap in the nuclear dye and the GFP signal as ligand concentration increases. (B) A cell with predominantly cytoplasmic fluorescence (left) and one with predominantly nuclear fluorescence (right) are shown with the areas where intensity measures should be taken and used for calculations highlighted. Blue line delineates the nucleus; the green line delineates the whole cell.

logarithmic units for the x axis. Fit the curve to standard formulas for EC_{50} calculations assuming a receptor–agonist interaction (Zhang and Danielsen, 1995). From this calculation, obtain the maximal value and the EC_{50} for the translocation process. The maximum corresponds to ligand efficacy; the EC_{50} defines ligand potency.

Note: from our experience, full agonists exhibit similar efficacies, as expected, while antagonists vary in their efficacy. A typical range for percentage nuclear fluorescence is between 20% in the absence of ligand and 75 to 80% at the highest concentrations of ligand.

Automated Measurements of Nuclear Translocation

The power of automation can be applied to measure translocation as a function of drug concentration. Among the imaging systems currently available with software for automated calculations of cytoplasmic to nuclear shuttling are the Discovery-1 from Universal Imaging, the IC 100 originally designed by Q3DM, the INCell analyzer from General Electrics, and the ArrayScan from Cellomics. The standard algorithms that can be purchased with these imaging systems allow for various degrees of end-user manipulation. When ready to begin drug screens, the investigator should test the various systems to see which one best fits the specific cell-based assay. We have had good results using both imaging and analysis tools of the IC 100 from Q3DM/Beckman, as well as the Discovery-1 system from Universal Imaging. We have not tested other systems directly.

When setting up automated imaging and data analysis, several parameters should be optimized. The choice of multiwell plates is an important component of both image quality and assay cost. Glass bottom, optic plastic and regular plastic plates can be tested unless the manufacturer of the imaging system clearly recommends a particular type of plate. Cell density and cell growth conditions for miniaturization will also need to be optimized. For the actual imaging, the best exposure time for each channel, the number of regions imaged per well, and the range for focus draft between channels will need to be determined. We have found it helpful to always focus on the DAPI channel without refocusing on the GFP channel. This ensures good focus even when the GFP signals are weak or when dust particles interfere with fluorescence detection. Once the parameters have been optimized, screens can proceed. Analyzed data should be generated from each run for the unbiased identification of hits, that is, of compounds, small molecules or other biologically active agents that cause translocation of the chimeric receptor into the nucleus (or that inhibit agonist-induced translocation). These hits should be retested for confirmation and then evaluated in secondary assays.

Built-in Secondary Assays

A cell-based screening protocol is not complete until hits are evaluated in secondary assays that test the ability of the hit to modulate a particular biological event in a manner that is distinct from the screening assay. Because the screening here is based on translocation of the receptor from the cytoplasm to the nucleus, an appropriate secondary assay would measure a downstream event in receptor action. We recommend building such an assay into the original stable cell line system for maximal relevance and ease of measurement. A good complement to the fluorescent-based primary translocation assay is a luciferase-based assay for transcriptional activity. When establishing stable cell lines as described earlier, a reporter construct can be introduced into the cells together with the expression vector for the chimeric receptor and a selection marker. Because the chimera contains the DNA-binding domain of the GR, the reporter gene (such as luciferase) should be driven by a promoter containing glucocorticoid response elements, that is, MMTV, 5xGRE, or tyrosine aminotransferase. Luciferase activity in response to the hits from the screen can be measured in a 96-well format in high- or low-throughput fashion, depending on the number of hits obtained. Luciferase assays should also be performed with individual hits in the presence of a known agonist to measure the potential antagonist activity of the hits. Alternative secondary assays amenable to automation include whole cell or *in vitro*-binding competition studies or measuring certain phenotypes specific to the activation of the receptor under study (such as apoptosis of a particular cell type by a ligand or morphological changes induced by ligand-activated differentiation).

Perspectives and Future Applications

The movement of proteins between compartments in the cell in response to biological signals can be measured using fluorescence technology and offers multiple opportunities for drug discovery applications. In the case of nuclear receptors, translocation from the cytoplasm to the nucleus triggered by ligand binding can be measured automatically by tagging the receptor fluorescently. This movement can be mimicked in nontranslocating nuclear receptors by designing glucocorticoid receptor chimeras as described in this chapter. To date, several orphan receptors await the identification of their cognate ligands, as do the over 200 nuclear receptors expressed in *Caenorhabditis elegans*. Conventional approaches to find these ligands have not yet succeeded. The strategy presented here could aid in catalyzing this discovery. In addition, modulators of many other proteins whose cellular localization is altered or regulated by signaling cascades could be targeted for fluorescence cell-based drug screens.

References

Botos, J., Xian, W., Smith, D. F., and Smith, C. L. (2004). Progesterone receptor deficient in chromatin binding has an altered cellular state. *J. Biol. Chem.* **279,** 15231–15239.

Crescenzi, A., Graziano, M. F., Carosa, E., Papini, E., Rucci, N., Nardi, F., Trimboli, P., Calvanese, E. A., Jannini, E. A., and D'Armiento, M. (2003). Localization and expression of thyroid hormone receptors in normal and neoplastic human thyroid. *J. Endocrinol. Invest.* **26,** 1008–1012.

de Lange, P., Koper, J. W., Huizenga, N. A., Brinkmann, A. O., de Jong, F. H., Karl, M., Chrousos, G. P., and Lamberts, S. W. (1997). Differential hormone-dependent transcriptional activation and -repression by naturally occurring human glucocorticoid receptor variants. *Mol. Endocrinol.* **11,** 1156–1164.

Hager, G. L. (1999). Studying nuclear receptors with green fluorescent protein fusions. *Methods Enzymol.* **302,** 73–84.

Kato, Y., Ying, H., Willingham, M. C., and Cheng, S. Y. (2004). A tumor suppressor role for thyroid hormone beta receptor in a mouse model of thyroid carcinogenesis. *Endocrinology* **145,** 4430–4438.

Mackem, S., Baumann, C. T., and Hager, G. L. (2001). A glucocorticoid/retinoic acid receptor chimera that displays cytoplasmic/nuclear translocation in response to retinoic acid: A real time sensing assay for nuclear receptor ligands. *J. Biol. Chem.* **276,** 45501–45504.

Martinez, E. D., Dull, A. B., Beutler, J. A., and Hager, G. L. (2006). High-content fluorescence-based screening for epigenetic modulators. *Methods Enzymol.* **414** (this volume).

Martinez, E. D., Rayasam, G. V., Dull, A. B., Walker, D. A., and Hager, G. L. (2005). An estrogen receptor chimera senses ligands by nuclear translocation. *J. Steroid Biochem. Mol. Biol.* **97**(4), 307–321.

Muscatelli, F., Strom, T. M., Walker, A. P., Zanaria, E., Recan, D., Meindl, A., Bardoni, B., Guioli, S., Zehetner, G., Rabl, G., *et al.* (1994). Mutations in the DAX-1 gene give rise to both X-linked adrenal hypoplasia congenita and hypogonadotropic hypogonadism. *Nature* **372,** 672–676.

Nishigori, H., Tomura, H., Tonooka, N., Kanamori, M., Yamada, S., Sho, K., Inoue, I., Kikuchi, K., Onigata, K., Kojima, I., Kohama, T., Yamagata, K., Yang, Q., Matsuzawa, Y., Miki, S., Seino, S., Kim, M. Y., Choi, H. S., Lee, Y. K., Moore, D. D., and Takeda, J. (2001). Mutations in the small heterodimer partner gene are associated with mild obesity in Japanese subjects. *Proc. Natl. Acad. Sci. USA* **98,** 575–580.

Shi, X. B., Ma, A. H., Xia, L., Kung, H. J., and Vere White, R. W. (2002). Functional analysis of 44 mutant androgen receptors from human prostate cancer. *Cancer Res.* **62,** 1496–1502.

Smith, C. L., Archer, T. K., Hamlin-Green, G., and Hager, G. L. (1993). Newly expressed progesterone receptor cannot activate stable, replicated mouse mammary tumor virus templates but acquires transactivation potential upon continuous expression. *Proc. Natl. Acad. Sci. USA* **90,** 11202–11206.

Tajima, T., Kitagawa, H., Yokoya, S., Tachibana, K., Adachi, M., Nakae, J., Suwa, S., Katoh, S., and Fujieda, K. (2000). A novel missense mutation of mineralocorticoid receptor gene in one Japanese family with a renal form of pseudohypoaldosteronism type 1. *J. Clin. Endocrinol. Metab.* **85,** 4690–4694.

Van Maldergem, L., Bachy, A., Feldman, D., Bouillon, R., Maassen, J., Dreyer, M., Rey, R., Holm, C., and Gillerot, Y. (1996). Syndrome of lipoatrophic diabetes, vitamin D resistant rickets, and persistent Mullerian ducts in a Turkish boy born to consanguineous parents. *Am. J. Med. Genet.* **64,** 506–513.

Walker, D., Htun, H., and Hager, G. L. (1999). Using inducible vectors to study intracellular trafficking of GFP-tagged steroid/nuclear receptors in living cells. *Methods* **19,** 386–393.

Wan, Y., Coxe, K. K., Thackray, V. G., Housley, P. R., and Nordeen, S. K. (2001). Separable features of the ligand-binding domain determine the differential subcellular localization and ligand-binding specificity of glucocorticoid receptor and progesterone receptor. *Mol. Endocrinol.* **15,** 17–31.

Yamamoto, Y., Wada, O., Suzawa, M., Yogiashi, Y., Yano, T., Kato, S., and Yanagisawa, J. (2001). The tamoxifen-responsive estrogen receptor alpha mutant D351Y shows reduced tamoxifen-dependent interaction with corepressor complexes. *J. Biol. Chem.* **276,** 42684–42691.

Zhang, S., and Danielsen, M. (1995). 8-Br-cAMP does not convert antagonists of the glucocorticoid receptor into agonists. *Recent Prog. Horm. Res.* **50,** 429–435.

[4] The Ligand-Independent Translocation Assay: An Enabling Technology for Screening Orphan G Protein-Coupled Receptors by Arrestin Recruitment

By Robert H. Oakley, Christine C. Hudson, Michael D. Sjaastad, and Carson R. Loomis

Abstract

Finding natural and/or synthetic ligands that activate orphan G protein-coupled receptors (oGPCRs) is a major focus in current drug discovery efforts. Transfluor® is a cell-based GPCR screening platform that utilizes an arrestin–green fluorescent protein conjugate (arrestin-GFP) to detect ligand interactions with GPCRs. The assay is ideally suited for oGPCRs because binding of arrestin-GFP to activated receptors is independent of the interacting G protein. Before embarking on a high-throughput screen, it is important to know that the target oGPCR can actually bind arrestin-GFP. This information was thought to be inaccessible, however, as arrestin-GFP recruitment is an agonist-driven process. This chapter describes an assay that enables GPCRs to be validated in Transfluor in the absence of ligand. This assay, termed the ligand-independent translocation (LITe) assay, utilizes a modified G protein-coupled receptor kinase to bypass the requirement of ligand for initiating arrestin-GFP translocation. Using the LITe assay, one can determine if an oGPCR binds arrestin-GFP and if the response is quantifiable by high-content screening instruments. In addition, the assay expedites the development and identification of oGPCR stable cell lines with the best Transfluor properties. In this way, the assay provides criteria for selecting the best oGPCRs to move forward for a Transfluor screening campaign. Moreover, the assay can be used for quality control purposes during the orphan receptor screen itself by providing positive translocation responses for calculation of Z prime values. In summary, the LITe assay is a powerful new technology that enables a faster and more reliable path forward in the deorphanization of GPCRs with Transfluor.

METHODS IN ENZYMOLOGY, VOL. 414   0076-6879/06 $35.00
DOI: 10.1016/S0076-6879(06)14004-5

Introduction

G protein-coupled receptors (GPCRs) comprise the largest family of cell-surface receptors. They span the plasma membrane a characteristic seven times and function to transduce extracellular cues into intracellular signaling events that regulate numerous physiological processes. These receptors represent one of the most important targets for drug discovery as approximately 50% of all marketed drugs regulate their activity (Howard, 2001). The GPCR superfamily contains approximately 720 members, yet only a third of these have a known ligand and function (Wise *et al.*, 2004). The remaining members are termed orphan receptors (oGPCRs) because their ligand and function are unknown. While many of the oGPCRs belong to the sensory class, approximately 160 represent potential therapeutic targets (Wise *et al.*, 2004). Finding natural or surrogate ligands that activate these oGPCRs is essential for understanding their physiology and is a major focus in current drug discovery programs.

Upon binding ligand, GPCRs undergo conformational changes that promote their interaction with heterotrimeric guanine nucleotide-binding proteins (G proteins) composed of α, β, and γ subunits. This interaction triggers the exchange of guanosine 5′-diphosphate (GDP) for guanosine 5′-triphosphate (GTP) on the $G\alpha$ subunit and the subsequent dissociation of the G protein from the receptor and the $G\alpha$ subunit from the $G\beta\gamma$ subunits. The released G proteins regulate distinct effectors either positively or negatively, resulting in changes in the level of intracellular second-messenger signaling molecules. Ligand binding not only initiates intracellular signaling cascades, but also triggers mechanisms that limit the magnitude and duration of the signaling event. This process, termed desensitization, protects cells from being overstimulated and is mediated by two protein families, G protein-coupled receptor kinases (GRKs) and arrestins (Claing *et al.*, 2002; Ferguson, 2001). The seven-member GRK family phosphorylates agonist-occupied receptors on serine and threonine residues located predominantly in the carboxyl-terminal tail to create high-affinity binding sites for arrestins. Arrestins comprise a family of four proteins that reside in the cytoplasm and translocate to and bind GRK-phosphorylated receptors at the plasma membrane. Binding of a single arrestin sterically prevents the receptor from activating additional G proteins, thereby terminating or "arresting" the receptor response. Arrestins subsequently target the desensitized receptor to clathrin-coated pits for endocytosis. Once internalized, GPCRs are either recycled back to the cell surface to respond again to agonist (resensitization) or degraded inside the cell (downregulation).

Transfluor technology is a cell-based GPCR screening platform that utilizes an arrestin–green fluorescent protein conjugate (arrestin-GFP) to detect ligand interactions with GPCRs (Oakley *et al.*, 2002). In the absence of agonist, the target receptor is inactive and arrestin-GFP is diffusely distributed in the cytoplasm. Upon activation of the receptor by agonist, arrestin-GFP translocates rapidly to the receptor at the plasma membrane and localizes with it in clathrin-coated pits. Two classes of GPCRs can be distinguished based on the stability of the receptor–arrestin complex (Oakley *et al.*, 2000, 2005). Class A or "pit-forming" receptors form transient complexes with arrestin-GFP that dissociate soon after the receptor is targeted to clathrin-coated pits. Class B or "vesicle-forming" receptors form stable complexes with arrestin-GFP that internalize as a unit into endocytic vesicles. The agonist-dependent redistribution of arrestin-GFP from a diffuse cytoplasmic pattern to a granular pattern at the plasma membrane (pits) or inside the cell (vesicles) provides a direct readout on the activation status of the target GPCR that is accurately quantitated by a variety of high-content screening instruments (Comley, 2005).

One of the unique features of Transfluor that makes it particularly well suited for screening oGPCRs is that the assay works for receptors independent of their interacting G protein (Oakley *et al.*, 2006). Published reports describe arrestin-GFP translocation to over 20 receptors that couple to $G\alpha_s$, over 20 receptors that couple to $G\alpha_i$, and over 30 receptors that couple to $G\alpha_q$. Therefore, screening an orphan GPCR with Transfluor, as opposed to traditional functional assays, requires no prior knowledge of the signaling pathway. Eliminating the need to determine the interacting G protein not only streamlines the drug discovery process but also may make it more productive. This is highlighted by reports describing GPCRs, such as the chemokine D6 receptor, that do not couple to classic G proteins but do bind arrestin-GFP (Galliera *et al.*, 2004). In fact, it has been suggested that the slow progress made using traditional assays in the deorphanization effort may indicate that many of the remaining oGPCRs signal through G protein-independent mechanisms (Wise *et al.*, 2004).

Before conducting a high-throughput screening campaign with Transfluor against a target orphan receptor, it is important to know that the receptor can bind arrestin-GFP and that the translocation response can be quantitated by the available automated image analysis system. However, without an agonist to activate the receptor (the case for all oGPCRs), how does one perform such assay validation? Faced with this issue, we developed an assay that allows evaluation of arrestin-GFP recruitment to receptors in the absence of ligand. This assay, termed the ligand-independent translocation (LITe) assay, utilizes a modified GRK to effectively take the place of ligand for initiating arrestin-GFP translocation to oGPCRs. This chapter describes the practice and utility of the LITe assay for oGPCRs to be screened with Transfluor.

Overview of the LITe Assay

Translocation of arrestin-GFP has been shown for many GPCRs to be dependent on both agonist occupancy and GRK phosphorylation. Several receptors, however, have been reported to recruit arrestin-GFP in the absence of ligand, including the 5-hydroxytryptamine 2C receptor (5HT2CR) and the human cytomegalovirus US28 receptor (Marion *et al.*, 2004; Miller *et al.*, 2003). Each of these receptors exhibits strong arrestin-GFP binding that is dependent on GRK-mediated phosphorylation but independent of agonist. GPCRs are thought to exist in an equilibrium between inactive and active conformations (Samama *et al.*, 1993). In the absence of agonist, most receptors spend the majority of their time in the inactive conformation. Occasionally, they isomerize into the active state and couple to G proteins to produce constitutive, or agonist-independent, activity. Binding of agonist markedly shifts the equilibrium by stabilizing the active conformation of the receptor. The unbound 5HT2C and US28 receptors are unique in that they spend the majority of their time in the active conformation, resulting in a high level of constitutive activity (Marion *et al.*, 2004; Miller *et al.*, 2003). These findings indicate that arrestin recruitment is not dependent on agonist binding per se, but on the adoption of the receptor active conformation. Without the stabilizing presence of the agonist, however, the active state for most GPCRs occurs too infrequently and is too short-lived for arrestin-GFP binding to be observed by fluorescence microscopy.

The GRK2 isoform of the GRK family resides in the cytoplasm of cells and is recruited to agonist-activated receptors by liberated $G\beta\gamma$ subunits (Pitcher *et al.*, 1998). To develop a ligand-independent translocation assay, we altered GRK2 by adding a CAAX motif (where C is cysteine, A is a small aliphatic residue, and X is an uncharged amino acid) to its carboxyl terminus. This motif directs the attachment of a 20-carbon lipid group (geranylgeranyl) to GRK2 and results in the insertion of the kinase in the plasma membrane (Inglese *et al.*, 1992). The mutant GRK2 (referred to as GRK2m) is no longer dependent on agonist-mediated release of free $G\beta\gamma$ subunits for recruitment to the cell surface. By positioning the kinase constitutively at the membrane (via the mutation) and in the vicinity of the GPCR (via overexpression), GRK2m is poised to immediately phosphorylate receptors that transition into the active conformation before they isomerize back to the inactive state. As a result, even GPCRs with very short-lived active states can be efficiently phosphorylated and observed to recruit arrestin-GFP in the complete absence of agonist. We call the GRK2m-induced translocation of arrestin-GFP to unoccupied receptors the LITe assay.

A comparison of the classic agonist-induced translocation response and the LITe assay translocation response is shown in Fig. 1 for cells stably

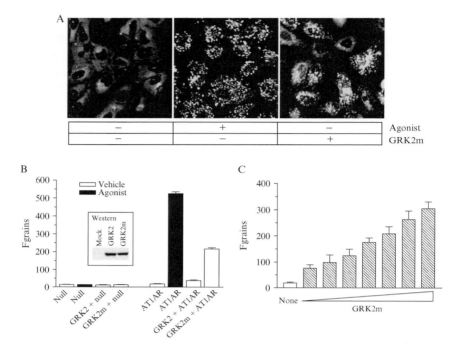

FIG. 1. Comparison of ligand-dependent and GRK2m-dependent translocation of arrestin-GFP to the AT1AR. Experiments were performed on U2OS cells stably expressing arrestin-GFP alone (null) or stably expressing both arrestin-GFP and the AT1AR (AT1AR). For the classic Transfluor assay, 100 nM angiotensin II (agonist) was added to cells for 30 min. For the LITe assay, GRK2m was transiently expressed using baculovirus, and the cells were analyzed the following day having never been exposed to the agonist. The INCell Analyzer 3000 was used for quantitation, and the reported parameter "Fgrains" is a measure of the fluorescent intensity of the spots resulting from localization of arrestin-GFP with the AT1AR in endocytic vesicles. (A) Confocal images showing distribution of arrestin-GFP in cells untreated (left image), treated with agonist (middle image), or transiently expressing GRK2m (right image). (B) Quantitation of arrestin-GFP translocation in cells untreated, treated with agonist, transiently expressing wild-type GRK2, or transiently expressing GRK2m. (Inset) Immunoblot of wild-type GRK2 and GRK2m. (C) Quantitation of arrestin-GFP translocation in cells transduced transiently with increasing amounts of GRK2m.

expressing arrestin-GFP and angiotensin II type 1A receptor (AT1AR). In the absence of agonist and GRK2m, arrestin-GFP is distributed evenly in the cytoplasm of the cells, indicating that the AT1AR, like most GPCRs, spends the majority of its time in the inactive conformation as no translocation is observed (Fig. 1A, left). Upon addition of a saturating concentration of agonist and stabilization of the receptor active state, arrestin-GFP translocates to the AT1AR at the plasma membrane and internalizes with it into endocytic vesicles (Fig. 1A, middle). Quantitation of the response measures approximately 550 Fgrains and represents a 20-fold increase over basal levels (~25 Fgrains) (Fig. 1B).

For the LITe assay, transient transfection of GRK2m results in a dramatic translocation of arrestin-GFP into vesicles with the AT1AR in the complete absence of agonist (Fig. 1A, right). Quantitation of the LITe assay response measures approximately 250 Fgrains and represents a 10-fold increase over basal levels (Fig. 1B). Importantly, the GRK2m-induced translocation response is dependent on overexpression of the AT1AR, as no measurable redistribution of arrestin-GFP is detected upon expression of GRK2m in cells stably expressing arrestin-GFP alone (Fig. 1B, compare values for *GRK2m + null* and *GRK2m + AT1AR*). In addition, the LITe assay response is dependent on the membrane-localized form of GRK2. When expressed at similar levels, GRK2m induces a much greater translocation response than wild-type GRK2 (Fig. 1B, compare values for *GRK2 + AT1AR* and *GRK2m + AT1AR*). Finally, the magnitude of the ligand-independent translocation response is dependent on the amount of over-expressed GRK2m, as a stepwise increase in GRK2m levels produces a stepwise increase in the LITe assay response (Fig. 1C).

Utility of the LITe Assay

Early Validation of oGPCRs in Transfluor

The LITe assay is ideally suited for investigating the arrestin-binding properties of oGPCRs, as no ligand is available for these receptors. Knowing whether and to what extent an oGPCR can bind arrestin-GFP is of great value not only for selecting oGPCRs to move forward for a high-throughput screen but also for ensuring that Transfluor screens are conducted on receptor targets that work well in the assay. Although most agonist-activated receptors recruit arrestin-GFP, there are a few reported exceptions such as the β_3-adrenergic receptor (Cao *et al.*, 2000). In addition, arrestin-GFP translocation to some receptors is weak and difficult to quantitate. For these reasons, we recommend that all oGPCR targets be validated in Transfluor using the LITe assay prior to conducting a large screening campaign.

Early validation studies with the LITe assay are accomplished easily and quickly in transiently transfected cells. Although many cell types support Transfluor, we typically employ either HEK293 cells for their high transfection efficiencies or U2OS cells for their superior image quality (Oakley *et al.*, 2006). To perform the LITe assay, we transiently cotransfect two dishes of cells. The first dish receives the target oGPCR, arrestin-GFP, and empty vector. The second dish receives the target oGPCR, arrestin-GFP, and GRK2m. Twenty-four hours after the transfection, cells are harvested and plated on 35-mm glass-bottom dishes. After an overnight incubation, we replace the culture medium with serum- and phenol-red free medium to reduce autofluorescence and then visualize the distribution of arrestin-GFP using a fluorescent microscope (63× objective and FITC filter set). We then compare the distribution of arrestin-GFP in cells expressing the target oGPCR in the absence and presence of GRK2m. In this way, the LITe assay on orphan receptors is similar to the classic agonist-induced translocation assay on known GPCRs, except that GRK2m takes the place of the agonist. A successful validation will show minimal arrestin-GFP recruitment in cells expressing the target oGPCR alone but a strong translocation response in cells coexpressing the target oGPCR and GRK2m.

Figure 2 shows the results of the LITe assay performed on four different oGPCRs. In the absence of GRK2m, no translocation response is observed for oGPCR1, oGPCR2, or oGPCR3 (Fig. 2, top). In each case, arrestin-GFP is distributed evenly in the cytoplasm of the cells. A weak vesicle response, however, is observed for oGPCR4, as small spots can be seen in the cytoplasm of the cells (Fig. 2, top). This suggests that oGPCR4 either possesses a significant degree of constitutive activity or that a natural ligand for this receptor is present in the medium. Upon coexpression of GRK2m, a profound redistribution of arrestin-GFP into vesicles is observed for each

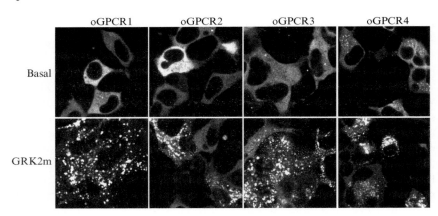

FIG. 2. LITe assay response for four oGPCRs expressed transiently in HEK293 cells. HEK293 cells were transfected transiently by calcium phosphate precipitation with arrestin-GFP, an oGPCR (oGPCR1, oGPCR2, oGPCR3, or oGPCR4), and either empty vector or GRK2m. Thirty-six hours later, the distribution of arrestin-GFP was visualized by fluorescence microscopy. (Top) Confocal images of the distribution of arrestin-GFP in cells expressing the oGPCR alone (basal). (Bottom) Confocal images of the distribution of arrestin-GFP in cells expressing both oGPCR and GRK2m.

of the orphan GPCRs (Fig. 2, bottom). These results in the LITe assay indicate that all four orphans strongly bind arrestin-GFP upon isomerization into the active conformation and suggest that each receptor will be a good candidate for a Transfluor screening campaign.

A negative or weak response in the LITe assay does not automatically disqualify an oGPCR as a target for a Transfluor screen. GPCRs can be genetically modified in their carboxyl-terminal tail to improve their response in Transfluor (Oakley *et al.*, 2006). The molecular motif mediating the stable binding of arrestin-GFP to vesicle-forming GPCRs, such as the AT1AR, has been identified and shown to be transferable (Oakley *et al.*, 1999, 2001). This motif is a cluster of serine/threonine residues positioned properly within the receptor carboxyl-terminal tail that serves as a primary site of GRK phosphorylation. Adding this motif to a receptor can convert a weak pit response into a strong vesicle response without significantly impacting receptor pharmacology. We have successfully employed this genetic alteration on many known and orphan receptors to better enable their detection in Transfluor.

Validation of oGPCR Stable Cell Lines for Transfluor Screening Campaigns

After validating an orphan receptor in Transfluor using the LITe assay in transiently transfected cells, we develop a stable cell line for each target receptor to be screened against a large compound library. Transfluor screening campaigns can be conducted on cells transiently or stably expressing the oGPCR of interest. The transient approach is achieved quickly but can result in more variability in the translocation response due to heterogeneity in receptor expression levels. Stables require a longer time investment but can result in individual clones that express a uniform level of receptor and give a more uniform translocation response. We utilize both approaches but favor stable expression of the target oGPCR for large screens that stretch over days and weeks. We have found the LITe assay to be instrumental for the rapid identification of the oGPCR stable clone with the best properties for a Transfluor screen.

The oGPCR stables are made in U2OS cells because of their superior adherence properties and image quality (Oakley *et al.*, 2006). We begin by transfecting U2OS cells stably expressing arrestin-GFP with the target oGPCR. Antibiotics are applied in order to select for the oGPCR-expressing cells, and the surviving cells are dilution plated to achieve individual clones. To quickly find the best oGPCR clone for an ensuing Transfluor screen, we utilize the LITe assay. Cells from each stable clone are plated into two wells of a 96- or 384-well plate. One well is left untreated for assessment of the basal translocation response, and the other well is transiently transfected with GRK2m. This can be accomplished with traditional lipid-based transfection procedures or by transducing the cells with baculovirus containing GRK2m. The advantage of the baculovirus gene delivery approach is that infection efficiencies greater than 95% can be achieved in U2OS cells. After an overnight incubation, the plates are analyzed on a high-content screening instrument. We then compare, qualitatively and quantitatively, the distribution of arrestin-GFP in the absence and presence of GRK2m for each clone.

To be used in a Transfluor high-throughput screen, the orphan receptor stable clone should meet two criteria. First, in the absence of GRK2m, little to no arrestin-GFP translocation should be observed in the cells. The reason for this requirement is that compounds that weakly activate the oGPCR will be better detected in a "clean" rather than "noisy" background. The top part of Fig. 3 shows images of appropriate basal responses for three different oGPCRs stably expressed in U2OS cells. We strongly recommend a visual inspection of the basal response to take advantage of the high-content value of Transfluor. Viewing the image not only confirms the absence of translocation for a clone receiving a low quantitative score, but also provides information on the overall health of the cells. Receptor clones with a large number of cells displaying vacuolated, rounded-up, and/or overly elongated morphology are discarded as these features predict poor growth and viability.

The second requirement of an oGPCR stable clone is that GRK2m expression induces a large, uniform translocation response that can be

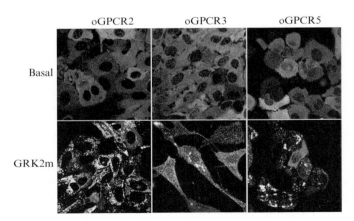

Fig. 3. LITe assay response for three oGPCRs stably expressed in U2OS cells. U2OS cells stably expressing arrestin-GFP and oGPCR2, oGPCR3, or oGPCR5 were plated in 35-mm glass-bottom dishes. For the LITe assay, the stable cell lines were either untreated (basal) or exposed to baculovirus containing GRK2m for 18 h. The distribution of arrestin-GFP was then visualized by fluorescence microscopy. (Top) Confocal images of the distribution of arrestin-GFP in each oGPCR stable in absence of GRK2m (basal). (Bottom) Confocal images of the distribution of arrestin-GFP in each oGPCR stable in the presence of GRK2m.

quantitated easily by the high-content screening instrument of choice. Clones in which the LITe assay response is heterogeneous across the population of cells or poorly quantitated are discarded. The lower part of Fig. 3 shows images of the LITe assay responses for the three different oGPCRs stably expressed in U2OS cells. In each case, almost all the cytoplasmic arrestin-GFP has redistributed into pits and/or vesicles, indicating that the translocation response is very strong. Additionally, both the magnitude and the phenotype of the response are uniform across most of the cells. Quantitation of the LITe assay response for each oGPCR stable measures approximately 300 Fgrains and represents a five- to eightfold increase over basal levels (Fig. 4). Thus, these oGPCR stables are excellent choices for a Transfluor screen, as the combination of a low basal response and a large GRK2m-induced response ensures that each cell line possesses a large signal window for detecting active compounds.

Quality Control for Transfluor Screening Campaigns

The actual Transfluor screen on the chosen oGPCR stable clone must be robust and reproducible such that active compounds can be reliably distinguished from the background noise. The most widely used parameter for evaluating the quality of a high-throughput screening assay is the Z prime value (Zhang et $al.$, 1999). The Z prime value is a statistical value used to determine assay robustness and reproducibility and is derived from the dynamic range and variability of positive and negative controls. A Z prime value of 0.5 or more indicates that data are of excellent quality. As the dynamic range becomes greater and the variability becomes smaller, the Z prime value approaches a perfect score of 1.0. The natural ligand provides the positive control for calculating Z prime values in screens for activators of known GPCRs. With no such ligand available for oGPCRs, Z prime values cannot be determined on the actual target receptor and an important quality control is lost.

We, however, have found that the LITe assay response can be used as a positive control for calculating Z prime values and evaluating the integrity of the oGPCR Transfluor screen. This is shown in Fig. 5 for a small screen of approximately 6000 compounds. U2OS cells stably expressing oGPCR3 are plated in 384-well plates. Wells 373–384 (column 24) are exposed to GRK2m-containing baculovirus and serve as the positive control. Wells 361–372 (column 23) are left untreated and serve as the negative control. After an overnight incubation, compounds are added to wells 1–360 (columns 1–22) and vehicle to wells 361–384. The plates are then analyzed on a high-content screening instrument. Figure 5A plots the basal and GRK2m-induced translocation responses for each of the 384-well plates. The basal

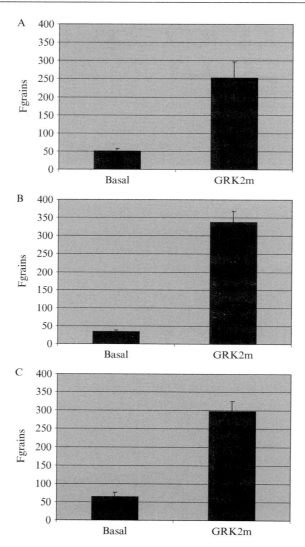

FIG. 4. Quantitation of the LITe assay response for three oGPCRs stably expressed in U2OS cells. U2OS cells stably expressing arrestin-GFP and oGPCR2 (A), oGPCR3 (B), or oGPCR5 (C) were plated in 384-well plates. For the LITe assay, wells were either untreated (basal) or exposed to baculovirus containing GRK2m for 18 h. The plates were then analyzed for arrestin-GFP translocation using the INCell Analyzer 3000.

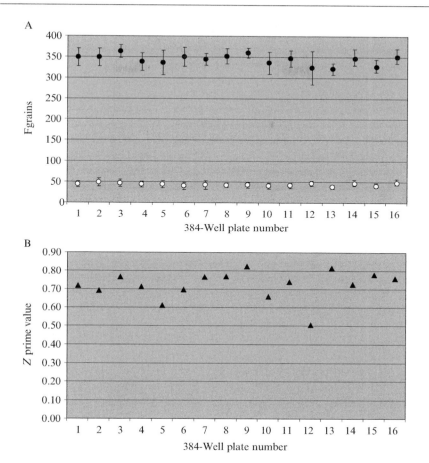

FIG. 5. LITe assay response provides quality control for oGPCR Transfluor screening campaigns. U2OS cells stably expressing arrestin-GFP and oGPCR3 were plated in 384-well plates for a Transfluor agonist screen. For the LITe assay, wells in column 23 of each plate were left untreated (basal) and wells in column 24 of each plate were exposed to baculovirus containing GRK2m. The following day compounds were added to remaining wells (columns 1–22) and plates were analyzed on the INCell Analyzer 3000 for arrestin-GFP translocation. (A) Quantitation of arrestin-GFP translocation in the absence ($\bigcirc$) and presence of GRK2m ($\bullet$) across 16 consecutive plates. (B) Calculated Z' prime values for each of the plates shown in A using the basal and GRK2m-induced responses for the negative and positive controls, respectively.

response (negative control) averaged 44 Fgrains, whereas the GRK2m-induced response (positive control) averaged 344 Fgrains, representing an eightfold signal to background ratio. The calculated Z prime values from each 384-well plate are shown in Fig. 5B and range from 0.51 to 0.82 with an overall average of 0.72, indicative of a high-quality screen. We have found similar high Z prime values for other oGPCRs and in large screening campaigns of greater than 100,000 compounds.

Conclusion

The LITe assay employs a modified GRK2 that constitutively resides at the plasma membrane to bypass the requirement of an agonist for inducing arrestin-GFP translocation to GPCRs. The assay has proven to be of great value for validating oGPCRs in Transfluor, for identifying the oGPCR stable clones with the best Transfluor properties, and for serving as a quality control during the Transfluor screen. In this way, the LITe assay ensures that the orphan receptor screening campaigns performed with Transfluor have the highest probability of success.

Acknowledgments

The authors thank Allen E. Eckhardt, Rachael D. Cruickshank, Diane M. Meyers, Richard E. Payne, Kirsten R. Wille, Jeremy L. Rouse, Shay Mullins, Michael T. Ouellette, Bryan W. Sherman, and Conrad L. Cowan for their experimental contributions to the LITe assay.

References

Cao, W., Luttrell, L. M., Medvedev, A. V., Pierce, K. L., Daniel, K. W., Dixon, T. M., Lefkowitz, R. J., and Collins, S. (2000). Direct binding of activated c-Src to the beta 3-adrenergic receptor is required for MAP kinase activation. *J. Biol. Chem.* **275,** 38131–38134.
Claing, A., Laporte, S. A., Caron, M. G., and Lefkowitz, R. J. (2002). Endocytosis of G protein-coupled receptors: Roles of G protein-coupled receptor kinases and ss-arrestin proteins. *Prog. Neurobiol.* **66,** 61–79.
Comley, J. (2005). High content screening: Emerging importance of novel reagents/probes and pathway analysis. *Drug Disc. World* **6,** 31–53.
Ferguson, S. S. (2001). Evolving concepts in G protein-coupled receptor endocytosis: The role in receptor desensitization and signaling. *Pharmacol. Rev.* **53,** 1–24.
Galliera, E., Jala, V. R., Trent, J. O., Bonecchi, R., Signorelli, P., Lefkowitz, R. J., Mantovani, A., Locati, M., and Haribabu, B. (2004). Beta-Arrestin-dependent constitutive internalization of the human chemokine decoy receptor D6. *J. Biol. Chem.* **279,** 25590–25597.
Howard, A. D., McAllister, G., Feighner, S. D., Liu, Q., Nargund, R. P., Van der Ploeg, L. H., and Patchett, A. A. (2001). Orphan G-protein-coupled receptors and natural ligand discovery. *Trends Pharmacol. Sci.* **22,** 132–140.
Inglese, J., Koch, W. J., Caron, M. G., and Lefkowitz, R. J. (1992). Isoprenylation in regulation of signal transduction by G-protein-coupled receptor kinases. *Nature* **359,** 147–150.
Marion, S., Weiner, D. M., and Caron, M. G. (2004). RNA editing induces variation in desensitization and trafficking of 5-hydroxytryptamine 2c receptor isoforms. *J. Biol. Chem.* **279,** 2945–2954.
Miller, W. E., Houtz, D. A., Nelson, C. D., Kolattukudy, P. E., and Lefkowitz, R. J. (2003). G-protein-coupled receptor (GPCR) kinase phosphorylation and beta-arrestin recruitment

regulate the constitutive signaling activity of the human cytomegalovirus US28 GPCR. *J. Biol. Chem.* **278,** 21663–21671.

Oakley, R. H., Barak, L. S., and Caron, M. G. (2005). Real time imaging of GPCR-mediated arrestin translocation as a strategy to evaluate receptor-protein interactions. *In* "G Protein Coupled Receptor-Protein Interactions" (S. R. George and B. F. O'Dowd, eds.), pp. 53–80. Wiley, New York.

Oakley, R. H., Cowan, C. L., Hudson, C. C., and Loomis, C. R. (2006). Transfluor provides a universal cell-based assay for screening G protein-coupled receptors. *In* "Handbook of Assay Development in Drug Discovery" (L. Minor, ed.), pp. 435–457. Dekker, New York.

Oakley, R. H., Hudson, C. C., Cruickshank, R. D., Meyers, D. M., Payne, R. E., Jr., Rhem, S. M., and Loomis, C. R. (2002). The cellular distribution of fluorescently labeled arrestins provides a robust, sensitive, and universal assay for screening G protein-coupled receptors. *Assay Drug Dev. Technol.* **1,** 21–30.

Oakley, R. H., Laporte, S. A., Holt, J. A., Barak, L. S., and Caron, M. G. (1999). Association of beta-arrestin with G protein-coupled receptors during clathrin-mediated endocytosis dictates the profile of receptor resensitization. *J. Biol. Chem.* **274,** 32248–32257.

Oakley, R. H., Laporte, S. A., Holt, J. A., Barak, L. S., and Caron, M. G. (2001). Molecular determinants underlying the formation of stable intracellular G protein-coupled receptor-beta-arrestin complexes after receptor endocytosis. *J. Biol. Chem.* **276,** 19452–19460.

Oakley, R. H., Laporte, S. A., Holt, J. A., Caron, M. G., and Barak, L. S. (2000). Differential affinities of visual arrestin, beta arrestin1, and beta arrestin2 for G protein-coupled receptors delineate two major classes of receptors. *J. Biol. Chem.* **275,** 17201–17210.

Pitcher, J. A., Freedman, N. J., and Lefkowitz, R. J. (1998). G protein-coupled receptor kinases. *Annu. Rev. Biochem.* **67,** 653–692.

Samama, P., Cotecchia, S., Costa, T., and Lefkowitz, R. J. (1993). A mutation-induced activated state of the beta 2-adrenergic receptor: Extending the ternary complex model. *J. Biol. Chem.* **268,** 4625–4636.

Wise, A., Jupe, S. C., and Rees, S. (2004). The identification of ligands at orphan G-protein coupled receptors. *Annu. Rev. Pharmacol. Toxicol.* **44,** 43–66.

Zhang, J. H., Chung, T. D., and Oldenburg, K. R. (1999). A simple statistical parameter for use in evaluation and validation of high throughput screening assays. *J. Biomol. Screen* **4,** 67–73.

[5]   High-Content Screening of Known G Protein-Coupled Receptors by Arrestin Translocation

By CHRISTINE C. HUDSON, ROBERT H. OAKLEY, MICHAEL D. SJAASTAD, and CARSON R. LOOMIS

Abstract

G protein-coupled receptors (GPCRs) have proven to be one of the most successful target classes for drug discovery. Accordingly, many assays are available to screen GPCRs, including radioactive-binding assays, second messenger signaling assays, and downstream reporter assays. One of

METHODS IN ENZYMOLOGY, VOL. 414   0076-6879/06 $35.00
                                                   DOI: 10.1016/S0076-6879(06)14005-7

the more novel approaches is the Transfluor technology, a cell-based assay that uses a detectable tag on a cytosolic protein, called arrestin, that is involved in the desensitization or inactivation of GPCRs. Monitoring the translocation of GFP-tagged arrestin from the cytosol to activated GPCRs at the plasma membrane measures the pharmacological effect of test compounds that bind the receptor target. Moreover, the Transfluor assay provides further, high-content information on the test compound itself and its effects on cell processes due to the fluorescent imaging of whole cells used in this screen. Screening known GPCRs with Transfluor against large compound libraries is best accomplished in cell lines stably expressing an optimum level of the target receptor. This chapter describes how to generate a clonal cell line stably expressing the known GPCR with suitable Transfluor properties. It then describes the steps involved in performing a Transfluor screen and discusses high content data resulting from the screen.

Introduction

G Protein-Coupled Receptors (GPCRs), G Protein-Coupled Receptor Kinases (GRKs), and Arrestins

G protein-coupled receptors comprise a large class of proteins that regulate many physiological functions such as sight, taste, smell, neurotransmission, cardiac output, and pain perception. In response to ligand binding, GPCRs activate heterotrimeric guanine nucleotide binding proteins (G proteins). Stimulation of the G protein results in its dissociation into an α subunit and a $\beta\gamma$ dimer that in concert initiates signaling cascades within the cell to produce a physiological response. Specifically, levels of the second messenger molecules—calcium and cyclic AMP—are altered, and transcription factors, such as NFAT and CREB, are activated to regulate gene expression. Agonist activation of GPCRs also begins the processes of receptor desensitization, internalization, and recycling (Ferguson, 2001). These processes are necessary to prevent overstimulation of the receptor in the presence of continuous agonist and are mediated by GRKs and arrestins.

G protein-coupled receptor kinases are a family of seven protein kinases responsible for phosphorylating GPCRs after agonist binding (Ferguson, 2001; Pitcher et al., 1998). GRK1 and GRK7 are expressed only in the eye and phosphorylate rhodopsin. GRK2 and GRK3, also known as βARK1 and βARK2, are expressed ubiquitously and are recruited to activated GPCRs by the $\beta\gamma$ subunits of the G proteins. GRK4, GRK5, and GRK6 are also expressed ubiquitously, but unlike GRK2 and GRK3, they are anchored at the plasma membrane by either palmitoylation or electrostatic interactions. Upon agonist activation of a GPCR, there is a conformational change resulting in a more open receptor configuration.

If a GRK is in close proximity to a GPCR in its open conformation, the GRK is able to phosphorylate serine and threonine residues on the intracellular loops and C-terminal tail of the GPCR (Fig. 1, step 1).

Arrestins are a family of four cytoplasmic, scaffolding proteins responsible for turning off the G protein-dependent signaling of an activated GPCR (Ferguson, 2001). Arrestin1, or visual arrestin, and arrestin4,

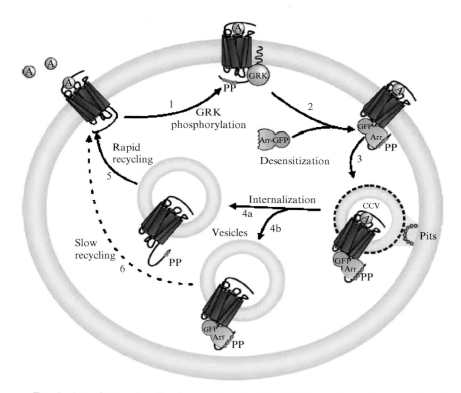

FIG. 1. Arrestin translocation to agonist-activated GPCRs and desensitization. Step 1: Upon agonist binding (A), GPCRs activate G proteins (not pictured) and change conformation to expose a site in the C-terminal tail for phosphorylation (PP) by GRK kinases. Step 2: Arrestin or a fusion protein of arrestin and GFP (Arr-GFP) binds the agonist-occupied, GRK-phosphorylated GPCR and occludes the receptor from binding and activating its G protein in a process termed desensitization. Step 3: Arrestin then targets the desensitized GPCR to clathrin-coated pits, wherein in step 4 the GPCR is internalized in clathrin-coated vesicles and delivered to endosomes. Step 4a: For some GPCRs, arrestin dissociates at or near the plasma membrane and is excluded from receptor-containing vesicles. Step 4b: In contrast, arrestin remains associated with other receptors and traffics with them into endocytic vesicles. Step 5: GPCRs that dissociate from β-arrestin at or near the plasma membrane are recycled rapidly, whereas in step 6, GPCRs that remain associated with arrestin are recycled slowly.

or cone arrestin, are only expressed in the eye. Arrestin2, or β-arrestin1, and arrestin3, or ß-arrestin2, are expressed ubiquitously. Arrestins translocate and bind to agonist-activated GPCRs that have been phosphorylated by GRKs (Fig. 1, step 2). Arrestins bind to the GPCR at intracellular sites that prevent the G protein α subunit from associating with the GPCR. Consequently, G protein-dependent signaling is terminated. Arrestin binding also targets the GPCR for endocytosis in clathrin-coated pits by acting as a scaffold linking the GPCR to internalization proteins such as clathrin and AP-2 (Fig. 1, step 3) (Goodman *et al.*, 1996; Laporte *et al.*, 2000). Some GPCRs bind arrestin weakly and dissociate from the arrestin once in clathrin-coated pits at the plasma membrane. These GPCRs are called class A GPCRs and recycle more quickly back to the cell surface to be available for agonist stimulation (Fig. 1, steps 4a and 5). Other GPCRs bind arrestin with high affinity and internalize with arrestin into endosomal vesicles. These GPCRs are called class B GPCRs and recycle more slowly (Fig. 1, steps 4b and 6) (Oakley *et al.*, 2000, 2005).

ArrestinGFP Translocation Assay

Because almost all GPCRs associate with arrestin upon agonist binding, one can measure GPCR activation, or inactivation, by monitoring the translocation of a fluorescently labeled arrestin from the cytosol to an activated GPCR at the plasma membrane (Barak *et al.*, 1997; Oakley *et al.*, 2002, 2006). In Transfluor technology, a green fluorescent protein (GFP) is fused to the carboxy terminus of arrestin, forming an arrestinGFP fusion protein. Both arrestinGFP and the target GPCR are overexpressed in the same cell. In the absence of an agonist, the arrestinGFP is distributed uniformly throughout the cytoplasm and is excluded from the nucleus. This can be visualized in live cells using a fluorescent microscope with the

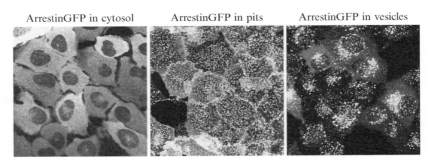

ArrestinGFP in cytosol   ArrestinGFP in pits   ArrestinGFP in vesicles

FIG. 2. Confocal microscopy of arrestinGFP translocation in cells. Images of live U2OS cells stably expressing a known GPCR and arrestinGFP were captured in real time before (left) and after 8 (middle) or 44 (right) min of stimulation with agonist.

appropriate filters for GFP excitation and emission (Fig. 2, left). Within seconds of agonist addition, the arrestinGFP translocates from the cytoplasm to the activated GPCRs and forms clathrin-coated pits at the plasma membrane. These pits resemble small fluorescent dots distributed evenly over the cell surface (Fig. 2, center). If a class A GPCR is stimulated, the pits will remain visible as long as the agonist is present. If a class B GPCR is stimulated, the arrestinGFP will internalize with the GPCRs from the pits to larger endocytic vesicles within several minutes. These vesicles resemble larger and brighter fluorescent dots and typically concentrate in the perinuclear region (Fig. 2, right). The magnitudes of these fluorescent pit and vesicle translocation responses of class A and B receptors can be quantitated on a variety of automated image analysis systems to distinguish different pharmacological profiles of GPCR ligands.

Stable Expression of a Known GPCR for the ArrestinGFP Translocation Assay

Transient Validation

In the development of a cell line stably expressing a known GPCR for the arrestinGFP translocation assay, the target GPCR is first validated in the Transfluor assay in transiently transfected cells. Several cell types have been used for the arrestin translocation assay, but U2OS osteosarcoma cells are recommended for screening campaigns because they adhere strongly to glass and plasticware and their flattened morphology yields high-quality images. We transfect U2OS cells stably expressing arrestinGFP with the target receptor. This is best accomplished using FuGENE6 on cells at 50% confluence. The following day, seed the cells into a 35-mm glass-bottom dish. After an overnight incubation, remove the growth medium and replace with 1 ml of serum and phenol red free-MEM with 10 mM HEPES. Incubate cells at 37° for 30 min to 1 h. During this incubation, examine the cells on a fluorescence microscope with a 20× or higher objective and the appropriate filter to visualize GFP. In the absence of agonist, neither pits nor vesicles should be present in the cells. If pits or vesicles are found, take careful note of their abundance and distribution. Stimulate cells with the agonist specific for the transfected GPCR. Within 2 to 45 min of stimulation, look for cells that have pits or vesicles.

Evaluation of Response

Prior to agonist addition, the arrestinGFP should be distributed uniformly throughout the cytoplasm of the cells and excluded from the nucleus (Fig. 2, left). After agonist addition, the arrestinGFP should be

Strong pit response   Strong vesicle response   Very weak pit response

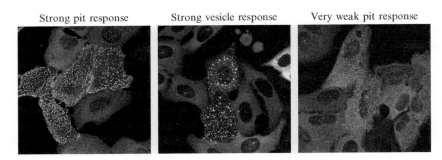

FIG. 3. ArrestinGFP translocation to transiently expressed GPCRs. U2OS cells stably expressing arrestinGFP were transiently transfected with three different GPCRs. Agonists were added, and confocal images of the translocation of arrestinGFP in the live cells were collected.

completely cleared from the cytoplasm and redistributed to clathrin-coated pits or endocytic vesicles (Fig. 3, left and middle). If the transfected receptor is a class A GPCR, arrestinGFP will only complex with agonist-occupied receptors in clathrin-coated pits at the plasma membrane. Pits are usually 0.5–2 μm in size (Fig. 3, left) and remain present on the surface of cells for at least 1 h in the continued presence of agonist. If the transfected receptor is a class B GPCR, arrestinGFP will traffic with the receptor from the clathrin-coated pits into endocytic vesicles. Vesicles are 3 to 5 μm in size (Fig. 3, middle) and remain present inside the cell for several hours even after agonist removal.

Occasionally, transient transfection of a GPCR yields either no arrestin GFP translocation in response to agonist or only a very weak translocation to tiny pits with the majority of the arrestinGFP remaining in the cytoplasm (Fig. 3, right). In these cases, the target receptor can be modified to improve or optimize the arrestinGFP translocation. The molecular motif that allows class B GPCRs to bind tightly to arrestinGFP and traffic together to endocytic vesicles has been identified and found to be transferable (Oakley *et al.*, 1999). Modifying a weak or nonresponding GPCR with this motif usually results in receptors that produce a robust Transfluor response that can be quantitated easily.

With a few GPCRs, arrestinGFP translocation is observed before the agonist is added to the cells. This basal translocation response may result from receptors having a high degree of constitutive activity or be caused by agonist already present in the culture media, especially in fetal bovine serum (FBS). While generally not problematic, basal activity may raise background levels in untreated cells and cause difficulty in developing or growing a cell line containing a GPCR that is under constant stimulation by its agonist. For this condition, culturing cells in reduced amounts of FBS or in a mixture with dialyzed FBS or charcoal/dextran-treated–FBS (stripped serum) is recommended to reduce undesirable levels of basal activity and improve cell growth.

Development of Stable Cell Lines for Known Receptors

Transfection and Drug Selection. After validating a GPCR in Transfluor using transiently transfected cells, we then develop a stable cell line for the receptor. Stable, clonal cell lines are advantageous for screening, because with optimal and uniform levels of receptor expression, there is less variability during screening and ideal screening statistics can be achieved. We begin by transfecting U2OS cells stably expressing arrestinGFP with the target receptor or mock DNA. The next day, the cells are split into a 35-mm glass-bottom dish for evaluating transfection efficiency and into a 15-cm dish for drug selection. If the transfection efficiency is acceptable, we add the appropriate selection antibiotic to the 15-cm dish and to the dish of cells that was transfected with mock DNA. This dish serves as a control to monitor the time needed to completely kill all nontransfected cells.

Evaluating Antibiotic-Resistant Cells (Parent Mix). The surviving antibiotic-resistant cells are known as the parent mix. Before continuing forward with stable development, we evaluate arrestinGFP translocation to the known receptor in the parent mix for two purposes. First is to ensure that the translocation response observed in transiently transfected cells is preserved in cells stably expressing receptors. Second is to evaluate the percentage of cells demonstrating the translocation response. Both results are obtained by seeding one 35-mm glass-bottom dish with 250,000 to 300,000 parent mix cells. After an overnight incubation at 37°, we assess the arrestinGFP translocation response in the absence and presence of a saturating concentration of the target receptor ligand. The magnitude of the agonist-induced response should be preserved, and the percentage of stimulated cells showing the response should have increased markedly as compared to that obtained after the initial transfection.

Clone Selection and Expansion. If the assessment of percentage responding cells in the parent mix is acceptable, then clonal cell populations are obtained by dilution plating in 15-cm gridded dishes with very low cell densities. In approximately 1.5 weeks, colonies on the 15-cm gridded dishes should have grown to an optimal size for picking, ~100 cells per colony. We isolate selected colonies using cloning cylinders and transfer each clone into 1 well of a 24-well plate. If the percentage of positive responding cells in the parent mix equaled 30 to 50%, then ~40 colonies are picked for screening. If the percentage of positive responding cells is less than 30%, then 70 to 100 colonies are picked. As clones grow to confluence in the 24-well plate, feed as necessary. When cells are 70 to 100% confluent in the wells, aspirate media and wash with phosphate-buffered saline (PBS). Add 100 μl of trypsin to the well and incubate at 37° until the cells round up. Then add 1 ml of media. Using a pipette, resuspend the cells and transfer 200 μl to duplicate wells of a 96-well, glass-bottom, black plate and seed the remaining cells in a six-well plate with 2 ml of media.

Identification of Positive Clones. Clones are initially screened in a 96-well glass-bottom plate after an overnight incubation. Replace seed media with 100 μl of phenol red-free MEM and 10 mM HEPES and incubate at 37° for 30 min to 1 h. Next, carefully add 25 μl of a 5× maximal dose of agonist to one of the pair of wells and 25 μl of dilution buffer to the other well. Incubate at 37° for 10 to 20 min. Examine the live cells on the fluorescence microscope and look for clones that meet the criteria described later. Identify three to eight of the best clones to further expand and characterize.

Ideal clones contain 100% of the cells responding to agonist by completely clearing the cytosol of arrestinGFP and redistributing it to pits

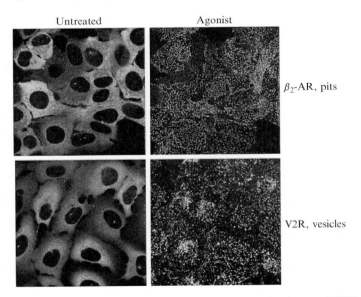

Untreated   Agonist

β_2-AR, pits

V2R, vesicles

Fig. 4. Clonal populations of Transfluor cell lines expressing known GPCRs. (Top) Confocal images of a clonal population of U2OS cells stably expressing the β_2-adrenergic receptor and arrestinGFP in the absence (untreated) or presence of isoproterenol (agonist). The β_2-adrenergic receptor is a class A receptor and produces a pit response. (Bottom) Confocal images of a clonal population of U2OS cells stably expressing the vasopressin 2 receptor and arrestinGFP in the absence (untreated) or presence of arginine vasopressin (agonist). The vasopressin 2 receptor is a class B receptor and produces a vesicle response. Modified from Oakley *et al.* (2002) with permission of Mary Ann Liebert.

or vesicles (Fig. 4, right). The arrestinGFP levels in untreated cells should be uniform and consistent within a clone and void of multiple localized concentrations or aggregates of the arrestinGFP (Fig. 4, left). Very little to no basal arrestinGFP translocation should occur in the absence of agonist. These clones should consist of healthy cells that have a regular, epithelial morphology. Avoid clones with cells that are elongated, rounded, unusually large, or have a high presence of vacuoles.

Choosing the Best Clone for the Screen

Seed the clones that were selected by visual assessment in triplicate rows or columns of a 96- or 384-well plate. Incubate at 37° overnight. The next day replace seed media with serum/phenol red free MEM with 10 mM HEPES (100 μl for a 96-well plate and 20 μl for a 384-well plate), and incubate at 37° for 30 min to 1 hr. Then to the triplicate rows or columns, add 25 μl for a 96-well plate or 10 μl for a 384-well plate of twelve 1:3 dilutions of a 5 or 3$\times$ stock of agonist, respectively, to generate a 12-point concentration–response curve. Incubate at 37° for 30 min. Fix the cells with 4% formaldehyde in PBS and an appropriate nuclear stain such as Hoechst or DRAQ5. Fix for 45 min at ambient temperature. Acquire images and quantitate the arrestinGFP translocation response for each of the clones on an image analysis instrument. Quantitate EC_{50} concentrations for the 12 point curve, and signal to background, signal to noise, and Z prime values for the minimum and maximum agonist concentration values for each clone (Zhang *et al.*, 1999).

Compare EC_{50} values that were collected for each clone from the imaging instrument to known EC_{50} values in the literature determined from signaling assays or to K_d values from binding experiments performed with the same agonist. Choose clones with Transfluor EC_{50} values similar to those found in the literature with other assays. Repeat concentration–response curves to ensure consistent results and run multiple agonists, if available, to check for expected pharmacology. Accordingly, if an antagonist screen is the goal, run concentration inhibition curves with known antagonists. Check IC_{50} values from the Transfluor assay with those reported in the literature for signaling assays or with K_i values from binding assays. Also consider how quickly the cells of a clone double and the kind of screening statistics each clone yields. Pick the clone that has a short doubling time and gives high signal to noise and Z prime values in the Transfluor assay. This ensures that the clone grows well, has low background and noise, and has a wide assay window for hits. Once the clone is decided upon, expand the cells and freeze down a large quantity of low passage cells for this clone to be able to expand from in the future.

Figure 5 shows quantitative data for three clonal cell lines of the same target GPCR. Concentration–response curves and the EC_{50} values of the clones (Fig. 5A) were in close proximity to each other and within a half-log of the literature K_d value for the agonist. Statistical analysis of the three clones (Fig. 5B) showed that all had excellent Z prime values and would provide an adequate assay window to identify hits. Clones differed only slightly in the average quantitation of their arrestinGFP translocation maxima and their basal noise. Because all three clones grew similarly in culture, clone B was chosen for the screen based on having the highest signal:noise and Z prime values.

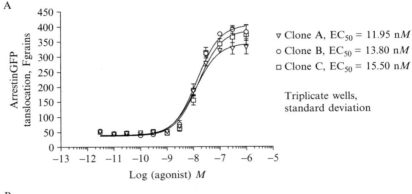

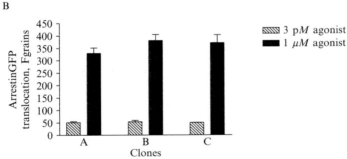

Clone	Ave min	SD min	N min	Ave max	SD max	N max	S:B	S:N	Z prime
A	51.33	4.04	3.00	329.67	22.23	3.00	6.42	12.32	0.72
B	54.33	5.03	3.00	381.00	23.64	3.00	7.01	13.51	0.74
C	50.67	0.58	3.00	371.33	31.94	3.00	7.33	10.04	0.70

FIG. 5. Quantitative data from the INCell 3000, generation 1, of positive, clonal Transfluor cell lines. (A) Concentration–response curves with EC50 values of three clonal populations of cells stably expressing a known GPCR and arrestinGFP. (B) Statistical analysis of the Transfluor response in the three clonal cell populations.

Screening Known Receptors with Transfluor Technology

Meeting the Demand of High-Throughput Screening with Transfluor

High-throughput screening assays need to be predictive of the pharmacology of a test compound on its target, simple and easy to perform, robust, and automatable. The arrestinGFP translocation assay has been shown to meet all of these requirements. First, it has been used successfully to determine the pharmacology of known agonists and antagonists for several GPCRs (Ghosh *et al.*, 2005; Oakley *et al.*, 2006). Next, the Transfluor assay is performed easily because it involves a few, basic steps (Fig. 6). In contrast to other high-content screening assays, which require primary and secondary antibody incubations and washes after cell fixation, no wash steps are necessary with Transfluor. In addition, all steps in the assay are automated easily by liquid-handling robotics such as MultiDrops and MiniTraks, and detection of assay results has been validated on multiple, automated fluorescent imaging platforms (Oakley *et al.*, 2006). Furthermore, the assay is very reproducible and gives excellent screening statistics, including Z prime values above 0.5, ensuring an optimal screening assay window.

Preparation Steps for Screening

Scale-up of Cells. The stable clone of cells expressing the target GPCR must be expanded to supply the cells needed for the actual screen. The number of compounds to be screened, the number of plates used in the screen, the number of cells seeded per well, the amount of excess void

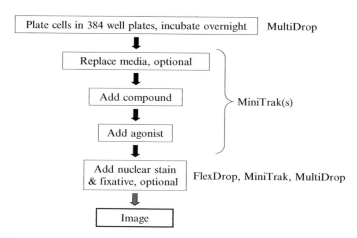

FIG. 6. Transfluor screening protocol steps with associated liquid-handling robotics.

volume of cell seeding solution, and the number of screening days all factor into the total number of cells necessary to complete the screen. To achieve this expansion of cells, low passage number cells are expanded into high surface area cell culture vessels such as roller bottles or cell factories. Cells in the high surface area vessels are fed as necessary until the cells are near confluence, at which time they can be harvested by trypsin or nonenzymatic dissociation reagents and seeded into either more high surface area vessels for further expansion or directly into the assay plates.

Optimization and Validation Experiments. Several important optimization and validation experiments need to be performed on the receptor stable prior to the screen. For example, the appropriate seeding density of U2OS cells expressing the GPCR target and arrestinGFP has to be obtained to ensure the highest-quality data and screening statistics; 5000 cells per well can be used as a starting density for a 384-well plate. In addition, incubation times and temperatures need to be optimized in regards to the arrestinGFP translocation response and constraints with automation. Furthermore, screening conditions for control agonists and/or antagonists need to be determined that address their cost, their final dimethyl sulfoxide concentrations in the screen, their possibility to nonspecifically bind to plasticware used in liquid handling, their stability over time and temperatures, and the concentration(s) to be used in the screen. Finally, a multiplate quality control test run should be done with all the optimized conditions to verify that appropriate screening statistics are achievable. If possible, a test run should be included with a subset of compounds from the larger library to mimic screening day conditions and confirm that all aspects of the screen, including the cellular response in the assay, liquid-handling robotics, and image acquisition and analysis, are all functioning correctly.

Screening a Known GPCR in the ArrestinGFP Translocation Assay

Liquid-Handling Assay Protocol. An outline of a typical Transfluor screening protocol is shown in Fig. 6. Liquid-handling robots are used to seed 384-well plates with an appropriate number of cells per well. The plates are incubated overnight at $37°/5\%$ CO_2. The next day the media in the cell assay plates is removed and replaced with assay buffer (serum and phenol red-free MEM with 10 mM HEPES). Test compounds are then transferred from compound library plates to the cell assay plates. For agonist screens, plates are incubated for 40 min at room temperature. The assay is then stopped by addition of a solution of formaldehyde and nuclear stain. For antagonist screens, after incubating the cells with compounds for 40 min, an appropriate concentration of agonist is added to each well for an additional incubation. The assay is then stopped as described earlier with the formaldehyde/nuclear stain addition.

Imaging Plates. After fixation, the assay plates can be sealed and imaged with the fixative still on them. Proper excitation and emission filters are selected to excite and detect the GFP fluorophore, and the fluorophore is used to stain nuclei of the cells. Neutral density filters, exposure times, camera gain, binning of pixel data, and the order of detecting fluorophores must be optimized to yield an unsaturated image that is highly resolved in the shortest period of time. The image size from a well also needs to be optimized for speed of acquisition and to ensure an adequate number of cells for the quantitation. In addition, a z axis offset of the objective may be set to help visualize the arrestinGFP translocation in the cells. This is especially useful for visualizing the pit response that is best seen at the surface where the cells sit on the plate bottom.

Data Analysis

Algorithms used by imaging platforms usually identify the number of cells present in the image(s) of a well by counting the number of fluorescently labeled nuclei. The algorithms then quantitate the number of fluorescent spots per well or per cell, the area and size of the spots, and/or the intensity of the spots. These data are used to quantitate the appearance of a pit or vesicle response if an agonist screen was performed or the absence of a pit or vesicle response in the case of an antagonist screen. Raw data can be used directly or it can be transformed further with software to calculate percentage inhibition or percentage response in relation to control wells. Mean and standard deviations are calculated from positive and negative controls on each plate and are then compared across all plates of the screen. Signal to background, signal to noise, and Z prime values are calculated from the positive and negative controls on each plate and averaged for the entire screen to determine the validity and accuracy of the screen. Responses from wells that meet a specified threshold are flagged as a hit by the software and can be reviewed visually.

Harvesting High Content Data

In addition to numerical data that are collected for measuring the arrestinGFP translocation, other information is also collected that can yield more high content information, such as the number of cells per well, the fluorescence intensity of the GFP or nuclei stain channels, overall cell morphology, and size/area of nuclei (Ghosh *et al.*, 2005; Oakley *et al.*, 2006). This additional, high content information can be harvested and used to automatically filter wells that have been flagged as hits into groups with similar characteristics. For example, by filtering the hits that have decreased number of cells in the well or smaller nuclei, one can separate out the wells that might have toxicity issues. Similarly, by filtering the hits that exceed a certain fluorescence intensity of the GFP channel and/or the nuclei stain channel, one can separate out the wells that have fluorescent compounds that could cause a false flagging of a well as a hit.

Figure 7 shows images of flagged wells that were automatically scored to rule out false hits quickly and prioritize the hits to further pursue. Figure 7A and B are active hits from an agonist and antagonist screen, respectively. These images are within numerical threshold limits for the fluorescence of their nuclei stain and the GFP signal. The images of these active hits are clear of artifacts, precipitates, and excess fluorescence. Cells appear to be in good health with no signs of starvation or toxicity. Figure 7C is an image of a false-positive fluorescent well in an antagonist screen. The fluorescence intensities in the GFP and nuclear stain channels are saturating and preclude the discernment of any other cellular details, such as the presence of pits or vesicles. Figure 7D is an image of a toxic, false-positive well for an antagonist screen. There are fewer cells per well and the cells are severely rounded. The size of the nuclei has shrunk and few, healthy cells are present that could show a Transfluor response. Figure 7E is an image of a

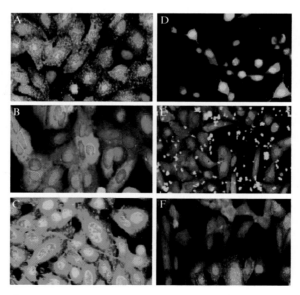

FIG. 7. Transfluor screening hits. (A) Active hit for agonist screen. (B) Active hit for antagonist screen. (C) Fluorescent false positive. (D) Toxic false positive. (E) Precipitate. (F) Miscellaneous; image is partially out of focus. Images from INCell 3000 generation 1.

well with a precipitate that appears as an excess of fluorescent spots throughout the well in either or both the GFP and nuclear stain channels. Precipitates can mask the arrestinGFP translocation response, cause toxicity to the cells, and appear as false-positive hits in an agonist screen. Figure 7F is an example of a flagged well that can be scored as miscellaneous. It is an image of an out-of-focus well. The quantitation for this well is inaccurate, as only a few cells in the well are in focus for measuring the arrestin GFP translocation response.

Conclusion

By exploiting the nearly universal process of GPCR desensitization to measure GPCR activation by arrestinGFP translocation in whole cells, Transfluor provides high content information on the test compound itself and its effects on cell processes. The Transfluor assay utilizes a cell line expressing a fluorescently tagged arrestin that resides in the cytosol of unstimulated cells. Upon transfection of a target GPCR and stimulation with a known agonist, the fluorescently tagged arrestin molecules translocate from the cytosol to the receptors at the plasma membrane, forming fluorescently labeled pits or endocytic vesicles. The redistribution of the fluorescence within whole cells can be measured by high-throughput, imaging platforms. Optimum screening conditions with the Transfluor assay are obtained with the generation and selection of a clonal cell line permanently expressing the target GPCR and arrestinGFP. ArrestinGFP translocation results are then collected easily using a simple Transfluor screening protocol. Imaging instruments used in the screening of known GPCRs for arrestinGFP translocation can make use of multiple data parameters, algorithms, and assays run in parallel to acquire high-content information on the test compounds, cells, and signaling events within the cells. This information can be used to automatically score the images of flagged wells, reduce the number of images to inspect visually, eliminate false-positives quickly, and prioritize hits for further pursuit. With the advent of running a second messenger GPCR assay in parallel with Transfluor (Turpin *et al.*, 2005), a new chapter opens in high-content screening of GPCRs in which researchers are able to measure activation of a GPCR by both signaling and desensitization assays being performed in one well.

Acknowledgments

We thank Richard E. Payne Jr. for his contributions to the stable cell line development section and Conrad L. Cowan, Shay Mullins, Michael Ouellette, and Bryan Sherman for their contributions including images to the screening section.

References

Barak, L. S., Ferguson, S. S., Zhang, J., and Caron, M. G. (1997). A beta-arrestin/green fluorescent protein biosensor for detecting G protein-coupled receptor activation. *J. Biol. Chem.* **272,** 27497–27500.

Ferguson, S. S. (2001). Evolving concepts in G protein-coupled receptor endocytosis: The role in receptor desensitization and signaling. *Pharmacol. Rev.* **53,** 1–24.

Ghosh, R. N., DeBiasio, R., Hudson, C. C., Ramer, E. R., Cowan, C. L., and Oakley, R. H. (2005). Quantitative cell-based high-content screening for vasopressin receptor agonists using Transfluor technology. *J. Biomol. Screen.* **10,** 476–484.

Goodman, O. B., Jr., Krupnick, J. G., Santini, F., Gurevich, V. V., Penn, R. B., Gagnon, J. H., Keen, J. H., and Benovic, J. L. (1996). Beta-arrestin acts as a clathrin adaptor in endocytosis of the beta2-adrenergic receptor. *Nature* **383,** 447–450.

Laporte, S. A., Oakley, R. H., Holt, J. A., Barak, L. S., and Caron, M. G. (2000). The interaction of beta-arrestin with the AP-2 adaptor is required for the clustering of beta 2-adrenergic receptor into clathrin-coated pits. *J. Biol. Chem.* **275,** 23120–23126.

Oakley, R. H., Barak, L. S., and Caron, M. G. (2005). Real time imaging of GPCR-mediated arrestin translocation as a strategy to evaluate receptor-protein interactions. *In* "G Protein Coupled Receptor-Protein Interactions" (S. R. George and B. F. O'Dowd, eds.), pp. 53–80. Wiley, New York.

Oakley, R. H., Cowan, C. L., Hudson, C. C., and Loomis, C. R. (2006). Transfluor provides a universal cell-based assay for screening G protein-coupled receptors. *In* "Handbook of Assay Development in Drug Discovery," (L. Miner, ed.), pp. 435–457. Dekker, New York.

Oakley, R. H., Hudson, C. C., Cruickshank, R. D., Meyers, D. M., Payne, R. E., Jr., Rhem, S. M., and Loomis, C. R. (2002). The cellular distribution of fluorescently labeled arrestins provides a robust, sensitive, and universal assay for screening G protein-coupled receptors. *Assay Drug Dev. Technol.* **1,** 21–30.

Oakley, R. H., Laporte, S. A., Holt, J. A., Barak, L. S., and Caron, M. G. (1999). Association of beta-arrestin with G protein-coupled receptors during clathrin-mediated endocytosis dictates the profile of receptor resensitization. *J. Biol. Chem.* **274,** 32248–32257.

Oakley, R. H., Laporte, S. A., Holt, J. A., Caron, M. G., and Barak, L. S. (2000). Differential affinities of visual arrestin, beta arrestin1, and beta arrestin2 for G protein-coupled receptors delineate two major classes of receptors. *J. Biol. Chem.* **275,** 17201–17210.

Pitcher, J. A., Freedman, N. J., and Lefkowitz, R. J. (1998). G protein-coupled receptor kinases. *Annu. Rev. Biochem.* **67,** 653–692.

Turpin, P., Loui, R., Quast, J., Kassinos, M., Guadet, L., Liao, J. F., Chan, W., Sportsman, R., Rickert, P., and Sjaastad, M. (2005). Combining FLIPR and Transfluor: A novel assay for the sequential analysis of calcium flux and receptor desensitization. *In* "Society for Biomolecular Screening," p. P06021. Geneva, Switzerland.

Zhang, J. H., Chung, T. D., and Oldenburg, K. R. (1999). A simple statistical parameter for use in evaluation and validation of high throughput screening assays. *J. Biomol. Screen.* **4,** 67–73.

[6]   Cell Imaging Assays for G Protein-Coupled Receptor Internalization: Application to High-Throughput Screening

By SEUNGTAEK LEE, BONNIE HOWELL, and PRIYA KUNAPULI

Abstract

There are a number of assays currently available to study G protein-coupled receptors (GPCRs), including ligand binding and functional assays. The latter category, albeit more complex, offers some obvious advantages over traditional ligand-binding assays. Functional cell-based assays typically include second messenger and reporter gene assays, which depend directly or indirectly on the cellular signaling cascade initiated upon receptor activation, respectively. More recently, cell imaging assays monitoring receptor trafficking are becoming increasingly popular. These assays, described in greater detail in this chapter, are independent of receptor signaling and are thus ideally suited for orphan receptors. In addition, these assays provide a valuable measure of receptor desensitization, an important feature for the use of GPCR agonists as potential therapeutic agents. The most popular GPCR imaging assays are based on the principles of receptor desensitization and internalization monitored directly or indirectly by green fluorescent protein.

Introduction

G protein-coupled receptors (GPCRs) constitute one of the largest families of druggable targets in the human genome (Bleicher, 2003; Drews, 2000). Not surprisingly, approximately 45% of currently available pharmaceutical drugs are targeted against this class of cell surface receptors (Bleicher *et al.*, 2003; Hopkins and Groom, 2002). The vast genomics effort has led to the identification of a number of GPCRs in the human genome, some of which remain "orphan" with unknown endogenous ligand and function (Civelli *et al.*, 2001; Marchese *et al.*, 1999). In this postgenomic era, a considerable research effort is aimed at deorphanizing these receptors and understanding their role in human physiology and their potential therapeutic value.

GPCRs constitute a superfamily of cell surface receptors with a common motif of seven membrane-spanning domains. Agonist stimulation initiates a cascade of signals that involve activation of the heterotrimeric GTP-binding proteins (G proteins) (Gautam *et al.*, 1998; Neer, 1995),

METHODS IN ENZYMOLOGY, VOL. 414   0076-6879/06 $35.00
DOI: 10.1016/S0076-6879(06)14006-9

resulting in second messenger-dependent modulation of various effector systems (Gilman, 1987) and feedback regulation of G protein coupling by receptor desensitization and receptor endocytosis (Krupnick and Benovic, 1998; Lefkowitz, 1993).

Functional assays to monitor the GPCR–ligand interaction are useful for overcoming some of the limitations of traditional ligand-binding assays. Commonly used functional assays for GPCRs include second messenger assays (for cAMP, IP_3, or intracellular Ca^{2+}) and reporter gene assays. More recently, the use of receptor trafficking assays is becoming increasingly popular for orphan receptors with unknown cellular signaling, as well as for the determination of receptor desensitization induced by agonists with the potential for use as therapeutic agents.

Elucidation of the mechanism of regulation of GPCR function by receptor desensitization in the 1990s laid the foundation for the receptor internalization/trafficking assay. This assay is unique in being independent of the second messenger signaling modulated by the receptor–ligand interaction. GPCR desensitization (waning of the receptor responsiveness with time) is mediated primarily by two protein families: G protein-coupled receptor kinases (GRKs) and arrestins (Ferguson, 2001). Agonist stimulation of GPCRs promotes the phosphorylation of serine/threonine residues located predominantly in the carboxyl-terminal tail and/or the third intracellular loop of the receptor by the family of GRKs. This in turn triggers the translocation of the arrestin family of proteins from the cytoplasm to the receptors at the plasma membrane. Arrestin binding to the activated and GRK-phosphorylated receptors effectively uncouples the receptor–G protein interaction, thereby terminating receptor signaling (Gurevich et al., 1995).

The arrestin proteins bound to the activated and phosphorylated receptor subsequently target GPCRs for endocytosis via clathrin-coated pits (Goodman et al., 1996; Krupnik et al., 1997; Oakley et al., 2002; Sterne-Marr and Benovic, 1995). Based on the affinity of the arrestin–GPCR interaction, receptor endocytosis results in either transient complexes of the GPCR/arrestin/clathrin-coated pits near the plasma membrane (for low-affinity interactions) or stable GPCR/arrestin complexes that internalize into intracellular vesicles (for high-affinity interactions) (Oakley et al., 1999, 2000, 2001). Thus, the movement of activated GPCRs from the plasma membrane into either pits or vesicles can be monitored as an indication of receptor activation and forms the basis of a functional high-content cell-based assay (Fig. 1).

Among the different GPCR functional cell-based assays currently available, the receptor trafficking assay offers value as a receptor "proximal" readout (unlike the reporter gene assays) and is particularly advantageous for receptors that lack a suitable radioligand for use in receptor-binding assays. In addition, this assay can be used for most GPCRs without a priori

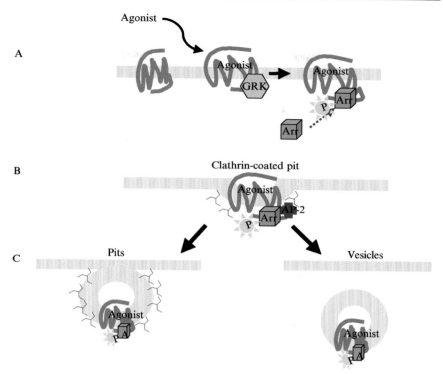

FIG. 1. Schematic representation of GPCR internalization process.

knowledge of the agonist-induced signaling cascade. Another advantage of this functional assay is the potential to identify different categories of ligands: classical competitive antagonists (also identified by the receptor-binding assay), as well as allosteric modulators (i.e., compounds that do not inhibit the binding of a radioligand to the receptor) that presumably inhibit receptor function by binding to sites on the receptor distinct from the agonist-binding pocket.

Monitoring GPCR Trafficking via Receptor–GFP

Overview

Two methods prevail to monitor GPCR trafficking in cells, both involving green fluorescent proteins (GFP). The most direct method is to express the GPCR of interest recombinantly as a chimera with GFP in the carboxyl-terminal tail and monitor the fluorescence localization in the cell upon receptor activation.

This chapter describes internalization of the chemokine receptor CXCR3, monitored by direct fusion of GFP to the carboxyl-terminal of the receptor tail. The human CXCR3 receptor cDNA was engineered with emerald GFP at the carboxyl-terminal tail of the receptor [stop codon and emerald GFP sequence separated by Gly(5)-Ala peptide linker]. The assay was developed to confirm CXCR3 receptor antagonists from a subset of compounds originally identified as potential receptor antagonists in a reporter gene β-lactamase (BLA) assay by high-throughput screening (HTS).

Assay Protocol

The homogeneous addition only assay protocol is detailed in Fig. 2. Cell images were analyzed with the Granularity Analysis algorithm (Fig. 3).

Kinetics of Receptor Internalization

Kinetics of CXCR3-GFP internalization observed by confocal microscopy using the INCell Analyzer 3000 indicated that within 5 to 10 min of stimulation by the agonist, I-TAC, small pits/vesicles approximately 1 μm in size are seen accumulating in the cytoplasm. Following longer incubations with agonist (30 min to 3 h), these small vesicles become less evident and a more prominent single, large, perinuclear vesicle (late endosome/lysosome) develops (Fig. 4). The size of the smaller vesicles is close to the resolution limit of the microscope (~0.6 μm), which also vary significantly

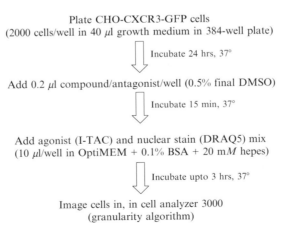

Plate CHO-CXCR3-GFP cells
(2000 cells/well in 40 μl growth medium in 384-well plate)

Incubate 24 hrs, 37°

Add 0.2 μl compound/antagonist/well (0.5% final DMSO)

Incubate 15 min, 37°

Add agonist (I-TAC) and nuclear stain (DRAQ5) mix
(10 μl/well in OptiMEM + 0.1% BSA + 20 mM hepes)

Incubate upto 3 hrs, 37°

Image cells in, in cell analyzer 3000
(granularity algorithm)

FIG. 2. Assay protocol for CXCR3-GFP receptor internalization assay.

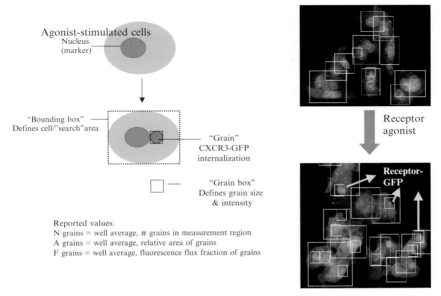

FIG. 3. Granularity analysis algorithm for CXCR3-GFP internalization assay.

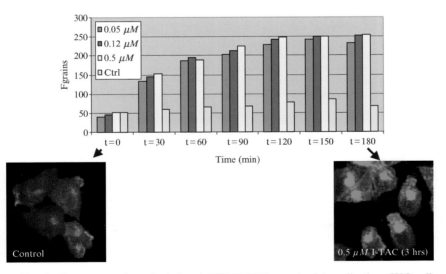

FIG. 4. Time course of agonist-induced CXCR3-GFP receptor internalization. CHO cells stably expressing CXCR3-GFP were stimulated with various (0, 0.05, 0.12, and 0.5 μM) concentrations of the agonist I-TAC and assayed as described for 0 to 180 min. Results are expressed as Fgrains based on granularity analysis algorithm as shown in Fig. 3. Confocal images from INCell 3000 are shown for control cells (no agonist stimulation) and for cells stimulated with 0.5 μM I-TAC for 3 h.

in number and size from cell to cell. Hence, larger vesicles are more practical to measure and quantitate.

Similar to the time course of agonist stimulation, the response (receptor trafficking from the plasma membrane to intracellular vesicle) is also dose dependent, as quantified in Fig. 5. Importantly, the dose-dependent response is inhibited in the presence of the receptor antagonist. The receptor trafficking assay monitored by receptor-GFP exhibits comparable sensitivity to other GPCR assays (Fig. 6).

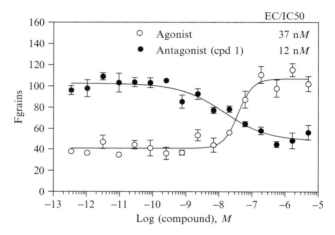

FIG. 5. Agonist and antagonist dose responses in CXCR3-GFP receptor internalization assay. Receptor internalization assays with CHO cells stably expressing CXCR3-GFP were conducted as described using various concentrations of the agonist I-TAC or various concentrations of an antagonist in the presence of 0.5 μM I-TAC (EC$_{90}$). Images were acquired and data quantified as described previously.

	IC50 (nM)			
Sample	Receptor binding (nM)	BLA (nM)	FLIPR (nM)	IN cell (nM)
Compound 2	640	462	196	510
Compound 3	690	271	269	540
Compound 4	520	289	173	607
Compound 5	1240	396	564	342
Compound 6	1800	981	206	1957
Compound 7	>50% inhibition @ 10 μM	113	43	250

FIG. 6. Comparison of assay sensitivity for the CXCR3 receptor. IC$_{50}$ values of six compounds were compared among receptor binding, reporter gene β-lactamase assay, second messenger Ca^{2+} assay measured by FLIPR, and the receptor internalization assay (INCell) for the CXCR3 receptor.

Monitoring GPCR Trafficking via Receptor–Arrestin–GFP

Overview

An indirect method to measure GPCR activation using localization of an arrestin–GFP chimera was initially developed by Norak Biosciences and commercialized as the Transfluor assay. In this assay, the arrestin–GFP fluorescence is localized in the cytoplasm as a diffuse signal when receptors are inactive at the plasma membrane and, upon receptor activation, the arrestin–GFP first translocates to the activated receptors at the plasma membrane and is subsequently internalized into small pits near the plasma membrane (as seen with the MRG-X1 receptor, case I described later) or larger vesicles (as seen with the NK1 receptor, case II described later) based on the affinity of the arrestin for the activated and phosphorylated receptor (Oakley *et al.*, 2002) (Figs. 1 and 7). An advantage of the Transfluor assay is that it does not require manipulation of the receptor sequence.

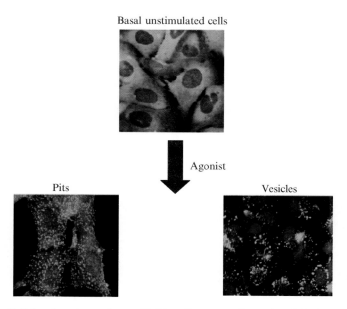

FIG. 7. Cellular phenotypes observed in Transfluor assay. Example cellular images of basal unstimulated cells, cells exhibiting small pits near the surface of the plasma membrane, and cells exhibiting larger internal vesicles upon agonist stimulation are shown.

Case I: MRG-X1 Receptor

Assay Protocol

U2OS-hMRGX1-βarrGFP stable cells are cultured in MEM supplemented with 10% heat-inactivated fetal bovine serum, 4 mM L-glutamine, 10 μg/ml gentamicin, 10 mM HEPES, 0.4 mg/ml G-418, and 0.4 mg/ml Zeocin. The assay protocol for an antagonist screen is shown in Fig. 8. Confocal microscopy is performed on the INCell Analyzer 3000 (GE Healthcare, Piscataway, NJ) using the granularity analysis and toxicity algorithms as shown in Fig. 9.

Data Analysis

In the INCell Analyzer 3000 imager, two excitation lines, 488 and 633 nm, are used to simultaneously excite β-arrestin–GFP and the DRAQ5 nuclear stain, respectively. Confocal images from the INCell 3000 are analyzed by the Raven software. Images of 200–300 cells from each well are captured and analyzed with the granularity analysis algorithm using the Raven software of IN Cell Analyzer 3000. Individual cells or nuclei are first identified by thresholding the red channel (DRAQ5). From the nucleus,

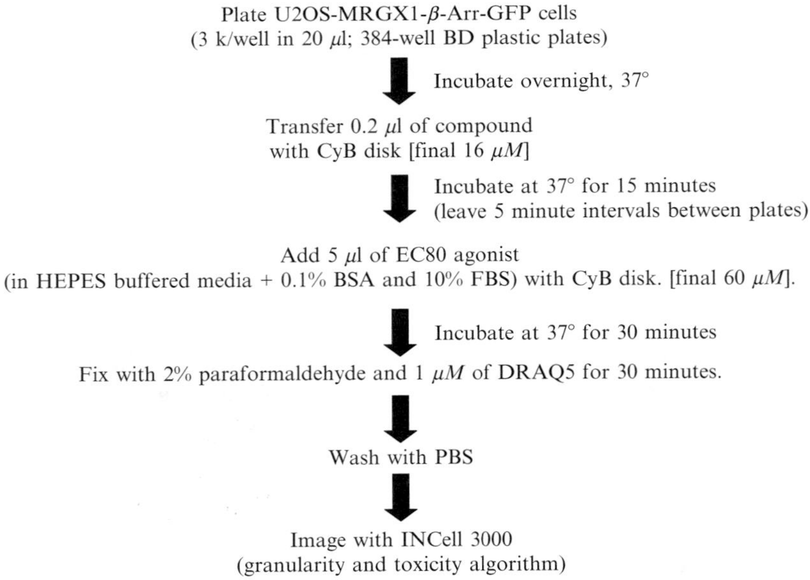

Plate U2OS-MRGX1-β-Arr-GFP cells
(3 k/well in 20 μl; 384-well BD plastic plates)

    Incubate overnight, 37°

Transfer 0.2 μl of compound
with CyB disk [final 16 μM]

    Incubate at 37° for 15 minutes
(leave 5 minute intervals between plates)

Add 5 μl of EC80 agonist
(in HEPES buffered media + 0.1% BSA and 10% FBS) with CyB disk. [final 60 μM].

    Incubate at 37° for 30 minutes

Fix with 2% paraformaldehyde and 1 μM of DRAQ5 for 30 minutes.

Wash with PBS

Image with INCell 3000
(granularity and toxicity algorithm)

FIG. 8. Transfluor assay protocol for MRG-X1 receptor internalization assay.

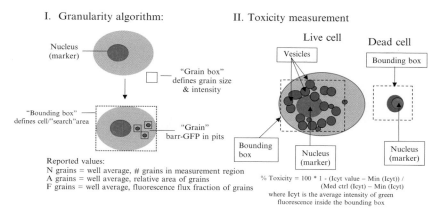

FIG. 9. Schematic representation of granularity and toxicity algorithms for the Transfluor receptor internalization assay.

a rectangular bounding box is dilated out to the edge of the cell with a dilation setting of 15 pixels. Within this bounding box, fluorescent spots of β-arrestin–GFP distribution are identified and outlined using an intensity gradient of 1.2 and a grain size of 4 pixels (Fig. 9). This granularity analysis algorithm is used to measure the number of fluorescent spots (Fgrains) based on fluorescent intensity and size of the grains, and the average value from all the cells in a well is used to obtain the "Fgrain" value for the well. The same granularity analysis algorithm is used to measure potential toxicity introduced by compounds based on the cellular morphology. The Icyt value is used to measure the average intensity of the green fluorescence inside the bounding box. Low Icyt values may indicate toxicity (resulting in either cell lysis or cells rounding up) displaying little or no GFP distribution in the cytoplasm and the plasma membrane, whereas high Icyt values may be due to the presence of a fluorescent compound.

Assay Characterization

INCell3000 analysis of MRG-X1 receptor trafficking indicates that in basal conditions, the β-arrestin–GFP fluorescence appears to be located primarily in the cytoplasm. Upon stimulation with increasing concentrations of the agonist BAM15, there is a corresponding increase in the punctate "pit" staining within the cell (Fig. 10). Agonist addition induces visible receptor internalization beginning from 1 μM BAM15 and is maximal at 100 μM BAM15. The receptor internalization quantified with the granularity analysis algorithm exhibits an EC_{50} of $\sim$14 μM (Fig. 10). Kinetics of receptor internalization indicates that this process plateaus

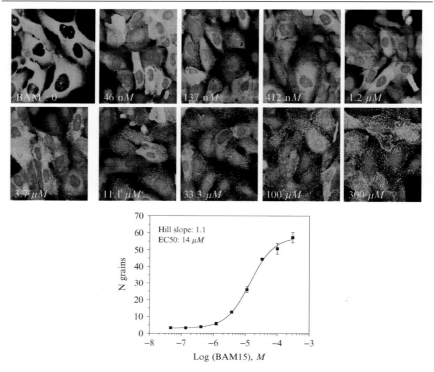

Fig. 10. Dose response of agonist-induced MRG-X1 receptor internalization. U2OS-MRGX1–βarr–GFP cells were assayed as described previously. Sample images from INCell 3000 (top) at increasing concentrations of the agonist BAM15 and a dose–response curve derived from processed data (bottom) are shown. Reprinted from Kunapuli, P., Lee, S., Zheng, W., Alberts, M., Kornienko, O., Mull, R., Kreamer, A., Hwang, J.-I., Simon, M. I., and Strulovici, B. Identification of small molecule antagonists of the human mass-related gene (MRG)-×1 receptor. *Anal. Biochem.* **351**, 50–61. Copyright (2006), with permission from Elsevier.

10 min after initiation of receptor stimulation and is stable for 60 min (Fig. 11).

Interestingly, the EC_{50} of the agonist response in the receptor-trafficking assay (14 μM) appears to be significantly right shifted compared to the BLA (55 nM) and second messenger FLIPR (8 nM) assays for the MRG-X1 receptor. However, it is important to bear in mind some key differentiating features in these assays, such as the cellular background (U2OS vs CHO), method of clone selection (traditional cloning vs single cell FACS based on functional response), and relationship to cellular signaling (receptor internalization is independent of cellular signaling, unlike the BLA or FLIPR assays). The lack of correlation between the agonist EC_{50} in the receptor-trafficking assay and the second messenger assay appears to be unique for this receptor and is

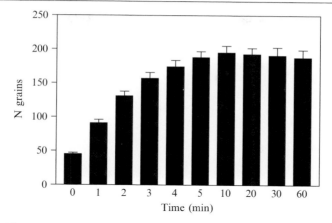

FIG. 11. Time course of receptor internalization for MRGX1 receptors in the Transfluor assay. U2OS-MRGX1–βarr–GFP cells were assayed as described with 0.3 mM BAM15 for 30 min at 37°. Data were processed as described previously. Reprinted from Kunapuli, P., Lee, S., Zheng, W., Alberts, M., Kornienko, O., Mull, R., Kreamer, A., Hwang, J.-I., Simon, M. I., and Strulovici, B. Identification of small molecule antagonists of the human mass-related gene (MRG)-×1 receptor. *Anal. Biochem.* **351,** 50–61. Copyright (2006), with permission from Elsevier.

exhibited by three different clones. This phenomenon has not been observed for other GPCRs (Howell, Lee, and Kunapuli, personal communication).

Assay Application

The MRG-X1 receptor-trafficking assay, monitored by β-arrestin–GFP, is used as a secondary/follow-up assay after HTS for MRG-X1 receptor antagonists using the BLA assay. In the absence of a suitable radioligand for a high-throughput receptor-binding assay using membranes from CHO-MRG-X1-BLA cells, the compounds identified from the primary screen are analyzed further for their ability to inhibit agonist-induced receptor internalization using the proximal, high-content assay. Consistent with the lower sensitivity of the receptor-trafficking assay as seen with BAM15, the IC_{50} of most compounds in this assay is right shifted compared to the BLA, FLIPR, or binding assays. In the presence of 10 μM concentration of each of these compounds and 60 μM BAM15 (EC_{80} concentration for receptor-trafficking assay), the cellular phenotype appears comparable to the basal, unstimulated state, with β-arrestin–GFP exhibiting diffuse cytoplasmic staining, implicating an uncoupled state from the receptor . The IC_{50} of some of these compounds is shown in Fig. 12. However, despite a lower assay sensitivity compared to the other functional assays, this high-content assay proved to be useful in assessing compound toxicity, which is often a common issue in cell-based assays. For example, in Fig. 13A, the sample

Sample #	IC50 (nM)			Binding
	BLA	FLIPR	InCell	
8	50	103	730	320
9	64	157	2800	ND
10	80	187	550	182
11	90	209	6782	ND
12	124	904	2200	1220
13	140	751	3600	607
14	163	220	2200	1430
15	165	334	4800	630
16	200	1500	5200	ND
17	209	492	10,000	ND
18	244	460	5200	1290
19	252	3300	8800	2400
20	290	1300	6300	1850
21	375	2700	29,000	1180
22	418	9800	13,500	2510
23	556	632	5600	1850

Fig. 12. Dose response for MRG-X1 receptor antagonist. The MRGX1 receptor internalization assay was performed as described. Cellular images from INCell 3000 confocal microscope (left) were analyzed with the granularity analysis algorithm (exhibited as percentage inhibition of 80 μM BAM15-induced signal) and toxicity algorithms (right) for compound exhibiting minimal toxicity (A) and compound exhibiting 66% toxicity at 80 μM (B). Reprinted from Kunapuli, P., Lee, S., Zheng, W., Alberts, M., Kornienko, O., Mull, R., Kreamer, A., Hwang, J.-I., Simon, M. I., and Strulovici, B. Identification of small molecule antagonists of the human mass-related gene (MRG)-×1 receptor. *Anal. Biochem.* **351,** 50–61. Copyright (2006), with permission from Elsevier.

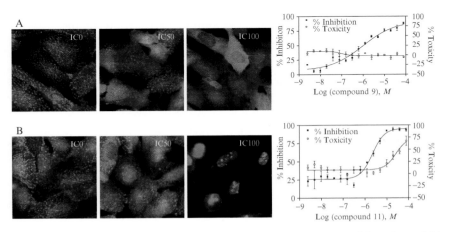

Fig. 13. Comparison of assay sensitivities for the MRG-X1 receptor. IC_{50} values of 16 compounds were compared among reporter gene β-lactamase assay, second messenger Ca^{2+} assay measured by FLIPR, receptor internalization assay, and whole cell receptor-binding assay for the MRG-X1 receptor.

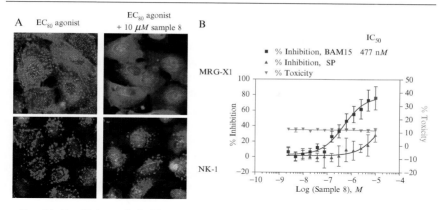

FIG. 14. Specificity analysis. The Transfluor assay was performed as described to determine the specificity of MRG-X1 antagonist for the MRG-X1 receptors. (A) Sample confocal images from INCell 3000 and (B) a dose–response curve of an MRG-X1 receptor antagonist are shown. Reprinted from Kunapuli, P., Lee, S., Zheng, W., Alberts, M., Kornienko, O., Mull, R., Kreamer, A., Hwang, J.-I., Simon, M. I., and Strulovici, B. Identification of small molecule antagonists of the human mass-related gene (MRG)-×1 receptor. *Anal. Biochem.* **351,** 50–61. Copyright (2006), with permission from Elsevier.

exhibits minimal (if any) toxicity even at 100 μM, as measured by a cell morphology-based toxicity algorithm, whereas in Fig. 13B, there is higher toxicity exhibited by the compound. In addition, some of these compounds were also tested against an unrelated class A GPCR, the neurokinin NK1 receptor, in a similar receptor-trafficking assay using U2OS-neurokinin receptor 1–β-arrestin–GFP cells. Most of these compounds were inactive or weak in inhibiting substance P-induced NK1 receptor internalization (Fig. 14), demonstrating specificity of the assay.

Case Study 2: NK1 Receptor Trafficking

Overview

The NK1–β-arrestin–GFP receptor internalization assay was used as a follow-up/hit funneling strategy for several thousand compounds originating from primary HTS using the BLA assay. The Transfluor assay was established in a semiautomated mode with various instruments for liquid handling. The granularity analysis and toxicity algorithms used for this analysis are similar to those described for the MRG-X1 Transfluor assay.

Assay Protocol

The Transfluor assay protocol for the U2OS-NK1–βarr–GFP cells in a semiautomated format is described in Fig. 15.

Assay Characterization

The kinetics of receptor internalization in U2OS-NK1–βarr–GFP cells upon stimulation with the agonist, substance P (SP), reveals that this process plateaus at ~45 min after receptor stimulation (Fig. 16). These cells appear to be tolerant to ~1% dimethyl sulfoxide (DMSO), an important parameter for compound screening (Fig. 17A). The assay appears tolerant to cell densities ranging from 2000 to 4000 cells/well, resulting in comparable responses with respect to assay window and assay sensitivity,

Plate U2OS-NK1-βarr-GFP cells
(3 k/well in 30 μl in 384 well plates)

↓ Incubate overnight, 37°, 5% CO_2

Cybi-disk   Transfer 0.15 μl of compounds from stock in DMSO
(final 0.43% DMSO)

↓ Incubate 10 min, 37°, 5% CO_2

Multidrop 1   Dispense 5 μl of 13 n M substance P (EC70)
(in HEPES buffered media + 0.1% BSA; final concentration 1.85 nM)

↓ Incubate 60 min, 37°, 5% CO_2

Multidrop 2   Fix with 2% formaldehyde and 0.5 μM DRAQ5 (nuclear stain)

↓ 30 min, RT

Embla ELX400   Remove fixative by aspiration.
plate crane, INCell 3000   wash 3× with PBS
image on INCell analyzer 3000 (using the plate crane)

FIG. 15. Transfluor assay protocol for NK1 receptor internalization assay.

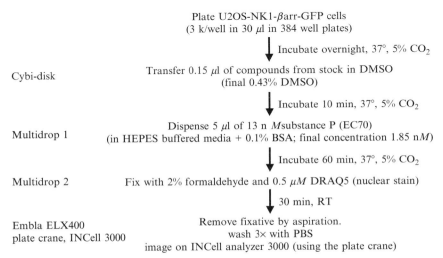

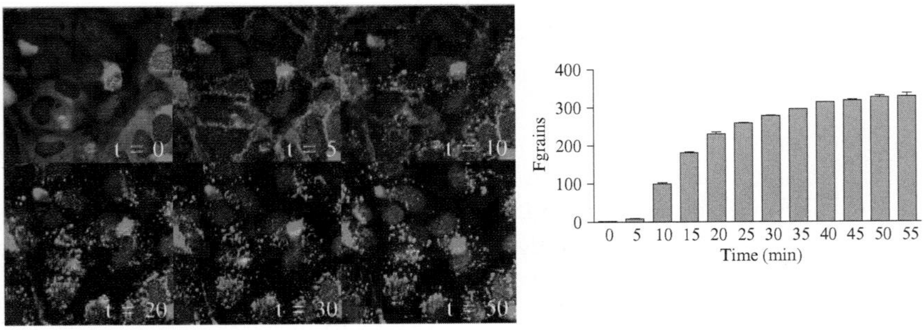

FIG. 16. Kinetics of NK1 receptor internalization. The NK1 Transfluor assay was conducted as described with 0.5 μM substance P (measured in Fgrains or relative grain intensity). Images taken at 0, 5, 10, 20, 30, and 50 min are shown.

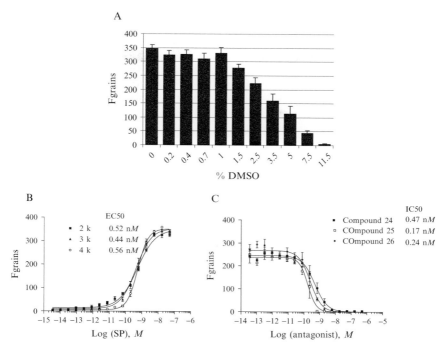

FIG. 17. Assay characteristics of the NK1 Transfluor assay. The NK1 Transfluor assay was conducted as described in the presence of various concentrations of DMSO and 0.5 μM substance P (A), in the presence of various cell densities and various concentrations of substance P (B), and in the presence of various concentrations of receptor antagonists with 0.5 μM substance P (C). The response was measured in Fgrains.

exhibiting an EC_{50} of 0.5 nM for SP (Fig. 17B). More importantly, receptor pharmacology (Fig. 17C) with known antagonists was also comparable for three compounds in this assay (and between second messenger and reporter gene assays, data not shown).

Screening

Whole plate assays using U2OS-NK1–βarr–GFP cells in the Transfluor assay show acceptable S/N and plate CVs ($\sim$6%) for screening (Fig. 18A). The assay plate format used for the follow-up screen of $\sim$3000 compounds in triplicate at two concentrations (8.5 and 2.1 μM, starting from 2 mM and 0.5 μM samples, respectively) is shown in Fig. 18C. Each of the assay steps was set up at a different instrument workstation (CyB Disk, Multidrop 1 and 2, ELX-405 and plate crane for the INCell 3000 as shown in Fig. 15).

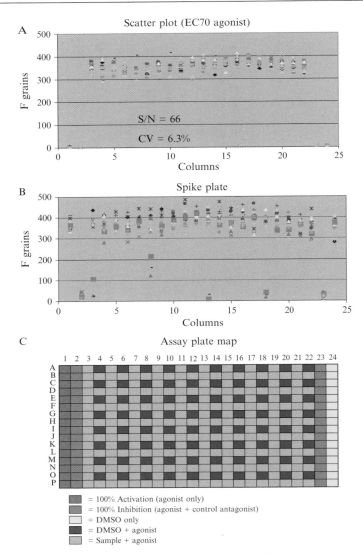

FIG. 18. Whole plate assay for NK1 receptor antagonists. (A) The whole plate assay with DMSO reveals acceptable S/N and plate CV (∼6%). The DMSO plate contained the NK1 receptor antagonist compound 25 in columns 1 and 2 and basal unstimulated cells in columns 23 and 24. Columns 1–22 were stimulated with 1.85 nM substance P. (B) The plate map of the spiked plate is similar to the plate map for the screen, with the exception that most wells between columns 3 and 22 contained DMSO. Some random wells were spiked with compound 25 at IC$_{50}$, IC$_{70}$, or IC$_{100}$. (C) Schematic plate map for assay plates in the screen. Column 1 was used for 100% activation with only agonist addition. Columns 2 and 23 were treated with

Compound stock concentration	2 mM	500 μM
Final compound concentration	8.5 μM	2.1 μM
# of compounds screened	2877	2877
Median > 30% inhibition	925 (32.2%)	336 (11.7%)
Toxic compounds	192 (20.8% of hits)	13 (3.9% of hits)
Fluorescent compounds	4 (0.4% of hits)	4 (1.2% of hits)
False positives (nonsample areas)	20 (2.2%)	0
Total actives in transfluor assay	709	319

FIG. 19. Screen summary. A total of 2877 compounds were screened in 384-well format at 8.5 and 2.1 μM concentrations in triplicate. Toxic compounds (based on cellular image and toxicity algorithm) were eliminated by visualizing the images of all of the hits.

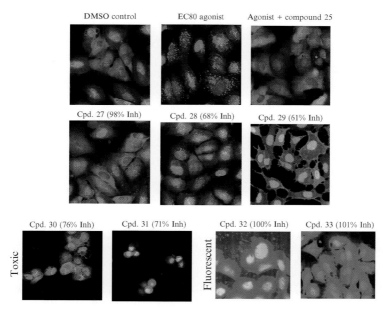

FIG. 20. Cellular phenotypes from NK1 antagonist screen. Example images of the cellular phenotype from the NK1 receptor antagonist screen using the Transfluor assay are shown.

agonist (1.85 nM final concentration of substance P) and compound 25 for 100% inhibition control. Column 24 was treated with DMSO and agonist. Quadrants 1, 3, and 4 were treated with sample compounds and agonist, while quadrant 2 was treated with DMSO as a control and agonist. Data were processed as described earlier.

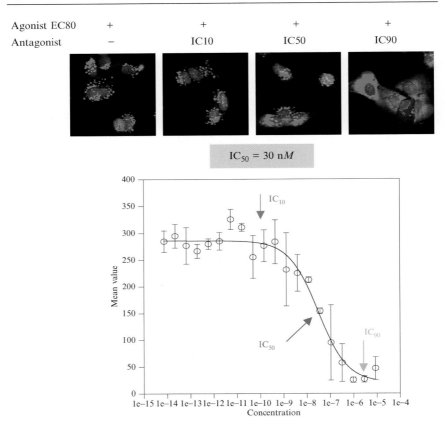

FIG. 21. Example compound dose response. Sample images (top) of compound 34 at IC_{10}, IC_{50}, and IC_{90} are shown with β-arrestion–GFP in the green channel and DRAQ5 in the red channel, and a IC_{50} curve of a 20-point titration (bottom) is shown for compound 24.

The assay took a total of $\sim$6 h to complete for $\sim$3000 compounds in triplicate at two compound concentrations. The image size for acquisition in INCell 3000 was set to 0.75×0.375 mm for two color channels: green for GFP and red for DRAQ5. The minimum cell count was set to 50 cells/well. The live image analysis took a total of 10.5 h to complete for all the assay plates. The images were analyzed by the granularity analysis algorithm. Fgrains refer to the fraction of cellular fluorescence present in the qualifying grains. Results from this screen are shown in Fig. 19, and sample images from different wells are shown in Fig. 20.

Among the confirmed compounds in the Transfluor assay, dose responses were performed for 72 compounds. An example titration with

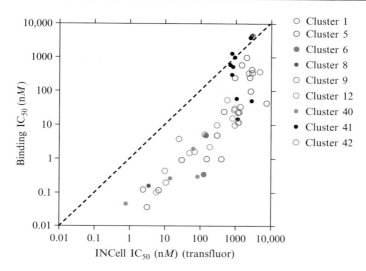

FIG. 22. Correlation plot for NK1 receptor antagonists: Comparison of the IC_{50} in the receptor binding and Transfluor assays among 72 NK1 receptor antagonists. IC_{50} values obtained from the INCell 3000 (TransFluor assay) are on the x axis, and IC_{50} values from binding data are plotted on the y axis. The compounds were clustered by their structure. Data based on Fgrain measurements using granularity algorithm v001.

corresponding cellular images is shown in Fig. 21. The IC_{50} values obtained were directly compared to those from receptor-binding experiments. Although there appeared to be a rightward shift in IC_{50} values from the Transfluor assay for these compounds, there was reasonable correlation with the binding data as shown in Fig. 22. Possible explanations for the rightward shift in the Transfluor assay could be due to differences in the assay protocol (e.g., Transfluor assay used 1.85 nM SP vs 0.1 nM SP in binding assay). In addition, in the binding assay, the compounds were preincubated with the cells for at least 30 min before agonist addition, thereby increasing the assay sensitivity for weak binders. Interestingly, for some of the compounds (e.g., cluster # 41), there is a better correlation between the receptor binding and Transfluor assays than others (e.g., cluster #6 and #40).

Acknowledgments

We thank Lin-Lin Shiao and Dr. Kathleen Sullivan for the CXCR3-GFP cell line. We are also grateful to Kevin Nguyen for his help in preparing the compound plates for dose–response studies and to Dr. James Inglese (current address: National Institutes of Health, MD) for valuable suggestions during the course of this work.

References

Bleicher, K. H. (2003). Hit and lead generation: Beyond high-throughput screening. *Nature Rev. Drug Disc.* **2,** 369–378.

Civelli, O., Nothacker, H. P., Saito, Y., Wang, Z., Lin, S. H. S., and Reinscheid, R. K. (2001). Novel neurotransmitters as natural ligands of orphan G-protein-coupled receptors. *Trends Neurosci.* **24,** 230–237.

Drews, J. (2000). Drug discovery: A historical perspective. *Science* **287,** 1960–1964.

Ferguson, S. S. (2001). Evolving concepts in G protein-coupled receptor endocytosis: The role in receptor desensitization and signaling. *Pharmacol. Rev.* **53,** 1–24.

Gautam, N., Downes, G. B., Yan, K., and Kisselev, O. (1998). The G protein bg complex. *Cell Signal* **10,** 447–455.

Gilman, A. G. (1987). G proteins: Transducers of receptor generated signals. *Annu Rev. Biochem.* **56,** 615–649.

Goodman, O. B., Jr., Krupnick, J. G., Santini, F., Gurevich, V. V., Penn, R. B., Gagnon, A. W., Keen, J. H., and Benovic, J. L. (1996). β-arrestin acts as a clathrin adaptor in endocytosis of the β2-adrenergic receptor. *Nature* **383,** 447–450.

Gurevich, V. V., Dion, S. B., Onorato, J. J., Ptasienski, J., Kim, C. M., Sterne-Marr, R., Hosey, M. M., and Benovic, J. L. (1995). Arrestin interactions with G protein-coupled receptors: Direct binding studies of wild type and mutant arrestins with rhodopsin, β2-adrenergic and M2 muscarinic cholinergic receptors. *J. Biol. Chem.* **270,** 720–731.

Hopkins, A. L., and Groom, C. R. (2002). The druggable genome. *Nature Rev. Drug Disc.* **1,** 727–730.

Krupnick, J. G., Goodman, O. B., Jr., Keen, J. H., and Benovic, J. L. (1997). Arrestin/clathrin interaction: Localizatin of the clathrin binding domain of non visual arrestins to the carboxy terminus. *J. Biol. Chem.* **272,** 15011–15016.

Krupnick, J. G., and Benovic, J. L. (1998). The role of receptor kinases and arrestins in G protein-coupled receptor regulation. *Annu. Rev. Pharmacol. Toxicol.* **38,** 289–319.

Lefkowitz, R. J. (1993). G protein-coupled receptor kinases. *Cell* **74,** 409–412.

Marchese, A., George, S. R., Kolalowski, L. F., Jr., Lynch, K. R., and O'Dowd, B. F. (1999). Novel GPCRs and their endogenous ligands: Expanding the boundaries of physiology and pharmacology. *Trends Pharmacol. Sci.* **20,** 370–375.

Neer, E. J. (1995). Heterotrimeric G proteins: Organizers of transmembrane signals. *Cell* **80,** 249–257.

Oakley, R. H., Hudson, C. C., Cruickshank, R. D., Meyers, D. M., Payne, R. E., Jr., Rhem, S. M., and Loomis, C. R. (2002). The cellular distribution of fluorescently labeled arrestins provides a robust, sensitive and universal assay for screening G protein-coupled receptors. *Assay Drug Dev. Tech.* **1,** 21–30.

Oakley, R. H., Laporte, S. A., Holt, J. A., Barak, L. S., and Caron, M. G. (1999). Association of beta-arrestin with G protein-coupled receptors during clathrin-mediated endocytosis dictates the profile of receptor resensitization. *J. Biol. Chem.* **274,** 32248–32257.

Oakley, R. H., Laporte, S. A., Holt, J. A., Barak, L. S., and Caron, M. G. (2001). Molecular determinants underlying the formation of stable intracellular G protein-coupled receptor-beta-arrestin complexes after receptor endocytosis. *J. Biol. Chem.* **276,** 19452–19460.

Oakley, R. H., Laporte, S. A., Holt, J. A., Caron, M. G., and Barak, L. S. (2000). Differential affinities of visual arrestin, beta arrestin 1, and beta arrestin 2 for G protein-coupled receptors delineate two major classes of receptors. *J. Biol. Chem.* **275,** 17201–17210.

Sterne-Marr, R., and Benovic, J. L. (1995). Regulation of G protein-coupled receptors by receptor kinases and arrestins. *Vitam. Horm.* **51,** 193–234.

[7] High-Throughput Confocal Microscopy for β-Arrestin–Green Fluorescent Protein Translocation G Protein-Coupled Receptor Assays Using the Evotec Opera

By Ralph J. Garippa, Ann F. Hoffman,
Gabriele Gradl, and Achim Kirsch

Abstract

Ligand-activated G protein-coupled receptors (GPCRs) are known to regulate a myriad of homeostatic functions. Inappropriate signaling is associated with several pathophysiological states. GPCRs belong to a ~800 member superfamily of seven transmembrane-spanning receptor proteins that respond to a diversity of ligands. As such, they present themselves as potential points of therapeutic intervention. Furthermore, orphan GPCRs, which are GPCRs without a known cognate ligand, offer new opportunities as drug development targets. This chapter describes a systems-based biological approach, one that combines *in silico* bioinformatics, genomics, high-throughput screening, and high-content cell-based confocal microscopy strategies to (1) identify a relevant subset of protein family targets, (2) within the therapeutic area of energy metabolism/obesity, (3) and to identify small molecule leads as tractable combinatorial and medicinal chemistry starting points. Our choice of screening platform was the Transfluor β-arrestin–green fluorescent protein translocation assay in which full-length human orphan GPCRs were stably expressed in a U-2 OS cell background. These cells lend themselves to high-speed confocal imaging techniques using the Evotec Technologies Opera automated microscope system. The basic assay system can be implemented in any laboratory using a fluorescent probe, a stably expressed GPCR of interest, automation-assisted plate and liquid-handling techniques, an optimized image analysis algorithm, and a high-speed confocal microscope with sophisticated data analysis tools.

Choice of Orphans

Introduction

G protein-coupled receptors (GPCRs) are a superfamily of seven transmembrane proteins that have long been established as fertile targets for exploitation within pharmaceutical programs for the development of new

METHODS IN ENZYMOLOGY, VOL. 414
0076-6879/06 $35.00
DOI: 10.1016/S0076-6879(06)14007-0

medicinals. In the decade from 1993 to 2002, compounds that had a mode of action relating to direct action on a GPCR (agonism, antagonism, inverse agonism, positive or negative modulation), for marketed compounds and for phase III entities, consistently occupied 23 to 27% of all target types addressed (Gong and Sjogren, personal communication, 2004). This list of target family proteins includes such well-known members such as kinases, proteases, ion channels, and nuclear receptors. GPCRs are involved in numerous physiological processes, and the ligands to these receptors have shown themselves to be clever mechanistic tools and often useful therapeutics. The entire family of known GPCRs has demonstrated an affinity for an unusually wide variety of ligands, from low molecular weight trace amines, to small peptides (<10 amino acids), large peptides (>10 amino acids), even to large phospholipids. Taking into account all of the seven transmembrane sequences identified in the human genome project, it is estimated that there are approximately 600 to 800 GPCRs, of which only 400 to 500 have known ligands. Therefore, the GPCRs with unknown ligands, or orphan GPCRs (oGPCRs), present a significant opportunity for investigators not only to elucidate their involvement in key biological processes, but also to mine them as drug targets. Characterization of oGPCRs would present opportunities for claiming intellectual property and also for developing first-to-market medicinals. However, oGPCR research can be hampered by the lack of literature-based data, making informed decisions on which oGPCR is most relevant within a disease area difficult, and by assay development issues.

We have taken a systems-based approach to identify a set of related oGPCR targets and small molecules with selective activity on these targets. To this end, we have assembled a diverse team of bioinformaticians, cell biologists, and peptide and small molecule chemists. As part of a general strategy for the team, we saw the need to incorporate bioinformatics, genomic, and combinatorial chemistry thinking in the process of generating new targets, leads, reagents, and expertise for a focused drug discovery effort. For example, an extensive list of anabolic and catabolic signaling molecules are known to exert their effect through binding of cell surface GPCRs in central nervous system effector pathways involved in the hypothalamic control of energy balance (Kieffer and Habener, 1999; Shimada *et al.*, 1998). Current pharmacological targets for class A and B known GPCRs include neuropeptide Y, melanocortin, orexin, and bombesin. However, if one examines the clustering of known GPCRs involved in energy balance, the clades within the genomic dendrogram reveal a number of oGPCR sequences with significant homology, that is, with cognate ligands as yet undiscovered. Since it was not technically feasible for us to screen hundreds of potential oGPCR targets for active compounds, as a first phase of the project we utilized an *in silico* bioinformatics approach to focus and filter the list of potential target candidates. A second phase involved checking these oGPCR sequences for expression in normal human and mouse hypothalamic tissues. A mouse homologue is important to move a compound forward in lead optimization in that rodent models of obesity and energy metabolism are prevalent and accepted within the field. Single target expression profiling (STEP) was accomplished by examining the oGPCR of interest, relevant to GAPDH expression, across ~50 organ-specific tissue samples. This procedure is well established for estimating the relative expression levels of a particular protein throughout body tissue sample in relatively high-throughput manners (Reidhaar-Olson et al., 2003). In evaluating candidate oGPCRs for this study we identified many with a ubiquitous expression pattern across all tissues queried. For oGPCR38, however, the expression of the receptor was more limited with expression identified in the brain and hypothalamic regions, fulfilling preset criteria for oGPCRs of interest. A handful of candidate oGPCRs fulfilled the criteria of exclusive and specific expression patterns within the brain and hypothalamic regions. Our third and final phase of oGPCR selection involved determining the differential expression of the oGPCR of interest within discrete anatomical nuclei of defined brain regions harvested from diet-induced obese (DIO) mice as compared to nonobese rodents. For this purpose, we employed laser capture microdissection techniques to extract RNA from numerous mouse brain (Kinnecom and Pachter, 2005). The regions of interest included the arcuate nucleus, the paraventricular nucleus, medial amygdala, dentate gyrus, and the anterior pituitary. As precedence, many of the known obesity-related GPCRs are differentially regulated in discrete brain regions (Ludwig et al., 2001). We used quantitative polymerase chain reaction (PCR) methods in order to estimate the degree of up- or downregulation of the oGPCRs of interest (Reidhaar-Olson et al., 2002). Without a priori knowledge of the oGPCR involvement with either the anabolic or the catabolic arm of central energy regulation, we examined both possibilities. At the end of our three-phase selection process, we were able to distill down our list of candidate oGPCRs to 12 out of a possible 200 sequences. Of these 12 oGPCRs, 2 were subsequently deorphanized in the literature: the prolactin-releasing peptide receptor (D'Ursi et al., 2002, Gu et al., 2004) and the neuropeptide W receptor (Shimomura et al., 2002). Both of these GPCRs have ties to energy metabolism, thereby further validating our approach. One oGPCR had considerable intellectual property issues associated with it and another oGPCR showed no evidence of having a mouse homologue. Therefore, we chose 8 oGPCR sequences to move forward in the program. Four of these were high-throughput screened (HTS) using a fluorescent imaging plate reader (FLIPR) assay for calcium flux readout using a promiscuous G-α16 protein background for facilitating the heterotrimeric G protein coupling (data not shown). Four other oGPCRs were marked for HTS using cell-based, high-speed microscopy-based, green fluorescent protein (GFP) translocation readout. All 8 oGPCR sequences were cloned for the full-length human receptor and subsequently transferred to a pCDNA3.1 cloning vector (see later).

Transfluor

The strategy has been partially validated during the past years because several of the orphan receptors have subsequently been deorphanized by others and demonstrated to regulate feeding behavior, for example, prolactin-releasing peptide receptor and neuromedin receptor (Bouchard *et al.*, 2004; Hanada *et al.*, 2004). We are, however, hampered by the ability to quickly develop high-throughput assays for these orphan GPCRs that couple to an appropriate G protein pathway for a reliable readout of receptor activation, such as calcium flux or changes in cAMP levels. Therefore, we were very interested in the Transfluor technology that is based on the interaction of β-arrestin and an activated GPCR initiating the process of receptor internalization. Multiple HCS instrument platforms are available for cellular imaging that allow for the quantification of the redistribution of β-arrestin–GFP when complexed with the overexpressed GPCR detecting the GPCR signaling. The feasibility of this approach has been shown by Norak Biosciences for over 50 known GPCRs where they have obtained quantitative data for receptor agonists and antagonists that compare with existing technologies. Advantages of this technology include a single readout for all GPCRs; regardless of the type of endogenous G protein coupling, it can be used with all orphan-GPCRs. Without the need to collate out those targets that do not couple to Gq, there is a drastic decrease in assay development time for all GPCR assays, and it is applicable to GPCRs of interest to any therapeutic area without modification. This assay format is HTS compatible and can be used to generate IC_{50}, EC_{50}, and SAR determinations. In addition, there are no expensive reagents required beyond the initial investment in the imaging platform. Therefore the choice of Transfluor offered us this relatively fast HCS/HTS screen on our predefined selected oGPCRs and has positioned us to establish patent claims on these targets.

Technology

For primary screening and secondary follow-up experiments of the potential "hit" compounds an Opera confocal imaging plate platform is used. The platform comprises the confocal microscope, the scheduling software, Bernstein, and acquisition and processing customized software for

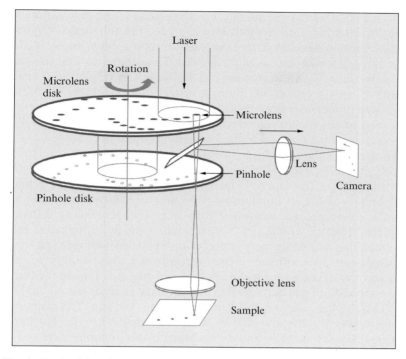

FIG. 1. Confocal imaging based on Nipkow disk technology. A collimated laser beam passes through a spinning disk containing microlenses. Each microlens focuses the incoming laser beam onto a corresponding pinhole, which is imaged into the sample. Fluorescence emitted by the sample in the focal plane is imaged back through the pinhole so that it can pass the pinhole disk and be detected with CCD cameras. Fluorescence from out-of-focus areas of the sample is not focused back through the pinhole but to a large extent is blocked by the pinhole disk. The disk contains approximately 20,000 pinholes with a diameter of approximately 50 μm.

data analysis. This instrument uses a Nipkow spinning disk to create confocal images composed from the approximately 1000 concurrent confocal beams (Fig. 1). The advantage of confocal imaging is that signals from areas of the image sample that are not in the focal plane are strongly attenuated, thus resulting in a very high signal-to-background ratio in the images. Of the four available light sources, 488- and 635-nm confocal excitation lines are used in this HCS assay in order to excite GFP from the β-arrestin–GFP fusion protein, GFP from Renilla mulleri (Prolume, Pinetop, AZ) DRAQ5 (the nuclear dye used to define both nuclear area and cytoplasmic area), respectively. Emission is detected simultaneously using two cameras in spectral bands at 535 and 690 nm. Although enabled, the third camera is not activated in these experiments. For the primary screening experiments, an air 20×/0.4 NA lens is used. However, for the follow-up of potential agonists accomplished by a 10-point dose–response experiment, a 20×/0.7 NA water immersion lens is implemented. To accomplish this, the immersion water supply is fully automated within the Opera platform, enabling automatic measurement of multiple plates without user intervention. The water immersion lens typically provides higher numeric aperture as compared to the air lenses, which results in higher image quality that is independent of the position of the focal plane within the sample.

To accommodate the high-throughput plate handling and reading requirements of the HCS/HTS assay, greater than 2000 384-well formatted microtiter plates, the Opera is interfaced with a Twister II plate handler (Caliper Life Sciences, Hopkinton, MA) and an external bar code reader (Fig. 2). These are all controlled with Bernstein scheduling software, a complete, dynamic scheduling software allowing the automated robotic plate handling and image acquisition of up to 80 384-well microtiter plates in a single run. This software proves versatile as it also handles the adding and removing of plates during an active processing run. Due to the fact that the time to acquire and perform the on-line image analysis takes 11 to 12 min per 384-well plate, we vary the number of plates per run. This allows us to accommodate (1) the transfer of data and images to the Roche data servers, (2) assay curator time to reevaluate the previous days run, and (3) reimage any out-of-focus wells or plate failures.

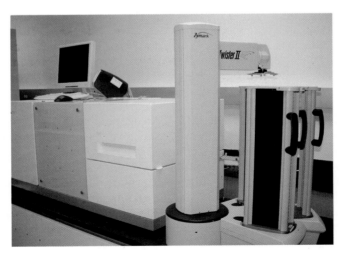

Fig. 2. Opera with Twister II plate robot and plate stacks on a Scala. An interface station with keyboard switch for the evaluation network is visible on top of the instrument.

The recorded images are analyzed and stored by a system of three parallel computers, using Acapella data analysis software. Numerical results from the analyzed images are stored in a hierarchical directory structure in XML files. For XML files, many tools for processing the information may be used, thus enabling a simple import of the results into different data processing, storage, and mining systems. For both the primary screening campaign and the follow-up dose–response experiments we use an XML transformation to convert the XML files into comma separated variable text files (csv) to import into ID Business Solutions, Inc. ActivityBase software (IDBS, Bridgewater, NJ). Furthermore, we use APlus screening data analysis software for follow-up dose–response assays that combine numerical screening data with compound information provided by supplementary data bases or imported via compound XML files. The analysis performed by APlus generates EC_{50} curves, percentage maximal responding efficacies, and a scoring based on the dose–response curve and positive/negative controls. We use this analysis to interpret what the specificity of compound agonist action is on one particular orphan GPCR versus others.

Image Analysis Algorithm

As with common HCS assays the image analysis is based on object detection and quantification. The Acapella application in this Transfluor assay is essentially an application quantifying receptor internalization. Representative images are shown defining distribution of the β-arrestin–GFP Prolume-labeled molecules and images depicting the nuclear stain, DRAQ5 (Fig. 3).

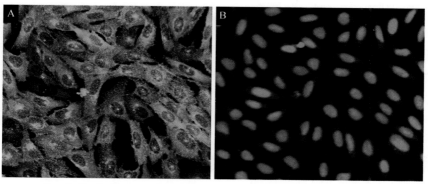

FIG. 3. Typical images showing GFP (A) and DRAQ5(B) fluorescence.

In the first steps of image analysis, nuclei and cell bodies are detected. For nuclei detection, red nuclear fluorescence is used (Fig. 4A). For cell body detection, GFP fluorescence could be used. However, in vesicle forming versions of the Transfluor assay we observed that virtually all of the GFP molecules were recruited to the vesicles so that no signal for cell detection was left. The DRAQ5 staining, however, was not an exclusive stain for DNA but showed some spurious staining in the cytoplasm. As a result, we could define both the nucleus and the cell body based on red nuclear fluorescence (Fig. 4B). The area that was analyzed further was defined using the identified cells. Cells detected using the DRAQ5 signal appeared smaller than cells visible in the GFP channel. Therefore, the cell shapes were dilated (Fig. 4C). The fluorescence distribution in the area covered by the cells was then analyzed in terms of object-oriented image analysis. In Acapella, an "object" is a collection of shapes or geometrical attributes combined with a number of numerical attributes, for example, a cell can be represented as an object that contains two geometrical attributes describing where the nucleus and the cell body are and a set of

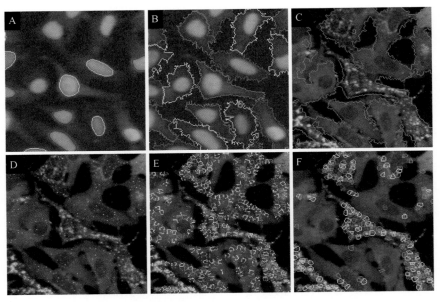

FIG. 4. Image analysis steps for the Transfluor assay. (A and B) Nucleus and whole cell body detection based on the DRAQ5 stain. (C) Area where GFP distribution is analyzed shown on the GFP image. (D and E) Initial set of local intensity maximums and corresponding reference areas for these maximums. (F) Reference regions of the selected maximums. Different colors help in identifying individual objects.

numbers such as area of the nucleus or intensity of GFP fluorescence in the cell body.

In analysis of the GFP fluorescence distribution, first the local intensity maximums were identified (Fig. 4D and E). This was the initial set of objects in the sequence of image analysis steps, and each maximum consisted of a single pixel in the image that was brighter than all surrounding pixels. These maximums were caused either by simple fluctuations in the fluorescence signal or by real aggregations of fluorescent molecules. The absolute intensity of the spots was not suitable for deciding if a spot was caused by a simple intensity fluctuation or not, as the average intensity of the cells containing the spots varied quite drastically. With a fixed intensity limit, some cells would have more maximums simply by the fact that more GFP-labeled molecules were expressed. Therefore, for each maximum, two other criteria were determined, enabling the selection of only those maximums, which were visually recognized as bright spots in the image without the artifacts created by different GFP expression levels in the different cells of the sample. With these two criteria, valid maximums were selected (Fig. 4F). These maximums were regarded as clathrin-coated pits or vesicles. No assumption had to be made in this selection process as to which type of fluorescence aggregation was observed.

Based on the selected maximums, the total number of qualified bright spots, the number of qualified bright spots normalized to the number of cells, or the total fluorescence contained in spots normalized to the total fluorescence in the cells was determined. All these parameters were used to determine the typical state of the whole cell population.

For the experiments described more outputs were added in order to get a better understanding of the processes in the assay. Adding such outputs required only a few lines of Acapella script code once the objects were identified. The first additional output employed quantifying nuclear area to add information about the compound effects on the cells. Nuclei of dead or apoptotic cells are typically smaller than nuclei of normal cells. Therefore the average size of nuclei was added to the list of outputs. Also the fluorescence signals in the nuclear channel are usually not influenced by the assay. Therefore the fluorescence intensity in this channel was used to determine if a compound was autofluorescent or not. Since two channels were being acquired, this translated into identifying whether compounds displayed fluorescence in the red or green channel comparing the same intensities over the nuclear area to the basal controls.

The second more assay-specific output added was the fraction of qualified maximums in the nuclear area. During assay development where a test library of compounds was screened it was observed that some compounds led to the formation of GFP aggregates in the nucleus (Fig. 5). Therefore

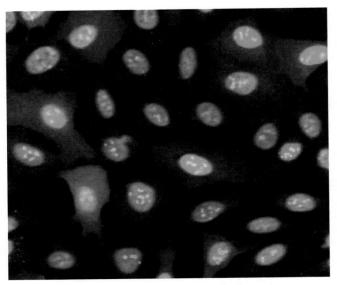

FIG. 5. Sample with GFP aggregates in the nucleus.

the fraction of the spots in the nucleus was added as an additional output. This output was subsequently used for flagging compounds that caused this kind of unusual GFP distribution. It was determined after subsequent experiments that this phenotype consequently led to cell death. These additional outputs demonstrated the virtues of high-content assays. While samples were analyzed with respect to the internalization of the receptor: β-arrestin complex quantifying the Transfluor assay, the simultaneous information on compound-specific properties, such as nuclear area, apoptosis, cytotoxicity, autofluorescence, and the induction of an unusual phenotype of GFP distribution, was obtained.

EC_{50} Fitting and Scoring

Numerical results of the dose–response confirmation screen were analyzed further using the APlus screening data analysis software. This is a client/server system for combining screening results with compound information from additional data sources. In the case of this experiment the compound information was provided using an XML file defining compound concentration per well. The location of the high and low control wells was derived from the plate layout. This information could also have been provided via the compound information path.

At first several screening metrics such as Z' or the signal-to-background ratio were determined (Zhang *et al.*, 1999). From the positive controls for each plate, G protein-related kinase (GRK) transfections, the full activity (100%) of the assay was determined and based on this, the activity of the individual wells was calculated. These activities were the basis for further analysis. Referencing to the controls on the same plate allowed for the compensation of potential drifts and so on.

In the next step for each compound the dose–response curve was fitted using various models. The best-fitting model was selected automatically by the software. Available models used were as follow.

- sigmoidal: classical dose response model
- linear: central part of the sigmoidal model
- exponentially converging/diverging: higher/lower end of the sigmoidal model
- constant positive/negative: constant high/low measurements
- unclassified: curves that cannot be described with any of the other models

For data generated, the sigmoidal, linear, and exponential models of EC_{50} values were generated.

In addition to the fitting model, the dynamic range was also reported. Dynamic range classes were none, medium, and full to extreme, where these classes describe how a dose–response curve compares to the positive and negative controls. As the potential compound's agonist response could be classified as full or partial, this means of categorizing the effects was practically constructive.

From all this information—the fitting model, the EC50 value, and the dynamic range—a score for each compound tested in a dose–response fashion was generated. These scores are in the range of 1 to 30 and describe the potential of a compound in a single value. The mathematical model for this scoring has been determined empirically before and was used successfully in many screening campaigns. Using this dynamic range measurement, compounds may be classified as to varying characteristic responses on the receptor internalization and β-arrestin–GFP redistribution process.

Assay Methods

Tissue Preparation

Three diet-induced obese (DIO) and three lean mice (C57Bl/6J) brains are used (Van Heek *et al.*, 1997). Using RNase-free conditions, the brain is removed rapidly, placed in a cryomold, covered with OCT compound

(Tissue Tek frozen tissue embedding medium Sakura Finetek, Torrance, CA), frozen in liquid nitrogen, and sectioned serially at 7 to 10 μm in a cryostat. Sections are mounted on plain (noncoated) clean microscope slides and placed on dry ice immediately. The sections are stored at $-70°$. The slides are fixed in 70% ethanol, rinsed in RNaase-free distilled water, stained in cresyl violet acetate (to identify the nuclei of interest), and dehydrated in graded ethanol (95% and absolute) with a final 5 min in xylene as the last step in the procedure. Slides are then placed in a desiccator for drying. Completely dried slides are used for laser capture microscopy.

Following the manufacturer's protocol, nuclei from the following hypothalamic regions are microdissected for further study: arcuate nucleus, dentate gyrus, dorsomedial hypothalamus, lateral hypothalamus, medial amygdala, median eminence, paraventricular nucleus, and ventromedial hypothalamus. These hypothalamic regions are identified as referenced in Franklin and Paxions (1997). After performing laser capture microdissection, the cap (with captured cells) is placed in a reagent tube containing 200 μl RNA denaturing buffer (GITC) and 1.6 μl β-mercaptoethanol. The tube is then inverted for 2 min to allow tissue to be digested off the cap. The reagent containing microdissected nuclei is then ready for RNA extraction, isolation, amplification, etc.

Laser Capture Microscopy

The PixCell II LCM system from Arcturus Engineering (Mountain View, CA) is used for laser capture microdissection. This system utilizes a low-power class IIIb invisible, infrared laser. Specially developed CapSure Macro Caps (Arcturus Engineering, Mountain View, CA) coated with thermoplastic film are placed over the tissue section. The cells of interest are positioned in the center of the field. A focused laser beam is pulsed over the cells of interest and the transfer film fuses to the selected cells. When the cap is lifted off, the selected cells remain adherent to the cap/film surface. The film on the under side of the CapSure is placed in direct contact with the reagent buffer. The cellular contents detach from the film into the reagent buffer and are ready for further processing.

Single Target Expression Profiling: Real Time Quantitative PCR

The snap-frozen human tissues used in the STEP analysis are obtained from Cooperative Human Tissue Network (CHTN, Eastern Division, NCI, Philadelphia, PA). Tumor tissues have been trimmed previously by licensed pathologists to limit potential cross-contamination. Total RNA is extracted from snap-frozen human tissues using "Ultraspecl RNA isolation kits" (Biotecx Laboratories, Houston, TX, BL10100) and purified further using RNeasy mini kits (Qiagen, Valencia, CA). Fifteen milligrams of total RNA is converted into double-stranded cDNA by reverse transcription (GIBCO BRL Life Technologies, Grand Island, NY) using the T7–T24 primer [5′-GGC CAG TGA ATT GTA ATA CGA CTC ACT ATA GGG AGG CGG (dT24)] and cleaned up by phenol/chloroform/isoamyl extraction using phase lock gel.

Master 384-well plates are generated containing 5 ng/ml double-stranded cDNA from 40 different tissue types (300 tissues total). Daughter plates are produced [final cDNA concentration: 40 pg/ml (200 pg/well)] via robotics. Duplex real-time PCR (target gene and GAPDH as reference gene) on 384-well optical plates is performed using TaqMan technology (Roche Molecular Systems, Inc., Alameda, CA) and analyzed on an ABI Prism PE7900 sequence detection system (Perkin-Elmer Applied Biosystems, Lincoln, CA), which uses the 5′ nuclease activity of *Taq* DNA polymerase to generate a real-time quantitative DNA analysis assay. PCR mix per well (25 ml) consists of commercially available, premixed GAPDH TaqMan primers/probe (PE), 900 nM each of 5′ and 3′ primers, and 200 nM TaqMan probe from each target gene, 200 pg cDNA and TaqMan universal PCR master mix (Perkin-Elmer Applied Biosystems). The following PCR conditions are used: 50° for 2 min and 95° for 10 min, followed by 40 cycles at 95° for 15 s and 62° for 1 min. The expression levels of target genes are normalized to reference gene levels and represented as relative expression (E) where DCt is the difference between reference and target gene cycles at which the amplification exceeds an arbitrary threshold. For reference gene values, GAPDH Ct values for each tissue are adjusted based on expression values of a panel of eight housekeeping genes in order to further improve normalization.

Development of Stable Cell Line

After the choice of the orphan candidate is made, oGPCR38 cDNA is cloned into the modified pCDNA3.1/Zeo(+) vector (Invitrogen Life Technologies, Carlsbad, CA). A PCR-generated IVS-IRES fragment using the pIRESpuro vector (Invitrogen Life Technologies, Carlsbad, CA) as a template is inserted into the *XbaI*-*Bst107I* sites. This vector is reused in a cassette-like manner to develop other oGPCR cell lines by insertion of the individual cDNAs. To construct the vector the following is performed.

1. Digest pCDNA3.1/Zeo(+) with *Xba*I and *Bst*1107 I. Isolate the fragment with the CMV promoter.
2. Using an IRES-containing vector (pIRESpuro), PCR amplify the IVS-IRES portion of this vector. Engineer an *Xba* I site at the 5′ end and incorporate about 20 5′ bases of the Zeo gene at the 3′ end.

3. Using the pCDNA3.1/Zeo vector as a template, PCS amplify the Zeocin-SV40 polyadenylation fragment incorporating about 20 3′ bases of IRES at the 5′ end and including a *Bst*1107I site at the 3′ end.

4. Set up overlapping PCR using amplicons generated from steps 2 and 3, digest with appropriate enzymes, and ligate with fragment from step 1.

5. The following restriction sites will no longer be unique when the construct is completed: *Hind*III, *Kpn*I, *Pst*I, *Apa*I, and 5′*Pme*I.

6. The remaining unique restriction sites in the MCS will be *Nhe*I, *Pme*I, *Afl*II, *Bam*HI, *Est*XI, *Eco*RI, *Eco*RV, *Bst*XI, *Not*I, *Xho*I, and *Xba*I.

7. The resulting vector is CMV–MCS–IVS–IRES–Zeo–SV40 PA.

The oGPCR DNA is inserted into the vector as a FLAG-oGPCR38 with a modified cytoplasmic tail region, referred to as "E" for enhanced activation. This modification of adding phosphorylation sites to the carboxy-terminal tail of the receptor has been shown to enhance the affinity of the receptor:β-arrestin interaction and has been described by Oakley *et al.* (2005a). The cells are maintained in DMEM 10% FCS (heat inactivated), 10 μg/ml gentamicin, 10 mM HEPES, 2 mM L-glutamine, 0.4 mg/ml Zeocin (to maintain selection of the oGPCR38), 0.4 mg/ml G418 (to maintain selection of the β-arrestin-GFP-Prolume tag), and penicillin/streptomycin. To identify the receptor expression during the cloning, propagation, and screening processes, cell aliquots are monitored for FACS analysis. To monitor receptor expression it is critical to remove the monolayers from the tissue culture flasks or the cell factories using EDTA. Although we use typical trypsinization procedures for propagating the cells, when cells are trypsinized 24 h prior to assay, FACS analysis results show an unexpectedly low receptor number due to the trypsin degrading the receptor at the surface of the cells. We monitor receptor expression on a twice-weekly basis using the Guava Technologies personal cell analysis instrument (Guava Technologies, Hayward, CA) employing an anti-FLAG M2 (mouse monoclonal) from Sigma and a secondary goat F(ab')$_2$ anti-mouse IgG-PE from Guava Technologies. This Express assay uses proprietary Guava Express software to quantify receptor expression along with a viability assessment. During the process of producing clones of oGPCR38, a minimum of 10 independent cell clones are characterized prior to identifying the clone for the HCS/HTS assay. We chose a clonal line (#216) for three reasons. One was due to the fact that the FACS analysis confirmed it as a highly responding population of receptor expressing cells. Second, we also identified #216 as the most responsive vesicle/pit forming clone responding to the GRK LITe assay (Oakley *et al.*, 2005b), achieving a S/B >10. Third, it also fulfilled the criteria of being a most stable clonal cell line during the course of assay development by reproducibly retaining receptor expression levels and maintaining the response to the GRK LITe assay over 22 passages. The propagated cells were maintained under the afore-mentioned standard culture conditions and used up to passage 21 for the screening assay and did not exceed this. We monitored receptor expression on a twice-weekly basis over the course of the screen, which confirmed that the receptor was maintained at >90% expression over the 20 days of screening runs.

HCS Assay Procedure

Materials

- Transfluor cell line(s)
- GRK LITe BacMam
- MEM media without phenol red
- HEPES
- HBSS
- Phosphate-buffered saline (PBS)
- Dimethy sulfoxide (DMSO)
- DRAQ5
- Formic acid

A generic description of the Transfluor assay methods used has been published by Hoffman (2005) among other literature and commercial methods (Ghosh *et al.*, 2005; Hudson *et al.*, 2002; Oakley *et al.*, 2002). The following methods for this screening campaign are similar with the following changes. We had been able to put into place a ligand-independent control using the GRK in a transiently expressed BacMam vector, referred to as the LITe assay (Ames *et al.*, 2004). This allowed us to monitor the 100% full activation of the assay with this transiently ex-pressed protein. Maximal transcription resulting in receptor activation and internalization occurred over the course of 16 to 20 h, allowing the β-arrestin:receptor internalization process to occur. The GRK virus is added 4 to 6 h after the double stable transfected U 2-OS oGPCR 38 cell line is plated onto 384-well microtiter plates. The multiplicity of infection (MOI) is optimized to 60, which translates into the addition of a 1- to 2-μl aliquot per control well depending on the viral stock potency. We chose to proceed with 16 wells of the positive GRK controls and 16 wells of the basal or untreated cells exposed to a maximum of 0.75% DMSO as the vehicle, as this is what the compounds were diluted in. Our choice of compound concentration was based upon experience running other cell-based screening assays, as well as the running the previous Transfluor assays where the results of hits and leads were quite feasible to follow up.

Compound plates are prepared previously as single-use plates containing a 2-μl DMSO aliquot of a 1 mM stock compound solution. Plates are diluted with 50 μl of HBSS containing 20 mM HEPES on a screening day using a Multidrop microplate dispenser with Titan stackers (Titertek, Huntsville, AL) for dispensing. These 384-well plates are bar coded to map the individual compounds per plate in IDBS's ActivityBase software. Typically, compounds are prealiquoted in only 352 of the 384 wells, leaving columns 1 and 2 open for control solutions. In compound plates, columns 1 and 2 are filled with 50 μl HBSS containing 20 mM HEPES and 4% DMSO.

Cell plates are formatted such that the first two columns are reserved for the controls: column 1 are GRK–LITe-treated positive controls and column 2 are basal vehicle controls. The 384-well cell plates (BD Falcon 384 well black/clear plates, Tissue Culture treated, reference #350504) are plated with 4000 Transfluor cells per well 20 h prior to assay. For the 16 wells in column 1, the 1- to 2-μl aliquot of the GRK BacMam is added 4 h after cell plating. Procedurally, 80 cell plates and 80 compound plates are matched up on a given daily screening day for executing the campaign. Four consecutive rounds of 20 compound and 20 cell plates are prepared and processed consecutively. Twenty compound plates are diluted as described at the same time as conditioned growth medium from 20 cell plates is replaced manually with 25 μl of prewarmed 37° MEM (without phenol red) containing 20 mM HEPES using a Multidrop from Titertek (Huntsville, AL). This replacement step is required to be done in a relatively quick time frame to avoid drying of the cells in the cell plates. Hereafter 6 μl of the compounds and controls is aliquoted into the cell plates using a programmed Tomtec Quadra 384 workstation (TOMTEC, Hamden, CT) in which tips are washed with DMSO followed by three washes in water and sonication to clean the tips between plates. The compound-containing cell assay plates are then left for 1 h incubation at room temperature. These plates are then quenched using a second dedicated stacker Mulitdrop where a 25-μl aliquot of 4% formic acid prepared in HBSS containing 20 mM HEPES and 2 μM DRAQ5 stain is dispensed. Cell assay plates are then lidded and remain in a laminar flow hood overnight. The following day, the plates are then washed with three 50-μl aliquots on a Titertek MAP-C2 liquid-handling assay processor with PBS. The final processing step dispenses a 25-μl aliquot to each well. The now-completed plates are heat sealed with plastic black sealers, and the run is placed into the Twister II Plate Handler stacker.

Image and Data Acquisition

The acquisition of 80 plates takes 12 h after which the assay curator reviews the plate-by-plate analysis to ensure that there were no plate errors in reading either whole plates due to occasional nonstandard plate specifications or errors due to no data acquisition, which occasionally is due to nonfocusing issues. These identified plates are rerun using the same algorithm collecting another field of view, which in 90% of the situations proves successful. Nonsuccessful acquisitions are then fully repeated in subsequent assays. A daily examination of the preliminary findings is done to verify that the images resulted in the expected phenotype defining the hit compounds.

Image Analysis Algorithm

The image analysis is implemented as Acapella script code. The relevant part of the algorithm is shown here:

```
nuclei_detection_g(Image = reference)
if (nuclei.count != 0)
  Cytoplasm_detection_b(cytoplasm = reference)
  WholeObjectRegion()
  set(WholeCells = Objects)
  set(SearchMask = WholeCells.ObjectRegion)

  set(NumberOfCells = nuclei.count)
  CalcIntensity(Image = signal, Cells = nuclei)
  set(GreenSignal = cells.intensity.mean)
  CalcIntensity(Image = reference, Cells = nuclei)
  set(RedSignal = cells.intensity.mean)
  set(NucleusArea = nuclei.area.mean)
  if (SearchMask.area != 0)
    spot_detection_c(signal, "ObjectRegion," ShowOutputParameters = no)
    // determine the number of the nucleus at this location
    calcintensity(SpotCenters, nuclei.index, Objects = spots)
    objectfilter(SpotCenters_intensity != 0)
    if (spots.count != 0)
      set(NuclearCenterFraction = objects.count/spots.count)
    end()
  end()
end()

output(IntegratedSpotSignalPerCellularSignal_BackgroundSubtracted,
  "Counts from centers per total counts")
output(NumberOfCells,"Number of cells")
```

```
output(NucleusArea, "Size of Nuclei")
output(GreenSignal, "Green Nuclear Fluorescence")
output(RedSignal, "Red Nuclear Fluorescence")
output(NumberOfSpots, "Number of translocation centers")
output(NuclearCenterFraction, "Fraction of translocation centers in
    the nucleus")
output(SpotsPerObject, "Centers per cell")
output(SpotsPerArea, "Centers per area")
```

First nuclei are detected using a single cell analysis procedure call. If nuclei are present, cells are detected using the reference signal channel (DRAQ5) and the complete cell is defined as mask where spots are searched. Some initial results are determined, which are based on the detected nuclei: the number of cells, intensity of the green and red signal in the nuclear area, and the average size of the nuclei.

If the search mask is not empty, spots are searched and then those spots that are in the nuclei area are selected and their fraction of all the spots is calculated. Finally, all results are output to the calling application. The main result is *counts from centers per total counts*. The result parameters *number of cells, size of nuclei, green nuclear fluorescence, red nuclear fluorescence,* and *fraction of translocation centers in the nucleus* are used to flag compounds and as additional information sources.

As per the previous section, the comma-separated variable files are processed by Abase software where the identification of potential compound hits is tabulated. Once identified, experiments are prepared to test the potential hit compounds under 10-point dose–response curves in orphan GPCR38, 7, and 105. For these experiments, compounds are prepared from 5 mM DMSO stock solutions and diluted in HBSS containing 20 mM HEPES at final concentrations in the assay of 30 μM to 1 nM in duplicate. In these cases, the compounds are evaluated on the oGPCR38 as well as two additional oGPCRs propagated and prepared in the same manner. This allowed us to evaluate the specificity of the potential agonist compounds in the same Transfluor system. Thereafter, other assays used to characterize GPCRs such as cAMP, calcium flux as measured by luminescence, FLIPR, and other HCS internalization assays were employed.

Screen Metrics and Results

Example of Experimental Findings

Using the aforementioned experimental paradigm, we were able to achieve a plate throughput of 80–100 clear-bottom 384-well plates per screening day. This allowed us to finish the entire primary screen of ~750,000 compounds within 5 weeks. The primary hit rate was 0.27%, that is, 2073 qualified hits based on *counts from centers per total counts* (see earlier discussion). We were confident that the BacMam GRK positive control was not overestimating the degree of vesiculation that would be seen in the Transfluor cell assay in the presence of active small molecule compounds. On a relative scale accounting for the translocated/internalized β-arrestin–GFP oGPCR complexes, the ~748,000 inactive compounds were clustered around 0.005 units, the BacMam GRK positive control displayed an average of 0.100 activation units, and active compounds ranged between 0.025 and 0.250 activation units. Therefore, in retrospect, the constitutively active GRK positive control was not overestimating the amount of receptor translocation in the presence of active compounds. Typically, at this point in a high-content, high-throughput cell-based screen, we conduct confirmation assays of the primary screening hits by reordering and rerunning each compound twice on three different cell plates. Compounds that maintain activity in two out of three retests are subjected to 10-point concentration response curves and selectivity assays using the oGPCR of interest (oGPCR38) and two other stably transfected oGPCR U-2 OS Transfluor cell lines (oGPCRs 7 and 105). The signal-to-background (S/B) ratio for oGPCR38 was 35, and the S/B ratio for oGPCRs 7 and 105 was 8.4 and 4.5, respectively. The S/B ratio for the latter two receptor screens was lower due to a level of residual basal activation in the resting state. The Z prime values for all three assays, the primary screen and the two counter screens, was >0.6. After identification of validated selective hit compounds, other mechanistic GPCR signal transduction assays, such as calcium flux and cAMP determination assays, were employed to investigate the specific heterotrimeric G protein-coupled pathway that was activated by the agonist compounds.

Preliminary findings on the validated hit compounds were verified by manual curation of the phenotypes displayed in the confocal images. Careful examination revealed hit compounds with the expected phenotype of β-arrestin–GFP:receptor complexes robustly distributed among a mixed population of clathrin-coated pits and internalized vesicles. However, a number of other expected and unexpected phenotypes were also identified.

i. Compounds that caused an activation event but exhibited a preference for either clathrin-coated pit morphology or internalized vesicle morphology via an unknown compound-specific mechanism.

ii. Compounds that elicited the expected pit or vesicle activation phenotype but also produced evidence of cytotoxicity, via either plasma membrane ruffling or cell shrinkage of overt cell loss.

iii. Compounds that caused a nuclear condensation event, as registered in the DRAQ5 (Biostatus Limited, Leicestershire, UK) nuclear stain channel for the object characteristic for *size of nuclei.*

iv. Compounds that produced *translocation centers* for the GFP
β-arrestin that reside within the nuclear mask area, an unexpected event
given the fact that a nuclear localization sequence was not factored into the
experimental design.

v. Compounds that precipitated out of solution and the resulting
particulate, aggregate, or crystalline-like structures were visualized

vi. Compounds flagged for autofluorescence in the red channel. These
compounds, if genuinely active, may still be potential drug development
candidates; however, a different mode of detection in the secondary assay
may be appropriate.

vii. Compounds flagged for autofluorescence in the green channel.
There was a great variety among the exact cellular localization of these
compounds, ranging from compounds that were distributed broadly within
the cytoplasm, to nuclear localization, to defined compartment/organellar
localization (presumably due to selective cellular uptake mechanisms).

viii. Compounds that caused spurious phenotypes, such as the
complete translocation of the β-arrestin–GFP to the actin fibrils within
the cell and a single characteristic foci in the nucleus. These compounds
may eventually become tools to help elucidate mechanism of β-arrestin
involvement in the cytoskeleton or in cell cycle regulation.

Of the validated active compounds, we found incidences where a compound was monoselective for only the oGPCR of interest. In other cases, a
compound was active on two out of three oGPCRs tested. There were also
compounds that activated all three of the Transfluor oGPCR cell lines.
There were even cases of oGPCR-specific cytotoxicity events, which may
have been due to a receptor-based mechanism or simply differences in the
U-2 OS cell background due to stable clonal selection. Most of the active
validated compounds displayed EC_{50} values in the low micromolar range;
however, a number of compounds were potent in the high nanomolar
range. The exact nature of the mechanism of each compound-induced
GFP β-arrestin activation event remains to be discovered. Notably, there
are literature reports of molecules that act through nonheterotrimeric G
protein-coupled pathways, possibly through extracellular signal-regulated
kinase activation or via arrestin-induced clustering of intracellular signaling
molecules (Ren *et al.*, 2005).

Concluding Remarks

We have described protocols for conducting high-throughput, fluorescent image-based confocal cell-based screens for the identification of surrogate small molecule chemistry as a starting point for identifying mechanistic

tools or for initiating medicinal drug discovery chemistry efforts toward a subset oGCPRs related to obesity and energy metabolism in humans. These selected oGPCRs were identified using a systems-based approach that combined the power of *in silico* bioinformatics, molecular biological techniques, genomics, stable green fluorescent protein expression, laser capture microdissection, quantitative expression profiling, high-speed automation-assisted laser confocal microscopy, and state-of-the-art live cell-based assay development. Such experimentation would not have been possible without the convergence of recent significant advancements within each of these fields of study. The use of GFP translocation events as a bona fide high-throughput readout opens the possibility of using similar translocation or activation events for addressing a wide variety of target classes in drug development and academic settings alike. In our hands, the Transfluor assay has been particularly effective in assisting in the elaboration of compounds to address oGPCRs and their signaling mechanisms, thus presenting a new avenue for receptor deorphanization strategies.

References

Ames, R., Fornwald, J., Nuthulaganti, P., Trill, J., Foley, J., Buckley, P., Kost, T., Wu, Z., and Romanos, M. (2004). BacMam recombinant baculoviruses in G protein-coupled receptor drug discovery. *Recept. Channels* **10**, 99–107.

Bouchard, L., Drapeau, V., Provencher, V., Lemieux, S., Chagnon, Y., Rice, T., Rao, D. C., Vohl, M., Tremblay, A., Bouchard, C., and Perusse, L. (2004). Neuromedin β: A strong candidate gene linking eating behaviors and susceptibility to obesity. *Am. J. Clin. Nutr.* **80**, 1478–1486.

D'Ursi, A., Albrizio, S., Di Fenza, A., Crescenzi, O., Carotenuto, A., Picone, D., Novellino, E., and Rovero, P. (2002). Structural studies on Hgr3 orphan receptor ligand prolactin-releasing peptide. *J. Med. Chem.* **45**, 5483–5491.

Franklin, K. B. J., and Paxions, G. (1997). "The Mouse Brain in Stereotaxic Coordinates." Academic Press, San Diego, CA.

Ghosh, R. N., DeBiasio, R., Hudson, C. C., Ramer, E. R., Cowan, C. L., and Oakley, R. H. (2005). Quantitative cell-based high-content screening for vasopressin receptor agonists using Transfluor technology. *J. Biomol. Screen* **10**, 476–484.

Gong and Sjogren, Personal communication.

Gu, W., Geddes, B. J., Zhang, C., Foley, K. P., and Stricker-Krongrad, A. (2004). The prolactin-releasing peptide receptor (GPR10) regulates body weight homeostasis in mice. *J. Mol. Neurosci.* **22**, 93–103.

Hanada, R., Teranishi, H., Pearson, J. T., Kurokawa, M., Hosoda, H., Fukushima, N., Fukue, Y., Serino, R., Fujihara, H., Ueta, Y., Ikawa, M., Okabe, M., Murakami, N., Shirai, M., Yoshimatsu, H., Kangawa, K., and Kojima, M. (2004). Neuromedin U has a novel anorexigenic effect independent of the leptin signaling pathway. *Nature Med.* **10**, 1067–1073.

Hoffman, A. F. (2005). The preparation of cells for high-content screening. *In* "Handbook of Assay Development in Drug Discovery" (A. F. Minor, ed.), pp. 227–242. Dekker, New York.

Hudson, C. C., Oakley, R. H., Cruickshank, R. D., Rhem, S. M., and Loomis, C. R. (2002). Automation and validation of the Transfluor technology: A universal screening assay for G protein-coupled receptors. *Proc. SPIE Int. Soc. Optic. Eng.* **4626,** 548–555.

Kieffer, T. J., and Habener, J. F. (1999). The glucagon-like peptides. *Endocr. Rev.* **20,** 876–913.

Kinnecom, K., and Pachter, J. S. (2005). Selective capture of endothelial and perivascular cells from brain microvessels using laser capture microdissection. *Brain Res. Protocols* **16,** 1–9.

Ludwig, D. S., Tritos, N. A., Mastaitis, J. W., Kulkarni, R., Kokkotou, E., Elmquist, J., Lowell, B., Flier, J. S., and Maratos-Flier, E. (2001). Melanin-concentrating hormone overexpression in transgenic mice leads to obesity and insulin resistance. *J. Clin. Invest.* **107,** 379–386.

Oakley, R. H., Barak, L. S., and Caron, M. G. (2005a). Real-time imaging of GPCR-mediated arrestin translocation as a strategy to evaluate receptor-protein interactions. *In* "G Protein Receptor-Protein Interactions" (M. G. George and S. R. Dowd, eds.). Wiley, New York.

Oakley, R. H., Cowan, C. L., Hudson, C. C., and Loomis, C. R. (2005b). Transfluor provides a universal cell-based assay for screening G protein-coupled receptors. *In* "Handbook of Assay Development in Drug Discovery" (C. R. Minor, ed.), pp. 435–457. Dekker, New York.

Oakley, R. H., Hudson, C. C., Cruickshank, R. D., Meyers, D. M., Payne, R., Jr., Rhem, S. M., and Loomis, C. R. (2002). The cellular distribution of fluorescently labeled arrestins provides a robust, sensitive, and universal assay for screening G protein-coupled receptors. *Assay Drug Dev. Technol.* **1,** 21–30.

Ren, X., Reiter, E., Ahn, S., Kim, J., Chen, W., and Lefkowitz, R. J. (2005). Different G protein-coupled receptor kinases govern G protein and β-arrestin-mediated signaling of V2 vasopressin receptor. *Proc. Natl. Acad. Sci. USA* **102**(5), 1448–1453.

Reidhaar-Olson, J. F., Braxenthaler, M., and Hammer, J. (2002). Process biology: Integrated genomics and bioinformatics tools for improved target assessment. *Targets* **1,** 189–195.

Reidhaar-Olson, J. F., Ohkawa, H., Babiss, L. E., and Hammer, J. (2003). Process biology: Managing information flow for improved decision making in preclinical R&D. *Preclinica* **1,** 161–169.

Shimada, M., Tritos, N. A., Lowell, B. B., Flier, J. S., and Maratos-Flier, E. (1998). Mice lacking melanin-concentrating hormone are hypophagic and lean. *Nature* **396,** 670–674.

Shimomura, Y., Harada, M., Goto, M., Sugo, T., Matsumoto, Y., Abe, M., Watanabe, T., Asami, T., Kitada, C., Mori, M., Onda, H., and Fujion, M. (2002). Identification of neuropeptide W as the endogenous ligand for orphan G-protein-coupled receptors GPR7 and GPR8. *J. Biol. Chem.* **277,** 35826–35832.

Van Heek, M., Compton, D. S., France, C. F., Tedesco, R. P., Fawzi, A. B., Graziano, M. P., Sybertz, E. J., Strader, C. D., and Davis, H. R., Jr. (1997). Diet-induced obese mice develop peripheral, but not central, resistance to leptin. *J. Clin. Invest.* **99,** 385–390.

Zhang, J., Chung, T. D. Y., and Oldenburg, K. R. (1999). A simple statistical parameter for use in evaluation and validation of high throughput screening assays. *J. Biomol. Screen.* **4,** 67–73.

[8]   G Protein-Coupled Receptor Internalization Assays in the High-Content Screening Format

By DOROTHEA HAASEN, ANDREAS SCHNAPP,
MARTIN J. VALLER, and RALF HEILKER

Abstract

High-content screening (HCS), a combination of fluorescence microscopic imaging and automated image analysis, has become a frequently applied tool to study test compound effects in cellular disease-modeling systems. This chapter describes the measurement of G protein-coupled receptor (GPCR) internalization in the HCS format using a high-throughput, confocal cellular imaging device. GPCRs are the most successful group of therapeutic targets on the pharmaceutical market. Accordingly, the search for compounds that interfere with GPCR function in a specific and selective way is a major focus of the pharmaceutical industry today. This chapter describes methods for the ligand-induced internalization of GPCRs labeled previously with either a fluorophore-conjugated ligand or an antibody directed against an N-terminal tag of the GPCR. Both labeling techniques produce robust assay formats. Complementary to other functional GPCR drug discovery assays, internalization assays enable a pharmacological analysis of test compounds. We conclude that GPCR internalization assays represent a valuable medium/high-throughput screening format to determine the cellular activity of GPCR ligands.

Introduction

Fluorescence microscopy has been widely employed in academic cell biology research as a nondestructive and sensitive technique to visualize subcellular structures and to monitor intracellular protein translocations. In recent years, the pharmaceutical industry has displayed an increasing preference to apply cell biological test systems in the early drug discovery process. Particularly, a novel technique generally referred to as high-content screening (HCS) has been introduced that combines high-resolution fluorescence microscopy with automated image analysis (Almholt *et al.*, 2004; Conway *et al.*, 1999; Ghosh *et al.*, 2000; Li *et al.*, 2003; Taylor *et al.*, 2001). The biomolecules of interest are labeled with different fluorophores that may be monitored in parallel at different wavelengths (multiplexing). Image analysis software quantifies the distribution and brightness of the

METHODS IN ENZYMOLOGY, VOL. 414   0076-6879/06 $35.00
          DOI: 10.1016/S0076-6879(06)14008-2

fluorophore-labeled biomolecules in the cells. In a kinetic mode, this enables the visualization of intracellular protein trafficking as a consequence of a pharmaceutical drug effect.

Apart from protein trafficking (Almholt *et al.*, 2004), HCS can provide information on the phosphorylation state of target proteins (Russello, 2004), on cellular proliferation (Bhawe *et al.*, 2004) or apoptosis (Steff *et al.*, 2001), on morphological changes such as neurite outgrowth (Simpson *et al.*, 2001), on modifications of the cytoskeleton (Giuliano, 2003; Olson and Olmsted, 1999), on cellular movements (Soll *et al.*, 2000), and on other overall changes of the fluorescence such as for the analysis of gap junctions (Li *et al.*, 2003).

By these means, HCS provides several advantages over normal high-throughput screening (HTS). Cellular HTS conventionally monitors the mean response of the whole cell population of a microtiter plate (MTP) well. In contrast, HCS can distinguish the individual response of many cells in an MTP well that may differ with respect to the differentiation, the stage of the cell cycle, the state of transfection, or due to natural variability. As a result, heterogeneous pharmaceutical drug effects on mixed cell populations may be analyzed in a single MTP well. "On-target" drug effects may be cross-correlated with other phenomena such as cellular toxicity (Wolff *et al.*, 2006). Compound artefacts such as cell lysis or compound autofluorescence may be detected. HCS permits work with endogenous targets and/or primary cells using specific antibodies or morphological image analysis. In this way, novel assay formats can be enabled that do not depend on an overall change of fluorescence or luminescence intensity from the whole MTP well.

High-Throughput Confocal Cellular Imaging Systems

HCS has benefited from the introduction of high-throughput confocal cellular imaging systems. In confocal optics, the spatial resolution in the vertical direction is improved dramatically by reducing background fluorescence from above or below the focal plane. This way, confocality enables the observation of cells that adhere to the bottom of a microtiter plate well without interference from dead cells, free fluorophore, or autofluorescent particles above the cellular layer. This increased optical resolution is particularly important to permit the visualization of biomolecule translocation processes across the complex subcellular membrane, vesicle, and organelle systems within eukaryotic cells.

In classic confocal optics (Wilson, 1990), the restriction to fluorescence emission from a specific focal plane is achieved by guiding the emitted light through a pinhole. Using this so-called "point scanning," the fluorescence emission from a femtoliter-sized observation volume within the focal plane is guided to the photon detector. However, the available confocal point scanning microscopes, which are based on this optical assembly, are generally too slow for drug screening applications.

Three fluorescence microscopic cellular imagers combine a high throughput with confocal optics (Zemanova *et al.*, 2003): the INCell Analyzer 3000 (Glaser, 2004; Haasen *et al.*, 2006) from GE Healthcare Biosciences (Little Chalfont, United Kingdom), the Opera (Eggeling *et al.*, 2003) from Evotec Technologies GmbH (Hamburg, Germany), and the Pathway Bioimager (Vanek and Tunon, 2002) from Becton-Dickinson Bioimaging Systems (Rockville, MD). To reduce measuring time, the Opera and the Pathway Bioimager employ a Nipkow disk (Nakano, 2002) to project fluorescence from several confocal volumes in parallel to a charged-coupled device (CCD) camera, whereas the INCell Analyzer 3000 employs line scanning (Glaser, 2004) through a confocal slit. The imaging and image analysis for the examples given in this chapter are described for the INCell Analyzer 3000.

The INCell Analyzer 3000 provides three laser lines for fluorescence excitation: two (364 nm, 488 nm) from an argon ion laser and one (647 nm) from a krypton laser (Fig. 1). The three laser lines are individually guided through neutral density filters but may be combined for parallel excitation. After reflection by a beam splitter they are autofocused via a 40× Nikon ELWD Plan Fluor/0.6 NA objective to the adherent cell layer at the bottom of the assay MTP well. The 40× Nikon extra-long working distance (ELWD) Plan Fluor/0.6 NA objective provides the combined benefits of a large field of view (FOV; 0.56 mm²) with a good optical resolution. Fluorescence emission light is collected through the same objective, passed through the aforementioned beam splitter, and guided through a confocal slit. The confocal slit serves to exclude emission light from above or below the focal layer. The light is split into up to three wavelength ranges, which permits simultaneous confocal imaging using three 12-bit −35° cooled CCD cameras. Prior to cellular imaging, a flat field correction for inhomogeneous illumination of the scanned area is carried out using an MTP well containing a homogeneous fluorophore solution.

GPCR Internalization Assays

GPCRs are seven-transmembrane helix proteins (Ji *et al.*, 1998), typically transmitting an extracellular signal into the cell by the conformational rearrangement of their helices and by the subsequent binding and activation of an intracellular heterotrimeric G protein (Perez and Karnik, 2005). In this way, GPCRs act as sensors of exogenous signals, which they transduce into

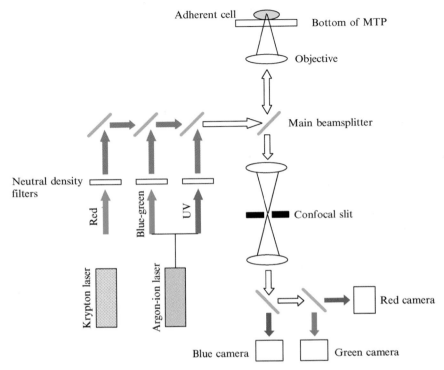

FIG. 1. Optical buildup of the IN Cell Analyzer 3000. The IN Cell Analyzer 3000 employs line scanning through a confocal slit. Two laser lines (364 and 488 nm) from an argon ion laser and one laser line (647 nm) from a krypton laser are guided individually through neutral density filters. All three laser lines may then be combined. After reflection by a beam splitter they are autofocused via a 40× Nikon extra-long working distance (ELWD) Plan Fluor/0.6 NA objective to the adherent cell layer in the assay MTP. Emission light is collected through the same objective, passed through the above beam splitter, and guided through a confocal slit. The light is then split into up to three wavelength ranges, which allows simultaneous confocal imaging using three 12-bit −35° cooled CCD cameras.

cytoplasmic signaling pathways. The first GPCRs to be cloned were bovine opsin (Nathans and Hogness, 1983) and the β-adrenergic receptor (Dixon et al., 1986). Since then, a large gene family of a further ∼2000 GPCRs has been reported, classified into more than 100 subfamilies according to sequence homology, ligand structure, and receptor function.

GPCRs are the most important class of therapeutic targets (Ma and Zemmel, 2002). Approximately 45% of all known pharmaceutical drugs are directed against transmembrane receptors (Drews, 2000), largely against GPCRs. GPCRs are involved in a broad diversity of physiological functions, such as pain perception, chemotaxis, neurotransmission, cardiovascular actions, and metabolism, and finding ways to modulate GPCR signaling remains a major focus of pharmaceutical research.

The interaction between GPCRs and their extracellular ligands has proven to be an attractive point of interference for therapeutic agents. For that reason, the pharmaceutical industry has developed biochemical drug discovery assays to investigate these ligand–GPCR interactions, such as scintillation proximity assays (Alouani, 2000) or the less frequently employed fluorescence polarization assays (Banks and Harvey, 2002; Harris et al., 2003) and fluorescence intensity distribution analysis assays (Auer et al., 1998; Zemanova et al., 2003). All the aforementioned biochemical binding assays rely on the competition of the test compound with a labeled reference ligand. An obvious disadvantage of these binding assays is the risk of missing noncompetitive, allosteric ligands. Further, the binding assay does not elucidate functional aspects of test compound activity, such as full/partial agonism, neutral antagonism, inverse agonism, or positive modulation. To expand compound testing in this direction, there is a need for functional high-throughput assays, possibly measuring GPCR activity in a more physiological, cellular background.

GPCR signal transduction mechanisms have been characterized in three major classes: Gq (phospholipase C), Gi, and Gs (inhibition and stimulation of cAMP production, respectively).

If the GPCR of interest signals via phospholipase C, the most broadly applied cell-based technique to measure GPCR activation is the Ca release assay, either measured in a fluorescent format using Ca-sensitive fluorophores (Sullivan et al., 1999) or in a luminescent format using aequorin and a chemiluminescent substrate (Dupriez et al., 2002). Alternatively, if the GPCR of interest signals via adenylate cyclase, the cytosolic cyclic adenosine monophosphate (cAMP) content may be determined using various detection technologies (Gabriel et al., 2003).

Apart from the aforementioned assays measuring the cellular signaling via G proteins, the functional activation of GPCRs may be monitored by agonist-induced receptor internalization (Milligan, 2003). The broad applicability of GPCR internalization assays (Fig. 2; scheme of GPCR internalization) is based on the common phenomenon of GPCR "desensitization" and has been demonstrated for numerous GPCRs (Ferguson, 2001; Krupnick and Benovic, 1998; Oakley et al., 2002). In the desensitization process, GPCR kinases (GRKs) phosphorylate agonist-activated GPCRs on serine and threonine residues. Arrestins are cytoplasmic proteins that are recruited to the plasma membrane by GRK-phosphorylated GPCRs (Barak et al., 1997).

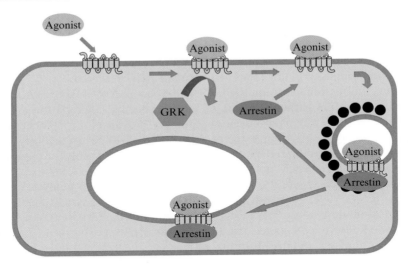

Fɪɢ. 2. Scheme of GPCR internalization assays. After agonist stimulation, the GPCR becomes phosphorylated by a GRK on its carboxy-terminal tail. Arrestin is recruited to the plasma membrane by the GRK-phosphorylated GPCR. Arrestin then targets the GPCR to clathrin-coated pits for endocytosis. Depending on the stability of the specific GPCR–arrestin interaction, arrestin is either released after the formation of clathrin-coated pits or cointernalized with the GPCR-loaded vesicles. The internalization process may be monitored by a fluorophore label on the agonist, on the N or C terminus of the GPCR, or on the arrestin.

Arrestins then uncouple the GPCR from the cognate G protein (Lohse *et al.*, 1992; Pippig *et al.*, 1993) and target the desensitized receptors to clathrin-coated pits for endocytosis (Goodman *et al.*, 1996; Laporte *et al.*, 2000).

In contrast to Ca release and cAMP assays, the internalization assay is independent of the individual GPCR intracellular signaling route. Thus, desensitization occurs independently of the associated G protein subclass or of the class of GPCR ligand. Further, the imaging-based GPCR internalization assays offer the general advantages of the HCS format as described earlier.

Internalization Assay and Image Analysis Protocols

Fluorophore-Labeled Ligand Internalization

One means to monitor the internalization of a GPCR is to cointernalize a specific, fluorophore-labeled ligand. In the following example protocol a tetramethylrhodamine-labeled endothelin-1 (TMR-ET-1) is used as a ligand for the endothelin A receptor (ET_AR). Physiologically, ET_AR is a GPCR that mediates the vasoconstrictive effects of the ET-1 peptide hormone (Bremnes *et al.*, 2000; Yanagisawa *et al.*, 1988).

Materials

Cell lines: hET_A-C1 (high expression level, B_{max} 60–150 pmol/mg protein) and hET_A-C2 (low expression level, B_{max} 17 pmol/mg protein) in CHO-K1 background (Euroscreen)

TMR-ET-1: Custom labeling (Evotec OAI, Hamburg, Germany) with 5-carboxy TMR on the side chain of the lysine in the ET-1 peptide CSCSSLMDKECVYFCHLDII. Dissolve lyophilized product in dimethyl sulfoxide (DMSO) and store in aliquots at $-80°$.

ET-1, unlabeled: Dissolve lyophilized product (Bachem Feinchemikalien, Bubendorf, Switzerland) in DMSO and store in aliquots at $-80°$.

Assay plate: black 96-well Viewplate, 800 μm plastic bottom, coated with collagen I (Perkin Elmer)

Complete medium: Ham's F12 supplemented with 10% fetal calf serum

High control: 4 nM TMR-ET-1, 0.6% DMSO in assay buffer

Low control: 4 nM TMR-ET-1 + 40 nM unlabeled ET-1, 0.6% DMSO in assay buffer

Assay buffer: Hank's balanced salt solution (HBSS)/HEPES (1.26 mM $CaCl_2$, 0.493 mM $MgCl_2$, 0.407 mM $MgSO_4$, 5.33 mM KCl, 0.441 mM KH_2PO_4, 4.17 mM $NaHCO_3$, 137.93 mM NaCl, 0.338 mM Na_2HPO_4, 5.56 mM D-glucose, 10 mM HEPES, pH 7.4), supplemented with 0.2% bovine serum albumin

Fixing solution: 2% formaldehyde, 10 mM HEPES in Ham's F12 medium

Phosphate-buffered saline (PBS): 2.67 mM KCl, 1.47 mM KH_2PO_4, 137.93 mM NaCl, 8.06 mM Na_2HPO_4

Ligand Cointernalization Protocol

1. Seed cells into assay plate using a multichannel pipette at a density of 25,000 cells in a volume of 100 μl volume per well in complete medium. Culture for 20 to 24 h at $37°$, 5% CO_2.
2. Remove complete medium and add 40 μl of ligand in HBSS/HEPES.
3. Incubate assay plate at room temperature.
4. Add 1 volume 2× concentrated fixing solution to wells at indicated time points ($t = 10$ min, 3.5 h, 8 h).
5. Fix for 30 min at room temperature.
6. Remove fixing solution and add 100 μl PBS.

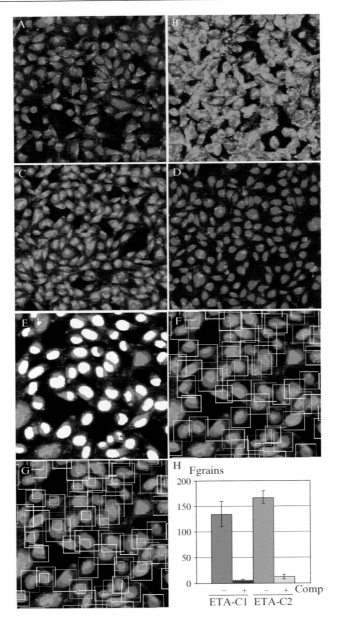

FIG. 3. TMR-ET-1 cointernalization by ETAR. CHO-K1 cells stably expressing medium levels of ETAR (ETA-C2) are stimulated with 4 nM TMR-ET-1 (A–C) or 4 nM TMR-ET-1 + 40 nM ET-1 (D) incubated at room temperature and fixed at 10 min (A), 3.5 h (B), and 8 h (C and D) poststimulation. Fixed cells are imaged on the IN Cell Analyzer 3000 and analyzed

Imaging. For this protocol, the fixed cells are imaged on the IN Cell Analyzer 3000 instrument employing the 488-nm laser line (121 mW; attenuated by a neutral density filter to 10%) combined with the 595BP60 emission filter for TMR-ET-1 and the 364-nm laser line (58 mW; attenuated by a neutral density filter to 50%) combined with the 450BP65 emission filter for Hoechst. Fluorescence emission is recorded sequentially in the green and in the blue channel. Prior to cellular imaging, a flat field correction for inhomogeneous illumination of the scanned area is carried out using an MTP well containing 50 nM Oregon Green and 20 μM Alexa Fluor 350 solution (Molecular Probes, Inc., Eugene, OR). All measurements are conducted at room temperature.

Figure 3 shows examples of representative images from wells fixed at different time points. The intracellular fluorescence of TMR-ET-1 increases over time (Fig. 3A–C) due to the formation of small endocytic vesicles that accumulate in larger structures close to the nuclear compartment. The cointernalization of labeled ET-1 can be competed with a 10-fold excess of unlabeled ligand (Fig. 3D).

Image Analysis and Quantification. The cointernalization of a fluorophore-labeled ligand with its specific GPCR upon receptor activation results in the formation of grain-like objects within the cell. Here, the granularity analysis module GRN0 of the IN Cell Analyzer 3000 analysis package is used as a method to identify, analyze, and quantify these grain-like structures. The granularity algorithm employs a two-color image composed of a marker channel and a signal channel. The marker channel image (blue) is employed to identify fluorescently stained cell nuclei, to classify each identified nucleus as an object (Fig. 3E), and to define a region of measurement by the dilation of the nuclear mask for each individual cell. In the signal channel image (green) the region of the dilated nuclear mask (Fig. 3F) is then analyzed for the presence of grains. The algorithm defines grains as distinct regions characterized by two parameters: a specific size and a defined intensity difference as compared to the region of the cell immediately surrounding the grains (Fig. 3G). The Fgrain value, which is the scaled measure of the fraction of cellular fluorescence

---

using the granularity algorithm GRN0 (E–H). Adapted analysis parameter settings are indicated by the nuclear mask (E), the measurement region of the dilated nuclear mask (large boxes; F and G), and the grain regions (small boxes; G). Images represent one-fourth (A–D) and one-sixteenth (E and F) of a full-size image (750 μm × 750 μm). Quantification of the grain structures (size: 20 pixels, intensity gradient: 2.0) is performed on the basis of Fgrain values for two stable cell lines expressing ETAR, ETA-C1 (high expression level, 60–150 pmol/mg protein), and ETA-C2 (medium expression level, 17 pmol/mg protein) stimulated with 4 nM TMR-ET-1 only (comp −) or 4 nM TMR-ET-1 + 40 nM ET-1 (comp +). Values are plotted as means of six images ± SD.

present in the qualifying grains, is used as the numeric output parameter. It is calculated as

$$\text{Fgrain} = \frac{\text{Sum of pixel values in test boxes corresponding to valid grains}}{\text{Sum of pixel values in dilated mask region}} \times 1000$$

The quantification of Fgrains by the granularity algorithm verifies that in two cell lines stably expressing different levels of ET_AR the co-internalization of TMR-ET-1 can be competed by unlabeled ET-1 (Fig. 3H). A 10-fold excess of unlabeled ligand reduced the Fgrain value from approximately 136 to 5 for the highly expressing cell line ETA-C1 and from approximately 167 to 12 for ETA-C2 with slightly reduced expression levels of ET_AR.

Fluorophore-Labeled Antibody (β2AR)

An alternative assay technology to follow the internalization of a GPCR is to cointernalize a specific antibody, directed either against an extracellular domain of the receptor or against an amino-terminal epitope tag. In the following example protocol the internalization of a vesicular stomatitis virus G (VSVG) epitope-tagged human β2AR expressed in HEK-293 cells is described. In this setup, the primary anti-VSVG-antibody is cointernalized with the receptor upon agonist stimulation and, in a second step, is detected by a fluorophore-labeled secondary antibody by immunofluorescence staining of fixed, permeabilized cells. In this way it is possible to determine the extent of agonist behavior. By way of illustration, the response of the β2AR to the partial agonist salmeterol (Sears and Lotvall, 2005) and the full agonist formoterol (Sears and Lotvall, 2005) is shown.

The molecular mechanisms of receptor tolerance summarized earlier have also been shown to be involved in the desensitization of β2AR (Ferguson, 2001; Lefkowitz, 1998; Su et al., 1980). In the context of β2AR the extent of functional desensitization induced by different β-adrenoceptor agonists is variable, and a correlation between agonist efficacy and receptor desensitization has been shown (Benovic et al., 1988; Clark et al., 1999; January et al., 1997, 1998).

Materials

Cell line: VSVG epitope-tagged β2AR (VSVG-β2AR) in HEK293 background (GE Healthcare, Cardiff, UK)

Assay plate: black 384-well 720-μm glass-bottom plate (MatriCal) coated with 5 μg/cm^2 poly-D-lysine (PDL)

Compound plate/predilution plate: 384-well MTP (REMP)

Complete medium: minimum essential medium (MEM) with Earle's salts and GlutaMAX I supplemented with 10% fetal calf serum (FCS) and 1% nonessential amino acids

High control: 10 μM isoproterenol (Sigma), 0.316% DMSO in complete medium

Low control: 0.316% DMSO in complete medium

Primary antibody: mouse monoclonal anti-VSVG antibody, clone P4D5 (Sigma)

Krebs–Ringer buffer (KRB): 125 mM NaCl, 4.8 mM KCl, 1.2 mM KH_2PO_4, 1.2 mM $MgSO_4$, 1.3 mM $CaCl_2$, 25 mM HEPES, pH 7.4

Fixing solution: 2% formaldehyde, 10 mM HEPES in KRB

PBS: 2.67 mM KCl, 1.47 mM KH_2PO_4, 137.93 mM NaCl, 8.06 mM Na_2HPO_4

Permeabilization solution: 0.1% saponin in PBS

Secondary antibody: anti-mouse AlexaFluor647 (Molecular Probes)

Antibody Cointernalization Protocol

1. Seed cells into an assay plate using a dispenser at a density of 10,000 cells in a volume of 20 μl per well in complete medium. Culture for 20 to 24 h at 37°, 5% CO_2.
2. Prepare an intraplate dilution series (1:3.16) of compounds in DMSO in a compound plate using an eight-tip pipettor.
3. Dilute compound DMSO solutions in complete medium (1:63.2) into a predilution plate using a 384-well pipetting head.
4. Add 20 μl primary antibody in complete medium to assay plate using a dispenser.
5. Bind antibody for 30 min at 37°, 5% CO_2.
6. Transfer 10 μl of prediluted compounds to assay plate using a 384-well pipetting head.
7. Incubate for indicated time point(s) at 37°, 5% CO_2.
8. Add 1 volume 2× concentrated fixing solution to assay plate.
9. Fix for 30 min at room temperature.
10. Wash assay plate three times with PBS.

Imaging. The fixed cells are imaged on the IN Cell Analyzer 3000 instrument employing the 647-nm laser line (114 mW) combined with the 695BP55 emission filter for Alexa Fluor 647 and the 364-nm laser line (60 mW; attenuated by a neutral density filter to 10%) combined with the 450BP65 emission filter for Hoechst. Fluorescence emission is recorded sequentially in the red and in the blue channel. Prior to cellular imaging, a flat field correction for inhomogeneous illumination of the scanned area is carried out using an MTP well containing 50 nM Cy5 solution (GE Healthcare Biosciences) and 50 μM Alexa Fluor 350 solution (Molecular Probes, Inc.). All measurements are conducted at room temperature.

Figure 4 shows representative images of antibody cointernalization in β2AR-expressing cells treated with a full agonist, formoterol, and a partial agonist, salmeterol. Results show that after stimulation with the full agonist, the red fluorescence representing the subcellular localization of β2AR moves from the plasma membrane to at first smaller vesicles (5 to 10 min) and then to larger vesicles (10 to 20 min) that accumulate in a compartment close to the nucleus at 20 to 30 min post stimulation. In contrast, the response to the partial agonist is delayed, starting with the formation of small vesicles at 10 min post stimulation and reaching a maximal response at 30 min post stimulation. Moreover, the partial response can be seen to apply to each individual cell, resulting in two receptor subpopulations: one that internalizes and another that remains localized at the plasma membrane. In the case of activation by either a full or a partial agonist, the total intracellular red fluorescence decreases at 60 min post stimulation and is reduced to almost background levels at 120 min post stimulation, most likely due to targeting to the lysosomal pathway and degradation.

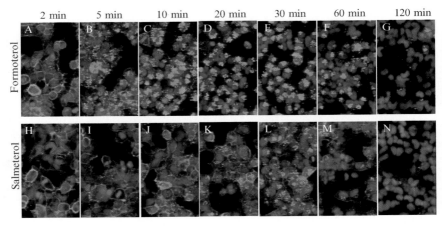

FIG. 4. Antibody cointernalization by β2AR. CHO-K1 cells stably expressing VSVG-tagged β2AR are allowed to bind the anti-VSVG antibody and are stimulated with formoterol (A–G) or salmeterol (H–N) and fixed at 2 min (A, H), 5 min (B, I), 10 min (C, J), 20 min (D, K), 30 min (E, L), 60 min (F, M), and 120 min (G, N) poststimulation. Fixed cells are subjected to immunofluorescence staining by a AF647-labeled secondary antibody and are imaged using the IN Cell Analyzer 3000. Images represent 1/27 of a full-size image (750 × 750 μm).

Image Analysis and Quantification. Corresponding to the example described in the preceding section for ligand cointernalization experiments, the granularity analysis module GRN0 of the IN Cell Analyzer 3000 analysis package is used as a method to identify, analyze, and quantify the internalized receptor in these experiments. In this case, the red channel image is defined as the signal channel. The GRN0 analysis parameters are optimized for nuclear morphology and the internalization response of the cell line is employed.

Changes in the granular appearance during the endocytic process provide a means to quantify the extent of the cellular response and to distinguish specific stages of endocytosis. Therefore, image analysis is performed setting different sizes and intensity gradients of grain-like objects. Grains are categorized in four subclasses according to their appearance (small, medium, large, and large bright), and their abundance is analyzed over the time course of the experiment (Fig. 5). As a representative example, upon activation by the full agonist formoterol, maximum values of 260 Fgrains are observed, as compared to an Fgrain value of 125 following activation by the partial agonist salmeterol. In case of the full agonist, all grain subclasses are present already 5 min post stimulation, reaching a maximum around 10 min for small, 20 min for medium, and 30 min for large and large bright vesicles (Fig. 5A). In contrast, in the case of the partial agonist, none of the vesicle subclasses are present at 5 min post stimulation, and all subclasses reach their maximum at 30 min post stimulation (Fig. 5B), indicating that the internalization response proceeds differently in comparison to full agonist stimulation.

In Fig. 6, the parameters for medium class vesicles are used to analyze the images from a dose–response experiment for the two agonists. Stimulation with the full agonist results in an efficacy comparable to the standard control isoproterenol (Fig. 6A), whereas the partial agonist reaches only 45% of the maximum response of the standard (Fig. 6B). In this way it can be shown that the antibody cointernalization approach is a sensitive assay system not only to determine EC_{50} values, but also to classify agonists according to their full or partial response.

Conclusions and Outlook

Both the labeled reference ligand internalization and the labeled GPCR internalization methods can produce stable HCS assays. Both formats provide advantageous, partially complementary features in the characterization of GPCR ligand test compounds. Thus, test compounds can be qualified as agonists or antagonists in the "labeled GPCR" internalization experiments. In the aforementioned assay protocol for the β2AR, the test

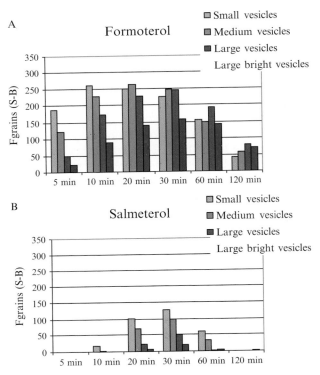

FIG. 5. Quantification of the internalization response of β2AR. Analysis of the grain structures is performed using the granularity analysis module GRN0. Grain objects are classified as follow: small (size: 5 pixels, intensity gradient: 1.5), medium (size: 7 pixels, intensity gradient: 2.0), large (size: 15 pixels, intensity gradient: 2.0), large bright (size: 15 pixels, intensity gradient: 2.5) vesicles. Fgrain values as a mean of two wells per experiment are plotted as S-B (signal from stimulated cells – background from control-treated cells).

compounds are analyzed for their agonistic properties. Alternatively, if the test compounds are preincubated with GPCR expressing cells, the absence of reference agonist-induced GPCR internalization can indicate antagonistic/inverse agonistic properties of the test compound. If the constitutive internalization and recycling rates of the GPCR are sufficient, the inverse agonists may be distinguished from the neutral antagonists by an enrichment of the receptor at the plasma membrane. As a further experimental modification to enable the identification of positive modulators, if the reference agonist is added at approximately EC_{10} concentration, a preincubation

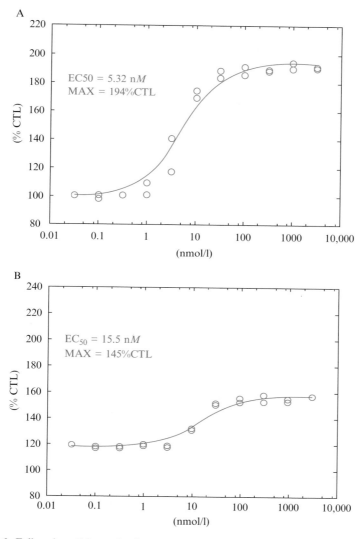

FIG. 6. Full and partial agonist dose response. β2AR-expressing cells are analyzed for medium size vesicles at 10 min poststimulation for formoterol as full agonist (A) and at 20 min poststimulation for salmeterol as partial agonist (B). Dose–response curves and EC_{50}/MAX values are calculated using a customized HTS software product. The assay window is defined by the response to isoproterenol and mock treatment, defining 200 and 100% CTL, respectively.

with a positive modulator compound will increase the internalization rate significantly.

A challenge in GPCR-labeling assays is that antibodies against the extracellular portion of the native receptor are not always available. Further, it is important that the GPCR-labeling antibody does not interfere with the ligand-binding site or produce a functional effect itself. To reduce the probability of such issues, an N-terminal tag may be attached to the GPCR of interest so that a tag-directed antibody may be used for GPCR detection.

In addition to the aforementioned GPCR-labeling assays, the "labeled ligand" protocol makes it possible to distinguish between orthosteric and allosteric GPCR-binding test compounds. Orthosterically acting test compounds block the binding of the labeled reference ligand to the cell surface and cannot therefore be analyzed further in "labeled ligand" assays. An allosterically acting test compound permits the binding of the labeled reference ligand to the plasma membrane-exposed GPCRs. If the mechanism of action is antagonistic, a preincubation of the allosteric test compound prevents cointernalization of the labeled reference ligand. If the mechanism is positive modulatory, a preincubation of the allosteric test compound leads to an increased rate of labeled ligand internalization. With regard to agonistic, allosteric test compounds, the interpretation of experimental results becomes more challenging: if added at the same time as the labeled reference ligand, depending on the ratio of kinetics, potencies, and efficacies between test compound and reference ligand, the test compounds may either increase or decrease the "labeled ligand" internalization rate.

The internalization assays described in this chapter employ an extracellular, ligand/GPCR-labeling step at the plasma membrane. Therefore, the initially increasing number and intensity of intracellular, fluorescent granules describe the rate of receptor internalization. After a few minutes, however, (i) the newly synthesized protein from the secretory pathway reaching the cell surface and (ii) recycling and/or (iii) degradation of ligand- or antibody-labeled GPCRs will be superimposed on the internalization kinetics. To help unravel these complex intracellular receptor trafficking pathways, it is possible to work with a second label that describes the overall distribution of the GPCR, such as a fluorescent protein label at the C terminus of the receptor (Xia et al., 2004) or to use a GPCR-detecting antibody on fixed and permeabilized cells. Such a second label enables observation of the ligand-induced net shift of local GPCR concentrations between the plasma membrane and intracellular compartments. Further, such labeling provides an impression of the GPCR distribution in the cells prior to ligand stimulation, thereby facilitating the overall investigation of test compound-induced changes.

The option to use fixed-cell internalization endpoint protocols, as described earlier for the model system of the β2A receptor, is advantageous for automated liquid handling and offline imaging. Thus, receptor internalization assays can provide robust medium and/or high-throughput screening formats and excellently complement the drug discovery tool spectrum for the GPCR target class.

Acknowledgment

We thank Paolo Meoni from Euroscreen S.A. (Brussels, Belgium) for providing the ETAR-overexpressing CHO cells that were used in this chapter.

References

Almholt, D. L., Loechel, F., Nielsen, S. J., Krog-Jensen, C., Terry, R., Bjorn, S. P., Pedersen, H. C., Praestegaard, M., Moller, S., Heide, M., Pagliaro, L., Mason, A. J., Butcher, S., and Dahl, S. W. (2004). Nuclear export inhibitors and kinase inhibitors identified using a MAPK-activated protein kinase 2 redistribution screen. *Assay Drug Dev. Technol.* **2,** 7–20.

Alouani, S. (2000). Scintillation proximity binding assay. *Methods Mol. Biol.* **138,** 135–141.

Auer, M., Moore, K. J., Meyer-Almes, F. J., Guenther, R., Pope, A. J., and Stoeckli, K. (1998). Fluorescence correlation spectroscopy: Lead discovery by miniaturized HTS. *Drug Discov. Today* **3,** 457–465.

Banks, P., and Harvey, M. (2002). Considerations for using fluorescence polarization in the screening of G protein-coupled receptors. *J. Biomol. Screen.* **7,** 111–117.

Barak, L. S., Ferguson, S. S., Zhang, J., and Caron, M. G. (1997). A beta-arrestin/green fluorescent protein biosensor for detecting G protein-coupled receptor activation. *J. Biol. Chem.* **272,** 27497–27500.

Benovic, J. L., Staniszewski, C., Mayor, F., Jr., Caron, M. G., and Lefkowitz, R. J. (1988). beta-Adrenergic receptor kinase: Activity of partial agonists for stimulation of adenylate cyclase correlates with ability to promote receptor phosphorylation. *J. Biol. Chem.* **263,** 3893–3897.

Bhawe, K. M., Blake, R. A., Clary, D. O., and Flanagan, P. M. (2004). An automated image capture and quantitation approach to identify proteins affecting tumor cell proliferation. *J. Biomol. Screen* **9,** 216–222.

Bremnes, T., Paasche, J. D., Mehlum, A., Sandberg, C., Bremnes, B., and Attramadal, H. (2000). Regulation and intracellular trafficking pathways of the endothelin receptors. *J. Biol. Chem.* **275,** 17596–17604.

Clark, R. B., Knoll, B. J., and Barber, R. (1999). Partial agonists and G protein-coupled receptor desensitization. *Trends Pharmacol. Sci.* **20,** 279–286.

Conway, B. R., Minor, L. K., Xu, J. Z., Gunnet, J. W., DeBiasio, R., D'Andrea, M. R., Rubin, R., DeBiasio, R., Giuliano, K., DeBiasio, L., and Demarest, K. T. (1999). Quantification of G-protein coupled receptor internatilization using G-protein coupled receptor-green fluorescent protein conjugates with the ArrayScantrade mark high-content screening system. *J. Biomol. Screen* **4,** 75–86.

Dixon, R. A., Kobilka, B. K., Strader, D. J., Benovic, J. L., Dohlman, H. G., Frielle, T., Bolanowski, M. A., Bennett, C. D., Rands, E., and Diehl, R. E. (1986). Cloning of the gene and cDNA for mammalian beta-adrenergic receptor and homology with rhodopsin. *Nature* **321,** 75–79.

Drews, J. (2000). Drug discovery: A historical perspective. *Science* **287,** 1960–1964.

Dupriez, V. J., Maes, K., Le Poul, E., Burgeon, E., and Detheux, M. (2002). Aequorin-based functional assays for G-protein-coupled receptors, ion channels, and tyrosine kinase receptors. *Receptors Channels* **8,** 319–330.

Eggeling, C., Brand, L., Ullmann, D., and Jager, S. (2003). Highly sensitive fluorescence detection technology currently available for HTS. *Drug Discov. Today* **8**, 632–641.

Ferguson, S. S. (2001). Evolving concepts in G protein-coupled receptor endocytosis: The role in receptor desensitization and signaling. *Pharmacol. Rev.* **53**, 1–24.

Gabriel, D., Vernier, M., Pfeifer, M. J., Dasen, B., Tenaillon, L., and Bouhelal, R. (2003). High throughput screening technologies for direct cyclic AMP measurement. *Assay Drug Dev. Technol.* **1**, 291–303.

Ghosh, R. N., Chen, Y. T., DeBiasio, R., DeBiasio, R. L., Conway, B. R., Minor, L. K., and Demarest, K. T. (2000). Cell-based, high-content screen for receptor internalization, recycling and intracellular trafficking. *Biotechniques* **29**, 170–175.

Giuliano, K. A. (2003). High-content profiling of drug-drug interactions: Cellular targets involved in the modulation of microtubule drug action by the antifungal ketoconazole. *J. Biomol. Screen* **8**, 125–135.

Glaser, V. (2004). An interview with Tim Harris, Ph.D. *Assay Drug Dev. Technol.* **1**, 403–408.

Goodman, O. B., Jr., Krupnick, J. G., Santini, F., Gurevich, V. V., Penn, R. B., Gagnon, A. W., Keen, J. H., and Benovic, J. L. (1996). Beta-arrestin acts as a clathrin adaptor in endocytosis of the beta2-adrenergic receptor. *Nature* **383**, 447–450.

Haasen, D., Wolff, M., Valler, M. J., and Heilker, R. (2006). Comparison of G-protein coupled receptor desensitization-related beta-arrestin redistribution using confocal and non-confocal imaging. *Comb. Chem. High Throughput Screen.* **9**, 37–47.

Harris, A., Cox, S., Burns, D., and Norey, C. (2003). Miniaturization of fluorescence polarization receptor-binding assays using CyDye-labeled ligands. *J. Biomol. Screen* **8**, 410–420.

January, B., Seibold, A., Allal, C., Whaley, B. S., Knoll, B. J., Moore, R. H., Dickey, B. F., Barber, R., and Clark, R. B. (1998). Salmeterol-induced desensitization, internalization and phosphorylation of the human beta2-adrenoceptor. *Br. J. Pharmacol.* **123**, 701–711.

January, B., Seibold, A., Whaley, B., Hipkin, R. W., Lin, D., Schonbrunn, A., Barber, R., and Clark, R. B. (1997). beta2-adrenergic receptor desensitization, internalization, and phosphorylation in response to full and partial agonists. *J. Biol. Chem.* **272**, 23871–23879.

Ji, T. H., Grossmann, M., and Ji, I. (1998). G protein-coupled receptors. I. Diversity of receptor-ligand interactions. *J. Biol. Chem.* **273**, 17299–17302.

Krupnick, J. G., and Benovic, J. L. (1998). The role of receptor kinases and arrestins in G protein-coupled receptor regulation. *Annu. Rev. Pharmacol. Toxicol.* **38**, 289–319.

Laporte, S. A., Oakley, R. H., Holt, J. A., Barak, L. S., and Caron, M. G. (2000). The interaction of beta-arrestin with the AP-2 adaptor is required for the clustering of beta 2-adrenergic receptor into clathrin-coated pits. *J. Biol. Chem.* **275**, 23120–23126.

Lefkowitz, R. J. (1998). G protein-coupled receptors. III. New roles for receptor kinases and beta-arrestins in receptor signaling and desensitization. *J. Biol. Chem.* **273**, 18677–18680.

Li, Z., Yan, Y., Powers, E. A., Ying, X., Janjua, K., Garyantes, T., and Baron, B. (2003). Identification of gap junction blockers using automated fluorescence microscopy imaging. *J. Biomol. Screen.* **8**, 489–499.

Lohse, M. J., Andexinger, S., Pitcher, J., Trukawinski, S., Codina, J., Faure, J. P., Caron, M. G., and Lefkowitz, R. J. (1992). Receptor-specific desensitization with purified proteins: Kinase dependence and receptor specificity of beta-arrestin and arrestin in the beta 2-adrenergic receptor and rhodopsin systems. *J. Biol. Chem.* **267**, 8558–8564.

Ma, P., and Zemmel, R. (2002). Value of novelty? *Nature Rev. Drug Discov.* **1**, 571–572.

Milligan, G. (2003). High-content assays for ligand regulation of G-protein-coupled receptors. *Drug Discov. Today* **8**, 579–585.

Nakano, A. (2002). Spinning-disk confocal microscopy: A cutting-edge tool for imaging of membrane traffic. *Cell Struct. Funct.* **27**, 349–355.

Nathans, J., and Hogness, D. S. (1983). Isolation, sequence analysis, and intron–exon arrangement of the gene encoding bovine rhodopsin. *Cell* **34**, 807–814.

Oakley, R. H., Hudson, C. C., Cruickshank, R. D., Meyers, D. M., Payne, R. E., Rhem, S. M., and Loomis, C. R. (2002). The cellular distribution of fluorescently labeled arrestins provides a robust, sensitive, and universal assay for screening of G protein-coupled receptors. *Assay Drug Dev. Technol.* **1**, 21–30.

Olson, K. R., and Olmsted, J. B. (1999). Analysis of microtubule organization and dynamics in living cells using green fluorescent protein-microtubule-associated protein 4 chimeras. *Methods Enzymol.* **302**, 103–120.

Perez, D. M., and Karnik, S. S. (2005). Multiple signaling states of G-protein-coupled receptors. *Pharmacol. Rev.* **57**, 147–161.

Pippig, S., Andexinger, S., Daniel, K., Puzicha, M., Caron, M. G., Lefkowitz, R. J., and Lohse, M. J. (1993). Overexpression of beta-arrestin and beta-adrenergic receptor kinase augment desensitization of beta 2-adrenergic receptors. *J. Biol. Chem.* **268**, 3201–3208.

Russello, S. V. (2004). Assessing cellular protein phosphorylation: High throughput drug discovery technologies. *Assay Drug Dev. Technol.* **2**, 225–235.

Sears, M. R., and Lotvall, J. (2005). Past, present and future: Beta2-adrenoceptor agonists in asthma management. *Respir. Med.* **99**, 152–170.

Simpson, P. B., Bacha, J. I., Palfreyman, E. L., Woollacott, A. J., McKernan, R. M., and Kerby, J. (2001). Retinoic acid evoked-differentiation of neuroblastoma cells predominates over growth factor stimulation: An automated image capture and quantitation approach to neuritogenesis. *Anal. Biochem.* **298**, 163–169.

Soll, D. R., Voss, E., Johnson, O., and Wessels, D. (2000). Three-dimensional reconstruction and motion analysis of living, crawling cells. *Scanning* **22**, 249–257.

Steff, A. M., Fortin, M., Arguin, C., and Hugo, P. (2001). Detection of a decrease in green fluorescent protein fluorescence for the monitoring of cell death: An assay amenable to high-throughput screening technologies. *Cytometry* **45**, 237–243.

Su, Y. F., Harden, T. K., and Perkins, J. P. (1980). Catecholamine-specific desensitization of adenylate cyclase: Evidence for a multistep process. *J. Biol. Chem.* **255**, 7410–7419.

Sullivan, E., Tucker, E. M., and Dale, I. L. (1999). Measurement of [Ca2+] using the fluorometric imaging plate reader (FLIPR). *Methods Mol. Biol.* **114**, 125–133.

Taylor, D. L., Woo, E. S., and Giuliano, K. A. (2001). Real-time molecular and cellular analysis: The new frontier of drug discovery. *Curr. Opin. Biotechnol.* **12**, 75–81.

Vanek, P. G., and Tunon, P. (2002). High-throughput single-cell tracking in real time. *Gen. Eng. News* **22**, 1–4.

Wilson, T. (1990). "Confocal Microscopy." Academic Press, New York.

Wolff, M., Haasen, D., Merk, S., Kroner, M., Maier, U., Bordel, S., Wiedenmann, J., Nienhaus, G. U., Valler, M. J., and Heilker, R. (2005). Automated high content screening for phosphoinositide 3 kinase inhibition using an AKT1 redistribution assay. *Comb. Chem. High Throughput Screen.* **9**, 339–350.

Xia, S., Kjaer, S., Zheng, K., Hu, P. S., Bai, L., Jia, J. Y., Rigler, R., Pramanik, A., Xu, T., Hokfelt, T., and Xu, Z. Q. (2004). Visualization of a functionally enhanced GFP-tagged galanin R2 receptor in PC12 cells: Constitutive and ligand-induced internalization. *Proc. Natl. Acad. Sci. USA* **101**, 15207–15212.

Yanagisawa, M., Kurihara, H., Kimura, S., Tomobe, Y., Kobayashi, M., Mitsui, Y., Yazaki, Y., Goto, K., and Masaki, T. (1988). A novel potent vasoconstrictor peptide produced by vascular endothelial cells. *Nature* **332**, 411–415.

Zemanova, L., Schenk, A., Valler, M. J., Nienhaus, G. U., and Heilker, R. (2003). Confocal optics microscopy for biochemical and cellular high-throughput screening. *Drug Discov. Today* **8**, 1085–1093.

[9]   Screening for Activators of the Wingless
Type/Frizzled Pathway by Automated
Fluorescent Microscopy

By KRISTEN M. BORCHERT, RACHELLE J. SELLS GALVIN,
LAURA V. HALE, O. JOSEPH TRASK,
DEBRA R. NICKISCHER, and KEITH A. HOUCK

Abstract

Development of means to screen primary human cells rather than established cell lines is important in improving the predictive value of cellular assays in drug discovery. We describe a method of using automated fluorescent microscopy to detect activators of the wingless type/Frizzled (Wnt/Fzd) pathway in primary human preosteoblasts. This technique relies on detection of endogenous β-catenin translocation to the nucleus as an indicator of pathway activation, requires only a limited number of primary cells, and is robust enough for automation and high-content, high-throughput screening. Identification of activators of the Wnt/Fzd pathway in human preosteoblasts may be useful in providing lead compounds for the treatment of osteoporosis.

Introduction

Developing effective means of screening primary human cells for phenotypic changes is an important goal in improving drug discovery research. Primary cells are more representative of *in vivo* physiology than established cell lines. During the process of cell line establishment, genetic and epigenetic changes occur that may influence the target being pursued in unknown ways (Baguley and Marshall, 2004; Horrocks *et al.*, 2003). Such changes may result in the generation of false-positive leads that are ineffective *in vivo*, as well as false negatives, eliminating potentially efficacious leads. Primary cells, particularly if cultured in the appropriate microenvironment, reduce the likelihood of such problems (Bissell *et al.*, 2002). However, as evidenced by the relative paucity of descriptions of use of primary human cells in screening campaigns, numerous obstacles exist to their successful implementation in screening assays. Some of these obstacles are beginning to be addressed. Although availability of primary cells previously required access to human tissue sources, a wide variety of cell types are now available from numerous sources. These cells can typically

METHODS IN ENZYMOLOGY, VOL. 414
Copyright 2006, Elsevier Inc. All rights reserved.

0076-6879/06 $35.00
DOI: 10.1016/S0076-6879(06)14009-4

be passaged only a limited number of times, however. Together with their cost, this puts practical limits on how they can be used in drug discovery. In addition, engineering of these cells is challenging in that introducing reporter genes with plasmid DNA is often difficult and immortal cell lines cannot be created without likely losing the primary cell phenotype. Use of lentiviral vectors is one possibility for dealing with this problem (Verhoeyen and Cosset, 2004). Alternatively, a means of addressing both the limited cell availability and difficulty introducing reporter gene systems is by using a high-content screening approach relying on measurement of phenotypic changes by automated fluorescence microscopy. This chapter illustrates this approach using a system to measure activation of the wingless type/Frizzled (Wnt/Fzd) pathway in primary human preosteoblast cells as a screen for compounds with potential bone anabolic activity.

Our interest in the Wnt/Fzd pathway in a preosteoblast cell arose from the clear involvement of the pathway in regulation of bone phenotype. The Wnt family of secreted lipoproteins consists of 19 family members that regulate cell–cell signaling controlling proper patterning during embryogenesis (Logan and Nusse, 2004). Maintenance of tissue phenotype in the adult is another function and dysregulation results in pathologies, including cancer (Sparks et al., 1998). Wnt binds to the seven-transmembrane domain receptor Fzd, a non-G-protein-coupled receptor family containing 10 members in human (Bhanot, 1996; Yang-Snyder, 1996). A coreceptor protein, low-density lipoprotein receptor-related protein (LRP) is required for signal transduction (Pinson 2000; Tamai et al., 2000). Wnt binding to Fzd/LRP results in the transmission of signal through disheveled (DSH) to a protein complex that includes axin, glycogen synthase kinase-3β (GSK3β), adenomatous polyposis coli (APC), and the transcription factor β-catenin (Nelson, 2004; Wnt home page) and results in inhibition of the phosphorylation of β-catenin by GSK3β. This causes inhibition of ubiquitin-mediated degradation of β-catenin and stabilization of cytoplasmic levels of the protein. The increasing cytoplasmic level leads to subsequent nuclear translocation of β-catenin. There it binds to the transcription factor T-cell-specific transcription factor/lymphoid enhancer-binding factor-1 (TCF/LEF), yielding a new transcription factor complex capable of regulating a wide variety of genes involved in growth and differentiation (Cadigan, 1997; Giles, 2003).

Involvement of the Wnt/Fzd pathway in bone formation was shown through a number of genetic studies in both human and rodent. In humans, mutations in the coreceptor LRP5 yielded either marked cortical bone thickening for activating mutations (Boyden, 2002; Little, 2002) or osteoporosis pseudoglioma syndrome for inactivating ones (Gong, 2001). Deletion of LRP5 in mice decreased osteoblast formation and function, resulting in osteopenia (Kato, 2002). Transgenic mice expressing the activating mutation had increased bone mass (Babij, 2003). A soluble, secreted decoy Wnt receptor, Fzd-related protein-1, caused increased trabecular bone formation when deleted in mice (Bodine, 2004). A pharmacological study of activation of the Wnt/Fzd pathway using inhibitors of GSK3β induced *de novo* bone formation and pronounced hypertrophy of osteoblasts (Kulkarni *et al.*, 2004).

Use of Primary Human Preosteoblasts

Although the Wnt/Fzd pathway could be manipulated pharmacologically through inhibition of GSK3β, this is a common node of a number of signal transduction pathways (Cohen and Goedert, 2004; Jope and Johnson, 2004). Thus, this may be too nonselective as a mechanism to safely induce increased bone formation as a treatment for osteoporosis. Because other known components of the Wnt/Fzd pathway are not typical druggable-type targets, we chose to pursue identification of activators of the pathway through a phenotypic screening approach without a specific molecular target selected a priori. With this approach, we thought it essential to choose a cell type representing the *in vivo* physiological state as accurately as possible so that relevant potential targets would be present. For these reasons, we developed an assay to measure activation of the Wnt/Fzd pathway through quantifying the nuclear translocation of β-catenin in primary human preosteoblast cells (Bi *et al.*, 2001). These cells were isolated as a subpopulation of stromal mesenchymal cells from the surface of femoral bone marrow and shown capable of being induced to differentiate into osteoblasts. Activation of the Wnt/Fzd pathway was quantified using the Cellomics (Pittsburgh, PA) Arrayscan IV automated fluorescent microscope system. The Cellomics ArrayScan system is described extensively in Williams *et al.* (2006).

Reagents and Materials

Human presosteoblast cells: Obtained by Dr. L.X. Bi, University of Texas Medical Branch, (Galveston, TX) from femoral long bone marrow in compliance with the National Institute of Health's requirements for human subjects (Bi, 2001)

Cell culture maintenance medium: MEM α (Gibco/Invitrogen, Carlsbad, CA) supplemented with 10% fetal calf serum (FCS) (Hyclone, Logan, UT) and 2 mM L-glutamine

Trypsin-versene (Biowhittaker, Walkersville, MD)

Dulbecco's phosphate-buffered saline (DPBS) (Biowhittaker)

Dimethyl sulfoxide (DMSO) (JT Baker, Phillipsburg, NJ)

96-well black viewplates (Perkin-Elmer, Wellesley, MA)

Indolemaleimide GSK3 inhibitor synthesized by Eli Lilly & Co. (Indianapolis, IN) (Engler *et al.*, 2004)

37% formaldehyde (Sigma-Aldrich, St. Louis, MO)

Hoescht 33352 (Sigma-Aldrich)

Triton X-100 and Tween-20 (Roche Biochemicals, Indianapolis, IN)

Mouse anti-β-catenin antibody (BD Biosciences, San Jose, CA)

Alexa-488 goat antimouse antibody (Molecular Probes/Invitrogen, Eugene, OR)

L cells and Wnt3A-expressing L cells [American Type Tissue Collection (ATCC) Manassas, VA]

G418 (Gibco/Invitrogen)

Centricon Plus 80 (polyethersulfone, molecular weight cutoff, 5000)

Preparation of L Cell Control- and Wnt3A-Conditioned Medium

1. Grow L cells to confluence in T-150 flasks.
2. Trypsinize and split 1:20 in 20 ml of medium without G418 in T-150 flasks.
3. Grow and condition medium for 4 days at 37° and 5% CO_2 in a humidified incubator.
4. Collect conditioned medium, sterile filter, and store at 4°. Refeed cells with 20 ml of medium.
5. Collect conditioned medium at 72 h, sterile filter, and combine with first collection.
6. Concentrate approximately five-fold with Centricon Plus 80 centrifugal filters, sterile filter, and store at 4°.

Primary Preosteoblast Cell Culture and Compound Screening

Due to the limited availability of these cells and the constraint to maximum passage number, cells should be expanded to provide a pool of cells sufficient for a screening campaign. These cells were found to be capable of being differentiated into osteoblasts until approximately passage 12 (data not shown). Therefore we expanded the cells and pooled, aliqouted, and cryopreserved a screening lot of cells at passage 5. Cells were found to respond best to Wnt3A-conditioned medium following at least 2 weeks of growth in culture after removal from cryopreservation. Therefore cells were maintained in T-150 flasks and passaged when nearly confluent for 2 to 4 weeks before being used in screening. Cells were not used past passage 11. Optimal cell density for measuring a response to Wnt3A-conditioned medium was determined to be approximately 2000 cells per well, which permitted a good compromise between low enough density to permit imaging of individual, segmented cells and high enough density to minimize the number of fields required to be imaged to achieve a statistically relevant population (Borchert *et al.*, 2005).

1. Thaw aliquots of frozen cells and plate in T-150 flasks in MEM α/10% FCS/2 mM glutamine. Incubate at 37° and 5% CO_2 in a humidified incubator.
2. Trypsinize and passage at 1:10 when nearly confluent. Cells are ready for screening after 2 weeks in culture. Do not use past passage 11.
3. Dilute to 20,000 cells/ml in MEM α/0.5% FCS/2 mM glutamine.
4. Plate 2000 cells/well with a Multidrop (Thermo Electron, Boston, MA) or other liquid dispenser in a black, 96-well Packard viewplate.
5. Incubate overnight at 37° and 5% CO_2.
6. Dilute compounds in 96-well polystyrene plates with MEM α to 10× desired screening concentration immediately before treating cells. The DMSO concentration should not exceed 0.25% in the assay plate. Include appropriate positive (10 μM GSK3 inhibitor or 2.5-fold concentrated Wnt3A-conditioned medium) and negative (2.5% DMSO or 2.5-fold concentrated control L cell-conditioned medium) controls at 10× final concentration.
7. Using Multimek 96 (Beckman Coulter, Fullerton, CA) or other suitable liquid handler, dispense 11 μl of compound or conditioned medium per well.
8. Return plates to incubator for 4 h before processing for immunofluorescence.

Immunofluorescent Staining

In order to identify the subcellular localization of β-catenin, we used the immunofluorescent staining approach and the Arrayscan automated fluorescent microscope for detection. To simplify the assay protocol, we initially tested a primary antibody against human β-catenin directly conjugated with FITC. Although we were able to observe the nuclear translocation of β-catenin upon stimulation with Wnt3A-conditioned medium, there was a very low signal-to-background due to a significant level of background staining. Attempts to reduce background were unsuccessful so an indirect immunofluorescent staining using an Alexa 488 goat antimouse antibody was implemented. While more cumbersome due to the required two staining and washing steps, this significantly enhanced the signal-to-background ratio. The following steps were automated using a Multimek for cell fixation and an MRD8 titrator (Titertek, Huntsville, AL) for immunofluorescent staining.

1. Following a 4-h incubation after compound treatment, aspirate medium from plates with Multimek and immediately add 100 μl 3.7% formaldehyde in DPBS, warmed to 37°, to fix cells; incubate 15 min at room temperature.
2. Using Multimek, wash cells twice with 100 μl DPBS.
3. Permeabilize cells with 100 μl 0.1% Triton X-100 in DPBS for 5 min.
4. Wash cells with 100 μl DPBS.
5. Add 50 μl of 1 μg/ml mouse anti-β-catenin in DPBS and incubate at room temperature for 1 h.
6. Wash cells with 100 μl blocking buffer (0.1% Tween-20 in DPBS).
7. Add 100 μl blocking buffer and incubate 15 min at room temperature.
8. Wash once with 100 μl DPBS.
9. Add 100 μl of DPBS containing 10 μg/ml Alexa 488 goat antimouse antibody and 2 μg/ml Hoescht 33342 and incubate 1 h at room temperature.
10. Wash once with blocking buffer and twice with DPBS.
11. Add 100 μl of DPBS and seal plates; read on ArrayScan immediately or store at 4° in the dark up to 4 weeks.

Imaging and Quantitation of Nuclear Translocation

Fixed and stained plates were imaged on the Cellomics ArrayScan IV using the nuclear translocation algorithm (Fig. 1). The algorithm identifies a cell by the nuclear Hoescht 33342 dye and draws a mask around the boundaries of the nucleus. A copy of this mask is then dilated as defined by the user to form the inner cytoplasm ring mask outside of the boundaries of the nucleus. Next a copy of this mask is expanded to form the outer cytoplasmic ring mask. Image analysis measurements, including size, shape, and fluorescent intensity of the target protein β-catenin as marked by Alexa488 in the nuclear and cytoplasm compartments, are made. The fluorescence difference between the nucleus and the cytoplasm is quantified and reported per individual cell or as average of the entire population in the field. Starting from the default protocol, the algorithm was modified to reduce artifacts induced by the fixing and staining technique and to accommodate the specific morphology of these cells. Specifically, the automated fixing and staining resulted in a significant number of cells that had folded over themselves, resulting in artificially high fluorescent readings. Also, some overlapping cells were seen with resulting high fluorescence.

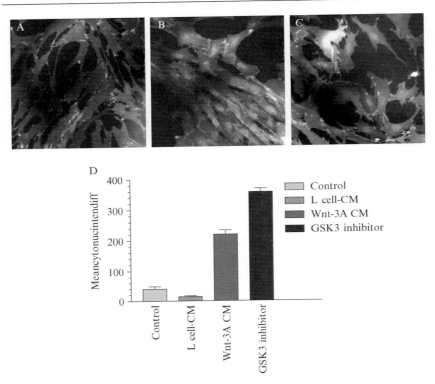

FIG. 1. Images of staining of β-catenin in human preosteoblasts treated with (A) control L cell-conditioned medium, (B) Wnt-3A-conditioned medium, or (C) GSK3 inhibitor. Quantitation of the degree of nuclear translocation of the β-catenin is shown in D.

We were unsuccessful in using morphologic parameters to exclude these cells from analysis but were able to do so setting upper limits on the total fluorescent intensity measurements in the target channel. Cells with folded nuclei were excluded by setting the upper limit of the parameter (nuclear fluorescence intensity/nuclear area) at that which would include approximately 99% of morphologically normal cells in a positive control population (treated with GSK3β inhibitor). Cells with overlapping cytoplasm were excluded by setting an upper limit on total fluorescence intensity in the target channel to a value that includes 99% of nonoverlapping cells in a negative control population. Finally, the cytoplasmic ring was set as a dilation of two pixels due to the relatively elongated shape of the cells and narrow band of cytoplasm that could be imaged outside the nucleus.

Translocation of β-catenin to the nucleus was reported as the difference in Alexa 488 fluorescence intensity between the nuclear area and the cytoplasmic area and was designated MeanNucCytoIntenDiff (Fig. 1D).

1. Set up ArrayScan IV with 10×/0.45NA Zeiss objective.
2. Set to default nuclear translocation algorithm and adjust parameters to measure 100 objects per well or up to a maximum of six fields per well.
3. Image positive and negative controls using the default nuclear translocation algorithm. Adjust algorithm as described earlier to exclude cells with undesirable morphologies.
4. Read plates.

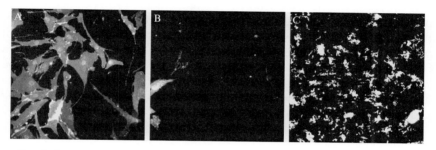

FIG. 2. Example of images of wells that were calculated as active by the cytoplasm to nuclear translocation algorithm illustrating (A) a true hit, (B) a false positive due to cytotoxicity, and (C) a false positive due to a fluorescent compound.

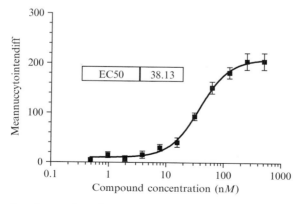

FIG. 3. Results of a confirmation assay done as a concentration–response curve for cytoplasm to nuclear translocation of β-catenin in primary human preosteoblasts.

Data Analysis

One of the advantages of high-content screening is the ability to look at a variety of result parameters that can be used to exclude false-positive results early in the screening process. For example, cytotoxicity often resulted in small, rounded-up cells with high fluorescence intensity or peculiar fluorescent accumulations in the target channel that the algorithm sometimes interpreted as an active well (Fig. 2). Generally, these wells had fewer than 100 objects counted per well and such wells are found easily by sorting data based on valid object count. Occasionally, compounds fluorescent in the target channel were noted that also resulted in false positives (Fig 2). These can be excluded by examining images from positive wells, in particular those with MeanNucCytoIntenDiff much greater than the positive control. Compounds with apparent real activity were confirmed in concentration–response format. As shown in Fig. 3, the resulting response curve gives reassurance as to normal pharmacological behavior of the compound and inspection of associated images can rule out artifacts. These hits were then advanced to additional testing, including counterscreening and mechanism of action studies.

Acknowledgments

We thank Dr. Liangxiang Bi for the human preosteoblast cells and Robin Gonyier, Ann McFarland, and Jonathan Gibbons for valuable technical assistance.

References

Babij, P., Zhao, W., Small, C., Kharode, Y., Yaworsky, P. J., Bouxsein, M. L., Reddy, P. S., Bodine, P. V., Robinson, J. A., Bhat, B., Marzolf, J., Moran, R. A., and Bex, F. (2003). High bone mass in mice expressing a mutant LRP5 gene. *J. Bone Miner. Res.* **18,** 960–974.

Baguley, B. C., and Marshall, E. S. (2004). *In vitro* modelling of human tumour behaviour in drug discovery programmes. *Eur. J. Cancer* **40,** 794–801.

Bhanot, P., Brink, M., Samos, C. H., Hsieh, J. C., Wang, Y., Macke, J. P., Andrew, D., Nathans, J., and Nusse, R. (1996). A new member of the frizzled family from Drosophila functions as a Wingless receptor. *Nature* **382,** 225–230.

Bi, L. X., Yngve, D., Buford, W. L., and Mainous, E. (2001). Isolation and characterization of preosteoblastic cells from human long bone marrow. *J. Bone Miner. Res.* **19**(Suppl. 1), SA050.

Bissell, M. J., Radisky, D. C., Rizki, A., Weaver, V. M., and Petersen, O. W. (2002). The organizing principle: Microenvironmental influences in the normal and malignant breast. *Differentiation* **70,** 537–546.

Bodine, P. V., Zhao, W., Kharode, Y. P., Bex, F. J., Lambert, A. J., Goad, M. B., Gaur, T., Stein, G. S., Lian, J. B., and Komm, B. S. (2004). The Wnt antagonist secreted frizzled-related protein-1 is a negative regulator of trabecular bone formation in adult mice. *Mol. Endocrinol.* **18,** 1222–1237.

Borchert, K. M., Galvin, R. J. S., Frolik, C. A., Hale, L. V., Gonyier, R., Trask, O. J., Nickischer, D. R., and Houck, K. A. (2005). Phenotypic screening for activators of the wnt/ frizzled pathway in primary human osteoblasts. *Assay Drug Dev. Technol.* **3**, 133–141.

Boyden, L. M., Mao, J., Belsky, J., Mitzner, L., Farhi, A., Mitnick, M. A., Wu, D., Insogna, K., and Lifton, R. P. (2002). High bone density due to a mutation in LDL-receptor-related protein 5. *N. Engl. J. Med.* **346**, 1513–1521.

Cadigan, K. M., and Nusse, R. (1997). Wnt signaling: A common theme in animal development. *Genes Dev.* **11**, 3286–3305.

Cohen, P., and Goedert, M. (2004). GSK3 inhibitors: Development and therapeutic potential. *Nature Rev. Drug Discov.* **3**, 479–487.

Engler, T. A., Henry, J. R., Malhotra, S., Cunningham, B., Furness, K., Brozinick, J., Burkholder, T. P., Clay, M. P., Clayton, J., Diefenbacher, C., Hawkins, E., Iversen, P. W., Li, Y., Lindstrom, T. D., Marquart, A. L., McLean, J., Mendel, D., Misener, E., Briere, D., O'Toole, J. C., Porter, W. J., Queener, S., Reel, J. K., Owens, R. A., Brier, R. A., Eessalu, T. E., Wagner, J. R., Campbell, R. M., and Vaughn, R. (2004). Substituted 3-imidazo[1,2-a]pyridin-3-yl-4-(1,2,3,4-tetrahydro-[1,4]diazepino-[6,7,1-hi]indol-7-yl)pyrrole-2,5-diones as highly selective and potent inhibitors of glycogen synthase kinase-3. *J. Med. Chem.* **47**, 3934–3937.

Giles, R. H., van Es, J. H., and Clevers, H. (2003). Caught up in a Wnt storm: Wnt signaling in cancer. *Biochim. Biophys. Acta* **1653**, 1–24.

Gong, Y., Slee, R. B., Fukai, N., Rawadi, G., Roman-Roman, S., Reginato, A. M., Wang, H., Cundy, T., Glorieux, F. H., Lev, D., Zacharin, M., Oexle, K., Marcelino, J., Suwairi, W., Heeger, S., Sabatakos, G., Apte, S., Adkins, W. N., Allgrove, J., Arslan-Kirchner, M., Batch, J. A., Beighton, P., Black, G. C., Boles, R. G., Boon, L. M., Borrone, C., Brunner, H. G., Carle, G. F., Dallapiccola, B., De Paepe, A., Floege, B., Halfhide, M. L., Hall, B., Hennekam, R. C., Hirose, T., Jans, A., Juppner, H., Kim, C. A., Keppler-Noreuil, K., Kohlschuetter, A., LaCombe, D., Lambert, M., Lemyre, E., Letteboer, T., Peltonen, L., Ramesar, R. S., Romanengo, M., Somer, H., Steichen-Gersdorf, E., Steinmann, B., Sullivan, B., Superti-Furga, A., Swoboda, W., van den Boogaard, M. J., Van Hul, W., Vikkula, M., Votruba, M., Zabel, B., Garcia, T., Baron, R., Olsen, B. R., and Warman, M. L. (2001). LDL receptor-related protein 5 (LRP5) affects bone accrual and eye development. *Cell* **107**, 513–523.

Horrocks, C., Halse, R., Suzuki, R., and Shepherd, P. R. (2003). Human cell systems for drug discovery. *Curr. Opin. Drug Discov. Devel.* **6**, 570–575.

Jope, R. S., and Johnson, G. V. (2004). The glamour and gloom of glycogen synthase kinase-3. *Trends Biochem. Sci.* **29**, 95–102.

Kato, M., Patel, M. S., Levasseur, R., Lobov, I., Chang, B. H., Glass, D. A., 2nd, Hartmann, C., Li, L., Hwang, T. H., Brayton, C. F., Lang, R. A., Karsenty, G., and Chan, L. (2002). Cbfa1-independent decrease in osteoblast proliferation, osteopenia, and persistent embryonic eye vascularization in mice deficient in Lrp5, a Wnt coreceptor. *J. Cell. Biol.* **157**, 303–314.

Kulkarni, N. H., Liu, M., Halladay, D. L., Frolik, C. A., Engler, T. A., Helvering, L. M., Wei, T., Kriauciunas, A., Martin, T. J., Sato, M., Bryant, H. U., Onyia, J. E., and Ma, Y. L. (2006). An orally bioavailable GSK3 alpha-beta dual inhibitor increases markers of cellular differentiation *in vitro* and bone mass *in vivo*. *J. Bone Miner. Res.* **21**, 910–920.

Little, R. D., Carulli, J. P., Del Mastro, R. G., Dupuis, J., Osborne, M., Folz, C., Manning, S. P., Swain, P. M., Zhao, S. C., Eustace, B., Lappe, M. M., Spitzer, L., Zweier, S., Braunschweiger, K., Benchekroun, Y., Hu, X., Adair, R., Chee, L., FitzGerald, M. G., Tulig, C., Caruso, A., Tzellas, N., Bawa, A., Franklin, B.,

McGuire, S., Nogues, X., Gong, G., Allen, K. M., Anisowicz, A., Morales, A. J., Lomedico, P. T., Recker, S. M., Van Eerdewegh, P., Recker, R. R., and Johnson, M. L. (2002). A mutation in the LDL receptor-related protein 5 gene results in the autosomal dominant high-bone-mass trait. *Am. J. Hum. Genet.* **70**, 11–19.

Logan, C. Y., and Nusse, R. (2004). The Wnt signaling pathway in development and disease. *Annu. Rev. Cell Dev. Biol.* **20**, 781–810.

Pinson, K. I., Brennan, J., Monkley, S., Avery, B. J., and Skarnes, W. C. (2000). An LDL-receptor-related protein mediates Wnt signaling in mice. *Nature* **407**, 535–538.

Sparks, A. B., Morin, P. J., Vogelstein, B., and Kinzler, K. W. (1998). Mutational analysis of the APC/beta-catenin/Tcf pathway in colorectal cancer. *Cancer Res.* **58**, 1130–1134.

Tamai, K., Semenov, M., Kato, Y., Spokony, R., Liu, C., Katsuyama, Y., Hess, F., Saint-Jeannet, J. P., and He, X. (2000). LDL-receptor-related proteins in Wnt signal transduction. *Nature* **407**, 530–535.

Verhoeyen, E., and Cosset, F. L. (2004). Surface-engineering of lentiviral vectors. *J. Gene Med.* **6**(Suppl. 1), S83–S94.

Williams, R. G., Kandasamy, R., Nickischer, D., Trask, O. J., Laethem, C., Johnston, P. A., and Johnston, P. A. (2006). Generation and characterization of a stable MK2-EGFP cell line and subsequent development of a high-content imaging assay on the Cellomics ArrayScan platform to screen for p38 mitogen-activated protein kinase inhibitors. *Methods Enzymol.* **414**(this volume).

[10] A Live Cell, Image-Based Approach to Understanding the Enzymology and Pharmacology of 2-Bromopalmitate and Palmitoylation

By Ivana Mikic, Sonia Planey, Jun Zhang, Carolina Ceballos, Terri Seron, Benedikt von Massenbach, Rachael Watson, Scott Callaway, Patrick M. McDonough, Jeffrey H. Price, Edward Hunter, and David Zacharias

Abstract

The addition of a lipid moiety to a protein increases its hydrophobicity and subsequently its attraction to lipophilic environments like membranes. Indeed most lipid-modified proteins are localized to membranes where they associate with multiprotein signaling complexes. Acylation and prenylation are the two common categories of lipidation. The enzymology and pharmacology of prenylation are well understood but relatively very little is known about palmitoylation, the most common form of acylation. One distinguishing characteristic of palmitoylation is that it is a dynamic modification. To understand more about how palmitoylation is regulated,

METHODS IN ENZYMOLOGY, VOL. 414
0076-6879/06 $35.00
DOI: 10.1016/S0076-6879(06)14010-0

we fused palmitoylation substrates to fluorescent proteins and reported their subcellular distribution and trafficking. We used automated high-throughput fluorescence microscopy and a specialized computer algorithm to image and measure the fraction of palmitoylation reporter on the plasma membrane versus the cytoplasm. Using this system we determined the residence half-life of palmitate on the dipalmitoyl substrate peptide from GAP43 as well as the EC_{50} for 2-bromopalmitate, a common inhibitor of palmitoylation.

Introduction

Many proteins are concentrated on the plasma membrane (PM) by virtue of their lipid modifications. Most data show that lipid modifications of proteins may well be the primary physical determinant for targeting to and retention of some proteins to membrane lipid microdomains such as synapses and caveolae (El-Husseini *et al.*, 2000; El-Husseini Ael *et al.*, 2002; Kanaani *et al.*, 2002; Loranger and Linder, 2002; Topinka and Bredt, 1998; Zacharias *et al.*, 2002). Fusion of green fluorescent protein (GFP) to small peptide substrates for lipid modification (e.g., the N-terminal 12–18 residues of GAP43) has been shown to be sufficient to localize the fusion proteins to the PM in the absence of any other targeting signal (Zacharias *et al.*, 2002). Similarly, mutagenic substitution of modifiable residues for ones that cannot be modified results in gross mislocalization and/or loss of function of the expressed proteins (Craven *et al.*, 1999; Hiol *et al.*, 2003b; Osterhout *et al.*, 2003; Wiedmer *et al.*, 2003). It has also been shown that different lipid moieties induce partitioning into different lipid environments or lipid microdomains of cells (Melkonian *et al.*, 1999; Moffett *et al.*, 2000; Zacharias *et al.*, 2002). Specific associations of proteins within such microdomains, whether mediated by attractive, protein–protein interactions, forced proximity, or both, are critical components in the architecture of cellular communication, and lipid modifications undoubtedly play an important role in the creation and modulation of such protein associations. Among the lipid modifications, palmitoylation is the most poorly understood and may hold the most promise as a druggable process in human health and disease. It was only a short time ago that the general assumption was that palmitoylation was likely exclusively an autocatalytic process. However, a family of enzymes has been shown to be able to mediate the process.

This chapter reviews, compares, and contrasts the various common forms of protein lipid modifications. It then describes how to go about constructing fluorescent protein (FP)-based reporters for lipid modification and live-cell based assays to study the processes that underlie protein lipid modification. It also describes advances in high-content or high-throughput microscopy that allow us to make very accurate measurements of even small changes in the subcellular distribution of lipid modified FPs. It also presents data demonstrating the utility of the methods described.

Types of Protein Lipid Modifications

Protein Prenylation

Prenyl and acyl groups are the most common forms of protein lipid modifications (Fig. 1). The two most common forms of prenylation are geranylgeranylation and farnesylation (Fig. 1A), whereas myristoylation and palmitoylation (Fig. 1B) are likely the most common forms of acylation. Most, if not all, of the biochemical steps regulating the prenylation of proteins have been deciphered (reviewed in Fu and Casey, 1999; Roskoski, 2003; Sinensky, 2000). In fact, the mechanistic pathway for farnesylation has been determined to the atomic level (Long *et al.*, 2002). This density of information

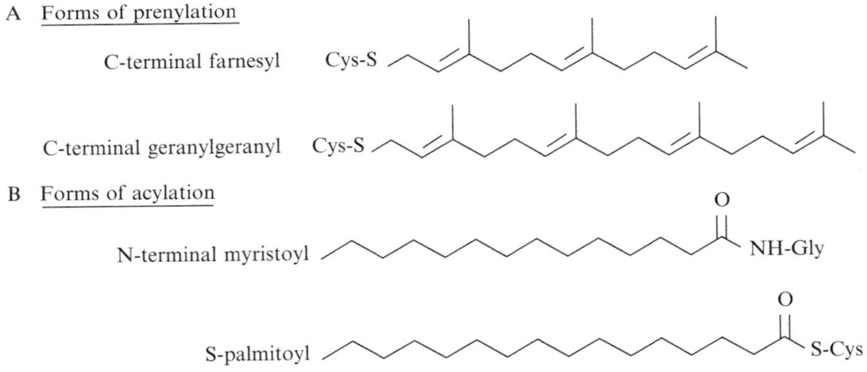

A   Forms of prenylation

C-terminal farnesyl   Cys-S

C-terminal geranylgeranyl   Cys-S

B   Forms of acylation

N-terminal myristoyl   NH-Gly

S-palmitoyl   S-Cys

FIG. 1. Lipids that modify proteins: (A) prenylation and (B) acylation. Each class of modification targets proteins to which they are attached to unique subcellular locales (Melkonian *et al.*, 1999; Moffett *et al.*, 2000; Zacharias *et al.*, 2002). This ability is likely due to their different chain lengths, degree of saturation and branching, and their physical position on the proteins. Both forms of prenylation occur via stable thioether bonds on the final cysteine of a "CAAX" box at the C terminus of a protein. Myristoylation occurs via a stable amide bond to the N-terminal glycine of a protein, whereas the addition of palmitate occurs most commonly via a labile, but regulated, thioester bond to the side chain of a free, reactive cysteine on the cytoplasmic side of the PM.

for prenylation is due in part to the fact that prenyltransferases have been fairly successful therapeutic targets for the treatment of several types of cancer (Dinsmore and Bell, 2003; Ghobrial and Adjei, 2002), underscoring the importance of protein lipidation for human health and disease.

Protein Acylation

Among the types of acylation, enzymatic processes regulating myristoylation have been characterized best and are reviewed elsewhere (Farazi et al., 2001; Rajala et al., 2000; Resh, 1999). Briefly, proteins that will become myristoylated begin with a consensus sequence Met-Gly-X-X-X-Ser/Thr. The start Met is cotranslationally, proteolytically removed and the myristate is added to the exposed N-terminal glycine via a stable amide bond. The formation of this bond is catalyzed by N-myristoyl transferase with a high degree of selectivity for 14-carbon myristate (Farazi et al., 2001; Rajala et al., 2000). N-terminal myristoylation often exists in combination with palmitoylation, which can take at least two forms: N-palmitoylation (a much less common form) and S-palmitoylation (the most common). N-Palmitoylation, first described for the protein sonic hedgehog (Pepinsky et al., 1998), is the addition of palmitic acid to the α-amide of Cys-24, which is exposed proteolytically to become the N-terminal residue of the functional protein. Addition of a palmitoyl group by an amide bond to the N-terminal glycine has been shown to occur on the heterotrimeric G protein, Gαs (Kleuss and Krause, 2003).

Myristoylation frequently occurs in close spatial and often temporal proximity to palmitoylation on a protein. Cysteine residues (C) in close proximity to a myristoylated N-terminal glycine (G) are often modified by palmitate (Dunphy and Linder, 1998; Resh, 1999), as is the case for the Src family of protein tyrosine kinases, e.g., Fyn (Liang et al., 2004) and Lyn (Honda et al., 2000), but can also occur up to 21 residues away in a de novo-designed, GFP fusion protein (Navarro-Lerida et al., 2002) and 26 residues from the N terminus in eNOS (Robinson and Michel, 1995).

Palmitoylation may also occur in conjunction with prenylation (either farnesylation or geranylgeranylation), as is the case with many of the Ras superfamilies of signaling proteins (Hancock, 2003; Hancock et al., 1989, 1990). A C-terminal "CaaX box" provides the substrate recognition motif for prenyltransferases. In this motif, C is cys, a is aliphatic, and X denotes the residue that specifies either farnesylation or geranylgeranylation. Prenylation at the C-terminal cysteine is followed by proteolytic cleavage of the next three amino acids and carboxymethylation of the newly exposed, prenylated cysteine. Cysteine residues immediately N-terminal to the prenylated cysteine residue are frequently palmitoylated. Examples of such dual modification can be found in H-Ras and N-Ras (illustrated here) (Apolloni *et al.*, 2000; Hancock, 2003; Hancock *et al.*, 1989).

Palmitoylation may also be the exclusive form of lipidation of a protein. In such cases, it is most common to find more than one palmitate attached to the protein [e.g., Gap43 (Liu *et al.*, 1993; Skene and Virag, 1989) and PSD-95 (Topinka and Bredt, 1998)]. Palmitoylated cysteines are often found in close proximity to basic residues, which provide additional membrane affinity by virtue of an electrostatic interaction with the negatively charged head groups of the membrane lipids.

While these are probably the most common lipid species to occupy cysteine, other acyl groups have been shown to take residence on thiol side chains of cysteines. The best studied examples (Liang *et al.*, 2001, 2002, 2004) use matrix assisted laser desorption, time of flight (MALDI-TOF) mass spectrometry to positively identify the adducts. These studies showed modification of the N terminus of Src family kinases where the N-terminal glycine is consistently modified by myristate, while the adjacent cysteine residue may be *S*-acylated by palmitate, palmitoleate, stearate, or oleate (Liang *et al.*, 2004).

The most common way to attach palmitate to a protein is via a labile thioester bond to the thiol side chain of a cysteine residue (Bernstein *et al.*, 2004) (Fig. 1), but palmitoylation can also occur in other ways, for example, on an N-terminal cysteine, as is the case with Hedgehog (Pepinsky *et al.*, 1998), a secreted signaling protein. An example of palmitate modifying the weaker -OH nucleophile of threonine occurs on the carboxyl terminus of a spider toxin (Branton *et al.*, 1993). The ε-amino group of lysine can also be modified by palmitate via an amide bond. This occurs in several secreted proteins, including a bacterial toxin (Stanley *et al.*, 1998). Little if anything is known about the enzymes that may modify lysines or threonines in eukaryotic cells. While these forms of palmitoylation are of great interest, they lie beyond the scope of this chapter and as such the main focus of this chapter is exclusively palmitoylated proteins where the linkage is via a thioester bond to the thiol side chain of cytosolic, accessible cysteine residues.

There is no obvious sequence of residues that constitute a motif for palmitoylation among cytoplasmic proteins that are palmitoylated other than a requirement for the free thiol side chain of cysteine. Proteins may be palmitoylated near either the N terminus or, less frequently, the C-terminus or any one or more sites in the middle. Some examples are listed in Table I.

Palmitate Turnover

The residence half-life of the palmitoyl group on proteins can be considerably shorter than the half-life of the proteins to which it is attached (Lane and Liu, 1997; Loisel *et al.*, 1996; Wolven *et al.*, 1997), suggesting that

TABLE I

PROTEIN PALMITOYLATION MOTIFS OF CYTOPLASMIC PROTEINS OCCUR THROUGHOUT THE
PROTEINS WITH NO DISCERNIBLE (AS OF YET) CONSENSUS SEQUENCE[a]

Protein	Relative location of the palmitoylation site within the protein	Palmitoylated sequence and the neighboring environment[b]	References
Gap43	N-terminal	N-MLCCMRRTKQV-	Liu et al. (1993); Skene and Virag (1989)
PSD95	N-terminal	N-MDCLCIVTTKKY-	Topinka and Bredt (1998)
RGS16	N-terminal	N-MCRTLAAFPTTCLE-	Hiol et al. (2003)
Yck2	C-terminal	-FFSKLGCC-COOH	Roth et al. (2002)
AtRac8 (Arabidopsis)	C-terminal	-GCLSNILCGKN-COOH	Lavy et al. (2002)
SNAP25	Internal	-[83]KFCGLCVCPCNKL[95]-	Lane and Liu (1997)
GAD65	Internal	-[27]RAWCQVAQKFTGGIGN-KLCALL[48]-	Kanaani et al. (2002)
GRK6	Internal	-[557]RQDCCGNCSDSEEE-LPTRL-[576]	Stoffel et al. (1994)

[a] After el-Husseini Ael and Bredt (2002) and Smotrys and Linder (2004).
[b] Palmitoylated residues are in bold. Basic residues that may contribute to PM binding are underlined; superscript numbers indicate the relative position in the protein.

the process should be mediated by enzymes in a manner analogous to kinases and phosphatases for phosphorylation. These examples of a fast (minutes) dynamic response also occur on a timescale that is shorter than what should be required to deplete or replenish a pool of available palmitate. Such observations strongly suggest and led many to hypothesize that palmitoylation is regulated enzymatically. Reversible palmitoylation of PSD-95 is regulated by the activity of neurons. Depalmitoylation promotes the removal of PSD-95 from synapses (El-Husseini Ael and Bredt, 2002). Several other interesting reports of the regulated turnover of palmitate support a role for palmitoylation in the regulation of the structure and function of synapses that is highly dynamic (El-Husseini Ael et al., 2002; Osten et al., 2000; Rocks et al., 2005).

Enzymes for Depalmitoylation

The relatively short residence half-life of palmitoylation of proteins is particularly suggestive of enzymatically mediated palmitate removal or depalmitoylation. Indeed, such proteins were among the first enzymes to be identified as having an influence on the palmitoylation state of proteins. PPT1 (Camp and Hofmann, 1993; Camp *et al.*, 1994) is a lysosomal hydrolase that participates in the degradation of palmitoylated proteins by deacylating cysteine thioesters; acyl protein thioesterase 1 (APT1), a cytoplasmic protein first biochemically characterized as an acyl thioesterase by Duncan and Gilman (1998), is a member of the serine hydrolase, α/β fold family of lysophospholipases that has several additional substrate and lipid specificities. Regulated removal of the palmitoyl group from proteins is critical in human health as defects in PPT1 result in a severe neurodegenerative disorder known as infantile neuronal ceroid lipofuscinosis (ICNL) (Vesa *et al.*, 1995). The role of PPT1 in ICNL was confirmed by targeted disruption of the gene in a mouse resulting in a model of the human disorder (Gupta *et al.*, 2001).

Palmitoylation increases the hydrophobicity and thus the affinity of the protein for membranes. The unique chemistry and dynamic nature of palmitoylation make it unique among the various forms of protein lipidation, providing an important mechanism for cells to integrate, sort, regulate, or even generate various cellular stimuli. In neurons, it is involved in protein targeting, endocytosis, vesicle fusion, neurotransmitter release (vesicle exocytosis), clustering, and modulation of ion channels in the plasma membrane, cell adhesion, and others. Palmitate is the most common lipid species to occupy cysteine residues but it is not the only one. Liang *et al.* (2001, 2002, 2004) used MALDI-TOF mass spectrometry to identify the lipid moieties on the N terminus of Src family kinases. For this kinase, the N-terminal glycine is consistently modified by myristate, but the adjacent cysteine residue is modified most frequently by palmitate but also by palmitoleate, stearate, or oleate (Liang *et al.*, 2004) with a frequency of residence on the protein that is apparently related to their abundance.

Role of Palmitoylation in Development

The involvement of palmitoylation in development is exemplified by GAP43 and paralemmin, which are localized to filopodia and growth cones (Liu *et al.*, 1991; Skene and Virag, 1989; Strittmatter, 1991, 1992; Zuber *et al.*, 1989a,b) where each has been shown to influence the formation of filopodia and the branching of dendrites and axons. Interestingly, it has also been shown that simply expressing the palmitoylation motifs of these proteins fused to GFP is sufficient to induce the changes seen when overexpressing the native forms of GAP43 and paralemmin (Gauthier-Campbell *et al.*, 2004). This suggests that the critical factor in the morphogenic transformation of the membrane is due to insertion of the tightly packing and rigid palmitates themselves. Similarly, examples of proteins involved in axon path finding and synapse formation are palmitoylated, including the cell adhesion molecule NCAM140 and the netrin-1 receptor DCC (deleted in colorectal cancer) (Herincs et al., 2005), both members of the NCAM family of proteins. Inhibiting palmitoylation disrupts raft localization of DCC, which is required for netrin-1-induced DCC-dependent ERK activation and netrin-1-dependent (Herincs et al., 2005) axon growth. Palmitoylated tetraspanins form unique lipid raft structures (called tetraspanin-enriched microdomains) that provide platforms for integrin-dependent cell adhesion (Yang et al., 2004).

During development, the ability of neurons to find their way to appropriate targets forming synaptic infrastructure is critical and, in many circumstances, dependent on the palmitoylation of key regulatory proteins. Subsequent to developmental morphogenic movements, formation, maintenance, and regulation of synaptic protein complexes are also dependent on palmitoylation. One of the first, most highly characterized and arguably most important palmitoylated synaptic protein is PSD-95, a PDZ domain-containing, scaffolding protein (Kim and Sheng, 2004). Retention of PSD-95 at the synapse is mediated by palmitoylation (Topinka and Bredt, 1998). A cycle of palmitoylation/depalmitoylation of PSD-95 facilitates the internalization of AMPA receptors (Ehrlich and Malinow, 2004) and presumably any other proteins bound to the scaffold at the time of internalization. In the case of the AMPA receptor, palmitoylation-related changes in the rate of internalization are closely tied to learning and memory in an in vitro preparation as well as during experience-driven synaptic strengthening by natural stimuli in vivo (Ehrlich and Malinow, 2004). The movement of ions through ionotropic channels sequestered at the synapse by palmitoylation-regulated scaffolding triggers exocytotic release of neurotransmitter. Not surprisingly, several of the critical components of the exocytotic apparatus are also palmitoylated, such as SNAP-25 (Hess et al., 1992) and members of the synaptotagmin family (Chapman et al., 1996; Heindel et al., 2003; Veit et al., 1996, 2000). The functional consequences of palmitoylation in these instances vary with the protein and range from sorting to presynaptic terminals (Han et al., 2005; Kang et al., 2004) to regulation of intracellular membrane flow and vesicle fusion (Fukasawa et al., 2004) in the case of Ykt6. Vesicles bound for the presynaptic membrane are loaded with the neurotransmitter. The neurotransmitter GABA is synthesized by gultamic acid decarboxylase 65 kDa (GAD65). Dual palmitoylation of GAD65 is required for moving the protein from the Golgi to presynaptic terminals. In this case, the palmitoylated cysteines are immediately adjacent to (1) a Golgi localization signal and (2) a membrane-anchoring motif (Kanaani et al., 2002). As is the case with palmitoylated Ras proteins, cellular targeting signals, in addition to the palmitate, operate in a hierarchical way to determine the subcellular residence of the protein. In the case of GAD65, the dynamic, reversible nature of palmitate provides an elegant switch to regulate protein function or subcellular distribution as opposed to an adjacent, "hardwired" short stretch of amino acid residues.

Cysteines: Primary Sites of Palmitoylation

The nature of cysteines, their thiol side chains, the types of reactions they undergo, and the nature of the bonds they form all figure very prominently into interest in palmitoylation. Cysteine residues are among the most nucleophilic agents in a cell (Bernstein *et al.*, 2004) and are the most common site of palmitoylation. However, palmitoylation can also occur in other ways, for example, on an amine of an N-terminal cysteine, as is the case with Hedgehog (Bijlsma *et al.*, 2004; Pepinsky *et al.*, 1998), a secreted signaling protein. An example of palmitate modifying the weaker -OH nucleophile of threonine occurs on the carboxyl terminus of a spider toxin (Branton *et al.*, 1993). The ε-amino group of lysine can also be modified by palmitate linked by an amide bond. This occurs in several secreted proteins, including a bacterial toxin (Basar *et al.*, 1999). Little is known about the enzymes that may catalyze the modification of lysines or threonines in eukaryotic cells. While these forms of palmitoylation are of great interest, they lie beyond the scope of this chapter and as such the main focus of this chapter is exclusively palmitoylated proteins where the linkage is via a thioester bond to the thiol side chain of a free, reactive cysteine residue.

Thiols of cysteine residues are also targets of other types of modifications (Di Simplicio *et al.*, 2003; Hershko and Ciechanover, 1998; Lipton *et al.*, 2002). The redox potential of different cellular compartments varies, but in general, in healthy cells, the cytoplasm is a reducing environment, meaning that solvent-exposed cysteine side chains are not typically disulfides (Sitia and Molteni, 2004). Thiols are the most potent nucleophilic amino acid residue side chains with rates of reaction that depend on the pK_a of the thiol, with the pK_a being dependent on the local environment of the side chain within the context of the whole protein. Unlike other residues with side chain nucleophilic moieties (-OH or -NH$_2$), thiol side chains undergo conjugations, redox, and exchange reactions (Di Simplicio *et al.*, 2003). Conjugation reactions include nitric oxide (NO) or *S*-nitrosylation, fatty acids such as palmitate, reactive oxygen species, and reactive nitrogen species forming bonds that are not susceptible to cleavage by hydroxylamine, a reagent used to quantitatively remove thioester-linked palmitate (Munro and Williams, 1999). Another source of thioester bonds in cells is the transient association between ubiquitin and the E1, E2, and certain E3 ubiquitination enzymes (Hershko and Ciechanover, 1998; Passmore and Barford, 2004). Based on the physical characteristics of ubiquitin-related thioesters it is highly unlikely that they are ever in a position to be palmitoylated.

Palmitoyl Acyl Transferases (PATs)

It is clear that palmitoylation is intricately tied to signaling pathways by a variety of mechanisms, giving the distinct appearance of being a regulated process. However, despite years of searching, only recently has the molecular identity of enzymes capable of palmitoylation been discovered. A family of at least 23 enzymes called palmitoyl acyl transferases (designated ZDHHC 1–23), capable of catalyzing the addition of palmitate to cysteines, was identified. Three of these have already been linked to human diseases. Erf2p and Akr1p were identified as the first PATs in yeast (Lobo et al., 2002; Roth et al., 2002). These were the first reports that provided the molecular identity of a protein associated with or having PAT activity. One very important piece of evidence also gleaned from these reports was the likelihood that the DHHC-CRD (Asp-His-His-Cys-cysteine-rich domain) motif was a signature for this type of activity. The identification of GODZ, a DHHC-CRD protein, as a PAT that specifically palmitoylates the $\gamma 2$ subunit of the $GABA_A$ receptor provided evidence that PATs exist in vertebrates (mammals) and that the DHHC-CRD motif is the hallmark of at least one type of enzyme that can palmitoylate proteins in vivo (Keller et al., 2004). One known, critical function of a DHHC-CRD protein is the transfer of palmitate to a cysteine residue substrate. When comparing the number of potentially palmitoylated proteins (hundreds to thousands?) to the current number of PATs (23), it is evident that at least some PATs will have more than a single substrate. The consequences of a nonfunctional PAT should, in theory, result in nonpalmitoylated substrates that are unable to function appropriately in their normal capacity. It is also probable that a particular substrate can be palmitoylated by more than one PAT, in which case, hypopalmitoylated substrates may be the result of a nonfunctional PAT. In any event, unless the palmitoylation-related "gain" (analogous to the gain in an electronic circuit) inherent in the signaling cascade that dictates one particular endpoint is known, the relationship between decremented palmitoylation and outcome cannot be determined a priori. Identification of PATs cleared a critical roadblock in our understanding of palmitoylation. Our ability to identify the substrates of PATs and to understand how PATs are regulated has continued to proceed, albeit slowly. Development of agonist and antagonist assays for the process will be important in the development of new pharmacological tools to study the process as well as pharmacological tools that are directly relevant to human health and disease.

Links between Palmitoylation and Disease

Immediately following their discovery, members of the DHHC-CRD family were linked to human diseases. Chromosome 22q11 is associated with a complex syndrome sometimes referred to as velocardiofacial syndrome or, more commonly, as 22q11 deletion syndrome (22q11DS). One of the several diseases associated with the syndrome is schizophrenia. Several genes and 72 identified single nucleotide polymorphisms (SNPs) (Bassett *et al.*, 2003; Karayiorgou *et al.*, 1995) have been identified in the affected region. Among the SNPs, rs175174 was tightly associated with susceptibility to schizophrenia (Mukai *et al.*, 2004). This SNP causes improper splicing of the gene ZDHHC8, retention of intron number 4, and introduction of an early stop codon resulting in early termination of translation. This sets up a situation where the truncated version behaves as a dominant-negative regulator of palmitoylation. Mukai *et al.* (2004) generated *Zddhc8* knock-out mice that expressed impaired sensorimotor gating, a fear of new spaces, and reduced motor activity (Jablensky, 2004). These mice also proved resistant to the NMDA receptor blocking drug MK801, which is known to activate locomotor activity, suggesting that ZDHHC8 affects behavior in part by interfering with glutamatergic transmission (Mukai *et al.*, 2004). These results correlate with other studies showing that palmitoylation modulates several neurotransmitter systems, including the glutamatergic system (Drisdel *et al.*, 2004; el-Husseini Ael and Bredt, 2002; Hayashi *et al.*, 2005).

HIP14 (huntingtin interacting protein 14) is a human DHHC-CRD-containing protein that interacts with htt, the protein that incurs hyperexpanded stretches of polyglutamine in Huntington's disease (Huang *et al.*, 2004; Singaraja *et al.*, 2002). Importantly, HIP14 expression is limited to the neurons (medium spiny neurons in the striatum) impacted most severely in the disease (Singaraja *et al.*, 2002), thereby providing a potential mechanism for the puzzling and important phenomenon of cell-type specificity so often observed in neurodegenerative diseases. ZDHHC15, one of the PATs that palmitoylates PSD-95 and GAP43 most efficiently (Fukata *et al.*, 2004), has been linked to nonsyndromic X-linked severe mental retardation (Mansouri *et al.*, 2005). There is a breakpoint in the regulatory portion of ZDHHC15 resulting in gene transcription failure. If any of the known mutations in PATs are acting as the sole genetic cause underlying the pathology of the disease they are associated with, it suggests that the degree of functional redundancy may be lower than it appears (Fukata *et al.*, 2004) or, alternatively, that specificity is the result of selective expression in a certain cell type or even a distinct subcellular domain.

These early examples of the profound consequences of misregulated palmitoylation illustrate the importance of the modification and suggest that other malfunctioning PATs will be associated with human disease.

Using Fluorescent Proteins to Study Protein Lipid Modifications

Over the last decade, fluorescent proteins have become an indispensable tool for understanding so many details in cell biology, signaling, and physiology, among others. Our understanding of the function and effects of lipid modifications of proteins is a prime example how enabling FPs have been. Substrates for the various forms of lipid modification are encoded by relatively short peptides and therefore easily fused to either the N or the C terminus of a FP. Such a fusion allows, at the very minimum, the ability to determine the effect of a lipid moiety on the subcellular distribution of an otherwise inert protein. In the case of *Aqueorea* FPs, there are no other known signal or targeting sequences existing in FPs that dictate a specific subcellular distribution; the monomeric proteins diffuse freely throughout the cytoplasm and nucleus unless directed to do otherwise by fusion with a targeting signal.

Monomeric Fluorescent Proteins: A Critical Feature for Studying Palmitoylation

Not all fluorescent proteins are created equally. Currently, *Aequorea* GFP is the only fluorescent protein known that is not an obligate homo-oligomer in its natural state; the GFP from *Renilla* is an obligate dimer (Ward, 1998; Ward and Cormier, 1979) and RFP (dsRED from coral) is an obligate tetramer (Baird *et al.*, 2000), as are all other characterized fluorescent proteins from corals. Quite some time before a rigorous determination of the homoaffinity was made, it was known that even *Aequorea* GFP dimerized to some degree in solution (Yang *et al.*, 1996). GFP crystallizes either as a dimer (Battistutta *et al.*, 2000; Palm *et al.*, 1997; Yang *et al.*, 1996), or as a monomer (Brejc *et al.*, 1997; Ormo *et al.*, 1996; Wachter *et al.*, 1997, 1998, 2000). In the dimeric crystal structure, the unit cell consists of two monomers associated in a slightly twisted, head-to-tail fashion via many hydrophilic contacts, as well as several hydrophobic contacts. The very different solvent conditions used by each group are sufficient to explain the differing results (Ward, 1998). Residing within a large hydrophobic patch, residues A206, L221, and F223 are sufficient to cause formation of the dimer at relatively low concentrations in solution and in living cells. However, changing these residues singly or in combination to positively charged residues such as A206K, L221K, and F223R effectively eliminated the interaction of the monomers (Zacharias *et al.*, 2001); the resulting monomeric GFPs are called mGFPs. To determine the strength of the interaction in solution, Zacharias *et al.* (2001) used sedimentation equilibrium analytical ultracentrifugation to characterize the affinity of GFPs with the wild-type interface as well as the mGFPs. In significant contrast to X-ray crystallography, the experimental conditions used in the analytical ultracentrifugation experiments approximated cellular physiological conditions and were able to provide definitive information (McRorie and Voelker, 1993) about the affinity of the complex. Other mutations are also thought to affect the state of GFP aggregation. F99S and M153T first described in relation to aggregation by Crameri *et al.* (1996) reduce obvious patches of surface hydrophobicity and could inhibit aggregation, but no dissociation constant has yet been determined for their Cycle3 mutant. Indeed, the triple mutant (F99S, M153T, V163A) (Crameri *et al.*, 1996) has a diffusion coefficient inside mammalian cells one order of magnitude higher than that of wild-type GFP, implying a corresponding reduction in binding to other macromolecules (Yokoe and Meyer, 1996). Because V163 points into the interior of the protein (Ormo *et al.*, 1996) and because F99S and M153 face outward, the latter two are most likely the culprits in wild-type GFP. However, the triple mutation did not alter the overall speed of fluorescence development at 37° compared to wild type (Crameri *et al.*, 1996). Mutations V163A and S175G together actually slow the final aerobic development of fluorescence (Siemering *et al.*, 1996), even though they improve greatly the yield of the properly matured protein.

The crystal structure of cyclized Cycle3 GFP was determined (Hofmann *et al.*, 2002). The authors found a crystallographic dimer interface different than those reported previously and concluded that various polar and nonpolar patches on the surface of GFP could serve to dimerize the proteins in ways previously undescribed. However, it is unlikely that the dimer interface described in this work exists under physiological conditions with great affinity in non-Cycle3 versions of GFP because it is clear from sedimentation equilibrium analytical ultracentrifugation experiments on GFP containing A206K (and other mutations) (Zacharias *et al.*, 2001) that there is virtually no remaining affinity when the more commonly observed dimer interface is altered. Further biophysical characterization of Cycle3 and mutants derived from it is clearly warranted.

The issue of GFP oligomerization is significant for several reasons. Most of the potential for trouble arises when GFP or its spectral mutants are fused to a host protein to track protein localization or expression or to measure interactions by fluorescence resonance energy transfer (FRET) (most often mCFP and mYFP). If GFP dimerizes in the context of being part of a fusion protein, it could also foreseeably dimerize the host protein to which it is fused. The situation could become even stickier if the host protein is itself an oligomer. When measuring the interactions of molecules by FRET, it is imperative that the fluorophores used to report the interactions do not themselves in any way influence, or worse yet, create the interactions being measured. Obviously, if the fluorophores have affinity for each other, then doubt is cast on the accuracy of any measurement made to the presumed interaction of the host proteins. The problems associated with GFP dimerization are most troublesome when measuring intermolecular FRET in a two-dimensional space such as a membrane (Dewey and Datta, 1989; Dewey and Hammes, 1980; Fung and Stryer, 1978; Snyder and Freire, 1982; Wolber and Hudson, 1979; Yguerabide, 1994; Zimet et al., 1995) where the interest in lipid modified proteins is greatest. In this situation, we found that wild-type GFPs were very likely to dimerize even when expressed at very low surface densities (Zacharias et al., 2001). Because the monomerizing mutations alter nothing but the homo-affinity of GFP, we recommend including them (preferably A206K) in all GFP expression constructs where dimerization is not desirable.

The issue of weak dimerization of *Aequorea* FPs is not as serious when the objective of the study is to look simply at proteins localized to the plasma membrane by lipid modification. In general, the subcellular distribution of dimerizing FPs on a gross level is indistinguishable from the monomeric versions. However, this is not the case with tetrameric coral proteins. The oligomeric state of many of them extends beyond the obligate tetramer to the formation of relatively high-affinity (<100 μM dissociation constants) dimers of tetramers, setting up a situation in which large subplasmlemmal lattices can form. We found that in most cell types, lipid-modified oligomeric versions of FPs give poorly defined PM localization, a very aggregated or patchy appearance with a great deal more fluorescence in perinuclear membranes.

High-Throughput Microscopy (HTM)/High-Content Screening (HCS) to Study Lipid-Modified Proteins

We have spent a lot of energy on the development and use of high-content screening (also called high-throughput microscopy because of the incredibly large amount of information that can be obtained from any number of single cells) for use as a tool for pharmacology, cell biology, and functional genomics. There is now a realized and growing need for the development of novel HCS algorithms to answer fundamental questions in cell biology and signaling. The use of such algorithms on our automated imaging system, the EIDAQ100 (later versions were called the IC100 after Beckman Coulter Instruments bought the EIDAQ100 and began development of the instrument for mass production), can provide highly quantitative analyses of cellular phenomena that could not be obtained easily using traditional microscopy techniques and analysis tools. The usefulness of HCS extends significantly beyond applications in drug discovery; the capability to objectively and quantitatively image millions of cells rapidly provides an unparalleled analytical advantage to cell biologists. Teasing apart signaling networks in live cells using image-based, functional assays is inherently more information dense than in traditional isolated, cell-free systems. Answering very basic biological questions using HCS also allows a seamless transition to an assay for an intermediate- to high-throughput screen of small molecule libraries in search of compounds with the potential to be therapeutic agents.

Morphometric Analysis of Palmitoylation with HCS

Existing HCS algorithms are capable of discriminating minute changes in many types of cell morphology or subcellular protein distribution (Conway *et al.*, 2001; Ghosh *et al.*, 2000; Minguez *et al.*, 2002; Morelock *et al.*, 2005). There is essentially no limitation on the types of morphology that can be used, alone or combinatorially, as criteria for a unique marker (Boland and Murphy, 2001; Price *et al.*, 1996, 2002; Roques and Murphy, 2002). Multiple criteria ranging from the location or concentration of a fluorophore in a cell (Boland and Murphy, 1999a,b, 2001; Boland *et al.*, 1998; Markey *et al.*, 1999; Murphy *et al.*, 2000; Price *et al.*, 2002; Roques and Murphy, 2002) to a physical change in the shape of a cell or a redistribution of cellular contents such as chromosomes, transcription factors (Ding *et al.*, 1998), microtubules (Minguez *et al.*, 2002), and membrane protrusions (e.g., neurites, ruffles) (Price *et al.*, 1996; Roques and Murphy, 2002) can be used individually or combined to enhance the sensitivity and accuracy (Boland and Murphy, 2001; Price *et al.*, 1996; Roques and Murphy, 2002). These algorithms provide flexible and unbiased reports of the existence and degree of interesting, visible changes in cells allowing for high-resolution determinations of pharmacological efficacy tied to such a change (Conway *et al.*, 2001; Ding *et al.*, 1998; Ghosh *et al.*, 2000). It is the exquisite sensitivity for subtle change and the objectivity of HCS that should make possible the use of cell lines for a wide variety of assays, including functional genomics, proteomics, and pharmacology.

HCS Machine Vision Algorithms to Quantify Reporter Density on the Plasma Membrane

We have focused on generating an application, a machine vision algorithm, for HCS that will accurately and objectively quantify the amount of a fluorescence localized to the plasma membrane of cells, as well as translocation between the cytoplasm and the PM in either direction. The approach we have used is to extract information on a per-cell basis rather than a field-based measurement, meaning that a much finer determination of the biological event can be extracted from original data. Algorithms that analyze images on a per-cell basis require significantly more effort to program and validate but our effort is being rewarded. The algorithm used to determine the plasma membrane localization (with respect to the amount of the fluorophore in question in any other subcellular domain) is named Thora (Vala Biosciences). Palmitoylation has served as one of the model biological systems during the development of Thora. Reporters of palmitoylation are FPs fused either to peptide substrates for palmitoylation or to whole proteins that are palmitoylated. In both cases, the palmitoylated versions of the proteins are localized to the PM. When palmitoylation is blocked or palmitates are removed pharmacologically, the reporter becomes displaced from the PM, diffusing to the cytoplasm, to a degree that depends on the efficiency of the block provided by the pharmacological tool. Upon displacement, the reporter diffuses to the cytoplasm where its accumulation can be measured and ratioed against the decrease in intensity on the plasma membrane. Increases in cytoplasmic abundance or decreases in the abundance of the reporters on the plasma membrane can also be compared to many other metrics that are determined automatically when the cell images are analyzed by algorithms such as Thora.

GAP43:YFP is Palmitoylated, Localized to the Plasma Membrane, and Is a Stereotypical Reporter of the Cellular Capacity for Palmitoylation

The N terminus of GAP43 (neuormodulin) is one of the most intensively studied models for acylation of a cytoplasmic protein (Arni et al., 1998; Fukata et al., 2004; Gauthier-Campbell et al., 2004; Hess et al., 1993; Liang et al., 2002; Skene and Virag, 1989). It is particularly useful for many reasons, including the fact that palmitoylation is the predominant form of lipidation and that the short N-terminal peptide substrate for palmitoylation can function efficiently in isolation of the remainder of the protein (Zacharias et al., 2002). The N terminus of GAP43 is doubly palmitoylated on two adjacent cysteine residues (Table I). When a 12-residue palmitoylation substrate peptide from the N terminus GAP43 is fused to the N terminus of YFP, this peptide, by virtue of its palmitoylation, traps YFP on the PM (Fig. 2). Conversely, if palmitoylation of GAP43:YFP is blocked or inhibited, the protein diffuses freely throughout the cell, including the nucleus (Fig. 2D), as is the case when GFP is not fused to any other peptide or protein (Fig. 2B). GAP43:YFP is free of such ancillary signals for PM targeting or any enzymatic activity present in the parent protein.

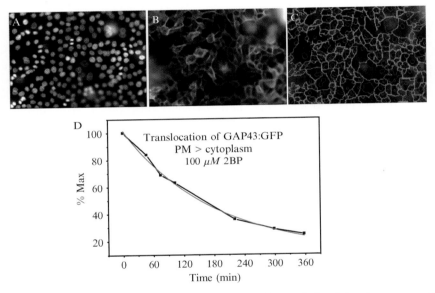

FIG. 2. Residence half-life of palmitate. Quantitative analysis of the time course of translocation of GAP43:YFP from the plasma membrane in response to 100 μM 2BP. MDCK cells stably expressing GAP43:YFP were exposed to 100 μM 2BP for 6 h. During this time, the same field of view of cells was imaged repeatedly in two channels: Hoechst/nucleus (A) and GFP/PM (B) on the EIDAQ100 (the three images, A, B, and C are at time 0). The PM mask (green lines) defined by Thora is shown (C); the nucleus is delimited by red lines; green lines mark the area identified as PM. By determining the average fluorescence intensity of the area defined by the PM mask (green in C), PM labeling was reduced by 80% from the maximum, or starting density during the 6-h period (D). Nonlinear least-squared fit of data to a single exponential decay curve (red line 4D) described data accurately and gave a decay constant, equivalent in this assay to the residence half-life of palmitate on this substrate of 179 min or ~3 h, the same as other published estimations for the residence of palmitate on cytosolic proteins (Lane and Liu, 1997; Wolven et al., 1997). Data were fit and the curve generated using MicroCal Origin.

Determination of the Residence Half-Life of Palmitate on GAP43:YFP Using HCS

Palmitoylation is a dynamic posttranslational modification. The thioester bond is labile, but regulated, and as a result, the turnover of palmitate residing on proteins is generally shorter than the rate of turnover of the protein that is modified by palmitate (El-Husseini Ael *et al.*, 2002; Lane and Liu, 1997; Loisel *et al.*, 1996; Wolven *et al.*, 1997). Using HCS we were able to determine the residence half-life of palmitate on GAP43:YFP expressed stably in MDCK cells (Fig. 2). While we expect that the residence half-life of palmitates will be context specific, the fact that we have identified a time that corresponds roughly to other measurements determined biochemically demonstrates the utility of the method. Another significant advantage provided by using this technology for such determinations is the fact that the measurement was generated in a living cell system over the course of 6 h rather than by incorporation of radiolabeled palmitate (tritiated palmitic acid; ^{3H}PA). Analysis of the incorporation of ^{3H}PA is relatively labor and cost intensive. The turn-around time for experiments using ^{3H}PA can be incredibly slow unless the system has incorporated a significant bias, such as stable overexpression of the substrate of interest. Even under ideal circumstances, the proteins must be separated by some means, typically SDS–polyacrylamide gel electrophoresis (one or two-dimensional SDS-PAGE) prepared for autofluorography and exposed to film for periods of up to 6 months. Typical times for a measurable autofluorographic signal from GAP43:YFP following incorporation of ^{3H}PA into the GAP43:YFP stable cell line following two-dimensional SDS-PAGE was 10 to 24 weeks. Using HCS to determine the state of palmitoylation provides mechanism for determining the kinetics for the palmitoylation of a substrate, as well as the activity of an associated PAT in the context of a living cell. Having the reporter of palmitoylation constitutively in the palmitoylated state lends itself best to an antagonist mode assay such as the ones illustrated in Figs. 2 to 5. In contrast, the disease states associated with PATs appear to be cases in which there is a state of hypopalmitoylation of substrates associated with a nonfunctional PAT. This suggests that in at least these cases, the assay mode needed for a drug discovery program would be agonist mode. There are several approaches to developing agonist mode assays for palmitoylation, each with advantages and drawbacks. Conceptually the simplest is to introduce a nonpalmitoylated, fluorescent substrate into the cytoplasm and track the translocation of the reporter from the cytoplasm to the plasma membrane as it becomes palmitoylated. Introduction of such a reporter on a large scale is somewhat problematic so methods to unmask a protected palmitoylation site fused to a stably expressed fluorescent reporter of palmitoylation are under development.

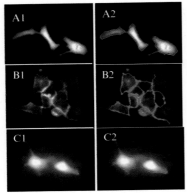

D

Image	Avg. intensity cytoplasm	Avg. intensity PM	Ratio
A	212.12	91.26	0.43
B	104.15	145. 65	1.40
C	181.83	89.57	0.49

FIG. 3. Determining the simple case of whether GAP43:YFP is on the PM or in the cytoplasm using the HCS algorithm. In this analysis, MDCK cells (same images as in Fig. 2) transiently transfected with (A) GFP alone, (B) GAP43:YFP, and (C) GAP43:YFP in the presence of 100 μM 2BP. Using masks that define the plasma membrane and the cytoplasm, the algorithm determined that the PM/cytoplasm ratio was significantly different between PM and cytoplasm localization (D). The localization of GFP alone (B) is described by a ratio similar to that of GAP43:YFP under conditions where its localization to the PM has been inhibited by incubation in 100 μM 2BP (C).

The analysis in Fig. 2 represents the case for a palmitoylated cytoplasmic protein. The residence half-life for palmitate on integral membrane (TM) proteins may be different (Loisel *et al.*, 1999). Because depalmitoylation of integral membrane proteins does not usually result in an immediate translocation of the protein from the PM into the interior of the cell, this particular assay is not suitable for quantifying depalmitoylation of TM-type proteins (e.g., G protein-coupled receptors [GPCRs]). Rather, depalmitoylation of TM proteins is likely to cause a lateral redistribution within the membrane into or out of lipid various lipid microdomains (Loisel *et al.*, 1999), in which case, monitoring changes in the associative properties of such protein in the two-dimensional space of the PM is better achieved using intermolecular FRET (Zacharias *et al.*, 2002).

Thora Can Measure Precisely the Subcellular Distribution of GAP43:YFP: IC$_{50}$ of 2BP

In a stepwise demonstration of the utility of Thora we show that it is able to make a simple binary decision of whether GAP43:YFP is on the PM or in the cytoplasm (Fig. 3) and subsequently that Thora can make fine

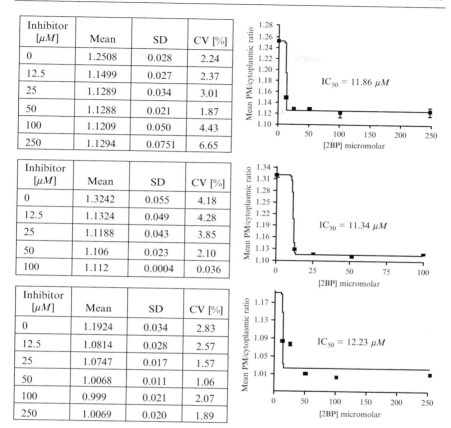

Inhibitor [μM]	Mean	SD	CV [%]
0	1.2508	0.028	2.24
12.5	1.1499	0.027	2.37
25	1.1289	0.034	3.01
50	1.1288	0.021	1.87
100	1.1209	0.050	4.43
250	1.1294	0.0751	6.65

Inhibitor [μM]	Mean	SD	CV [%]
0	1.3242	0.055	4.18
12.5	1.1324	0.049	4.28
25	1.1188	0.043	3.85
50	1.106	0.023	2.10
100	1.112	0.0004	0.036

Inhibitor [μM]	Mean	SD	CV [%]
0	1.1924	0.034	2.83
12.5	1.0814	0.028	2.57
25	1.0747	0.017	1.57
50	1.0068	0.011	1.06
100	0.999	0.021	2.07
250	1.0069	0.020	1.89

FIG. 4. Determination of the IC_{50} of 2BP using HTM. Inhibiting palmitoylation of GAP43: YFP with 2BP causes translocation of the reporter from the PM to the cytoplasm, allowing the determination, in live cells, of the IC_{50} of 2BP (the concentration at which 2BP reaches half its maximal effect) for this substrate. The IC_{50} determination was done simultaneously in three separate plates; 16 individual wells of cells were exposed to each concentration of 2BP. MDCK cells stably expressing GAP43:YFP were plated at 10,000 cells per well in three 96-well imaging plate (Costar) and allowed to adhere for at least 6 h. Pairs of columns were exposed to 2BP at concentrations ranging from 0 to 250 μM (16 wells/concentration). After a 2-h exposure, cells were imaged on the EIDAQ100 (Q3DM/Beckman Instruments) at 32°. The IC_{50} determined from each of these plates was approximately the same, even though the range of ratio values was shifted slightly for each plate (Fig. 3 tables and graphs). The PM distribution of GAP32:YFP was analyzed using Thora. PM fluorescence was ratioed against the fluorescence in the cytoplasm and plotted versus the concentration of 2BP. Z' for the assays was 0.2 to 0.22. The mean, SD, and CV (%) were cumulative determinations from individual cells in all wells. The identity of the PAT specific for GAP43 has not been confirmed but it is likely that HIP14 may contribute to its state of palmitoylation (Huang *et al.*, 2004). Even though the IC_{50} values are almost identical, information density around the IC_{50} value is low. For this reason, we increased the number of 2BP concentration, surveyed around the IC_{50} to make a finder determination.

Inhibitor [μM]	Mean	SD
0	1.34	0.052
0.1	1.32	0.057
0.5	1.27	0.061
1	1.32	0.044
2.5	1.29	0.078
5	1.29	0.089
10	1.24	0.11
15	1.09	0.02
20	1.003	0.191
50	0.916	0.106
100	0.872	0.170

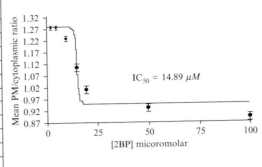

FIG. 5. Determination of the IC_{50} of 2BP using HTM. Because IC_{50} determinations in Fig. 3 had very few data points surrounding the IC_{50}, we increased the number of concentrations surveyed around the IC_{50} value to fill the gap. The IC_{50} value obtained was almost the same; however, the confidence in the value is increased dramatically when the value is supported by a greater concentration of data in this area.

determinations (Figs. 4 and 5) of the quantity of fluorescence on the PM, providing a level of sophistication great enough to make incremental determinations of the localization of the reporter. Two important questions are answered here. Data in Fig. 3 illustrate that the algorithm is sufficient to ensure success in an assay that is attempting to determine the simple question of whether the probe has moved from the membrane. Additionally, the second level of sophistication, the fractional localization of the reporter to the PM, provides an unprecedented degree of precision at high throughput that makes it possible to determine when there has been only a partial redistribution of a reporter (Figs. 4 and 5). This provides a tool that allowed us to determine the IC_{50} value (concentration at which the effectiveness of the pharmacological reagent is at half-maximum) of 2BP, the most commonly used, most specific inhibitor of 2BP known from the literature. 2BP has been used at many different concentrations to inhibit palmitoylation, apparently without regard to its efficacy. The expense and time involved in determining the IC_{50} of 2BP using radioactive methods and biochemistry would be significant. HCS offers a relatively inexpensive and reproducible way to determine this value.

Once we were sure that simple determinations could be made using artificially induced extreme conditions (i.e., all PM versus all cytoplasm), we focused on determining the IC_{50}, a more delicate measurement. Our first data set illustrates that the reproducibility among cells, wells, and plates is fairly good. We were able to generate almost identical IC_{50} values in all three plates. The first set of experiments illustrates that there is a ratio change in the PM/cytoplasm ratio when using this technique. The second important point clearly illustrated is the importance of having many data points surrounding the IC_{50}. If the value for a compound is not known, then a standard, broad range of concentrations is a good place to start. Once a general idea of where the IC_{50} will lie is obtained it is important to repeat such experiments using a larger number of concentrations hovering at either side of the IC_{50}, as in Fig. 5. Our narrowed, more focused attempt revealed that our original IC_{50} values were fairly accurate given the paucity of information in the IC_{50} value range of the concentrations used in Fig. 4. Data in Fig. 5 also provide a greater deal of confidence in the IC_{50} value.

Determination of the Compatibility of the Cellular Reporter System with Dimethyl Sulfoxide (DMSO)

Data from experiments using 2BP were done by diluting a concentrated stock solution of 100 mM 2BP in DMSO (a "1000×" stock solution) to the final concentrations indicated. Under these conditions, exposing the cells to 100 μM 2BP resulted in an exposure to 0.1% DMSO. It is important in experiments like this to maintain a constant final concentration of DMSO. The DMSO concentration should be below the level where there are detectable effects of DMSO. Rows of DMSO-only controls (0 and 1%) should be included on the same plate at least until the assay has been completely validated. Figure 6 illustrates a dose–response experiment for DMSO alone using concentrations ranging from 0.01 to 10%. From these data it is clear that even at DMSO concentrations much higher than normally used in a HTS screen of a small molecule library, there is little, if any, effect on this particular reporter system; it is encouraging that despite its common use in such experiments, we found no mention in the literature of DMSO having any effect on palmitoylation.

Cytotoxic Effects of Antagonists of Palmitoylation

As discussed previously, the single best characterized and the most commonly used pharmacological inhibitor of palmitoylation is 2BP. However, there are few reports about the effects of 2BP other than its ability to block palmitoylation. One report (Cnop et al., 2001) has shown a correlation between the concentration of palmitate and 2BP in cytotoxicity of preparations of rat β-cell preparations and that the mechanism of toxicity was due to necrosis and apoptosis. It has also been suggested that palmitate-related cytotoxicity may be due to the impairment of mitochondrial membrane (de Pablo et al., 1999). We frequently observed cellular

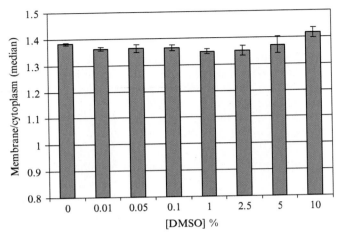

FIG. 6. DMSO does not have an effect on the displacement of GAP43:YFP from the plasma membrane. Cells were plated as in Figs. 4 and 5. DMSO was added to the cells at a range of concentrations that greatly exceeds what is introduced with the addition of 2BP. Even at 100 μM 2BP, the final percentage of DMSO is 0.1%.

toxicity that appeared to be correlated to the cell plating density when each well was exposed to the same concentration (100 μM) of 2BP (Fig. 7). It is likely that lower cell density results in a greater effective cellular membrane concentration of the 2BP, elevating it to the point where toxicity can be observed easily; the lipophilic 2BP will partition out of the aqueous culture medium and into cellular membranes—fewer cells will concentrate available 2BP to a greater degree than the same amount being distributed among more cells. This finding is critical in the development of assays to characterize PATs because cell plating density and homogeneity are critical, standard parameters that must be known to conduct all cell-based assays. Likewise, it is critical to understand any toxic effects of compounds that modulate palmitoylation. Specifically, in the toxicity "mechanism-based" meaning, is there something inherent in inhibiting all PATs or specific PATs that is responsible for cell death or is there some alternative reason, such as disruption of normal membrane integrity or the uncontrolled induction of one or more lipid-mediated signaling pathways in a cell, that is responsible for cell death? With efficacy and molecular mechanism of action, toxicity is a critical issue in the development of a high-throughput screen to find small molecule modulators of palmitoylation or

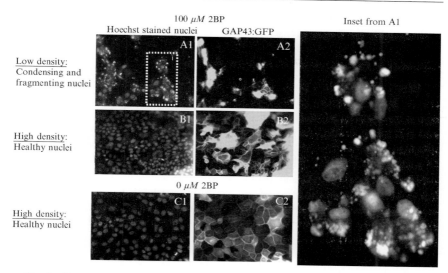

FIG. 7. Cytotoxic effects of 2BP. MDCK cells stably expressing GAP43:YFP were plated at (A) low (1000 cells/well) and (B and C) high/normal density (10,000 cells per well) and then exposed to 100 μM 2BP (A and B) or no 2BP (C). The low-density plating (A1) shows widespread nuclear condensation and fragmentation suggestive of cells undergoing apoptosis (Robertson *et al.*, 2000); see also the inset from A1 for a better view of the cells. This morphological change from normal nuclear integrity (see B1 and C1) occurred in as few as 6 h following addition of 2BP to the low-density culture. Cells at high-density culture were resistant to such changes for at least 18 h following exposure to 2BP (B and C). Interestingly, 2BP successfully caused translocation of GAP43:YFP from the PM in cells plated at high and low density (A2 and B2). Cells not treated with 2BP (C) had normal nuclei and GAP43:YFP remained localized to the PM (C2). Cells were allowed to adhere for at least 6 h before any treatment. 2BP (100 mM, in DMSO) was added to culture medium with 2% serum and vortexed just prior to addition to the cells in culture.

any other high-throughput screen. Such information will provide a critical foundation for monitoring the toxic effects of compounds that modulate palmitoylation, the most critical factor, closely following efficacy, of a high-throughput screen for small molecule modulators of palmitoylation. Observations like this are easy to make when developing an HCS, live-cell or image-based assay. Most cell biologists will be familiar with many important changes in cell morphology that differ from normal and as such should be very vigilant for such occurrences. Using a vision-based system such as the EIDAQ100 to make these determinations provides an unprecedented opportunity to make better decisions about what constitutes a healthy assay.

Considerations for Designing HCS, Cell-Based Assays with an Emphasis on Palmitoylation

Choosing a Palmitoylated Protein or Peptide

Many of the points made here can be applied generally to other types of assays, but we present them with lipid-modified proteins in mind. The first step is developing a biosensor system that will report on the activity of known and unknown modulators of the ability of a cell to palmitoylate a specific substrate. Our preferred substrate for palmitoylation is the N-terminal 12 or 18 residues of GAP43. This substrate represents only one of many possible substrates for palmitoylation; other substrates (see Table I) can be substituted but GAP43 is one of the most widely used and most extensively characterized of all substrates for palmitoylation. We utilize many other substrates of palmitoylation to answer specific questions about individual proteins, but for general questions, we use GAP43 most commonly.

Choosing a Fluorescent Protein

If fluorescent protein-based sensors are to be used, the choices are now significantly greater than they were a few years ago. This choice is made much in the same manner as choosing a small molecule fluorophore to label a particular cellular component. The excitation and emission spectra should be unique from other existing or future coexisting fluorophores. The fluorescent proteins should be bright (highest possible extinction coefficient and quantum yield) and should be truly monomeric. In general, the more red shifted the FP is, the better, as there is less cellular autofluorescence at redder wavelengths. Among these factors, perhaps only the oligomeric state of the FP is a unique consideration. Many frequently encountered FPs from corals are obligate tetramers and will cause a distinct patchy distribution that does not necessarily reflect the endogenous situation for the experimental peptide (e.g., GAP43).

Creation of Stable Reporter Cell Lines

There are also many factors that must be taken into account when choosing a cell line for developing these assays, ranging from the organism of origin (e.g., human, mouse, or dog) to the morphology of the cells. The fluorescent signal from a stable line expressing any FP-based reporter must be robust. Generating a robust cell line is not always a simple matter. In a practical sense the challenge is to express enough of the reporter so that image contrast is sufficient while retaining low image exposure times and, at the same time, not expressing so much of the reporter that normal cellular processes are overwhelmed. While the EIDAQ100 can and often does use brightly stained nuclei to establish an initial focal plane, the second image, often the FP-based reporter, must be bright enough to generate sufficient contrast so that the EIDAQ100 autofocus system is able to refocus if necessary. Large signal-to-noise ratios are also helpful in the subsequent analysis of any image data using Thora where image contrast is a key factor in the identification and quantification of the PM. In general, stable lines expressing the minimum amount of reporter necessary for efficient image acquisition and analysis are the best. This must be determined empirically for each line.

Drift is a problem encountered occasionally in such reporter cell lines. In our experience with reporter lines expressing lipid-modified FPs, the most common type of drift is a slow diminution of the fluorescent signal over many passages level but never so far in the subcellular distribution. For this reason it is important to expand any useful reporter cell lines and freeze down many tubes of the cells at early passages.

Cell Plating

Determining and maintaining optimal plating conditions are the most critical parameters for reducing the noise in a cell-based assay system. Automated cell dispensation systems such as the Multidrop systems from Thermo Electron, among others, are not prohibitively expensive and will soon pay for themselves by reducing error and frustration. The optimal cell densities must be determined on a per-assay basis. For all lipid-modified FPs studied to date, the subcellular distribution is defined most clearly when the cells are at or near confluence or in islands of tens to hundreds of cells. Within such areas, the best definition is at sites of cell–cell contact. For palmitoylation, islands of less than 10 cells provide sufficiently distinct morphology and high enough contrast to give robust measurements.

Choice of Cell Type

The choice of cell type for studying lipid-modified proteins is absolutely critical, as the variability of the expression pattern of such proteins can vary substantially among the cell type. For palmitoylated proteins, we find that MDCK and U2OS work very well, whereas HeLa, COS, and HEK are poor, each for different reasons. Both MDCK and U2OS are flat and, when confluent relatively columnar with high, fairly vertical walls between cells creating strong, bright, distinct boundaries where the lipid-modified FPs accumulate. These cells have small nuclei relative to their cytoplasm compartment, leaving sufficient room for clear observation of the reporter signal when it moves between the plasma membrane and the cytoplasm. Cell types other than MDCK and U2OS may work well for similar applications but such determinations must be made empirically. Generally the

cells should be flat, column shaped, have relatively small nuclei, be sufficiently adherent to a wide variety of substrates, and remain monolayers
rather than growing into amorphous piles as the cells reach confluence.

Control Compounds

One of the primary impediments to developing assays for protein palmitoylation has been the near absence of specific pharmacological tools to
characterize the system. It has been known for quite some time that only
very few compounds inhibit palmitoylation of proteins in tissue and cell
lines; these include cerulenin (DeJesus and Bizzozero, 2002; Hiol *et al.*,
2003a; Lawrence *et al.*, 1999), polyunsaturated fatty acids, and 2BP (Webb
et al., 2000). Of these compounds, 2BP is used most frequently and successfully and is now perceived as the pharmacological benchmark agent to
understanding palmitoylation (Webb *et al.*, 2000). The molecular mechanism of action of 2BP is not known but it is believed it may act by forming an
irreversible bond with PATs. The molecular mechanism of action of cerulenin is believed to be by the inhibition of fatty acid synthase (Menendez
et al., 2004), specifically 3-ketoacyl ACP synthase, a fairly early point in the
palmitoylation pathway well upstream of the physical addition of palmitate
to a protein. Cerulenin inhibition of fatty acid synthesis is also known to
induce apoptosis in a variety of cancer cells in as little as 2 h (Thupari *et al.*,
2001; Zhou *et al.*, 2003).

Evaluation of Assay Quality by Z'

Z' is a dimensionless statistical characteristic and is used for comparison and evaluation of assay quality (Zhang *et al.*, 1999). The Z' factor is

$$Z' = 1 - \frac{(3\sigma_{+control} + 3\sigma_{-control})}{\mu_{+control} - \mu_{-control}}$$

and can be calculated using only control data. A Z' factor equal to 1.0 is a
perfect assay. $Z' > 0.5$ indicates a good assay. This coefficient takes into
account the dynamic range of the assay signal, as well as the variation
associated with the reference control measurements. Because the Z' factor
is dimensionless, it is suitable for comparison between assays.

Troubleshooting

Data presented in Fig. 3 illustrate that the variability of the IC_{50}
determination for 2BP plate-to-plate basis is remarkably small. Likewise,
the standard deviations (σ in the formula just given) and %CVs are very

good. However, the Z' factors were poor. The main problem with these data is that the change in fluorescence intensity under the mask identified by the algorithm as being the PM does not drop significantly during the period of time the cells were exposed to 2BP. The most likely explanation for this is that the time of exposure to 2BP was not long enough to achieve the maximum level of depalmitoylation of the GAP43:YFP reporter. What this means in a practical sense is that a significant proportion of the GAP43:YFP reporter did not translocate from PM to the cytoplasm, resulting in a small change in the PM/Cyt ratio. This hypothesis is supported by the experiment shown in Fig. 8 where after 6 h approximately an 80% reduction in overall PM fluorescence intensity was obtained. Our rationale for shortening the exposure time to determine the IC_{50} for 2BP was that cells were not able to withstand a 6-h exposure to 250 μM 2BP; after 2 h cells in wells at this concentration were dying and floating away. Together these results suggest that we have started at a point where the instrument and algorithm noise are at a sufficiently low level and the variability of the biology is remarkably low. The primary problem being that the dynamic range of the assay is too small. In our experience, this component of the Z' equation is the one most frequently being inadequately robust when developing cell-based assays with FP reporters of some activity.

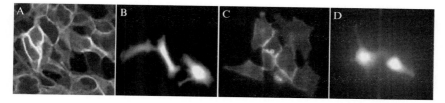

FIG. 8. GAP43:YFP is a reporter of palmitoylation. (A) Stably expressed GAP43:YFP is localized to the PM of MDCK cells, illustrating the remarkable homogeneity in the expression pattern that has been achieved in this stable line. (B) GFP not fused to any subcellular targeting motif is expressed throughout the interior of the cell, including the nucleus, illustrating that GFP alone has no known inherent targeting signals. (C) Transiently transfected GAP43:YFP is also expressed on the plasma membrane of cells except when palmitoylation is inhibited (D) by preincubation of transiently transfected cells in 2-bromopalmitate (2BP) (100 μM) (Webb *et al.*, 2000), illustrating that palmitoylation of the adjacent cysteines on the 12-residue peptide from the N terminus of GAP43 (NH2-MLCCMRRTKQV-YFP) fused to the N terminus of YFP is sufficient to retain the protein at the PM (Arni *et al.*, 1998; Liu *et al.*, 1993); there is nothing else inherent in the GAP43 peptide or YFP that will localize this protein to the PM. Cells in B, C, and D are representative of many having the same morphology.

System Error Identification

In addition to plate-to-plate variability of measurement such as the IC_{50} for 2BP, we will determine the day-to-day stability of repeated measurements of the same plates. Such measurements will illustrate the consistency and stability of the hardware and software components of our assay system. Based on previous experience with this instrument we believe that the variability or noise generated from this component will be minimal (Moran, 2005). Day-to-day and plate-to-plate determinations of the consistency and stability of the biological component will also be performed. Plating cells and dispensing reagents in a consistent manner will be enabled using dedicated multichannel pipettors and a Multidrop 384 automated cell-dispensing system.

Data Tracking

All variability from all sources should be presented in a tabulated matrix in a way that will clearly illustrate the source and extent of all variabilities. Such tabulation provides a convenient tracking system allowing one to manage assay development. Information tracking can be established easily using MS Excel spreadsheets. Data and analyses in the spreadsheets can be hyperlinked easily to original imaging data, allowing one to drill down to individual original data points, including images of single cells and eventually link out to individual compounds. This connectivity will enhance our ability to identify strengths and weaknesses of each step in the process efficiently.

Image Organization and Analysis

This aspect of HTM or HCS is certainly one of the most problematic. There is a great deal of effort being made to develop programs that allow researchers to catalog and store image and associated data. The enormous volume of data that can be generated in a very short time was almost unmanageable just a few years ago. The expense of collecting and saving data from a few experiments was almost prohibitive. However, the price of storage media has dropped to the point where it is no longer the issue it once was. The bigger problem is managing image data so that subsequent analyses within and between data sets are efficient and that archived data are not lost or overlooked due to poor documentation and organization.

Even modest use of HTM or HCS will generate, in a short time, very large amounts of data. A plan to manage these data should be integrated into the experimental design at the earliest stages. HCS systems often include software that aids in the organization of data, but these are most frequently platform specific. Efforts are also being made at the commercial level to fill the need for efficient data management systems. The most significant open source effort to our knowledge is the open microscopy environment (Goldberg *et al.*, 2005). This increasingly comprehensive system offers many advantages, including cross-platform compatibility. The project already has significant momentum, becoming an important community resource that is worth supporting.

Conclusions

Palmitoylation is a key determinant of the subcellular localization of proteins to regions of membrane specialization, including lipid rafts and synapses. Many synaptic proteins are palmitoylated and are the anchor for the formation of multimolecular signaling complexes, as is the case with PSD-95. Palmitoylation is a dynamic modification, making it unique among the various lipid modifications and a posttranslational modification that likely works like other reversible modifications such as phosphorylation. The process has been shown to be mediated by a family of enzymes, PATs, that contain ZDHHC-CRD motifs. Links between PATs and human diseases are still tentative but represent promising avenues for understanding the molecular basis for diseases such as Huntington's disease, mental retardation, and schizophrenia. Understanding the changes that occur in the substrates for PATs in disease states will be fertile ground for the development of HTS assays to discover pharmacological agents to correct the problem. The assay that may be the most important is an agonist mode assay, the more difficult of the forms to design.

References

Apolloni, A., Prior, I. A., Lindsay, M., Parton, R. G., and Hancock, J. F. (2000). H-ras but not K-ras traffics to the plasma membrane through the exocytic pathway. *Mol. Cell. Biol.* **20,** 2475–2487.

Arni, S., Keilbaugh, S. A., Ostermeyer, A. G., and Brown, D. A. (1998). Association of GAP-43 with detergent-resistant membranes requires two palmitoylated cysteine residues. *J. Biol. Chem.* **273,** 28478–28485.

Baird, G. S., Zacharias, D. A., and Tsien, R. Y. (2000). Biochemistry, mutagenesis, and oligomerization of DsRed, a red fluorescent protein from coral. *Proc. Natl. Acad. Sci. USA* **97,** 11984–11989.

Basar, T., Havlicek, V., Bezouskova, S., Halada, P., Hackett, M., and Sebo, P. (1999). The conserved lysine 860 in the additional fatty-acylation site of Bordetella pertussis adenylate cyclase is crucial for toxin function independently of its acylation status. *J. Biol. Chem.* **274**, 10777–10783.

Bassett, A. S., Chow, E. W., AbdelMalik, P., Gheorghiu, M., Husted, J., and Weksberg, R. (2003). The schizophrenia phenotype in 22q11 deletion syndrome. *Am. J. Psychiatr.* **160**, 1580–1586.

Battistutta, R., Negro, A., and Zanotti, G. (2000). Crystal structure and refolding properties of the mutant F99S/M153T/V163A of the green fluorescent protein. *Proteins* **41**, 429–437.

Bernstein, L. S., Linder, M. E., and Hepler, J. R. (2004). Analysis of RGS protein palmitoylation. *Methods Mol. Biol.* **237**, 195–204.

Bijlsma, M. F., Spek, C. A., and Peppelenbosch, M. P. (2004). Hedgehog: An unusual signal transducer. *Bioessays* **26**, 387–394.

Boland, M. V., Markey, M. K., and Murphy, R. F. (1998). Automated recognition of patterns characteristic of subcellular structures in fluorescence microscopy images. *Cytometry* **33**, 366–375.

Boland, M. V., and Murphy, R. F. (1999a). After sequencing: Quantitative analysis of protein localization. *IEEE Eng. Med. Biol. Mag.* **18**, 115–119.

Boland, M. V., and Murphy, R. F. (1999b). Automated analysis of patterns in fluorescence-microscope images. *Trends Cell Biol.* **9**, 201–202.

Boland, M. V., and Murphy, R. F. (2001). A neural network classifier capable of recognizing the patterns of all major subcellular structures in fluorescence microscope images of HeLa cells. *Bioinformatics* **17**, 1213–1223.

Branton, W. D., Rudnick, M. S., Zhou, Y., Eccleston, E. D., Fields, G. B., and Bowers, L. D. (1993). Fatty acylated toxin structure. *Nature* **365**, 496–497.

Brejc, K., Sixma, T. K., Kitts, P. A., Kain, S. R., Tsien, R. Y., Ormo, M., and Remington, S. J. (1997). Structural basis for dual excitation and photoisomerization of the Aequorea victoria green fluorescent protein. *Proc. Natl. Acad. Sci. USA* **94**, 2306–2311.

Camp, L. A., and Hofmann, S. L. (1993). Purification and properties of a palmitoyl-protein thioesterase that cleaves palmitate from H-Ras. *J. Biol. Chem.* **268**, 22566–22574.

Camp, L. A., Verkruyse, L. A., Afendis, S. J., Slaughter, C. A., and Hofmann, S. L. (1994). Molecular cloning and expression of palmitoyl-protein thioesterase. *J. Biol. Chem.* **269**, 23212–23219.

Chapman, E. R., Blasi, J., An, S., Brose, N., Johnston, P. A., Sudhof, T. C., and Jahn, R. (1996). Fatty acylation of synaptotagmin in PC12 cells and synaptosomes. *Biochem. Biophys. Res. Commun.* **225**, 326–332.

Cnop, M., Hannaert, J. C., Hoorens, A., Eizirik, D. L., and Pipeleers, D. G. (2001). Inverse relationship between cytotoxicity of free fatty acids in pancreatic islet cells and cellular triglyceride accumulation. *Diabetes* **50**, 1771–1777.

Conway, B. R., Minor, L. K., Xu, J. Z., D'Andrea, M. R., Ghosh, R. N., and Demarest, K. T. (2001). Quantitative analysis of agonist-dependent parathyroid hormone receptor trafficking in whole cells using a functional green fluorescent protein conjugate. *J. Cell Physiol.* **189**, 341–355.

Crameri, A., Whitehorn, E. A., Tate, E., and Stemmer, W. P. (1996). Improved green fluorescent protein by molecular evolution using DNA shuffling. *Nat. Biotechnol.* **14**, 315–319.

Craven, S. E., El-Husseini, A. E., and Bredt, D. S. (1999). Synaptic targeting of the postsynaptic density protein PSD-95 mediated by lipid and protein motifs. *Neuron* **22**, 497–509.

de Pablo, M. A., Susin, S. A., Jacotot, E., Larochette, N., Costantini, P., Ravagnan, L., Zamzami, N., and Kroemer, G. (1999). Palmitate induces apoptosis via a direct effect on mitochondria. *Apoptosis* **4**, 81–87.

DeJesus, G., and Bizzozero, O. A. (2002). Effect of 2-fluoropalmitate, cerulenin and tunicamycin on the palmitoylation and intracellular translocation of myelin proteolipid protein. *Neurochem. Res.* **27,** 1669–1675.

Dewey, T. G., and Datta, M. M. (1989). Determination of the fractal dimension of membrane protein aggregates using fluorescence energy transfer. *Biophys. J.* **56,** 415–420.

Dewey, T. G., and Hammes, G. G. (1980). Calculation on fluorescence resonance energy transfer on surfaces. *Biophys. J.* **32,** 1023–1035.

Di Simplicio, P., Franconi, F., Frosali, S., and Di Giuseppe, D. (2003). Thiolation and nitrosation of cysteines in biological fluids and cells. *Amino Acids* **25,** 323–339.

Ding, G. J., Fischer, P. A., Boltz, R. C., Schmidt, J. A., Colaianne, J. J., Gough, A., Rubin, R. A., and Miller, D. K. (1998). Characterization and quantitation of NF-kappaB nuclear translocation induced by interleukin-1 and tumor necrosis factor-alpha: Development and use of a high capacity fluorescence cytometric system. *J. Biol. Chem.* **273,** 28897–28905.

Dinsmore, C. J., and Bell, I. M. (2003). Inhibitors of farnesyltransferase and geranylgeranyl transferase-I for antitumor therapy: substrate-based design, conformational constraint and biological activity. *Curr. Top. Med. Chem.* **3,** 1075–1093.

Drisdel, R. C., Manzana, E., and Green, W. N. (2004). The role of palmitoylation in functional expression of nicotinic alpha7 receptors. *J. Neurosci.* **24,** 10502–10510.

Duncan, J. A., and Gilman, A. G. (1998). A cytoplasmic acyl-protein thioesterase that removes palmitate from G protein alpha subunits and p21(RAS). *J. Biol. Chem.* **273,** 15830–15837.

Dunphy, J. T., and Linder, M. E. (1998). Signalling functions of protein palmitoylation. *Biochim. Biophys. Acta* **1436,** 245–261.

Ehrlich, I., and Malinow, R. (2004). Postsynaptic density 95 controls AMPA receptor incorporation during long-term potentiation and experience-driven synaptic plasticity. *J. Neurosci.* **24,** 916–927.

El-Husseini, A. E., Topinka, J. R., Lehrer-Graiwer, J. E., Firestein, B. L., Craven, S. E., Aoki, C., and Bredt, D. S. (2000). Ion channel clustering by membrane-associated guanylate kinases. Differential regulation by N-terminal lipid and metal binding motifs. *J. Biol. Chem.* **275,** 23904–23910.

El-Husseini Ael, D., and Bredt, D. S. (2002). Protein palmitoylation: A regulator of neuronal development and function. *Nature Rev. Neurosci.* **3,** 791–802.

El-Husseini Ael, D., Schnell, E., Dakoji, S., Sweeney, N., Zhou, Q., Prange, O., Gauthier-Campbell, C., Aguilera-Moreno, A., Nicoll, R. A., and Bredt, D. S. (2002). Synaptic strength regulated by palmitate cycling on PSD-95. *Cell* **108,** 849–863.

Farazi, T. A., Waksman, G., and Gordon, J. I. (2001). The biology and enzymology of protein N-myristoylation. *J. Biol. Chem.* **276,** 39501–39504.

Fu, H. W., and Casey, P. J. (1999). Enzymology and biology of CaaX protein prenylation. *Recent Prog. Horm. Res.* **54,** 315–342; discussion 342–343.

Fukasawa, M., Varlamov, O., Eng, W. S., Sollner, T. H., and Rothman, J. E. (2004). Localization and activity of the SNARE Ykt6 determined by its regulatory domain and palmitoylation. *Proc. Natl. Acad. Sci. USA* **101,** 4815–4820.

Fukata, M., Fukata, Y., Adesnik, H., Nicoll, R. A., and Bredt, D. S. (2004). Identification of PSD-95 palmitoylating enzymes. *Neuron* **44,** 987–996.

Fung, B. K., and Stryer, L. (1978). Surface density determination in membranes by fluorescence energy transfer. *Biochemistry* **17,** 5241–5248.

Gauthier-Campbell, C., Bredt, D. S., Murphy, T. H., and El-Husseini Ael, D. (2004). Regulation of dendritic branching and filopodia formation in hippocampal neurons by specific acylated protein motifs. *Mol. Biol. Cell* **15,** 2205–2217.

Ghobrial, I. M., and Adjei, A. A. (2002). Inhibitors of the ras oncogene as therapeutic targets. *Hematol. Oncol. Clin. North Am.* **16,** 1065–1088.

Ghosh, R. N., Chen, Y. T., DeBiasio, R., DeBiasio, R. L., Conway, B. R., Minor, L. K., and Demarest, K. T. (2000). Cell-based, high-content screen for receptor internalization, recycling and intracellular trafficking. *Biotechniques* **29,** 170–175.

Goldberg, I. G., Allan, C., Burel, J. M., Creager, D., Falconi, A., Hochheiser, H., Johnston, J., Mellen, J., Sorger, P. K., and Swedlow, J. R. (2005). The Open Microscopy Environment (OME) Data Model and XML file: Open tools for informatics and quantitative analysis in biological imaging. *Genome Biol.* **6,** R47.

Gupta, P., Soyombo, A. A., Atashband, A., Wisniewski, K. E., Shelton, J. M., Richardson, J. A., Hammer, R. E., and Hofmann, S. L. (2001). Disruption of PPT1 or PPT2 causes neuronal ceroid lipofuscinosis in knockout mice. *Proc. Natl. Acad. Sci. USA* **98,** 13566–13571.

Han, W., Rhee, J. S., Maximov, A., Lin, W., Hammer, R. E., Rosenmund, C., and Sudhof, T. C. (2005). C-terminal ECFP fusion impairs synaptotagmin 1 function: Crowding out synaptotagmin 1. *J. Biol. Chem.* **280,** 5089–5100.

Hancock, J. F. (2003). Ras proteins: Different signals from different locations. *Nature Rev. Mol. Cell. Biol.* **4,** 373–384.

Hancock, J. F., Magee, A. I., Childs, J. E., and Marshall, C. J. (1989). All ras proteins are polyisoprenylated but only some are palmitoylated. *Cell* **57,** 1167–1177.

Hancock, J. F., Paterson, H., and Marshall, C. J. (1990). A polybasic domain or palmitoylation is required in addition to the CAAX motif to localize p21ras to the plasma membrane. *Cell* **63,** 133–139.

Hayashi, T., Rumbaugh, G., and Huganir, R. L. (2005). Differential regulation of AMPA receptor subunit trafficking by palmitoylation of two distinct sites. *Neuron* **47,** 709–723.

Heindel, U., Schmidt, M. F., and Veit, M. (2003). Palmitoylation sites and processing of synaptotagmin I, the putative calcium sensor for neurosecretion. *FEBS Lett.* **544,** 57–62.

Herincs, Z., Corset, V., Cahuzac, N., Furne, C., Castellani, V., Hueber, A. O., and Mehlen, P. (2005). DCC association with lipid rafts is required for netrin-1-mediated axon guidance. *J. Cell Sci.* **118,** 1687–1692.

Hershko, A., and Ciechanover, A. (1998). The ubiquitin system. *Annu. Rev. Biochem.* **67,** 425–479.

Hess, D. T., Patterson, S. I., Smith, D. S., and Skene, J. H. (1993). Neuronal growth cone collapse and inhibition of protein fatty acylation by nitric oxide. *Nature* **366,** 562–565.

Hess, D. T., Slater, T. M., Wilson, M. C., and Skene, J. H. (1992). The 25 kDa synaptosomal-associated protein SNAP-25 is the major methionine-rich polypeptide in rapid axonal transport and a major substrate for palmitoylation in adult CNS. *J. Neurosci.* **12,** 4634–4641.

Hiol, A., Caron, J. M., Smith, C. D., and Jones, T. L. (2003a). Characterization and partial purification of protein fatty acyltransferase activity from rat liver. *Biochim. Biophys. Acta* **1635,** 10–19.

Hiol, A., Davey, P. C., Osterhout, J. L., Waheed, A. A., Fischer, E. R., Chen, C. K., Milligan, G., Druey, K. M., and Jones, T. L. (2003b). Palmitoylation regulates regulators of G-protein signaling (RGS) 16 function. I. Mutation of amino-terminal cysteine residues on RGS16 prevents its targeting to lipid rafts and palmitoylation of an internal cysteine residue. *J. Biol. Chem.* **278,** 19301–19308.

Hofmann, A., Iwai, H., Hess, S., Pluckthun, A., and Wlodawer, A. (2002). Structure of cyclized green fluorescent protein. *Acta Crystallogr. D Biol. Crystallogr.* **58,** 1400–1406.

Honda, Z., Suzuki, T., Kono, H., Okada, M., Yamamoto, T., Ra, C., Morita, Y., and Yamamoto, K. (2000). Sequential requirements of the N-terminal palmitoylation site and

SH2 domain of Src family kinases in the initiation and progression of FcepsilonRI signaling. *Mol. Cell. Biol.* **20,** 1759–17571.

Huang, K., Yanai, A., Kang, R., Arstikaitis, P., Singaraja, R. R., Metzler, M., Mullard, A., Haigh, B., Gauthier-Campbell, C., Gutekunst, C. A., Hayden, M. R., and El-Husseini, A. (2004). Huntingtin-interacting protein HIP14 is a palmitoyl transferase involved in palmitoylation and trafficking of multiple neuronal proteins. *Neuron* **44,** 977–986.

Jablensky, A. (2004). Resolving schizophrenia's CATCH22. *Nature Genet.* **36,** 674–675.

Kanaani, J., el-Husseini Ael, D., Aguilera-Moreno, A., Diacovo, J. M., Bredt, D. S., and Baekkeskov, S. (2002). A combination of three distinct trafficking signals mediates axonal targeting and presynaptic clustering of GAD65. *J. Cell Biol.* **158,** 1229–1238.

Kang, R., Swayze, R., Lise, M. F., Gerrow, K., Mullard, A., Honer, W. G., and El-Husseini, A. (2004). Presynaptic trafficking of synaptotagmin I is regulated by protein palmitoylation. *J. Biol. Chem.* **279,** 50524–50536.

Karayiorgou, M., Morris, M. A., Morrow, B., Shprintzen, R. J., Goldberg, R., Borrow, J., Gos, A., Nestadt, G., Wolyniec, P. S., Lasseter, V. K., *et al.* (1995). Schizophrenia susceptibility associated with interstitial deletions of chromosome 22q11. *Proc. Natl. Acad. Sci. USA* **92,** 7612–7616.

Keller, C. A., Yuan, X., Panzanelli, P., Martin, M. L., Alldred, M., Sassoe-Pognetto, M., and Luscher, B. (2004). The gamma2 subunit of GABA(A) receptors is a substrate for palmitoylation by GODZ. *J. Neurosci.* **24,** 5881–5891.

Kim, E., and Sheng, M. (2004). PDZ domain proteins of synapses. *Nature Rev. Neurosci.* **5,** 771–781.

Kleuss, C., and Krause, E. (2003). Galpha(s) is palmitoylated at the N-terminal glycine. *EMBO J.* **22,** 826–832.

Lane, S. R., and Liu, Y. (1997). Characterization of the palmitoylation domain of SNAP-25. *J. Neurochem.* **69,** 1864–1869.

Lavy, M., Bracha-Drori, K., Sternberg, H., and Yalovsky, S. (2002). A cell-specific, prenylation-independent mechanism regulates targeting of type II RACs. *Plant Cell* **14,** 2431–2450.

Lawrence, D. S., Zilfou, J. T., and Smith, C. D. (1999). Structure-activity studies of cerulenin analogues as protein palmitoylation inhibitors. *J. Med. Chem.* **42,** 4932–4941.

Liang, X., Lu, Y., Neubert, T. A., and Resh, M. D. (2002). Mass spectrometric analysis of GAP-43/neuromodulin reveals the presence of a variety of fatty acylated species. *J. Biol. Chem.* **277,** 33032–33040.

Liang, X., Lu, Y., Wilkes, M., Neubert, T. A., and Resh, M. D. (2004). The N-terminal SH4 region of the Src family kinase Fyn is modified by methylation and heterogeneous fatty acylation: role in membrane targeting, cell adhesion, and spreading. *J. Biol. Chem.* **279,** 8133–8139.

Liang, X., Nazarian, A., Erdjument-Bromage, H., Bornmann, W., Tempst, P., and Resh, M. D. (2001). Heterogeneous fatty acylation of Src family kinases with polyunsaturated fatty acids regulates raft localization and signal transduction. *J. Biol. Chem.* **276,** 30987–30994.

Lipton, S. A., Choi, Y. B., Takahashi, H., Zhang, D., Li, W., Godzik, A., and Bankston, L. A. (2002). Cysteine regulation of protein function—as exemplified by NMDA-receptor modulation. *Trends Neurosci.* **25,** 474–480.

Liu, Y., Fisher, D. A., and Storm, D. R. (1993). Analysis of the palmitoylation and membrane targeting domain of neuromodulin (GAP-43) by site-specific mutagenesis. *Biochemistry* **32,** 10714–10719.

Liu, Y. C., Chapman, E. R., and Storm, D. R. (1991). Targeting of neuromodulin (GAP-43) fusion proteins to growth cones in cultured rat embryonic neurons. *Neuron* **6,** 411–420.

Lobo, S., Greentree, W. K., Linder, M. E., and Deschenes, R. J. (2002). Identification of a Ras palmitoyltransferase in *Saccharomyces cerevisiae*. *J. Biol. Chem.* **277,** 41268–41273.

Loisel, T. P., Adam, L., Hebert, T. E., and Bouvier, M. (1996). Agonist stimulation increases the turnover rate of beta 2AR-bound palmitate and promotes receptor depalmitoylation. *Biochemistry* **35,** 15923–15932.

Loisel, T. P., Ansanay, H., Adam, L., Marullo, S., Seifert, R., Lagace, M., and Bouvier, M. (1999). Activation of the beta(2)-adrenergic receptor-Galpha(s) complex leads to rapid depalmitoylation and inhibition of repalmitoylation of both the receptor and Galpha(s). *J. Biol. Chem.* **274,** 31014–31019.

Long, S. B., Casey, P. J., and Beese, L. S. (2002). Reaction path of protein farnesyltransferase at atomic resolution. *Nature* **419,** 645–650.

Loranger, S. S., and Linder, M. E. (2002). SNAP-25 traffics to the plasma membrane by a syntaxin-independent mechanism. *J. Biol. Chem.* **277,** 34303–34309.

Mansouri, M. R., Marklund, L., Gustavsson, P., Davey, E., Carlsson, B., Larsson, C., White, I., Gustavson, K. H., and Dahl, N. (2005). Loss of ZDHHC15 expression in a woman with a balanced translocation t(X;15)(q13.3;cen) and severe mental retardation. *Eur. J. Hum. Genet.* **13,** 970–977.

Markey, M. K., Boland, M. V., and Murphy, R. F. (1999). Toward objective selection of representative microscope images. *Biophys. J.* **76,** 2230–2237.

McRorie, D., and Voelker, P. (1993). "Self-associating Systems in the Analytical Ultracentrifuge." Beckman Instruments Inc., Fullerton, CA.

Melkonian, K. A., Ostermeyer, A. G., Chen, J. Z., Roth, M. G., and Brown, D. A. (1999). Role of lipid modifications in targeting proteins to detergent-resistant membrane rafts. Many raft proteins are acylated, while few are prenylated. *J. Biol. Chem.* **274,** 3910–3917.

Menendez, J. A., Vellon, L., Mehmi, I., Oza, B. P., Ropero, S., Colomer, R., and Lupu, R. (2004). Inhibition of fatty acid synthase (FAS) suppresses HER2/neu (erbB-2) oncogene overexpression in cancer cells. *Proc. Natl. Acad. Sci. USA* **101,** 10715–10720.

Minguez, J. M., Giuliano, K. A., Balachandran, R., Madiraju, C., Curran, D. P., and Day, B. W. (2002). Synthesis and high content cell-based profiling of simplified analogues of the microtubule stabilizer (+)-discodermolide. *Mol. Cancer Ther.* **1,** 1305–1313.

Moffett, S., Brown, D. A., and Linder, M. E. (2000). Lipid-dependent targeting of G proteins into rafts. *J. Biol. Chem.* **275,** 2191–2198.

Moran, T. J., Hunter, E., Morelock, M., Heynen, S., Laris, C., Thieleking, M., Akong, M., Mikic, I., Callaway, S., Zacharias, D., and Price, J. H. (2005). NF B nuclear translocation validation and performance in high throughput cell imaging. Submitted for publication.

Morelock, M. M., Hunter, E. A., Moran, T. J., Heynen, S., Laris, C., Thieleking, M., Akong, M., Mikic, I., Callaway, S., Zacharias, D. A., and Price, J. H. (2005). Statistics of assay validation in high throughput cell imaging of NFkB nuclear translocation. *Assay Drug Dev. Technol.* **3**(5), 483–499.

Mukai, J., Liu, H., Burt, R. A., Swor, D. E., Lai, W. S., Karayiorgou, M., and Gogos, J. A. (2004). Evidence that the gene encoding ZDHHC8 contributes to the risk of schizophrenia. *Nature Genet.* **36,** 725–731.

Munro, A. P., and Williams, D. L. H. (1999). Reactivity of nitrogen nucleophiles towards S-nitrosopenicillamine. *J. Chem. Soc. Perkin Trans.* **2,** 1989–1993.

Murphy, R. F., Boland, M. V., and Velliste, M. (2000). Towards a systematics for protein subcelluar location: Quantitative description of protein localization patterns and automated analysis of fluorescence microscope images. *Proc. Int. Conf. Intell. Syst. Mol. Biol.* **8,** 251–259.

Navarro-Lerida, I., Alvarez-Barrientos, A., Gavilanes, F., and Rodriguez-Crespo, I. (2002). Distance-dependent cellular palmitoylation of de-novo-designed sequences and their translocation to plasma membrane subdomains. *J. Cell Sci.* **115,** 3119–3130.

Ormo, M., Cubitt, A. B., Kallio, K., Gross, L. A., Tsien, R. Y., and Remington, S. J. (1996). Crystal structure of the Aequorea victoria green fluorescent protein. *Science* **273,** 1392–1395.

Osten, P., Khatri, L., Perez, J. L., Kohr, G., Giese, G., Daly, C., Schulz, T. W., Wensky, A., Lee, L. M., and Ziff, E. B. (2000). Mutagenesis reveals a role for ABP/GRIP binding to GluR2 in synaptic surface accumulation of the AMPA receptor. *Neuron* **27,** 313–325.

Osterhout, J. L., Waheed, A. A., Hiol, A., Ward, R. J., Davey, P. C., Nini, L., Wang, J., Milligan, G., Jones, T. L., and Druey, K. M. (2003). Palmitoylation regulates regulator of G-protein signaling (RGS) 16 function. II. Palmitoylation of a cysteine residue in the RGS box is critical for RGS16 GTPase accelerating activity and regulation of Gi-coupled signalling. *J. Biol. Chem.* **278,** 19309–19316.

Palm, G. J., Zdanov, A., Gaitanaris, G. A., Stauber, R., Pavlakis, G. N., and Wlodawer, A. (1997). The structural basis for spectral variations in green fluorescent protein. *Nature Struct. Biol.* **4,** 361–365.

Passmore, L. A., and Barford, D. (2004). Getting into position: The catalytic mechanisms of protein ubiquitylation. *Biochem. J.* **379,** 513–525.

Pepinsky, R. B., Zeng, C., Wen, D., Rayhorn, P., Baker, D. P., Williams, K. P., Bixler, S. A., Ambrose, C. M., Garber, E. A., Miatkowski, K., Taylor, F. R., Wang, E. A., and Galdes, A. (1998). Identification of a palmitic acid-modified form of human Sonic hedgehog. *J. Biol. Chem.* **273,** 14037–14045.

Price, J. H., Goodacre, A., Hahn, K., Hodgson, L., Hunter, E. A., Krajewski, S., Murphy, R. F., Rabinovich, A., Reed, J. C., and Heynen, S. (2002). Advances in molecular labeling, high throughput imaging and machine intelligence portend powerful functional cellular biochemistry tools. *J. Cell Biochem. Suppl.* **39,** 194–210.

Price, J. H., Hunter, E. A., and Gough, D. A. (1996). Accuracy of least squares designed spatial FIR filters for segmentation of images of fluorescence stained cell nuclei. *Cytometry* **25,** 303–316.

Rajala, R. V., Datla, R. S., Moyana, T. N., Kakkar, R., Carlsen, S. A., and Sharma, R. K. (2000). N-myristoyltransferase. *Mol. Cell. Biochem.* **204,** 135–155.

Resh, M. D. (1999). Fatty acylation of proteins: New insights into membrane targeting of myristoylated and palmitoylated proteins. *Biochim. Biophys. Acta* **1451,** 1–16.

Robertson, J. D., Orrenius, S., and Zhivotovsky, B. (2000). Review: Nuclear events in apoptosis. *J. Struct. Biol.* **129,** 346–358.

Robinson, L. J., and Michel, T. (1995). Mutagenesis of palmitoylation sites in endothelial nitric oxide synthase identifies a novel motif for dual acylation and subcellular targeting. *Proc. Natl. Acad. Sci. USA* **92,** 11776–11780.

Rocks, O., Peyker, A., Kahms, M., Verveer, P. J., Koerner, C., Lumbierres, M., Kuhlmann, J., Waldmann, H., Wittinghofer, A., and Bastiaens, P. I. (2005). An acylation cycle regulates localization and activity of palmitoylated Ras isoforms. *Science* **307,** 1746–1752.

Roques, E. J., and Murphy, R. F. (2002). Objective evaluation of differences in protein subcellular distribution. *Traffic* **3,** 61–65.

Roskoski, R., Jr. (2003). Protein prenylation: A pivotal posttranslational process. *Biochem. Biophys. Res. Commun.* **303,** 1–7.

Roth, A. F., Feng, Y., Chen, L., and Davis, N. G. (2002). The yeast DHHC cysteine-rich domain protein Akr1p is a palmitoyl transferase. *J. Cell Biol.* **159,** 23–28.

Siemering, K. R., Golbik, R., Sever, R., and Haseloff, J. (1996). Mutations that suppress the thermosensitivity of green fluorescent protein. *Curr. Biol.* **6,** 1653–1663.

Sinensky, M. (2000). Recent advances in the study of prenylated proteins. *Biochim. Biophys. Acta* **1484,** 93–106.

Singaraja, R. R., Hadano, S., Metzler, M., Givan, S., Wellington, C. L., Warby, S., Yanai, A., Gutekunst, C. A., Leavitt, B. R., Yi, H., Fichter, K., Gan, L., McCutcheon, K., Chopra, V., Michel, J., Hersch, S. M., Ikeda, J. E., and Hayden, M. R. (2002). HIP14, a novel ankyrin domain-containing protein, links huntingtin to intracellular trafficking and endocytosis. *Hum. Mol. Genet.* **11**, 2815–2828.

Sitia, R., and Molteni, S. N. (2004). Stress, protein (mis)folding, and signaling: The redox connection. *Sci. STKE* 2004, pe27.

Skene, J. H., and Virag, I. (1989). Posttranslational membrane attachment and dynamic fatty acylation of a neuronal growth cone protein, GAP-43. *J. Cell Biol.* **108**, 613–624.

Smotrys, J. E., and Linder, M. E. (2004). Palmitoylation of intracellular signaling proteins: Regulation and function. *Annu. Rev. Biochem.* **73**, 559–587.

Snyder, B., and Freire, E. (1982). Fluorescence energy transfer in two dimensions: A numeric solution for random and nonrandom distributions. *Biophys. J.* **40**, 137–148.

Stanley, P., Koronakis, V., and Hughes, C. (1998). Acylation of *Escherichia coli* hemolysin: A unique protein lipidation mechanism underlying toxin function. *Microbiol. Mol. Biol. Rev.* **62**, 309–333.

Stoffel, R. H., Randall, R. R., Premont, R. T., Lefkowitz, R. J., and Inglese, J. (1994). Palmitoylation of G protein-coupled receptor kinase, GRK6: Lipid modification diversity in the GRK family. *J. Biol. Chem.* **269**, 27791–27794.

Strittmatter, S. M. (1992). GAP-43 as a modulator of G protein transduction in the growth cone. *Perspect. Dev. Neurobiol.* **1**, 13–19.

Strittmatter, S. M., Valenzuela, D., Vartanian, T., Sudo, Y., Zuber, M. X., and Fishman, M. C. (1991). Growth cone transduction: Go and GAP-43. *J. Cell Sci. Suppl.* **15**, 27–33.

Thupari, J. N., Pinn, M. L., and Kuhajda, F. P. (2001). Fatty acid synthase inhibition in human breast cancer cells leads to malonyl-CoA-induced inhibition of fatty acid oxidation and cytotoxicity. *Biochem. Biophys. Res. Commun.* **285**, 217–223.

Topinka, J. R., and Bredt, D. S. (1998). N-terminal palmitoylation of PSD-95 regulates association with cell membranes and interaction with K+ channel Kv1.4. *Neuron* **20**, 125–134.

Veit, M., Becher, A., and Ahnert-Hilger, G. (2000). Synaptobrevin 2 is palmitoylated in synaptic vesicles prepared from adult, but not from embryonic brain. *Mol. Cell. Neurosci.* **15**, 408–416.

Veit, M., Sollner, T. H., and Rothman, J. E. (1996). Multiple palmitoylation of synaptotagmin and the t-SNARE SNAP-25. *FEBS Lett.* **385**, 119–123.

Vesa, J., Hellsten, E., Verkruyse, L. A., Camp, L. A., Rapola, J., Santavuori, P., Hofmann, S. L., and Peltonen, L. (1995). Mutations in the palmitoyl protein thioesterase gene causing infantile neuronal ceroid lipofuscinosis. *Nature* **376**, 584–587.

Wachter, R. M., Elsliger, M. A., Kallio, K., Hanson, G. T., and Remington, S. J. (1998). Structural basis of spectral shifts in the yellow-emission variants of green fluorescent protein. *Structure* **6**, 1267–1277.

Wachter, R. M., King, B. A., Heim, R., Kallio, K., Tsien, R. Y., Boxer, S. G., and Remington, S. J. (1997). Crystal structure and photodynamic behavior of the blue emission variant Y66H/Y145F of green fluorescent protein. *Biochemistry* **36**, 9759–9765.

Wachter, R. M., Yarbrough, D., Kallio, K., and Remington, S. J. (2000). Crystallographic and energetic analysis of binding of selected anions to the yellow variants of green fluorescent protein. *J. Mol. Biol.* **301**, 157–171.

Ward, W. W. (1998). Biochemical and physical properties of green fluorescent protein. *In* "GFP Green Fluorescent Protein Properties, Applications and Protocols" (M. A. Chalfie and S. Kain, eds.), Vol. 1, pp. 45–75. Wiley-Liss, New York.

Ward, W. W., and Cormier, M. J. (1979). An energy transfer protein in coelenterate bioluminescence: Characterization of the Renilla green-fluorescent protein. *J. Biol. Chem.* **254,** 781–788.

Webb, Y., Hermida-Matsumoto, L., and Resh, M. D. (2000). Inhibition of protein palmitoylation, raft localization, and T cell signaling by 2-bromopalmitate and polyunsaturated fatty acids. *J. Biol. Chem.* **275,** 261–270.

Wiedmer, T., Zhao, J., Nanjundan, M., and Sims, P. J. (2003). Palmitoylation of phospholipid scramblase 1 controls its distribution between nucleus and plasma membrane. *Biochemistry* **42,** 1227–1233.

Wolber, P. K., and Hudson, B. S. (1979). An analytic solution to the Forster energy transfer problem in two dimensions. *Biophys. J.* **28,** 197–210.

Wolven, A., Okamura, H., Rosenblatt, Y., and Resh, M. D. (1997). Palmitoylation of p59fyn is reversible and sufficient for plasma membrane association. *Mol. Biol. Cell* **8,** 1159–1173.

Yang, F., Moss, L. G., and Phillips, G. N., Jr. (1996). The molecular structure of green fluorescent protein. *Nature Biotechnol.* **14,** 1246–1251.

Yang, X., Kovalenko, O. V., Tang, W., Claas, C., Stipp, C. S., and Hemler, M. E. (2004). Palmitoylation supports assembly and function of integrin-tetraspanin complexes. *J. Cell Biol.* **167,** 1231–1240.

Yguerabide, J. (1994). Theory for establishing proximity relations in biological membranes by excitation energy transfer measurements. *Biophys. J.* **66,** 683–693.

Yokoe, H., and Meyer, T. (1996). Spatial dynamics of GFP-tagged proteins investigated by local fluorescence enhancement. *Nature Biotechnol.* **14,** 1252–1256.

Zacharias, D. A., Violin, J. D., Newton, A. C., and Tsien, R. Y. (2002). Partitioning of lipid-modified monomeric GFPs into membrane microdomains of live cells. *Science* **296,** 913–916.

Zhang, J. H., Chung, T. D., and Oldenburg, K. R. (1999). A simple statistical parameter for use in evaluation and validation of high throughput screening assays. *J. Biomol. Screen* **4,** 67–73.

Zhou, W., Simpson, P. J., McFadden, J. M., Townsend, C. A., Medghalchi, S. M., Vadlamudi, A., Pinn, M. L., Ronnett, G. V., and Kuhajda, F. P. (2003). Fatty acid synthase inhibition triggers apoptosis during S phase in human cancer cells. *Cancer Res.* **63,** 7330–7337.

Zimet, D. B., Thevenin, B. J., Verkman, A. S., Shohet, S. B., and Abney, J. R. (1995). Calculation of resonance energy transfer in crowded biological membranes. *Biophys. J.* **68,** 1592–1603.

Zuber, M. X., Goodman, D. W., Karns, L. R., and Fishman, M. C. (1989a). The neuronal growth-associated protein GAP-43 induces filopodia in non-neuronal cells. *Science* **244,** 1193–1195.

Zuber, M. X., Strittmatter, S. M., and Fishman, M. C. (1989b). A membrane-targeting signal in the amino terminus of the neuronal protein GAP-43. *Nature* **341,** 345–348.

[11] High-Resolution, High-Throughput Microscopy Analyses of Nuclear Receptor and Coregulator Function

By VALERIA BERNO,* CRUZ A. HINOJOS,* LARBI AMAZIT, ADAM T. SZAFRAN, and MICHAEL A. MANCINI

Abstract

Steroid nuclear receptors are ligand-dependent transcription factors that have been studied since the early 1960s by principally biochemical and reporter assay approaches. From these studies an elegant and complex model of nuclear receptor transcription regulation has been developed. Inherent to both biochemical and reporter assay approaches is the generation of averaged responses and it is not generally considered that individual cells could exhibit quite varied responses. In some cases, recent microscopic single-cell analyses provide markedly different responses relative to traditional approaches based on population averaging and underscore the need to continue refinement of the current model of nuclear receptor-regulated transcription. While single-cell analyses of nuclear receptor action have been hindered by the predominantly qualitative nature of the approach, high-throughput microscopy is now available to resolve this issue. This chapter demonstrates the utility of high-throughput microscopic analyses of nuclear receptor and nuclear receptor coregulator function. The ability of high-throughput microscopy to generate physiologically appropriate test populations by filtering based on morphological and protein of interest expression criteria is demonstrated. High-resolution, high-throughput microscopy is illustrated that provides quantitative subcellular information for both androgen and estrogen receptors. Efforts are ongoing to develop model systems that provide additional multiplex data and with refined image analyses to achieve true high-content imaging screens.

High-Throughput Microscopy (HTM)

High-throughput microscopy for biological purposes may be broadly defined as an automated image acquisition system coupled to cell identification and cell morphological detection/measurement algorithms. Automated image acquisition and the multiple aspects of cellular and subcellular morphology measured and catalogued at the individual cell level enable the potential generation of millions of measurements per day and impart the

---

*Primary contributors.

METHODS IN ENZYMOLOGY, VOL. 414
0076-6879/06 $35.00
DOI: 10.1016/S0076-6879(06)14011-2

high-information content inherent to this methodology. A significant benefit of the morphological measurements is the ability to filter (or gate) the imaged cell population to obtain a more homogeneous subpopulation for analysis. For example, apoptotic and mitotic cells may be morphologically identified and excluded from the final population to be analyzed. The ability to filter cells is especially beneficial when studying exogenously expressed proteins by allowing the exclusion of cells that are over- or underexpressing the protein relative to the endogenous level. Further, the numerous measurements enable the multiplexing of numerous morphological features in order to rapidly identify correlations and underscore the large analytical potential of HTM (Perlman *et al.*, 2004).

Technological innovations continue to improve HTM as developers advance imaging systems, design specialized algorithms to measure specific subcellular phenomena, and create more intuitive and comprehensive software tools for data mining. For example, algorithms were developed that specifically measure the fractional localization of a protein between the cytoplasm and the nucleus. These algorithms were applied to quantitate the cytoplasmic to nuclear translocation of the transcription factor NF-κB–p65 subunit upon proinflammatory cytokine stimulation by two separate HTM systems (Ding *et al.*, 1998; Morelock *et al.*, 2005). The successful matching of NF-κB biology with proper analytical tools was an early and highly extrapolatable HTM analysis tool that formed the basis of many translocation assays. Successful HTM analysis is therefore predicated upon utilizing or developing appropriate model systems.

Steroid Nuclear Receptors and Coregulators

Steroids are secreted endocrine hormones that are involved in a broad range of physiological functions such as sexual differentiation, growth and development, metabolism, and maintenance of homeostasis. In response to trophic signals, the adrenal glands (cortisol and aldosterone producing) or gonads (progesterone, testosterone, and estradiol producing) secrete steroids into the blood stream (Singh and Kumar, 2005). Steroid signals are transduced through binding to steroid nuclear receptors (NR). NRs are a family of transcription factors that include androgen (AR), estrogen (α and β isotypes, ERα and ERβ), glucocorticoid (GR), progesterone (A and B isoforms, PRA and PRB), and mineralocorticoid receptors (Evans, 1988). Steroid binding to these receptors affects transcription regulation of specific genes. Thus, steroids play a key role in the development or maintenance of multiple physiological phenotypes by altering transcription through NRs; not surprisingly, aberrant regulation of steroid signaling may result in pathophysiological conditions. The Nuclear Receptor Signaling Atlas (www.nursa.org) is a useful online resource for the entire NR community.

Nuclear receptors share common structural domains that are critical for the regulation of transcription (Kumar and Thompson, 1999). These domains include the DNA-binding domain (DBD), the ligand-binding domain (LBD), and activation functions 1 and 2 domains (AF-1, AF-2). Binding of ligand to the LBD causes a conformational change that promotes DNA promoter binding through the DBD. The receptor interacts as a dimer with DBD-specific hormone response elements to either inhibit or promote transcription of a set of responsive genes through the recruitment of coregulators. In the inactive or repressed state, the promoter-localized NR directly or indirectly recruits corepressor proteins such as histone deacetylases and other nuclear receptor corepressors. These proteins generate a transcriptionally prohibitive environment through histone modifications (e.g., acetylation, methylation) and chromatin condensation. Alternatively, binding of agonist to the NR reduces the association with corepressor proteins and increases the recruitment of multiple coactivator proteins such as steroid receptor coactivators (SRC-1, -2, and -3), chromatin-remodeling proteins, basal transcription factors, and RNA pol II. Specificity is generated by (1) the functional redundancy of many of the coregulators, resulting in an exponential number of protein combinations within a complex; (2) promoter-specific recruitment of coregulators; and (3) modulation of coregulators by signal transduction pathways. Thus, transcription is regulated through the NR and ligand-dependent recruitment to promoters of functionally distinct proteins.

Nuclear receptors each have unique properties that either lend to or detract from HTM analysis. The subcellular localization of each NR is a critical feature in any imaging assay. The AR, GR, MR, and PR members of the NR family are principally located in the cytoplasm in the absence of their ligands. Here, the nonliganded NR associates with chaperone proteins, such as heat-shock protein 90 (Pratt *et al.*, 2004). Addition of ligand induces NR homodimerization and a cytoplasm to nuclear translocation facilitated by dynamic interactions with chaperone proteins to the liganded receptor. The cytoplasmic NR translocation to the nucleus is analogous to NFκB translocation and is, therefore, conducive for study by existing HTM capabilities. Therefore, novel NR ligands may be screened and analyzed by HTM for their ability to induce NR cytoplasm to nuclear translocation (see later). In contrast, the ER and PRA are principally localized to the nucleus in the absence of ligand preventing HTM analysis of translocation (Press *et al.*, 1989; Lim *et al.*, 1999), although one could envision screens to examine effectors that lead redistribution to the cytoplasm. For nuclear resident proteins, novel HTM methods and model systems are required to study functional responses to ligand (see later).

We have adopted HTM as an analytical tool to study steroid nuclear receptor and nuclear receptor coregulator biology at the single-cell level. The critical aspects of experimental setup necessary for successful and reliable HTM analysis of AR and ER and one NR coregulator, SRC-3, are discussed. Each protein has distinct properties that require different model systems and application of HTM for analysis. The following sections describe these model systems, their analyses by HTM, and representative results.

General Methods

Cover Glass Preparation

Acid Etching Coverslips. Within a fume hood, heat coverslips in a loosely covered glass beaker (can use foil) in 1 *M* HCl at 50 to 60° for 4 to 16 h. An entire box of coverslips may be processed at once. Next pour off and save the acid and wash the coverslips extensively with sterile dH_2O. At this point, the coverslips may be rinsed with ethanol and laid out separately and left to dry on a sheet of 3MM paper under ultraviolet light. When dry, place the coverslips in a sterile culture dish for storage. We use 12-mm circular coverslips, which are conveniently sized for 24-well cell culture plates for growth and processing.

Coating with Poly-D-Lysine

Some cells do not adhere well to acid-etched cover slips and require an additional treatment. We coat acid-washed cover slips with poly-D-lysine; this can also be done in bulk using 100-mm cell culture dishes. If an entire box of coverslips is to be coated, divide them among four dishes. Prepare 1 mg/ml poly-D-lysine, reconstituted in sterile phosphate-buffered salilne (PBS), and filter sterilize (use a 10-ml syringe and syringe filter). Use about 10 ml per dish, sealing the dishes with Parafilm. One dish at a time may be coated and the poly-D-lysine transferred to the next dish when the first one is finished, and so on. Rock or rotate the dishes a minimum of 30 min. Remove and save the poly-D-lysine; it can be reused three or four times if refiltered and stored at −20°. Wash the coverslips with sterile dH_2O a minimum of five times in order to completely remove leftover poly-D-lysine. Next, rinse with ethanol, separate, and store as described in the acid-washing section. The coated coverslips may also be stored in ethanol, sealing the dishes with Parafilm until separating.

For glass-bottom well plates we also coat with poly-D-lysine or hormone-free stripped-dialyzed fetal bovine serum. The protocol for acid etching and poly-D-lysine coating well plates is similar to that of coverslips. The differences are that the acid etching should occur for a maximum of 4 h at 50° in order to minimize damage to the adhesive present in the plate. Care should also be taken not to let the poly-D-lysine dry within the well as this will cause an uneven surface for the cells to grow upon. Alternatively, the glass-bottom plates may be coated with stripped-dialyzed serum overnight at 37° within a cell culture incubator in order to prevent drying of the serum onto the cover glass. After incubation, the excess serum is removed; the plates are ready to be seeded with cells.

Cell Culture and Experimental Setup

Grow cells in the appropriate growth media and serum (usually 5% fetal bovine serum) to 90 to 95% confluency. Refresh the growth media 1 day prior to subculturing the cells onto coverslips or well plates following standard trypsin digestion protocols. For NR experiments, the subcultured cells are grown in medium that contains stripped and dialyzed serum that lacks hormones that can obfuscate results. The appropriate seeding density is dependent on many factors and must be determined empirically for each experiment (see specific examples). The cells are grown for 24 to 48 h on the cover glass until ligand treatment or transfection of expression vector(s).

DNA Transfection

Perform transfections by standard calcium phosphate precipitation protocols or by commercially available lipid-based transfection reagents following the manufacturer's protocol. If cells are not to be transfected then perform the ligand treatments and proceed to the fixation and immunolabeling protocol. Remove the medium containing the transfection reagent and expression vector and replace with medium containing the appropriate ligand and control treatments. In our experience, empirically evaluating a variety of transfection reagents and protocols is required for optimal results, which can be significantly cell line specific.

Formaldehyde Fixation and Immunolabeling of Cells

Antibody labeling of endogenous proteins must be optimized to achieve maximal signal-to-noise levels of the fluorescent signals. For cells expressing fluorescent protein fusions, the signal can be increased by immunolabeling the fluorescent protein and/or the protein of interest. For example, cells expressing green fluorescent protein (GFP) fused to the estrogen receptor (GFP-ERα) can be labeled using an ERα antibody or GFP antibody and a green fluorescing secondary antibody. We have found that double labeling can decrease the exposure time manyfold.

Initial Fixation

1. Remove culture media by aspiration or pipetting. Wash cells one to two times with ice-cold PBS (Ca^{2+} and Mg^{2+} containing).

2. Remove PBS and fix the cells with 4% formaldehyde on ice for 30 min.

3. Remove formaldehyde fix, discard in toxic waste bottle, and wash three times with PEM buffer, about 1 to 5 min per wash.

4. Quench autofluorescence: add PEM buffer to preweighed sodium borohydride (final concentration of 1 mg/ml). Mix and add to coverslips. Incubate for 5 min at room temperature, replace with fresh $PEM/NaBH_4$, and incubate another 5 min. Check coverslips to make sure they are not floating on top of the buffer due to the expected bubbling action generated. Alternatively, add 100 mM of ammonium chloride for 10 min at room temperature.

5. Wash with PEM twice, 5 min each time.

6. Permeabilize cell membranes: incubate cells at room temperature 30 min with PEM containing 0.5% Triton X-100. The amount of time in detergent is generally antigen/antibody specific; ~5 min is sometimes sufficient, but to ensure full penetration of antibodies into dense chromatin, we have found 30 min is required.

7. Wash three times with PEM, about 1 to 5 min per wash. If immunolabeling, wash once with Tris-buffered saline plus Tween 20 (TBS-T) buffer after the PEM washes.

Note: If working with cells that are not going to be immunolabeled (such as cells expressing XFP-fused proteins), proceed to the 4′,6-diamidino-2-phenylindole (DAPI) counterstaining step. Instead of using TBS-T, you may dilute the DAPI in PEM to 1 μg/ml working concentration. Incubate with DAPI a minimum of 3 min; again, this should be determined empirically for different cell lines.

Immunolabeling

1. Block nonspecific binding sites with 5% nonfat powdered milk in TBS-T buffer plus 0.02% sodium azide. Blocking time can be anywhere from 30 min to 1 h at room temperature or overnight at 4°. One percent bovine serum albumin in TBS-T can be substituted for powdered milk.

2. Remove blocking buffer and add primary antibody diluted in blocking buffer. Incubate 30 min at 37°, 1 to 2 h at room temperature, or overnight at 4°.

3. Remove primary antibody (generally, the primary antibody can be saved, frozen, and reused several times). Wash four to five times with blocking buffer, 1 min or more per wash.

4. Add fluorochrome-coupled secondary antibody diluted in blocking buffer. Incubate 30 min at room temperature protected from light. We

have found the Alexa dyes from Molecular Probes to be outstanding in terms of brightness and stability (www.probes.com). We routinely dilute the secondary antibodies manyfold more than recommended. Remove and save secondary antibody as it also can be reused. Wash five times with TBS-T and then wash with PEM.

Postfixation and Quenching

As high-magnification, high-resolution imaging systems became more sensitive, we found that antibody-labeled proteins sometimes exhibited minor movements, even through the eyepiece (at 100×). Presumably this was due to partial unfixing from formaldehyde cross-linking and could cause some blurriness in the images. Thus, we began introducing a post-fixation step to ensure stabilization of the antigen–antibody complexes for high-resolution work or, importantly, for longer-term storage of slides (weeks at 4°).

Note: The cells now need to be shielded from light.

1. Fix 10 to 30 min in 4% formaldehyde in PEM buffer.
2. Remove fix (discard in toxic waste) and wash three times in PEM, 2–5 min per wash.
3. Quench autofluorescence as described initially.
4. Wash with PEM twice, 5 min each time.
5. Wash two to three times with TBS-T.
6. Counterstain DNA with 1 μg/ml of DAPI diluted in TBS-T for a minimum of 3 min. Note: DAPI staining is a critical step in achieving successful HTM imaging. For imaging systems that do not use reflection of cell substratum, for example, focusing on the cells directly, a bright high-contrast DAPI stain of nuclei is necessary and is inherently useful in subsequent cell identification. Hoeschst or other DNA stains may also be used.
7. Remove DAPI and add TBS-T.
8. Mount the coverslips on the slide with an antifade mounting medium and seal with nail polish if necessary. For well plates add weak fixative (PEM or PBS and 0.4% formaldehyde) and store at 4° until imaging. Because formaldehyde is autofluorescent, the weak fixative must be replaced with either PEM or PBS buffer prior to imaging. We have found that the newly released Slow Fade or Prolong "Gold" from Molecular Probes is particularly useful to prevent photobleaching, especially in the green and red channels.

Reagents and Supplies

- PBS containing Ca^{2+} and Mg^{2+} (PBS) and TBS-T buffers: formulations of these buffers can be found in common protocol books.

- PEM buffer: 80 mM potassium PIPES, pH 6.8, 5 mM EGTA, pH 7.0, and 2 mM MgCl$_2$.

Filter sterilize and store at 4°.

- Formaldehyde: Use EM-grade formaldehyde (stock solutions range from 16 to 20% and are provided in glass ampoules). Working solution is 4% formaldehyde diluted in PEM buffer. Cover opened ampoules with Parafilm and store at 4°. We abandoned preparation of our own formaldehyde from paraformaldehyde powder to improve fixation consistency and for safety consideration.
- Triton X-100: For ease of use, make a 10% weight/volume stock solution of the Triton X-100 in PEM buffer and aliquot to microcentrifuge tubes covered with foil or amber-colored tubes to protect from light. Store the stocks at −20°. The working solution is 0.5% Triton X-100 diluted in PEM buffer. Diluted Triton is unstable. Do not refreeze the aliquots (store in a refrigerator a maximum of 1 week).
- DAPI: Make a 1-mg/ml stock solution in PEM buffer and aliquot to microcentrifuge tubes covered with foil or amber-colored tubes to protect from light. Store at 4°.

Imaging

High-resolution HTM analyses for all studies described are performed using the Cell Lab IC 100 Image Cytometer (IC100) platform and Cytoshop Version 2.1 analysis software (Beckman Coulter). The imaging platform consists of (1) Nikon Eclipse TE2000-U inverted microscope (Nikon; Melville, NY) with Chroma 82000 triple band filter set (Chroma; Brattleboro, VT), (2) Hamamatsu ORCA-ER digital CCD camera (Hamamatus; Bridgewater, NJ), and (3) Photonics COHU progressive scan camera (Photonics; Oxford, MA). The objective used is a Nikon S Fluor 40×/0.90NA. The objective is chosen as the minimum required to obtain subcellular resolution. The 12-bit images are generated with 1 × 1 binning (1344 × 1024 pixels; 6.5 μm^2 pixel size).

Nuclear Receptor Coregulator SRC-3

The nuclear receptor coactivator SRC-3 belongs to the p160/SRC (steroid receptor activator) family that also includes SRC-1 and SRC-2 (McKenna *et al.*, 1999). Members of this family directly bind to nuclear receptors and are intricately involved in receptor regulation. SRCs assist in activating transcription through their two transcription activation domains as well as acting as a scaffold for other transcription-related proteins.

SRC-3 is strongly suggested to play an oncogenic role in cancer: the SRC-3 gene is amplified in ~10% of human breast cancers, and SRC-3 mRNA and protein expression are elevated in a majority of tumors. *In vivo* mouse studies also show that overexpression of SRC-3 alone is sufficient to initiate tumorigenesis. At the subcellular level, the majority of SRC-3 is localized within the nucleus but translocates rapidly between the nucleus and the cytoplasm (Amazit *et al.*, 2003). At the molecular level, SRC-3 contains several serine/threonine phosphorylation sites. Standard reporter assays show that phosphorylation of SRC-3 regulates its transcription activity (Wu *et al.*, 2005). The ability of HTM and single-cell analysis to generate a more homogeneous and physiologically appropriate cell population for the study of subcellular phenomena is a great advantage over traditional reporter and biochemical assays. HTM will allow for the study of SRC-3 phosphorylation in regulating its subcellular distribution, cyto/nuclear shuttling, and interaction with transcription proteins at visible promoter arrays. To analyze SRC-3 by single-cell, microscopy techniques rapidly, the protein and its phosphorylation mutant variants are fused with green fluorescent protein and transiently expressed in tissue culture cell lines. In experiments with transiently expressed SRC-3 and, in general, all exogenously expressed proteins, we find that it is imperative to select cells that have near-endogenous levels of expression in order to evaluate natural functions. This section describes how to use HTM to properly select cells for analysis based on cell viability and expression of exogenous SRC-3.

Protein Expression Level Assay

Experimental Setup

Cell Culture, Treatment, and Labeling. Cell culture and transfection of HeLa cells are performed as described in earlier. Cells may be grown on either coverslips or glass-bottom well plates. Particular to SRC-3, optimal expression time posttransfection for pEGFP-SRC-3 expression vector has been determined to be 48 h and the cell density at the time of fixation is 50 to 70%. In addition, HeLa cells are mock transfected in parallel to pEGFP-SRC-3 transfection. After fixation both mock transfected and transfected cells are immunolabeled for SRC-3 as described earlier. The secondary antibody is conjugated to Alexa 555 (Molecular Probes).

Imaging, Data Extraction and Filtering, and Data Analysis. Both pEGFP-SRC-3-transfected and mock-transfected cells are imaged using the 40×/0.90 NA high-resolution objective. The 40× objective is chosen, in part, based on the balance between field size (number of cells imaged) and resolution. The high resolution will be required in subsequent studies to allow observation of subnuclear patterns such as the nuclear speckling associated with transcriptionally active SRC-3, as well as the androgen and estrogen receptors. The mock-transfected cells (containing only endogenous SRC-3) are imaged for DAPI and red immunofluorescence. The pEGFP-SRC-3-transfected cells are imaged for DAPI, red immunofluorescence, and GFP, maintaining the same exposure time for the red channel.

Cells are identified and extracted by generating nuclear area masks using a nonlinear least-squares optimized image filter to create marked object–background contrast, followed by automatic histogram-based thresholding. Estimating and subtracting the mean background image intensity, determined dynamically on a per-image basis, correct effects due to background fluorescence. The correlated channel mask is computed as an intersection between the threshold correlated channel image (the threshold level is computed dynamically using a proprietary background level estimation method, Beckman Coulter), a Voroni tessellation polygon, and a circle of user-defined radius from the nuclear centroid. The resulting imaged cell population is biologically heterogeneous (Fig. 1A).

Threshold filters standard to Cytoshop are applied to increase the homogeneity of the imaged cell population by removing aberrant cells. Three filters are applied simultaneously to the total imaged cell population:

1. DNA content outlier: removes apoptotic and mitotic cells based on DAPI intensity (Fig. 1B)
2. DNA cluster: removes cell clusters that could not be resolved as individual cells (Fig. 1B).
3. Cells acquired with two channels: removes cells that do not contain data for at least two user-defined channels (Fig. 1B). Filter application to the mock-transfected and GFP-SRC-3 expressing imaged cell populations produces morphologically homogeneous and viable interphase cell subpopulations (Fig. 1B).

Examination of the mean nuclear fluorescence of endogenous (red channel) SRC-3 in mock-transfected cells indicates a broad range of expression (Fig. 2A). Similar examination (red channel) of GFP-SRC-3 expressing cells indicates a higher maximal range of expression as expected. Also, overexpression of SRC-3 causes the formation of nonphysiological globules (or aggregates, foci, etc.) of the SRC-3 protein (Fig. 2B). To minimize total SRC-3 overexpression artifacts, the maximum expression in GFP-SRC-3 expressing cells is conservatively set to three times the

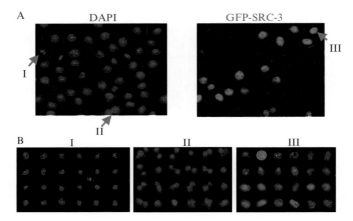

FIG. 1. Morphological filtering of total imaged cell population. Two channels are imaged in a single plane using a 40×/0.90 NA objective: channel 0 (DAPI) and channel 1 (pEGFP-SRC-3). (A) A representative field of Hela cells transiently expressing GFP-SRC-3 is presented. Contrast measurements identify DAPI-stained nuclei and Voroni tessellation polygons (red boundaries) define the cell boundaries. Present within this field are (I) apoptotic or mitotic cells, (II) unresolved nuclei clusters, and (III) interphase cells differentially expressing GFP-SRC-3. (B) Galleries of I, II, and III cells are presented. For gallery I, a DNA content algorithm is applied to identify the apoptotic and mitotic cells. For gallery II, a DNA cluster algorithm is applied to identify the clusters of nonresolved nuclei. Keep cells with two channels filter are combined with the previous filters and applied to the total cell population to generate a more homogeneous population (gallery III) of interphase cells differentially expressing GFP-SRC-3.

maximum expression found in mock-transfected cells. The corresponding fluorescence value in the green channel is identified and used as the maximum fluorescence value in a custom filter for GFP-SRC-3 expressing cells not immunolabeled and used for subsequent cell filtering (Fig. 2C). All cells within this range do not exhibit SRC-3 globule formation; importantly, antibody labeling of SRC-3 (and SRC-1 and SRC-2, data not shown) also fail to show large intranuclear foci. It is important to note that the expression range values are not absolute and differ between experiments due to biological, experimental, and optical variability.

Androgen Receptor

This section discusses the HTM procedure for studying translocation and nuclear distribution of one specific NR, the androgen receptor. The main function of androgenic signaling is male sex differentiation, pubertal

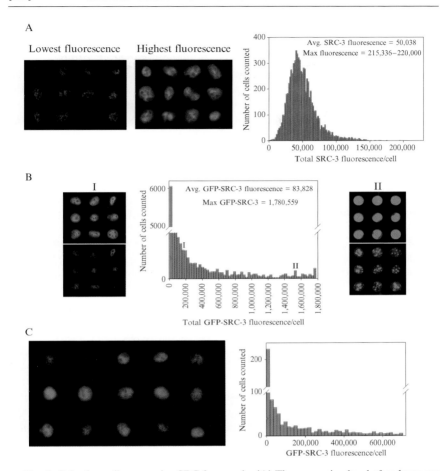

FIG. 2. Selecting cells expressing SRC-3 properly. (A) The expression level of endogenous SRC-3 in HeLa cells is examined by immunofluorescence. Two channels are imaged in a single plane using a 40×/0.90 NA objective: channel 0 (DAPI) and channel 1 (SRC-3) and the resulting cell population morphologically filtered as in Fig. 1. Two galleries featuring low- and high-expressing cells are presented. A histogram is presented that demonstrates the broad range of SRC-3 immunofluorescence with a maximum fluorescence of ~220,000. (B) HeLa cells transiently expressing GFP-SRC-3 and examined for total SRC-3 (immunofluorescence) and exogenous SRC-3 expression are presented. Gallery I demonstrates the lower range of SRC-3 expression. Gallery II demonstrates the upper range of SRC-3 expression. Note the appearance of GFP-SRC-3 globules (or foci, aggregates, etc.) in mid to high SRC-3 expressing cells. To remove cells overexpressing GFP-SRC-3, the upper limit of total expression is set to three times the maximum expression found in mock-transfected cells. The corresponding maximum GFP-SRC-3 fluorescence is identified and used as a custom filter for cells transiently expressing GFP-SRC-3 and are not immunolabeled. (C) A GFP-SRC-3 expression histogram after morphological and expression filtering and a representative image gallery of HeLa cells transiently expressing GFP-SRC-3 are presented. Note the absence of subnuclear GFP-SRC-3 globules.

changes, and maintenance of the male phenotype. Testosterone mediates these functions through its receptor, AR. As noted earlier, AR resides mainly in the cytoplasm in the absence of ligand associated with heat shock proteins (HSPs). Upon agonist binding to AR, HSPs are shed, homodimerization occurs, and the receptor is phosphorylated and translocates into the nucleus. Once inside the nucleus, AR becomes organized into discrete, but unstable foci based on live cell photobleaching experiments. Activated AR then binds androgen response elements and regulates transcription through the recruitment to the promoter of transcriptional coregulators and basal transcriptional machinery (Gao *et al.*, 2005). Interestingly, AR antagonists also cause nuclear translocation but without the accompanying subnuclear hyperspeckled distribution and transcriptional activity, indicating that translocation and hyperspeckling events are separable from transcription. Disruption of AR regulation has a major role in human disease (Gao *et al.*, 2005; Marcelli *et al.*, 2006). Mutations in AR are known to cause androgen insensitivity syndrome (AIS), which can result in a genetic male appearing phenotypically female. AR regulation of transcription also plays an important role in prostate cancer, which kills ~30,000 American men each year, second only to lung cancer. Although not causative, prostate cancer growth depends on improper regulation of AR signaling. The aberrant regulation is thought to occur by AR overexpression, AR mutations, and/or changes in AR phosphorylation (Eder *et al.*, 2001; Edwards and Bartlett, 2005). The high-throughput microscopy approach allows for the rapid determination of the subcellular distribution of the hundreds of AR mutants found in association with either prostate cancer or AIS in response to large libraries of chemical compounds. This has the potential to further define mutant molecular phenotypes and identify new compounds to regulate AR activity.

Protein Translocation and Nuclear Variance Assays

Experimental Setup

Cell Culture, Treatment, and Labeling. HeLa cells for AR translocation experiments are cultured, treated, and processed as described earlier with slight modifications. HeLa cells are subcultured into either 35- or 60-mm culture dishes depending on the experiment. After 24 h of growth in stripped-FBS containing media (DMEM), cells are transfected using optimized amounts of pEGFP-AR expression vector or pEGFP-AR mutant vectors and carrier DNA (pBluescript). The cells incubate with the lipid/DNA mixture for 6 h after which the cells are subcultured again onto the optical bottom multiwell plate at a density of 2×10^4 cells/cm^2. Higher cell densities are avoided due to the fact that it is prohibitive to accurate cytoplasmic measurements. Twenty-four hours after transfection, ligand treatments are initiated; ligands are serially diluted in stripped-dialyzed FBS containing media to generate a dose response. The length of time cells are exposed to ligand can vary depending on experimental questions, although we have found that a 2-h incubation is ideal for cytological studies of AR localization. At 2 h of 10 nM ligands, nuclear translocation is generally complete. Ligand treatment is stopped by washing cells with ice-cold PBS and 4% formaldehyde fixation. Fixation protocol is as described earlier.

Imaging, Data Extraction, and Data Analysis. For the DAPI channel, background fluorescence subtraction is applied, and a nucleus size range of 30 to 300 μm^2 is imposed. For the AR channel, background fluorescence subtraction is automatic, no size limit is imposed, and the extraction radius is set to ~20% greater than the average nuclear radius. In effect, only a ring of cytoplasm around the nucleus is examined. This abolishes the variation due to cells too closely spaced and errors in calculating the true cytoplasmic space for each cell.

Threshold gates similar to those described for SRC-3 are applied to each of the data sets to first extract a subpopulation of single cell objects with a biologically relevant level expression of AR. The expression level range is determined by comparative studies between transiently transfected HeLa cells and LNCaP cells, a prostate cancer-derived cell line that expresses AR. This generates the final, homogeneous population. After segmentation, a number of features are measured for each object; here we focus on two: cellular distribution and subnuclear patterns, such as the nuclear speckling associated with transcriptionally active AR (Fig. 3A).

The degree of nuclear translocation is represented by the fraction localized in nucleus (FLIN) measurement (Morelock *et al.*, 2005); the progressive reorganization of GFP-AR into a speckled pattern is captured by the nuclear variation (NVAR) measurement, which is the statistical variation in pixel brightness of the channel of interest in the nucleus. Data analysis is achieved by exporting the database files pertaining to the final gated populations to common analysis programs such as Excel and Sigma-Plot/SigmaStat. Complete analysis of HTM results is beyond the scope of this chapter. Average FLIN and standard deviation among a cell population of more than thousands of cells per condition were calculated from the Excel sheet associated with extraction data.

HTM analysis is able to define distinct responses for WT AR and AR C619Y, a DNA-binding domain mutant, to various ligands. Response curves correlating the efficiency of nuclear translocation (FLIN) of

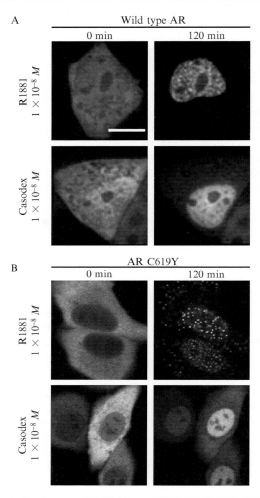

FIG. 3. High-resolution images of GFP-AR and GFP-AR C619Y. HeLa cells transiently transfected with GFP-AR (A) or GFP-AR C619Y (B) were incubated with either the synthetic AR agonist R1881 (1×10^{-8} M) or the AR antagonist Casodex (1×10^{-6} M). Cells are imaged both at 0 min (immediately before ligand addition) and at 120 min ligand treatment. R1881 induces nuclear translocation and a nuclear hyperspeckled pattern with wild-type AR. For the mutant AR C619Y, R1881 also induces nuclear translocation, but a much more dramatic focal pattern within the nucleus. For both forms of AR, Casodex is able to induce a degree of nuclear translocation, but with no hyperspeckling or foci formation.

AR-WT after administration of ligand demonstrated that nuclear translocation occurs more efficiently in response to R1881 than hydroxyflutamide (OHF), Casodex (Cas), and 17β-estradiol (E2; Figs. 3A and 4A). NVAR analysis reveals that R1881 induces the maximal degree of hyperspeckled intranuclear foci (Fig. 3A), with moderate reorganization from high levels of OHF and E2 (Fig. 4B). Importantly, AR antagonists OHF and Cas are unable to induce the same degree of nuclear speckling associated with R1881 (and are transcriptionally inactive). Highlighting the comparison between WT and mutant ARs, AR C619Y demonstrates similar ligand responses with the greatest difference being a significantly higher NVAR response to R1881, confirming the tendency of this mutant to form larger "aggregates" within the nucleus (Figs. 3B and 4) (Marcelli *et al.*, 2006; Nazareth *et al.*, 1999).

Estrogen Receptor

Estrogens are major promoters of cell proliferation in both normal and neoplastic breast epithelium. ERα and ERβ mediate the primary actions of estrogens at target sites around the body. These two receptors are encoded by different genes and exhibit tissue and cell-type specific expression. Augmentation of cell proliferation and an increased risk of uncontrolled cell growth and cancer is one outcome of excessive stimulation of the ERα pathway due to increased estrogen or ERα levels (Singh and Kumar, 2005). Approximately 70% of breast cancer patients are positive for elevated ERα expression at diagnosis. These patients are suitable candidates for hormone therapy that aims to block estrogen stimulation of breast cancer cells. Several antiestrogens have been developed and used as anticancer therapeutic agents and these include 4-hydroxytamoxifen (4HT) and ICI 182,780 (ICI) (Osborne *et al.*, 2000). Major goals of translational science in this arena include defining mechanisms of ligand action in regulating transcription and identifying new and tissue-specific ligands for disease treatment.

The molecular mechanisms of action for estrogens and antiestrogens at the single cell level are still being determined (Hinojos *et al.*, 2005). Single cell studies of ERα show that its subcellular distribution is predominantly nuclear, unlike AR, thus minimizing the direct usefulness prohibiting the application of cytoplasmic/nuclear fractional localization algorithms. In the absence of ligand, ERα is distributed in a diffuse nuclear pattern. The addition of ligand causes a redistribution of ERα into discrete foci. However, the nuclear redistribution of ERα does not always correlate well with transcription; for example, E2 and tamoxifen both induce subnuclear redistribution; however, in HeLa, the latter does not induce transcription. Therefore, we

A

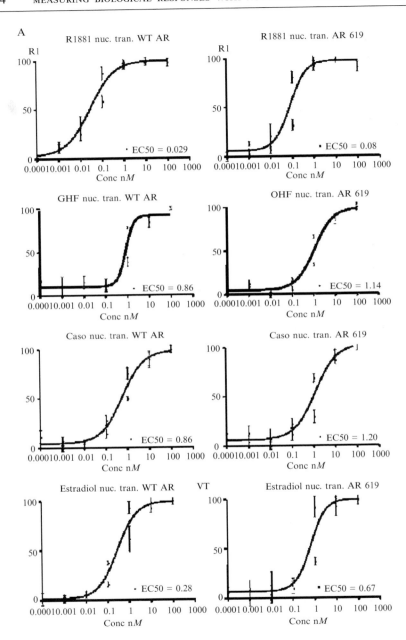

B

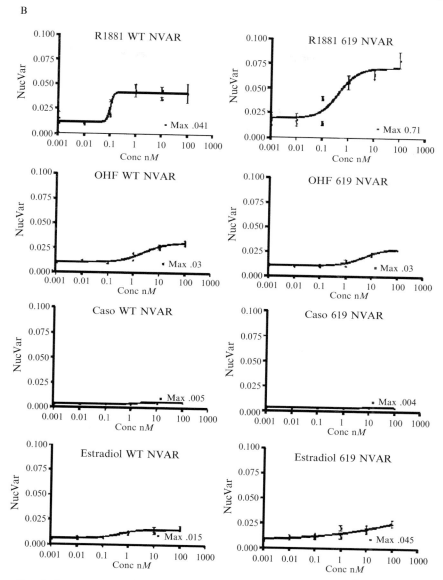

FIG. 4. Quantifying subcellular localization and patterning of AR using high-resolution HTM. HeLa cells transiently transfected with GFP-AR and GFP-AR C619Y were treated in an 11-point dose–response manner using an agonist (R1881), two antagonists (o-hydroxy-flutamide, OHF; Casodex, Caso), and a non-AR steroid (estradiol). Experiments were done in a 96-well format and in triplicate. Nuclear translocation is quantified by using the fraction localized in nucleus (FLIN) measurement (A). Degree of nuclear speckling or foci formation is quantified using the nuclear variation (NVAR) measurement (B).

have developed a model system (PRL-HeLa cell line) that is conducive to HTM analysis in order to study ERα transcriptional properties at the single cell level. We discuss the model system and the HTM procedure for studying the chromatin remodeling/transcription function of ERα.

PRL-HeLa is a cell line specifically engineered for the single cell study of ER function (Sharp *et al.*, 2006). PRL-HeLa cells contain multiple genomic integrations of a replicated prolactin (PRL) promoter/enhancer. The multiple integrations (PRL array) are spatially confined and are visualized by the accumulation of fluorescently tagged ERα, coregulators and modified histones, and DNA FISH. The promoter/enhancers within the PRL array controls the expression of a fluorescent reporter protein targeted to cytoplasmic peroxisomes (dsRED2-Serine-Lysine-Leucine [SKL]). Addition of ERα agonist (17β-estradiol, E2) causes a rapid (within minutes) visible decondensation of the array relative to no ligand; further, antiestrogens ICI or 4HT induce a marked condensation. Importantly, the chromatin condensation/decondensation is a direct reflection of ERα-dependent coregulator recruitment and histone modification, integral features of the generally accepted model of NR regulation of transcription. Reflecting the physiological significance of the cell line and in further accord with the model of NR function, large arrays (E2-treated cells) have more (>3-fold) dsRED2-SKL transcripts at the array relative to no ligand, whereas small arrays (ICI- or 4HT-treated cells) are transcriptionally repressed (>10 fold reduction) relative to no ligand. It follows that array size in this cell line expressing ERα is an indicator of receptor transcription functionality in response to estrogenic or antiestrogenic ligand. In this manner, PRL-HeLa affords the ability to measure multiple aspects of ERα transcriptional function by HTM that is impossible in conventional cell lines.

Foci Identification and Chromatin Remodeling Assay

Experimental Setup

Cell Culture, Treatment, and Labeling. PRL-HeLa cells are cultured, treated, and processed as described earlier. Specifically, the cells are seeded onto coverslips at a density of 1–1.25 × 10^5 cells/cm^2. After 48 h in hormone-free media, cells are transiently transfected with pEGFP-ERα expression vector and carrier DNA (pBluescript) using Transfectin (Bio-Rad). The carrier to test vector DNA amount ratio is 1:1 and the lipid to DNA ratio is 2:1. Cell incubation with the DNA and transfection reagent typically occurs overnight in hormone and phenol red-free media (DMEM). Replacing the media with prewarmed ligand containing media ends the transfection. Ligand incubation occurs for 2 h and is ended by rinsing with ice-cold PBS and fixation (see General Methods).

Imaging, Data Extraction and Filtering, and Data Analysis. Imaging of PRL-HeLa transiently expressing pEGFP-ERα is performed using the high numerical aperture $40\times$, 0.9 NA objective. Exposure time for pEGFP-ERα is $\sim$2 s. A single Z section is imaged for DAPI and GFP. To maximize the number of arrays imaged, the Z section for pEGFP-ERα is offset 1 μm from the DAPI focal plane. Depending on transfection efficiency, up to $\sim$100 fields per coverslip are imaged to obtain an unfiltered cell population of at least 5000 cells. The aggregate algorithm is automatically applied to the unfiltered cell population to identify and quantify the pEGFP–ERα foci. The aggregate identification parameters are 350 (maximum foci area in pixels), 30 (object scale), and 5 (minimum peak height). These parameters were determined to be optimal for identifying arrays in cells treated with either estrogenic (large arrays) or antiestrogenic (small arrays) ligands and for excluding nonarray identified foci. To further enhance the specificity and precision of the analysis a homogeneous cell population must be generated and is visually inspected manually in subpopulation galleries for quality control.

The unfiltered cell population has a high degree of heterogeneity (GFP-ERα expression level and identified foci, cell clusters, cell cycle, etc.) (Fig. 5A). To increase the cell population homogeneity, standard morphological and custom GFP-ER expression filters are applied as described earlier. An additional filter (number of aggregates identified per cell <10) is applied in order to remove cells with a high number of nonarray foci identified by the aggregate algorithm. The resultant filtered cell subpopulation predominantly contains in-focus and correctly identified arrays (Fig. 5B) for each treatment condition ($n > 100$). In this manner, cell filtering of a morphologically diverse cell population successfully achieved generation of a homogeneous cell population suitable for analysis of PRL array size. Quantitative cell morphology data and array measurements are contained in a data base file format.

Results of the ER chromatin-remodeling assay indicate a basal array size (no ligand control) in pixels of 70.8 ± 5.3 (Fig. 5C). Treatment with E2 causes a spatial expansion of the array (112.9 ± 3.8). Treatment with antiestrogens and breast cancer drugs 4HT or ICI causes array contraction (31.0 ± 2.5 and 29.1 ± 1.7, respectively). Array size for each treatment is significantly different than control ($p < 5 \times 10^{-6}$). Array size and HTM

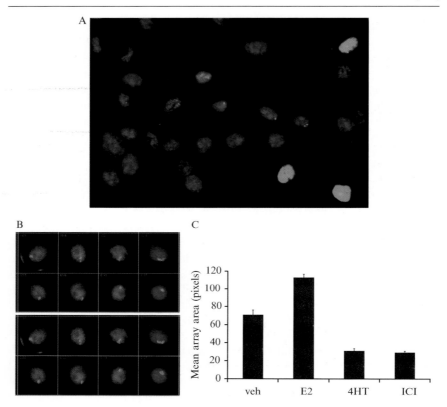

FIG. 5. Quantitative PRL array analyses. PRL-HeLa cells transiently expressing GFP-ER were treated (veh, E2, 4HT, or ICI at 10 nM for each ligand) for 2 h, fixed, and DAPI stained. Two channels are imaged: channel 0 (DAPI) and channel 1 (GFP-ER). (A) A representative field of E2-treated cell images is presented. Mitotic and apoptotic cells, nuclei clusters, out-of-focus cells, and cells overexpressing GFP-ER are filtered out to generate the final cell subpopulation to be analyzed as described in the text (B, top row). An aggregate identification algorithm is used to identify and quantify the GFP-ER-targeted PRL array. (B, bottom row) Results of algorithm application to identify and mask (red) the PRL array. (C) Results are presented as the mean array area in pixels (n >100 for each treatment) and indicate an ER ligand-dependent regulation of chromatin remodeling.

analysis are also used to rapidly generate EC_{50} and IC_{50} values for various ERα ligands (data not shown). PRL array size is a reliable indicator of ERα ligand-dependent regulation of transcription and is quantified rapidly by HTM. Thus, PRL-HeLa and HTM have successfully demonstrated the capability to quantitatively measure distinct aspects of ERα transcription function and can now be scaled to a 96-well plate format for true high-resolution HTM screening of ER ligand effects on chromatin remodeling. To this end, we have developed a viral-based system for high-efficiency introduction of GFP-ER and a more homogeneous subcloned PRL-HeLa line, which will be highly amenable to microtiter plate analyses.

Conclusions

Realization of high-resolution HTM as a rapid analytical method for the multiplex study of protein function is fast approaching. Driving this development is the generation of model systems such as PRL-HeLa, which allows the simultaneous study of numerous aspects of ER regulation of transcription, including nuclear translocation, DNA array targeting, chromatin modeling, coregulator recruitment, and transcription readout. While strides have been made in reducing the biological heterogeneity of cell populations to be analyzed, efficient pattern recognition algorithms still require development to overcome subcellular heterogeneity and also to take additional advantage of assays that incorporate multichannel labeling. Maturation of the field will allow for the shift of HTM research from primarily reflecting feasibility studies to true large-scale screening studies. For nuclear receptors, these studies will include mechanistic approaches such as chemical and siRNA library screening interference to define critical effectors and molecular players, respectively, in the complex signaling pathway of ligand- or nonligand-based NR activation and transcription regulation.

References

Amazit, L., Alj, Y., Tyagi, R. K., Chauchereau, A., Loosfelt, H., Pichon, C., Pantel, J., Foulon-Guinchard, E., Leclerc, P., Milgrom, E., and Guiochan-Mantel, A. (2003). Subcellular localization and mechanism of nucleocytoplasmic trafficking of steroid receptor coactivator-1. *J. Biol. Chem.* **278**, 32195–32203.

Ding, G. J., Fischer, P. A., Boltz, R. C., Schmidt, J. A., Colaianne, J. J., Gough, A., Rubin, R. A., and Miller, D. K. (1998). Development and use of a high capacity fluorescence cytometric system. *J. Biol. Chem.* **273**, 28897–28905.

Eder, I. E., Culig, Z., Putz, T., Nessler-Menardi, C., Bartsch, G., and Klocker, H. (2001). Molecular biology of the androgen receptor: From molecular understanding to the clinic. *Eur. Urol.* **40**, 241–251.

Edwards, J., and Bartlett, J. M. (2005). The androgen receptor and signal-transduction pathways in hormone-refractory prostate cancer. 1. Modifications to the androgen receptor. *BJU Int.* **95**, 1320–1326.

Evans, R. M. (1988). The steroid and thyroid hormone receptor superfamily. *Science* **240**, 889–895.

Gao, W., Bohl, C. E., and Dalton, J. T. (2005). Chemistry and structural biology of androgen receptor. *Chem. Rev.* **105**, 3352–3370.

Hinojos, C. A., Sharp, Z. D., and Mancini, M. A. (2005). Molecular dynamics and nuclear receptor function. *Trends Endocrinol. Metab.* **16**, 12–18.

Kumar, R., and Thompson, E. B. (1999). The structure of the nuclear hormone receptors. *Steroids* **64,** 310–319.

Lim, C. S., Baumann, C. T., Htun, H., Xian, W., Irie, M., Smith, C. L., and Hager, G. L. (1999). Differential localization and activity of the A- and B-forms of the human progesterone receptor using green fluorescent protein chimeras. *Mol. Endocrinol.* **13,** 366–375.

Marcelli, M., Stenoien, D. L., Szafran, A. T., Simeoni, S., Agoulnik, I. U., Weigel, N. L., Moran, T., Mikic, I., Price, J. H., and Mancini, M. A. (2006). Quantifying effects of ligands on androgen receptor nuclear translocation, intranuclear dynamics, and solubility. *J. Cell. Biochem.*

McKenna, N. J., Lanz, R. B., and O'Malley, B. W. (1999). Nuclear receptor coregulators: Cellular and molecular biology. *Endocr. Rev.* **20,** 321–344.

Morelock, M. M., Hunter, E. A., Moran, T. J., Heynen, S., Laris, C., Thieleking, M., Akong, M., Mikic, I., Callaway, S., DeLeon, R. P., Goodacre, A., Zacharias, D., and Price, J. H. (2005). Statistics of assay validation in high throughput cell imaging of nuclear factor kappaB nuclear translocation. *Assay Drug Dev. Technol.* **3,** 483–499.

Nazareth, L. V., Stenoien, D. L., Bingman, W. E., 3rd, James, A. J., Wu, C., Zhang, Y., Edwards, D. P., Mancini, M., Marcelli, M., Lamb, D. J., and Weigel, N. L. (1999). A C619Y mutation in the human androgen receptor causes inactivation and mislocalization of the receptor with concomitant sequestration of SRC-1 (steroid receptor coactivator 1). *Mol. Endocrinol.* **13,** 2065–2075.

Osborne, C. K., Zhao, H., and Fuqua, S. A. (2000). Selective estrogen receptor modulators: Structure, function, and clinical use. *J. Clin. Oncol.* **18,** 3172–3186.

Perlman, Z. E., Slack, M. D., Feng, Y., Mitchison, T. J., Wu, L. F., and Altschuler, S. J. (2004). Multidimensional drug profiling by automated microscopy. *Science* **306,** 1194–1198.

Pratt, W.B, Galigniana, M. D., Morishima, Y., and Murphy, P. J. (2004). Role of molecular chaperones in steroid receptor action. *Essays Biochem.* **40,** 41–58.

Press, M. F., Xu, S. H., Wang, J. D., and Greene, G. L. (1989). Subcellular distribution of estrogen receptor and progesterone receptor with and without specific ligand. *Am. J. Pathol.* **135,** 857–864.

Sharp, Z. D., Mancini, M. G., Hinojos, C., Dai, F., Berno, V., Szafran, A., Smith, K. P., Lele, T., Ingber, D., and Mancini, M. A. (2006). Estrogen receptor-α exchange and chromatin dynamics are ligand- and domain-dependent. *J. Cell. Sci.* In press.

Singh, R. R., and Kumar, R. (2005). Steroid hormone receptor signaling in tumorigenesis. *J. Cell. Biochem.* **96,** 490–505.

Wu, R. C., Smith, C. L., and O'Malley, B. W. (2005). Transcriptional regulation by steroid receptor coactivator phosphorylation. *Endocr. Rev.* **26,** 393–399.

[12]   Tracking Individual Proteins in Living Cells Using Single Quantum Dot Imaging

By Sébastien Courty, Cédric Bouzigues, Camilla Luccardini, Marie-Virginie Ehrensperger, Stéphane Bonneau, and Maxime Dahan

Abstract

Single quantum dot imaging is a powerful approach to probe the complex dynamics of individual biomolecules in living systems. Due to their remarkable photophysical properties and relatively small size, quantum dots can be used as ultrasensitive detection probes. They make possible the study of biological processes, both in the membrane or in the cytoplasm, at a truly molecular scale and with high spatial and temporal resolutions. This chapter presents methods used for tracking single biomolecules coupled to quantum dots in living cells from labeling procedures to the analysis of the quantum dot motion.

Introduction

Developments in cell biology are increasingly driven by the emergence of new imaging techniques. Advances in fluorescence microscopy, which allow biological processes to be imaged in real time, have had a particularly significant impact. Progress in cell imaging has mainly stemmed from the advent of genetically encoded markers (Lippincott-Schwartz and Patterson, 2003; Tsien, 2005), which have unquestionably revolutionized the way proteins can be studied in live cells. Thanks to the green fluorescent protein (GFP) and its many variants, important mechanisms such as molecular diffusion, gene expression, protein–protein interactions, or cell signaling can now be detected optically (Lippincott-Schwartz and Patterson, 2003; Shav-Tal *et al.*, 2004; Tsien, 2005). The need to analyze them quantitatively has spurred a large effort involving techniques from cell and molecular biology, biochemistry, physics, and computer science.

The imaging techniques based on fluorescent proteins, however, suffer from some limitations. First, the photodegradation of the probes limits the observation time. Second, the optical sensitivity is usually limited to the simultaneous detection of several molecules and does not truly reach the level of single molecules. New physical techniques have pushed the boundaries of detection by enabling the observation of individual bio-molecules *in vitro* (Weiss, 1999) and *in vivo* (Harms *et al.*, 2001; Iino *et al.*,

METHODS IN ENZYMOLOGY, VOL. 414   0076-6879/06 $35.00
              DOI: 10.1016/S0076-6879(06)14012-4

2001; Schutz *et al.*, 2000; Seisenberg *et al.*, 2001; Ueda *et al.*, 2001). These experiments have already shown that SM measurements provide valuable information that remains hidden in conventional measurements.

The introduction of semiconductor quantum dots as biological markers represents a new step for single molecule detection in live cells (Jaiswal and Simon, 2004; Medintz *et al.*, 2005; Michalet *et al.*, 2005). Due to their remarkable brightness and superior photostability (compared to conventional dyes), quantum dots can be used as ultrasensitive fluorescent probes allowing long-term acquisition with a good signal-to-noise ratio (SNR). They render possible the investigation of cellular processes with high spatial and temporal resolutions at a truly molecular scale. The ultrasensitive imaging methods using quantum dots have already shown their potential for the study of membrane proteins in living cells (Dahan *et al.*, 2003) and open up exciting prospects for the study of many complex biological processes.

Semiconductor Quantum Dots as Biological Probes

By combining a relatively small size and unique photophysical properties (Fig. 1), quantum dots (QDs) present a high interest as detection probes for biological imaging (Bruchez *et al.*, 1998; Chan and Nie, 1998).

- QDs are small inorganic nanoparticles with a diameter comprised between 2 and 10 nm and composed of different semiconductor materials. Following their synthesis in organic solvent, they can be

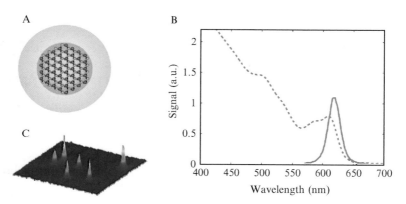

Fig. 1. (A) Structure of a functionalized quantum dot. The inorganic nanoparticle is covered by a layer of organic material used to solubilize and functionalize the QD, (B) absorption (dotted line) and emission spectra (plain line) of a QD sample, and (C) fluorescence image of single quantum dots deposited on a glass coverslip.

chemically modified for water solubilization and further functionalization (e.g., with Fab fragments or streptavidin).

- They possess both a large absorption and a narrow emission spectrum whose peak position is determined by the size of the semiconductor core and can be adjusted precisely during their synthesis by inorganic chemistry. As a consequence, QD samples with distinguishable emission wavelength can be excited with a single laser line, which make them ideal for multicolor detection (Wu *et al.*, 2003).
- Their photostability, which is far superior to that of conventional fluorophores (e.g., fluorescent proteins or organic dyes), overcomes the traditional limitation of photodegradation in fluorescent labeling.
- The synthesis and engineering of nanoparticles with different semiconductor materials and structures have expanded the range of possible emission wavelengths from the visible, ideally suited for immunofluorescence, to the red and infrared regions, more appropriate for imaging in tissues and animals (Kim *et al.*, 2004).

The potential of QDs as biological probes has now been demonstrated in many cases, and their applications, *in vitro* either in live cells or in organisms, are multiplying rapidly (Michalet *et al.*, 2005; Medintz *et al.*, 2005).

Single Molecule Imaging in Living Cells

A large choice of probes, such as fluorophores, quantum dots or beads, is now available to perform single molecule experiments in live cells. Selecting the right one depends on a number of physical and biochemical properties that should be evaluated for each system. Important factors are (i) the photostability, (ii) the extinction coefficient, (iii) the size, and (iv) the stoichiometry with which the detection probe can be bioconjugated to a biomolecule of interest (ideally, it should be achieved in a 1:1 ratio). A figure of merit, based on the value of these parameters for the different probes, is summarized in Table I.

For single molecule tracking, quantum dots are now probably the most favorable probe as they combine a relatively small size (radius ~10 nm, including the surface organic groups) with a remarkable brightness and a superior photostability, allowing long-term acquisition with a good signal-to-noise ratio. The high SNR primarily comes from the extinction coefficient ε of the QDs, which is on the order of $10^6\ M^{-1}\ 1\ cm^{-1}$ in the visible, 10 to 100 times higher than that of standard organic fluorophores (dyes or GFP). This translates directly into the number of photons N detected for a single emitting QD in a given acquisition time τ:

TABLE I

FIGURE OF MERIT FOR DIFFERENT PROBES USED IN SINGLE MOLECULE TRACKING EXPERIMENTS

Probe	Size	Photostability	Extinction coefficient	Stoichiometry
Organic fluorophores	+++	+	++	++
(Cy3, Cy5, Alexi, etc.)	1 nm	~1–10 s	~10^5	
GFP	++	−	+	+++
	2–4 nm	~ 100 ms	~10^4–10^5	
Quantum dots	+	+++	+++	+
	~10–20 nm	>20 mn	~10^6	
Beads	−	∞	Scattered light	−
	40 nm – 1 μm			

Photostability corresponds to the typical time during which a probe can be imaged continuously in a single molecule experiment. The extinction coefficient is expressed in $M^{-1}cm^{-1}$. "Stoichiometry" is an indication of how easily a probe can be conjugated to a single biomolecule with a ratio of 1:1.

$$N = \alpha q \left(\frac{2303\varepsilon}{N_A}\right)\left(\frac{I\lambda}{hc}\right)\tau$$

where α is the detection efficiency (~0.1–5%), q is the QD emission quantum yield (ca. 30–80%), λ is the QD emission wavelength, h is the Planck constant ($h = 6.63\ 10(-34)$ J.s), c is the speed of light, I is the excitation power per surface unit, and N_A is the Avogadro number. For the same input intensity I, one detects 10 to 100 times more fluorescence photons with a QD than with a fluorophore. With QDs, high values of N usually imply that the photon noise (also known as shot noise) is the primary source of noise in the detection system. It dominates other contributions, such as the background photon noise $\sigma = \sqrt{N}$ (due to scattered light and cellular autofluorescence), and the electronic noise (due to readout and dark current). In this context, the SNR:

$$SNR = \frac{N}{\sqrt{\sigma^2 + \sigma_b^2 + \sigma_e^2}} \approx \frac{N}{\sigma}$$
$$\approx \sqrt{N}$$

Using seminconductor QDs as fluorescent probes, individual protein receptors have been tracked in the membrane of live neurons for durations longer than 20 min (Dahan *et al.*, 2003). This ultrasensitive imaging method has been extended to study the dynamics of single protein motors in living cells (Courty *et al.*, 2006). This chapter describes methods for tracking single biomolecules coupled to quantum dots in living cells, including labeling procedures, optical microscopy setup and analysis of the QD motions.

Single Quantum Dot Tracking of Membrane Proteins

Understanding the structure and the dynamic organization of the plasma membrane is a major challenge in cell biology. To address these issues, single particle tracking is often considered a tool of choice, as it provides information with a spatial resolution on the order of 10 nm (see later), much higher than conventional confocal imaging; measuring the trajectories of single membrane molecules is therefore a way to directly probe the properties of the membrane with high spatial accuracy (Saxton and Jacobson, 1997).

In the fluid-mosaic model proposed by Singer and Nicolson (1972), the membrane is a homogeneous viscous fluid in which lipids and proteins undergo Brownian diffusion. Although this model has proven extremely fruitful in understanding some properties of the membrane, it is considered too simple. Nowadays, the membrane is often described as compartmentalized with obstacles (barriers, fences, pickets) hindering the free diffusion (Murase et al., 2004). In addition, there has been active debate about the structure, the dynamics, and the role of small lipid-rich membrane microdomains, known as lipid rafts.

Single molecule tracking experiments have also been used to understand the membrane organization on a larger length scale. Several experiments have focused on the dynamics of receptors for neurotransmitters, either excitatory (Borgdorff and Choquet, 2002; Tardin et al., 2003) or inhibitory (Dahan et al., 2003; Meier et al., 2001), as the lateral diffusion may play an important role in the regulation of the number of receptors at synapses. The first single QD tracking experiments were in fact performed on glycine receptors in live neurons (Dahan et al., 2003), and a key point in these experiments was the ability to observe receptors in the synaptic cleft, a domain previously inaccessible when using micrometer-sized latex beads as labels (Borgdorff and Choquet, 2002; Meier et al., 2001).

The following protocol is used to label individual membrane GABA receptors in cultured neurons. The concentrations of labeling reagents (primary and secondary antibodies, quantum dots) indicated correspond to conditions found appropriate in our experiments. For a different biological system, the immunolabeling conditions may vary depending on the antigens and the antibodies and should be adjusted accordingly. When testing for the right dilution, cells should be imaged in live conditions, as fixation might induce an additional autofluorescent background that would complicate the detection of individual QDs.

Solutions and Materials

Culture Reagents and Buffers

1. Minimum essential medium (MEM; Sigma-Aldrich Corp., St. Louis, MO)
2. Neurobasal medium supplemented with B27, glutamine (Invitrogen, Carlsbad, CA), and 25 μl antibiotics
3. L-15 medium (Invitrogen) supplemented (for a 30-ml final volume) with 15 μl antibiotics, 30 μl progesterone 10 μg/ml, and glucose to 3.6 μg/ml
4. B27 (Invitrogen) ready to use
5. Glutamine (Invitrogen) diluted in water to 200 mM
6. Poly-ornithine (Sigma-Aldrich Corp.)
7. Fetal veal serum (FVS) (Invitrogen)
8. Deoxyribonuclease (DNase) (Sigma-Aldrich Corp.)
9. Bovine serum albumin (Sigma-Aldrich Corp.)
10. Antibiotics: penicillin 10,000 UI and streptomycin 10,000 μg/ml (Invitrogen)
11. Glucose (Sigma-Aldrich Corp.)
12. Vectashield (Vector Laboratories, Burlingame, CA).
13. Phosphate-buffered saline (PBS) 10$\times$: 80.06 g NaCl, 2.02 g KCl, 2.05 g KH_2PO_4, 23.3 g Na_2HPO_4 in 1 liter of deionized water
14. Incubation buffer: A 10% orthoboric acid solution (10 g of orthoboric acid dissolved in a 100-ml water solution adjusted to pH 8 and filtered on 0.22-μm pores) and a 10% bovine serum albumin solution are added to water in respective volume proportions of 1/40 and 1/5. In order to obtain a final osmolarity around 260 mOsm, compatible with live cells, the incubation buffer has to be supplemented with 15% in volume of 1.43 M sucrose.
15. Paraformaldehyde 4%: 1 g of paraformaldehyde is dissolved in 22.5 ml of magnetically agitated water at 80°. If solution remains unclear after agitation, a droplet of 1 M NaOH is added. Solution is then completed with 2.5 ml of PBS10$\times$ and stored at $-20°$.

Antibodies and Quantum Dots

1. Primary antibody against γ2 subunit of $GABA_A$ receptor raised in rabbit (Euromedex, Mundolsheim, France) at 0.6 mg/ml stored at $-20°$ in deionized water
2. Fab fragment of biotinylated goat antirabbit (Fab-GaR-Biot) (Jackson Immunoresearch Laboratories, West Grove, PA) stored at $-20°$ in 50-50 glycerol-water buffer at 0.5 mg/ml
3. Streptavidin-coated quantum dots (2 μM) emitting at 605 nm (Quantum Dots Corp., Hayward, CA) stored at 4°

Methods

Cells and Medium Preparation

1. Coverslips are coated with polyornithine for one night and for a few hours with L15 medium supplemented with FVS (5% in volume) and NaHCO$_3$ 7% (2.5% in volume) before cell plating. Spinal cords of E14 rat embryo are dissected in PBS-glucose, incubated for 15 min at 37° in 1 ml of PBS supplemented with 20 μl of trypsine-EDTA 2.5%, and smoothly dissociated in DNase supplemented L15 medium. Spinal neurons are then plated on coverslips at concentrations between 5×10^4/ml and 2×10^5/ml in B27-neurobasal medium and stored in sterile boxes in an incubator at 37° under 5% of CO$_2$ until use for quantum dot staining.

2. In order to work with neurons in ambient air, it is necessary to use a specific medium, referenced here as MEM-Air. MEM powder is diluted in 800 ml of water, which is then completed with 6 g glucose, 4.5 ml of NaHCO$_3$ 7%, and 10 ml HEPES and sterilized by filtration on 0.22-μm pores. MEM-Air is completed immediately before experiments with 3 ml of deionized water 1 ml B27, 500 μl natrium pyruvate, and 500 μl glutamine for a 50-ml final volume.

Labeling for Single Molecule Experiments

1. Remove a coverslip from the culture boxes. Place it on a surface maintained at 37° and incubate with 500 μl of MEM-Air.
2. Aspirate the medium and add 100 μl of primary antibody (anti-γ2, 1:100) on the coverslip for 10 min at 37°.
3. Discard the primary antibody and wash the sample three (or more) times in 500 μl MEM-air.
4. Add 100 μl of secondary antibody (Fab-GaR-Bio, 1:200) for 10 min at 37°.
5. Discard the secondary antibodies and rinse the sample three times in 500 μl MEM-air.
6. Prepare the quantum dots at 1:2000 to 1:10,000 in the incubation buffer.
7. Incubate the coverslip for 1 min with QDs and then wash at least five times in MEM-air to remove free QD remaining in solution.

Single Quantum Dot Tracking of Intracellular Proteins

The high sensitivity achieved in single molecule imaging of membrane proteins suggests that QDs could be employed to accomplish a more challenging task: tracking the motion of individual intracellular proteins. QDs have already been used successfully to visualize biological processes in the cytoplasm of live cells but so far not at the single molecule level. For instance, QDs have been microinjected in Xenopus embryos (Dubertret *et al.*, 2002) in order to trace the cell lineage but the number of nanoparticles was superior to 10^9 per injected cell. Lidke *et al.* (2004) also used QD-tagged epidermal growth factor (EGF) to directly image the internalization of EGF receptors and the subsequent signaling pathways.

Reaching single molecule sensitivity in the imaging of intracellular biomolecules raises several difficulties and has been achieved in very few experiments. First, QDs have to enter the cell cytoplasm and reach their molecular target. Second, the fluorescence of individual QDs has to be detected in a noisy environment due to the autofluorescence of intracellular compartments and organelles. Finally, the motion in the cytosol is likely to be three dimensional compared to bidimensional diffusion in the membrane. It makes SM tracking more complex than in membrane experiments and will likely require new computational methods to extract and analyze trajectories from imaging data.

The following approach has been used successfully to image the motion of biotinylated kinesin motors coupled to QDs in the cytoplasm of living HeLa cells (Courty *et al.*, 2006). The protocol can be adapted easily to detect other intracellular biomolecules.

Solutions and Materials

Culture Reagents and Buffers

1. Hypertonic medium (Influx-pinocytic cell-loading reagent, Invitrogen)
2. Hypotonic medium (prepared by combining D-MEM medium, without serum, and sterile deionized water in a 6:4 ratio)
3. Air-MEM medium (D-MEM medium without phenol red, supplemented with 33 mM glucose, 2 mM glutamine, and 10 mM HEPES)
4. D-MEM medium (Gibco, Invitrogen, Carlsbad, CA) with 4500 mg/liter D-glucose, 110 mg/liter sodium pyruvate, and nonessential amino acids, without L-glutamine
5. Glucose (Sigma-Aldrich Corp.)
6. FVS (Invitrogen)
7. Glutamine (Invitrogen) diluted in water to 200 mM
8. Penicillin-streptomycin (Gibco, Invitrogen) contains 10,000 units of penicillin [base] and 10,000 μg of streptomycin [base]/ml utilizing penicillin G [sodium salt] and streptomycin sulfate in 0.85% saline
9. PBS 10×: 80.06 g NaCl, 2.02 g KCl, 2.05 g KH_2PO_4, 23.3 g Na_2HPO_4 in 1 liter of deionized water
10. Circular glass coverslips (18 mm, Karl Hecht KG, Sondheim, Germany)

Quantum Dots

1. Commercial streptavidin-coated quantum dots (2 μM stock) emitting at 655 nm (Quantum Dots Corp.) stored at 4°
2. MicroSpin SR-400 column (GE Health Care) stored at 4°

Methods

Purification of Quantum Dots Streptavidin Conjugates (QD-SAVs)

Commercial QD-SAVs emitting at 655 nm need to be purified from free avidin present in the stock solution as shown on the HPLC chromatogram (Fig. 2). QD-SAVs are purified from free avidin by using microSpin SR-400 columns, as described in the following procedure.

1. Resuspend the resin in the microspin by vortexing for a few minutes.
2. Loosen the cap one-fourth turn and snap off the bottom closure.
3. Place the microspin column in a 1.5-ml Eppendorf tube for support.
4. Prespin the column at 735g for 1 min.
5. Place the microspin column in a new 1.5-ml low-binding Eppendorf tube (LoBind). Remove and discard the cap.
6. Load the quantum dots solution slowly (e.g., 50 μl at 100 nM) to the top center of the resin, being careful not to disturb the bed. Check with an ultraviolet lamp.
7. Spin the column at 735g for 2 min, and collect the purified quantum dots in the low binding support tube.
8. Measure the new concentration of purified QD-SAVs with a spectrophotometer (extinction coefficient at 405 nm $= 5.7 \times 10^6 \, M^{-1} \, | \, cm^{-1}$.

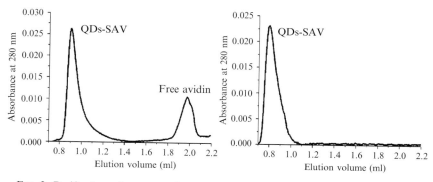

FIG. 2. Purification of quantum dots streptavidin conjugate. HPLC chromatograms of quantum dots streptavidin conjugate before (left) and after (right) purification on a microSpin SR-400 column.

Coupling of Biotinylated Proteins to QD-SAVs

For single molecule experiments, QD-SAVs are coupled at a final concentration of 1 nM with biotinylated proteins (ideally in a 1:1 ratio) in hypertonic solution for about 10 min at room temperature, before internalization in cultured cells. The hypertonic solution is prepared as described in the Invitrogen procedure (I-14402).

1. Prewarm 5 ml of D-MEM medium, without serum to 37°.
2. Melt the PEG (waxy solid on top of sucrose crystals) contained in the tube of influx-pinocytic cell-loading reagent by heating the tube in hot water (around 80°) for at least 2 min.
3. Remove the cap of the tube and quickly add 4.7 ml of 37° D-MEM medium without serum.
4. Replace the cap and vortex the tube vigorously to dissolve the sucrose crystals completely.
5. Once the solution is homogeneous, add 250 μl of serum and 50 μl of 1 M HEPES buffer, pH 7.4, in the tube.
6. Replace the cap and mix by vortexing the tube for a few minutes.

Internalization of QD-Protein Constructs in Live HeLa Cells

After coupling QD-SAVs to biotinylated proteins, the reaction product, QD-Ps, is loaded into live cells via a pinocytic process.

1. Cultured cells are plated at a density of 1.5×10^5 cells/cm^2 on 18-mm glass coverslips and maintained in D-MEM medium containing 10% fetal calf serum at 37° in a 5% CO$_2$ atmosphere.
2. Cells are incubated 10 min at 37° in the hypertonic medium along with QD-Ps.
3. After osmotic lysis of pinocytic vesicles, induce by incubating the cells at 37° for exactly 2 min in a hypotonic culture medium; QD-Ps are released homogeneously in the cell cytosol.
4. Cells are left for 10 min in Air-MEM medium before imaging. The osmotic lysis of pinocytic vesicles in the hypotonic solution does not alter the viability of cultured cells and does not result in lysosomal enzyme release. By comparison with other internalization techniques, such as microinjection, this method presents the main advantage that a large number of cells are loaded simultaneously in the same conditions.

Imaging of Single Conjugated Quantum Dot in Living Cells

Optical Microscope

The experiments are carried out using a standard inverted microscope (Olympus, IX71) mounted with an oil objective (Olympus × 60 NA = 1.4

PlanApo Ph3). An immersion objective should always be used with a numerical aperture of at least 1.2 and preferably 1.3 to 1.45. QDs are excited with a 100-W mercury lamp (excitation filter XF1074 525AF45, dichroic mirror DRLP 515, emission filters XF3304 605NB20 or 655WB20, Omega Opticals, Brattleboro, VT). Fluorescence images are collected with a Coolsnap ES (Roper Scientific) or a Peltier-cooled Micromax EB512FT CCD camera (Roper Scientific, Trenton, NJ). The pixel size p of the camera and the magnification M of the objective should be such that p/M is no larger than $\lambda/2NA$. In our case, $p = 13 \ \mu$m and $M = 60$ such that $p/M = 217$ nm. If p/M is too small, each individual spot will spread over too many pixels, leading to a decrease in SNR.

Calibration of the Optical System for Single Molecule Detection

To evaluate the ability of the optical system (objective, filter, camera) to detect individual quantum dots, a simple procedure can be followed.

1. Prepare a clean coverslip by rinsing it successively with methanol, acetone, and water. Dry it with clean compressed air or nitrogen.
2. Deposit a drop of QDs diluted at 1 nM in water. After 5–10 min, rinse with water and dry with clean compressed air or nitrogen.
3. Mount the coverslip on the microscope and focus on the surface. To facilitate focusing, a mark from a fluorescent pencil can be added on the proper side of the coverslip.
4. The signal of individual QDs corresponds to diffraction-limited spots and should be easily observable with the camera or in the eyepieces. The spots have a spatial extension given by $\lambda/2NA$. Individual QDs are identified by their fluorescence intermittency, that is, by the random succession of periods with high- and low-emission intensity.
5. If necessary, the blinking rate can be decreased by reducing the excitation intensity with a neutral density. As this will diminish the emission signal accordingly, a compromise should be found among detection sensitivity, acquisition time, and blinking.

Single Molecule Imaging in Live Cells

1. The coverslip is mounted on a cell chamber, maintained either at room temperature or at $37°$.
2. By means of a phase-contract objective, select a zone of interest in a cell. This selection can be also achieved with less sensitivity with bright-field imaging.
3. Turn off the transmission light and turn on the fluorescence excitation. Individual spots can be detected with the camera or in the eyepieces. If necessary, refocus the microscope slightly.

4. A sequence of images can then be acquired (Figs. 3 and 4). In our experiments, we perform two different types of recordings: time lapse or streaming. In both cases, the excitation illumination is controlled by a computer-driven mechanical shutter. For streaming acquisition, the shutter is open permanently, allowing a continuous recording at an acquisition rate varying usually between 10 and 30 Hz. The maximum number of images in the sequence is determined by the storage space in the RAM or the hard drive of the computer. The acquisition rate is usually limited by the readout time of the camera and not by the brightness of the probe (single static QDs can be detected with 5 or 10 ms acquisition time).

5. The sequences of fluorescence contain fluorescence spots coming either from single QDs or from small aggregate. The blinking of QDs

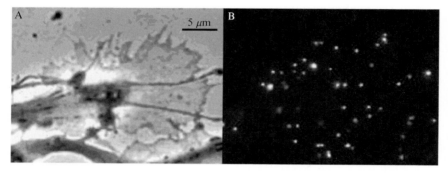

Fig. 3. Membrane labeling of single QDs in a nerve growth cone. (A: bright field image, B: fluorescence image).

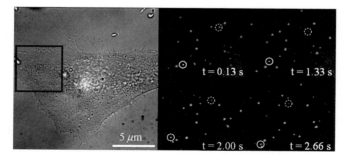

Fig. 4. Intracellular labeling of single QDs. (Left) Bright-field image of a HeLa cell in which QDs coupled to cytoplasmic proteins have been internalized. (Right) Tracking of individual biomolecules (two examples circled in plain and dashed lines) corresponding to the area marked on the left side.

provides a simple criterion to identify single molecules, as the probability for all the QDs forming an aggregate to simultaneously blink is small.

Data Processing

Acquisition of fluorescence image stacks is performed with MetaView (Universal Imaging, Downingtown, PA). Tracking and statistical analysis are carried out with custom-made software using Matlab (Mathworks, Natick, MA).

Analysis of Single QD trajectories

One of the main issues of single particle tracking experiments is extracting biological information from recorded trajectories. This section presents some of the techniques that have been used in order to measure quantities such as the diffusion coefficient or velocity.

Localization Accuracy of Single Quantum Dots

The value of the SNR has a direct consequence on the accuracy with which individual spots can be localized (Bobroff, 1986; Ober *et al.*, 2004; Thompson *et al.*, 2002; Yildiz *et al.*, 2003). Because QDs are much smaller than the excitation wavelength $\lambda\sigma$, the shape of the fluorescence spots corresponds to the point spread function (PSF) of the optical system. Its width w is given by the optical resolution of the microscope, that is, $w \approx \lambda/(2*NA)$ where NA is the objective numerical aperture. For nanoparticles emitting around 600 nm and an oil-immersion objective with a NA of 1.3, this corresponds to about 230 nm. However, by fitting the PSF with a two-dimensional Gaussian curve, its center can be pinpointed with a much higher accuracy, on the order of w/SNR when the detection sensitivity is limited by the photon noise (Thomson *et al.*, 2002). For SNR close to 30, as observed frequently with individual QD-tagged membrane receptors ($\tau \sim 50$–100 ms), the accuracy can be as low as 5–10 nm.

Image Processing and Extraction of Single QD Trajectories

The technical details of our image processing method have been presented in Bonneau *et al.* (2004). Efficient tracking routines can also be found as plugins for imageJ (http://rsb.info.nih.gov/ij/).

Measurement of the Point Spread Function

Our method for the detection and tracking of quantum dots requires measurement of the PSF, which is the response of the optical system to a point source. Because quantum dots are much smaller than λ, they can be

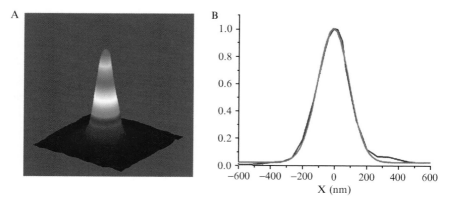

FIG. 5. (A) Experimental point spread function (PSF) of a single quantum dot. (B) The PSF (blue line) can be fitted by a Gaussian curve (red line) and its center can be localized with a high accuracy, on the order of λ/SNR.

reliably considered as a point source, and an image of a single QD is a good approximation of PSF (Fig. 5).

Detection of Spots

Fluorescent spots are detected on each frame of the stack by a method of correlation. A measure of similarity between real fluorescent image and the PSF (known from calibration images) is computed around local maxima of a small region of the fluorescence image. Correlation maxima, selected above a threshold depending on our estimation of the noise in the fluorescence image, reveal the position of QD-tagged molecules.

Individual Trajectories

Once QDs are detected on each frame, spots coming from the same QD have to be linked from frame to frame. Assuming that receptors are freely diffusing, we computed the probability for a spot, detected at a given position on a frame to be at another position on the following frame. The highest likelihood for a spot of the following frame provides the correct reconstruction of the trajectory. Spots of a given frame are then associated to the trajectory they are the most likely to belong to.

Blinking

Blinking brings additional difficulties for the tracking of single QDs, as the spots can disappear temporarily for random durations. A simple approach to account for blinking consists in computing continuous trajectories and then manually associating those corresponding to the same QD.

To reduce blinking rate, it is recommended to perform experiments with an excitation as low as possible.

Software

General tracking routines can be found as plugins for imageJ (http://rsb. info.nih.gov/ij/). We are currently implementing new tracking algorithms more suited for the emission properties of semiconductor QDs (Bonneau *et al.*, 2005). In the future, the corresponding software is intended to be freely available for potential users. More information can be found on our web site: http://www.lkb.ens.fr/recherche/optetbio/.

Mean Square Displacement Function

Once a trajectory $[x(t), y(t)]$ is determined, a way to extract physical information is the computation of $\rho(t)$, mean square displacement (MSD) (Saxton and Jacobson 1997). It is determined by the following formula

$$\rho(n\tau) = \frac{1}{N-n} \sum_{i=1}^{N-n} \left[\left(x((i+n)\tau) - x(i\tau) \right)^2 + \left(y((i+n)\tau) - y(i\tau) \right)^2 \right]$$

where τ is the acquisition time and N is the total number of frames. Through computation of the MSD (Fig. 6), the nature of the molecular motions can be analyzed. A diffusive motion is revealed by a MSD varying linearly with a slope $4D$ (where D is the diffusion coefficient). In practice, the parameter D is obtained by fitting the first five points of the MSD with a linear curve. When an additional directed motion (with velocity v) is

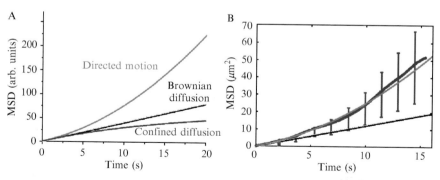

FIG. 6. (A) Theoretical curves of mean square displacement (MSD) as a function of time for directed (red line), Brownian (black line), and confined (blue line) motions. (B) Experimental MSD for a directed motion with velocity v and diffusion coefficient D. The red curve is a parabolic adjustment $4Dt + v^2t^2$ and the black linear curve corresponds to the initial slope $4Dt$.

present, $\rho(t)$ is in average equal to $4Dt + v^2 t^2$. In the opposite, if the motion is confined, $\rho(t)$ should in average exhibit a negative curvature and saturate toward a finite value equal to the surface area explored during the movement.

Statistical Errors

Because the MSD is calculated on trajectories with a finite number of points, it is inherently affected by statistical errors. $s(\tau)$ the variance of $\rho(\tau)$ can be either estimated by computer simulations or, in the case of Brownian diffusion, computed by the following formula (Qian *et al.* 1991):

$$\sigma(\tau) = 4D\tau \left[(4n^2(N-n) + 2(N-n) + n - n^3)/6n(N-n)^2 \right]^{1/2}$$

for $n \leq N/2$

$$\sigma(\tau) = 4D\tau \left[1 + \left((N-n)^3 - 4n(N-n)^2 + 4n - (N-n) \right)/6n^2(N-n) \right]^{1/2}$$

for $n \geq N/2$

Because $\sigma(\tau)$ increases rapidly with n, great care should be taken to account for these statistical fluctuations, especially when looking for deviation to simple Brownian diffusion (Qian *et al.*, 1991).

Acknowledgments

We are grateful to A. Triller, S. Lévi, P. Rostaing, B. Riveau, C. Charrier, C. Hanus, L. Cohen, Y. Bellaiche, and G. Cappello for their contribution to the experiments and for many valuable discussions.

References

Bobroff, N. (1986). Position measurement with a resolution and noise limited instrument. *Rev. Sci. Instrum.* **57**, 1152–1157.

Bonneau, S., Cohen, L., and Dahan, M. (2004). A multiple target approach for single quantum dot tracking. *In* "Proceedings of the IEEE International Symposium on Biological Imaging (ISBI 2004)," p. 664.

Bonneau, S., Dahan, M., and Cohen, L. (2005). Single quantum dot tracking based on perceptual grouping using minimal paths in a spatio-temporal volume. *IEEE Transact. Image Proc.* **14**, 1384–1395.

Borgdorff, A. J., and Choquet, D. (2002). Regulation of AMPA receptor lateral movements. *Nature* **417**, 649–653.

Bruchez, M., Jr., Moronne, M., Gin, P., Weiss, S., and Alivisatos, A. P. (1998). Semiconductor nanocrystals as fluorescent biological labels. *Science* **281**, 2013–2016.

Chan, W. C., and Nie, S. (1998). Quantum dot bioconjugates for ultrasensitive nonisotopic detection. *Science* **281**, 2016–2018.

Courty, S., Luccardini, C., Bellaiche, Y., Cappello, G., and Dahan, M. (2006). Single quantum dot imaging of kinesin motors in living cells. **6**(7), 1491–1495.

Dahan, M., Levi, S., Luccardini, C., Rostaing, P., Riveau, B., and Triller, A. (2003). Diffusion dynamics of glycine receptors revealed by single-quantum dot tracking. *Science* **302,** 442–445.

Dubertret, B., Skourides, P., Norris, D. J., Noireaux, V., Brivanlou, A. H., and Libchaber, A. (2002). *In vivo* imaging of quantum dots encapsulated in phospholipid micelles. *Science* **298,** 1759–1762.

Harms, G. S., Cognet, L., Lommerse, P. H., Blab, G. A., Kahr, H., Gamsjager, R., Spaink, H. P., Soldatov, N. M., Romanin, C., and Schmidt, T. (2001). Single-molecule imaging of l-type Ca2+ channels in live cells. *Biophys. J.* **81,** 2639–2646.

Iino, R., Koyama, I., and Kusumi, A. (2001). Single molecule imaging of green fluorescent proteins in living cells: E-cadherin forms oligomers on the free cell surface. *Biophys. J.* **80,** 2667–2677.

Jaiswal, J. K., and Simon, S. M. (2004). Potentials and pitfalls of fluorescent quantum dots for biological imaging. *Trends Cell Biol.* **14,** 497–504.

Kim, S., Lim, Y. T., Soltesz, E. G., De Grand, A. M., Lee, J., Nakayama, A., Parker, J. A., Mihaljevic, T., Laurence, R. G., Dor, D. M., Cohn, L. H., Bawendi, M. G., and Frangioni, J. V. (2004). Near-infrared fluorescent type II quantum dots for sentinel lymph node mapping. *Nature Biotechnol.* **22,** 93–97.

Lippincott-Schwartz, J., and Patterson, G. H. (2003). Development and use of fluorescent protein markers in living cells. *Science* **300,** 87–91.

Medintz, I. L., Uyeda, H. T., Goldman, E. R., and Mattoussi, H. (2005). Quantum dot bioconjugates for imaging, labeling and sensing. *Nature Mater.* **4,** 435–446.

Meier, J., Vannier, C., Serge, A., Triller, A., and Choquet, D. (2001). Fast and reversible trapping of surface glycine receptors by gephyrin. *Nature Neurosci.* **4,** 253–260.

Michalet, X., Pinaud, F. F., Bentolila, L. A., Tsay, J. M., Doose, S., Li, J. J., Sundaresan, G., Wu, A. M., Gambhir, S. S., and Weiss, S. (2005). Quantum dots for live cells, *in vivo* imaging, and diagnostics. *Science* **307,** 538–544.

Murase, K., Fujiwara, T., Umemura, Y., Suzuki, K., Iino, R., Yamashita, H., Saito, M., Murakoshi, H., Ritchie, K., and Kusumi, A. (2004). Ultrafine membrane compartments for molecular diffusion as revealed by single molecule techniques. *Biophys. J.* **86,** 4075–4093.

Ober, R. J., Ram, S., and Ward, E. S. (2004). Localization accuracy in single-molecule microscopy. *Biophys. J.* **86,** 11185–11200.

Qian, H., Sheetz, M. P., and Elson, E. L. (1991). Analysis of diffusion and flow in two-dimensional systems. *Biophys. J.* **60**(4), 910–921.

Saxton, M. J., and Jacobson, K. (1997). Single-particle tracking: Applications to membrane dynamics. *Annu. Rev. Biophys. Biomol. Struct.* **26,** 373–399.

Schutz, G. J., Kada, G., Pastushenko, V. P., and Schindler, H. (2000). Properties of lipid microdomains in a muscle cell membrane visualized by single molecule microscopy. *EMBO J.* **19,** 892–901.

Seisenberger, G., Ried, M. U., Endress, T., Buning, H., Hallek, M., and Brauchle, C. (2001). Real-time single-molecule imaging of the infection pathway of an adeno-associated virus. *Science* **294,** 1929–1932.

Shav-Tal, Y., Singer, R. H., and Darzacq, X. (2004). Imaging gene expression in single living cells. *Nature Rev. Mol. Cell. Biol.* **5,** 855–861.

Singer, S. J., and Nicolson, G. L. (1972). The fluid mosaic model of the structure of cell membranes. *Science* **175,** 720–731.

Tardin, C., Cognet, L., Bats, C., Lounis, B., and Choquet, D. (2003). Direct imaging of lateral movements of AMPA receptors inside synapses. *EMBO J.* **22,** 4656–4665.

Thompson, R. E., Larson, D. R., and Webb, W. W. (2002). Precise nanometer localization analysis for individual fluorescent probes. *Biophys. J.* **82,** 2775–2783.

Tsien, R. Y. (2005). Building and breeding molecules to spy on cells and tumors. *FEBS Lett.* **579,** 927–932.

Ueda, M., Sako, Y., Tanaka, T., Devreotes, P., and Yanagida, T. (2001). Single-molecule analysis of chemotactic signaling in Dictyostelium cells. *Science* **294,** 864–867.

Yildiz, A., Forkey, J. N., McKinney, S. A., Ha, T., Goldman, Y. E., and Selvin, P. R. (2003). Myosin V walks hand-over-hand: Single fluorophore imaging with 1.5-nm localization. *Science* **300,** 2061–2065.

Weiss, S. (1999). Fluorescence spectroscopy of single biomolecules. *Science* **283,** 1676–1683.

Wu, X., Liu, H., Liu, J., Haley, K. N., Treadway, J. A., Larson, J. P., Ge, N., Peale, F., and Bruchez, M. P. (2003). Immunofluorescent labeling of cancer marker Her2 and other cellular targets with semiconductor quantum dots. *Nature Biotechnol.* **21,** 41–46.

[13] Development and Application of Automatic High-Resolution Light Microscopy for Cell-Based Screens

By Yael Paran, Irena Lavelin, Suha Naffar-Abu-Amara,
Sabina Winograd-Katz, Yuvalal Liron,
Benjamin Geiger, and Zvi Kam

Abstract

Large-scale microscopy-based screens offer compelling advantages for assessing the effects of genetic and pharmacological modulations on a wide variety of cellular features. However, development of such assays is often confronted by an apparent conflict between the need for high throughput, which usually provides limited information on a large number of samples, and a high-content approach, providing detailed information on each sample. This chapter describes a novel high-resolution screening (HRS) platform that is able to acquire large sets of data at a high rate and light microscope resolution using specific "reporter cells," cultured in multiwell plates. To harvest extensive morphological and molecular information in these automated screens, we have constructed a general analysis pipeline that is capable of assigning scores to multiparameter-based comparisons between treated cells and controls. This chapter demonstrates the structure of this system and its application for several research projects, including screening of chemical compound libraries for their effect on cell adhesion, discovery of novel cytoskeletal genes, discovery of cell migration-related genes, and a siRNA screen for perturbation of cell adhesion.

METHODS IN ENZYMOLOGY, VOL. 414   0076-6879/06 $35.00
Copyright 2006, Elsevier Inc. All rights reserved.   DOI: 10.1016/S0076-6879(06)14013-6

Introduction

One of the primary challenges of biological research in the "post-genomic era" is a comprehensive understanding of the concerted action of multiple genes and their protein products in performing basic life processes. A fundamental requirement for studying such issues is the characterization of gene expression patterns, modification and subcellular localization of proteins and elucidation of their involvement in multiprotein complexes, and signaling networks within cells. It is noteworthy that in addition to their synthesis and spatial organization, protein assemblies are highly dynamic and undergo continuous rearrangement. Apparently many of these complex processes cannot be monitored effectively using purely biochemical approaches, which are well designed for assessing specific molecular interactions, yet they lack spatial resolution and are usually based on the collective behavior of cell populations. Such approaches fail to follow dynamic events in individual cells due to cell-to-cell heterogeneity and limited synchrony.

Many of the cellular features just mentioned can be monitored by advanced light microscopy, which has developed in recent years into an increasingly powerful tool for studying molecular and cellular events. Information about cells, which is attainable by microscopy, includes, for example, cell morphology, architectural organization of organelles, monitoring of gene expression (using promoter reporters), and localization of proteins (using fluorescently tagged fusion proteins). In addition, time-lapse recording, combined with fluorescence tracking, photobleaching, or photoactivation, can shed light on dynamic molecular processes in live cells.

A most compelling challenge is to harvest such quantitative structural and functional information from large numbers of samples. This need is a natural development of large genetic screens, where the effect of gene overexpression or suppression is tested using a wide variety of model systems, including yeast (Huh *et al.*, 2003), Drosophila (Boutros *et al.*, 2004), and human cells (Sachse *et al.*, 2005). Similar requirements are apparent in pharmacological screens, where libraries of chemical compounds are tested for their physiological effects (Clemons, 2004; Stockwell, 2004). In order to adapt the throughput to the number of screened samples, automation in microscopy is required, both in the acquisition and in the analysis stages. Moreover, each screen should provide specific information about a rather narrow slice of the multidimensional space characterizing the specific cellular systems. To resolve many features in a large number of samples, exhaustive screens are needed, which further emphasizes the need for suitable high-throughput and automated technology (Abraham *et al.*, 2004).

The implementation of advanced microscopy-based screening requires that the interpretation of data will be not only accurate, but also as fast as the rate of acquisition, and preferably online. Thus, quantitative analysis needs to deal with cell-to-cell variability based on an analysis of sufficiently large numbers of cells. This requirement further complicates the screening process, due to both the need to process highly elaborate images and the need to acquire and analyze a large number of images, sufficient for reliable statistical evaluation.

In view of these needs, many automatic microscope systems have been developed, some of which are available commercially (Smith and Eisenstein, 2005). Moreover, it is noteworthy that computerized analyses of microscope images have been amply described over the years, including the development of algorithms that can be applied to extract textural and morphological parameters from high-throughput experiments (Conrad *et al.*, 2004; Mitchison, 2005; Murphy, 2005; Perlman *et al.*, 2004, 2005; Tanaka *et al.*, 2005; Yarrow *et al.*, 2003; Zhou *et al.*, 2005). However, the capacity to perform fast and automated data acquisition at high resolution, using high numerical aperture (NA) objectives (e.g., $\times 60$ NA $= 0.9$), combined with automated analysis pipeline, is still rather limited.

This chapter describes the design of a fully automated, high-throughput, high-resolution multiwell plate-scanning microscope platform and its application in several cell-based screening projects. We focus on cytoskeletal and cell adhesion sites, which play a central role in dynamic cellular events, including cell migration, tissue formation, and transmembrane signaling. Due to their importance, these cell structures are also attractive targets for a variety of drugs (Peterson and Mitchison, 2002). The system described here can acquire up to two images per second using objective magnifications up to $\times 100/0.9$ and focusing before each acquisition by a fast and robust laser-based AutoFocus device (Liron *et al.*, 2006). An automated image analysis pipeline has been assembled with project-specific analysis routines and applied for identifying morphological and fluorescence intensity-related changes induced by genetic or chemical perturbations.

Technical Description

Microscope

The automated systems (Fig. 1) are based on IX71 and IX81 microscopes (Olympus, Japan). Microscope automation is provided by a ProScan system (Prior, Cambridge, UK) consisting of a stage, focus, objectives and dichroic cube changers, excitation and emission filter wheels, and shutters. Images are taken using a Quantix camera (Photometrics, Tucson, AZ),

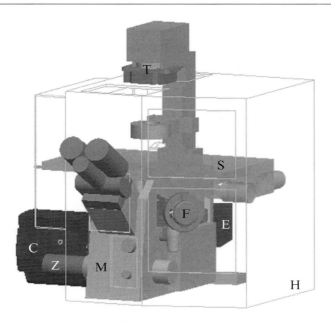

FIG. 1. A schematic representation of the high-resolution automated plate-screening microscope system. M, microscope body, Olympus IX81; S, XY stage and plate holder; Z, focus motor; C, camera with emission filter wheel; E, epi-illumination, including Hg arc lamp, shutter, and fluorescence excitation filter wheel; T, transmitted light halogen lamp and shutter; F, AutoFocus, optoelectronics; H, environmental chamber. Controlling electronics, computer, xyz joystick, monitor, and keyboard are not shown.

equipped with a Marconi (EEV) CCD57 back-illuminated chip with very high quantum efficiency in the visible and the near-IR (>80% quantum efficiency [QE] in 450- to 650-nm wavelength range). A manual focus jog and stage joystick are used for system alignment and initial setting of the screen.

For live cells screening, microscopes are equipped with a home-built Plexiglas box heated to 37° by an air circulator. CO_2 can be supplied, when necessary, via a small chamber encasing the multiwell plate. To allow gas exchange, without increased evaporation of the medium, the wells are covered by a Luminex BioFolie foil (Hanover, Germany).

The laser AutoFocus is a key component for the automated capabilities of the microscope. High-resolution imaging necessitates precise axial setting of the focus before each image is acquired, yet at minimal extra time and sufficient robustness to be repeated during long screens with negligible failure rates. These features were designed into our system (Liron et al., 2006).

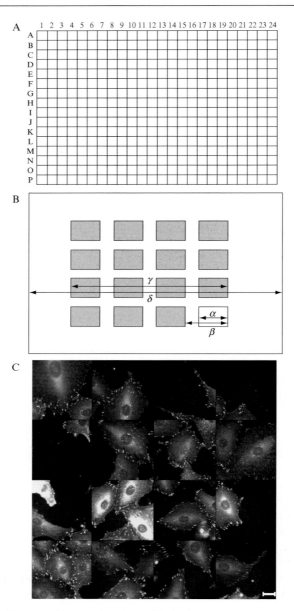

FIG. 2. (A) Plate graphic user interface (GUI). Graphics are used during automated acquisition to display the progress of the screen, for interactive visualization (mouse click on a well in a plate GUI displays montage of all images acquired in the well, and click on a field within a well displays the corresponding image in full resolution), and for display of analysis

Image Acquisition Software

The acquisition program, Resolve6D, organizes six-dimensional image files (i.e., *X, Y, Z,* wavelength, time, and position in the multiwell plate). It uses the UCSF Linux-based Image Visualization Environment and Priism software (IVE, http:/msg.ucsf.edu/IVE) for image display and Graphic User Interface. A command-line interpreter runs computer control sequences on all microscope functions, including objective changes, stage, focus, shutters, and filters, and allows unlimited depth of looping on all variables. Loops on wells are set by a selection of regions within all plate formats (96, 384, 1536 round or square wells). Within each well the imaged fields are defined by setting a fraction of the well area and a fill factor (>1 implies partially overlapping fields for matching image tiles into continuous carpets, see Fig. 2). The scanning program then chooses the acquisition path, minimizing distances between successive fields to reduce the time spent on AutoFocusing. One can also manually select a point list for revisiting interesting cells and record multiple time-lapse movies, as well as automatically depicting points of interest. For example, a list of coordinates for all the fluorescent cells can be extracted automatically from low-magnification images, which then may be revisited at high magnification. This multimagnification imaging capability can accelerate screens by factors of 10 or more when informative high-magnification images are scarce (~10%), such as in case of low-efficiency transfection.

Image Analysis

We developed an integrated solution for a data processing pipeline consisting of several analysis steps or modules. Image processing is highly image dependent. Development of open architecture is therefore a key issue in this stepwise design. The challenge is to build a generalized platform capable of incorporating a wide and adaptable range of image interpretation schemes and statistical analyses. We want to deal with low-, medium-, and high-magnification images of different cell types and with various assays based on morphology and subcellular distributions of tagged molecules and organelles. The modules are as follow.

Image Database Module

Microscope-based screens can produce Terabytes of high-resolution image data. We adopted the MRC file format (http://www.msg.ucsf.edu/IVE/

scores (coded in rainbow colors). (B) In-well imaged fields GUI. The fields distribution is determined by two parameters: filling factor = α/β and fraction of the well covered by images (γ/δ), as marked. (C) Montage of images acquired in one well. Scale bar: 20 μm.

IVE4_HTML/IM_ref2.html), which includes a header for global information about the experimental parameters, extended header for individual image information, and multiple images per file. This allows compact directory contents (file name annotates owner, date, and plate), with back-and-forth links between each condition or treatment (via plate/well numbers) and the corresponding images. The image files are associated with additional files created by the various analysis steps (e.g., montages, segmented object parameters, statistical scores, see later). Files are stored from the acquisition computers onto mass storage and accessed for analysis.

Interactive Visualization Module

Accessible visualization is a primary requirement for the researcher, for monitoring the acquisition, for inspecting controls, and for qualitatively evaluating effects. In addition, display of selected images, following automated computerized analysis, allows focusing on interesting samples. We base our visualization on whole well montages (tiling all images acquired in a well) called upon by plate and well numbers or via a plate–graphic interface (Fig. 2). A full-resolution display for each of the tiles can be called. Display windows have FIFO structure, keeping last displayed N images (N can be set), therefore allowing fast play of image sequences for the purpose of comparison.

Image Processing Module

Image processing is a mature field, nevertheless, biological image processing faces special needs that need to be addressed. The first requirement in microscopy is correction of the inhomogeneous illumination field. It can be achieved by a fiber optical scrambler spatially stabilizing the illumination intensity profile (Kam *et al.*, 1993). However, the tight well walls perturb this profile, imposing development of alternative solutions. A second problem encountered in biological imaging is background, originating from optoelectronics, autofluorescence, nonlocalized diffuse labeling, and out-of-focus contributions. We use various processing schemes such as linear and nonlinear filtration, local contrast enhancement, and rolling-ball background estimates to "flatten" images prior to segmentation (Lichtenstein *et al.*, 2003). It should be emphasized that the same image may be processed differently for segmentation of whole cell extents or for intracellular organelles. Simple and fast automated contrast-enhancement techniques are extremely useful also for visualization of the acquired images.

Image Segmentation and Connected Component Analysis Module

Segmentation and connected component analysis is the most common, but also most difficult approach to interpret image contents in terms of "objects." Over the years we have developed and applied a number of segmentation algorithms based on adaptive thresholds: WaterShed (Zamir *et al.*, 1999), fiber recognition (Lichtenstein *et al.*, 2003), and multiscale algorithms (Sharon *et al.*, 2001). The binary masks define objects, which are characterized by quantitative morphological and image intensity parameters (such as total and mean fluorescent intensity, area, axial ratio from best-fitted ellipsoid). Multicolor correlation analysis yields quantitative colocalization parameters (e.g., with a labeled organelle) (Kam *et al.*, 2001). For cells expressing fluorescently tagged proteins, one can segment the overall cell body by the higher diffuse fluorescence compared to the cell-free surrounding. Specific tags, labeling intracellular structures, can be used to evaluate proliferation, survival, and death and to recognize, at low resolution, cell locations for further analyses. Detailed quantification parameters for intracellular features, such as the nucleus and its substructures, cytoskeletal fibers, endoplasmic reticulum, Golgi, and mitochondria, serve as excellent reporters for cell phenotyping and identification of anomalous morphologies.

Statistical Module

A major feature of cell-based screens is the large cell-by-cell variability. We rely on statistics to find an optimal balance between the desired high throughput and the required high sensitivity and resolution. We feed the outputs of the quantitative parameters for all segmented individual objects into analyses, which are capable of accumulating data from multiple images (one or many wells) and rejecting outliers. The module exports cumulative characterization as histograms, scatter plots, percentiles, and so on for each parameter and for all the analyzed wells.

Multiparametric Scoring of Changes in Treated versus Control Cells

This module compares each "test well" to controls and supplies scores, nominating the wells that were affected by the treatment. Such scores are commonly defined in nonparametric statistical methods by comparing two lists of variables (e.g., Komologov–Smirnov and rank sum tests). For our screen analysis purpose, we extended these tests for histograms. In addition, we utilized comparison of median and other percentiles, which were found useful to characterize far from normal distributions and cases where few extreme events are to be weighted heavily or median behavior shifts are expected. We also compared two parameter domains (quadrants), similar to definition of subpopulations in flow cytometry scatter plots (Boddy *et al.*, 2001). We have good grounds to propose that multiparameter scores will depict variable changes that a single parameter test may overlook. We therefore apply different multiparametric schemes for scoring differences between an experimental well and controls, including ranking and clustering of multiparameter vectors (Zamir *et al.*, 2005). The purpose of this process is to grade all wells according to the degree of changes and to characterize the parameters mostly aberrant compared to the controls. This analysis links directly to the original images, supplying convenient visualization to survey the results (detailed description of the analysis is described elsewhere).

Biological Screens

Screening projects are aimed at quite diverse goals. They can address a wide variety of cellular processes and/or structures and use a wide range of genetic or chemical perturbations. This section describes several automatic, cell-based screens for a variety of research projects, including screens for adhesion- and cytoskeleton-perturbing chemical compounds, a search for novel structural cellular proteins, and analysis of cell migration.

Perturbation of Cell Adhesions

A common application of automated microscopic systems is cell-based screening of cell-perturbing agents. This section describes two kinds of perturbation screens, both targeting focal adhesions in cultured cells: a chemical compounds screen and a siRNA screen. Both screens used reporter cell lines expressing fluorescently tagged paxillin, a prominent component in the focal adhesion (FA) complex. It is noteworthy that FAs are involved in the anchoring of cells to the substrate, as well as in the transduction of adhesion-induced signals and regulation of actin organization within cells.

Chemical Compounds Screens

A search for novel drugs that might interfere with cellular functions has been approached in recent years by screening diverse chemical libraries and their modifications. Examination of the drug effect on live cells, including effects on cellular morphology and relocation of structural and signaling proteins, requires microscope-based imaging at high resolution. The higher the image resolution, the more sensitive the readout of drug effects on cellular functions manifested as subtle cell architectural attributes. Here we examine the effects of the chemical compound library on FAs. REF52 cells expressing stably paxillin-yellow fluorescent protein (YFP) served as reporter cells. Cells are cultured in multiwell plates and treated by compounds from a library of 10,000 low molecular weight, chemically defined molecules (ChemDiv Inc., San Diego, CA) using a single compound per well. Following treatment, cells are fixed and plates are screened for effects on FAs on the automated microscope

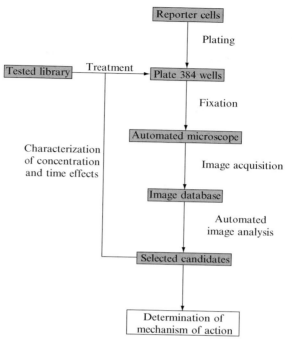

FIG. 3. Flowchart of the perturbation screen process.

at high resolution (Fig. 3). Acquired images are subjected to quantitative analysis, which yields more than 100 compounds with interesting cytoskeletal and FA-perturbing effects (see database of the screen results at http://www.weizmann.ac.il/mcb/Geiger/Screening.html). Major perturbations include FA elongation, reduction in FA size, loss of FAs at the cell center, abnormal paxillin localization, and overall cytoskeletal collapse (Fig. 4). The automated microscope is also utilized to study, in detail, the time- and concentration-dependent effects of selected compounds and their mechanism of action using labeling of cytoskeletal and FA-related proteins (Fig. 3).

Protocol for Cell-Based FA Perturbation Screening

1. REF52 cells, stably expressing paxillin-YFP, are seeded [800 cells/well in 50 μl Dulbecco's Modified Eagle's Medium (DMEM) with 10% fetal calf serum (FCS) and antibiotics] and cultured in optical bottom, 384-well plates (Greiner bio-one GmbH, Fricjkenhausen, Germany. F-bottom, μClear, black, tissue-culture treated).

2. Following a 24-h incubation, cells are treated with chemical compounds at a concentration of 10 μM (diluted from stock solution with DMEM containing 10% FCS and antibiotics).

3. Following a 90-min incubation, cells are fixed by dipping the plate in 3% paraformaldehyde bath for 25 min and then washing three times with phosphate-buffered saline (PBS).

4. Plates are examined by an automated microscope (objective $\times$ 60/ 0.9, 1-s exposure). In each well 16 images are acquired.

5. The image database is subjected to processing, segmentation, and analysis, which highlight the selected hits for significant FA perturbations.

siRNA Screen

To explore the mechanism involved in focal adhesion formation, individual signaling and focal adhesion complexes were knocked down using siRNA technology (Schutze, 2004). For that purpose, HeLa JW cells, expressing paxillin-YFP, were transfected with a library of "SMARTpools" RNA duplexes, each consisting of a mixture of four siRNAs, targeting the same gene (Dharmacon, USA). The details concerning this screen are

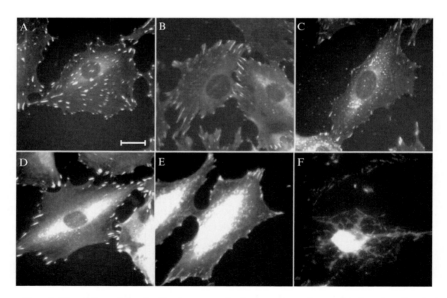

FIG. 4. Examples of focal adhesion perturbation patterns observed using the chemical compounds screen. (A) Untreated REF52 cells expressing paxillin-YFP. (B) Focal adhesion (FA) elongation. (C) Increase in number and decrease in size of FA. (D) Loss of FA at the cell center (usually associated with larger peripheral FA). (E) Increase in FA signal intensity and high amount of unlocalized paxillin-YFP. (F) Major cell damage. Scale bar: 20 μm.

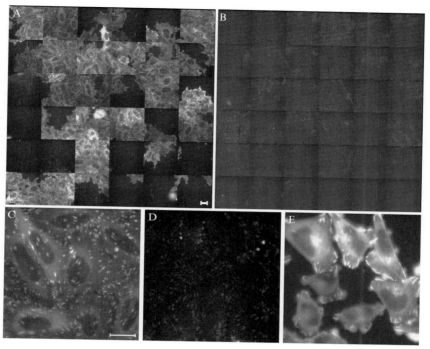

Fig. 5. Examples of images from siRNA screen. (A) Montage of images of untransfected reporter cells (HeLa cells expressing YFP-paxillin). (B) Montage of images of reporter cells transfected with siRNA targeting paxillin. (C) Single image of untransfected reporter cells. (D) Single image of cells transfected with siRNA targeting paxillin. (E) Single image of cells transfected with siRNA targeting talin. Scale bar: 20 μm.

beyond the scope of this chapter; however, the use of automatic microscopy was mandatory for screening the effects of a large number of siRNA duplexes. Thus, following treatment with the specific siRNA, cells were fixed and their FAs recorded using the automated microscope (Fig. 3). Major cytoskeletal changes detected in this screen include reduction in FA number and/or size, changes in FA localization within the cells, and overall effect on cell spreading and polarization. Figure 5 shows images of non-transfected cells, cells transfected with siRNA targeting paxillin, and cells transfected with siRNA targeting talin. Transfection with paxillin leads to a decrease in focal adhesion intensity as expected from the decrease in the expression of the YFP-tagged paxillin. The image of cells treated with talin siRNA is an example of the dramatic change in FA distribution and morphology following the treatment (detailed description of this screen will be published elsewhere).

Protocol for the Screening of siRNA Library

1. HeLa JW cells, stably expressing paxillin-YFP, are seeded in fibronectin-coated 384-well microplates (250 cells/well in 50 μl DMEM 10% FCS without antibiotics).

2. Following a 24-h incubation, cells are transfected with 100 nM siRNA using oligofectamine (Invitrogen, USA). SiRNA is diluted in DMEM without serum to 10 μl final volume.

3. Oligofectamine is diluted 1:10 in medium. After 5 min 10 μl of the Oligofectamine–DMEM mixture is mixed with each siRNA and the mixture is incubated for 20 min.

4. Medium (30 μl) is removed from each well with HeLa cells, and 5 μl of the siRNA Oligofectamine mix is added to the cells.

5. Cells are fixed 72 h after transfection with 3% paraformaldehyde and examined by the automatic microscope system.

6. Each plate contains, in addition, wells with cells transfected with siControl pool (a nontargeting siRNA), cyclophylin siRNA, and siTOX for transfection efficiency control. Transfection efficiency (calibrated to 80 to 90% in our case) is assessed by cell survival after siTOX transfection.

7. Plates are examined by the automated microscope (objective $\times$ 60/0.9, 1.5-s exposure). In each well 36 images are acquired and analyzed to extract FA affective genes.

A cDNA screen for Identifying Novel Structural Cellular Proteins Based on Localization of YFP-Fusion Proteins

One of the applications of the automated microscope system is a visual screen that aims at the identification of novel proteins, according to their subcellular localization. For that purpose, a normalized, oriented cDNA library was constructed with poly(A)-primed cDNA, prepared from mixed mRNA of rat brain tissue and rat kidney fibroblast (RKF) cell line, randomly fragmented to an average insert size of 500 bp. These fragments were cloned into pLPCX retroviral vector (Clontech) using *Sfi*I sites upstream to the YFP cDNA. The cDNA libraries contain $0.5–2 \times 10^6$ clones, more than 50% of which have cDNA insertions.

To screen the expression library we infected REF52 cells with pools of clones generated by subdividing the library. Those clones that contained upstream open reading frames out of frame with YFP were excluded from the subsequent analysis by FACS sorting. Cells expressing YFP fusion proteins were plated 5 cells per well in 96-well plates. Each well is then "backed up" and frozen in a 96-well plate for future investigation, as well as plated into 384 plates with optical plastic bottoms for screening for

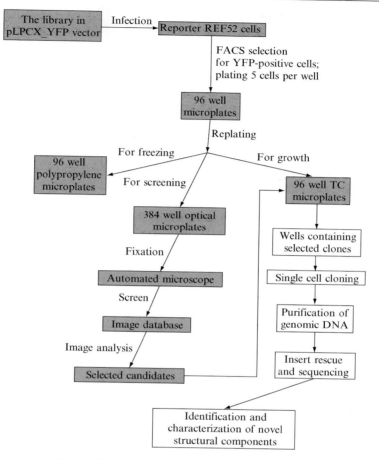

Fɪɢ. 6. Flowchart of the protein localization screen.

distinct patterns of YFP fluorescence by the automated microscope. Cells from wells displaying patterns of interest were then subjected to single-cell cloning and rescreened to isolate the clone responsible for the distinct pattern. Genes of interest were cloned, sequenced, and subjected to further characterization (Fig. 6). About 4% of total infected cells had visible cellular localization of the exogenous proteins representing cell structures such as nucleus, Golgi complex, mitochondria, and others. As expected, a significant percentage of the cDNA products localized to the cytosol (20% of the proteins showed definite localization) and the nucleus (15%), and

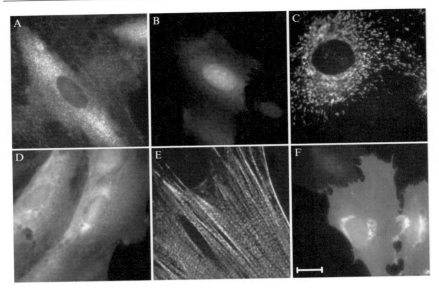

FIG. 7. Examples of subcellular localizations of YFP-fusion proteins observed in localization screening. (A) Vesicules, (B) nuclear, (C) mitochondrial, (D) cytoplasmic, (E) stress fibers, and (F) membrane. Scale bar: 20 μm.

about 20% were found to be associated with the secretory pathway (see database of the screen results at http://www.weizmann.ac.il/mcb/Geiger/Screening.html). Figure 7 shows a few examples of the subcellular localizations observed in the experiments.

Protocol for cDNA Screening

1. cDNA library (10 mg) in the pLPCX_YFP vector are used for infection of 3×10^6 target REF52 cells. Infection efficiency is titrated to about 80% to prevent insertion of multiple inserts.

2. Three days after infection, cells expressing YFP are selected by FACS and plated in a 96-well plate (5 cells/well) and mixed with naive cells (500 cell/well) for cells with growth problems at low confluence. Cells are then cultured to 100% confluence.

3. Puromycin (2 μg/ml) is added to select colonies of the infected cells (5 colonies per well). Cells are cultured to 80 to 100% confluence.

4. Cells from each well are replated to three plates: (1) 384-well plate with optical bottom for screening; (2) 96-well polypropylene plate for immediate freezing in liquid nitrogen ("back-up"); and (3) 96-well regular cell culture plate for following cell growth ("growing plate").

5. The next day cells in the 384-well plate are fixed by dumping the plate in a 3% paraformaldehyde bath for 25 min and are then washed three times with PBS.

6. The plate is transferred to the automated microscope (objective × 60/0.9, exposure 1.5 s, 25 to 36 fields per well).

7. Montages are prepared for visual selection of clones displaying YFP-labeled subcellular patterns of interest (positive clones).

8. Cells from wells in the 96-well "growing plate," corresponding to 384-plate wells containing positive clones, are transferred to new dishes for single-cell cloning and rescreening for isolation of clones responsible for the distinct patterns.

9. Genomic DNA from each single positive clone is purified and the cDNA inserts are rescued by polymerase chain reaction for sequencing, identification, and future characterization.

Screening for Genes Affecting Cell Migration

Cell migration is essential for the development and maintenance of normal tissues and organs; furthermore, enhanced migratory activity is common in malignant cells and is believed to be involved in metastasis. In order to identify proteins involved in or affecting cancer-related cell migration, we developed a quantitative high-throughput method for assessing cell motility. Many methods have been employed to measure two- and three-dimensional cell migration (Boyden, 1962; Hallab *et al.*, 2000; Lee *et al.*, 2000; Tamura *et al.*, 1998; Zigmond, 1977). However, high-throughput screening for cell motility demands the use of assays that record time-dependent changes in cell position, such as time-lapse movies, wound healing, or phagokinetic track formation (Albrecht-Buehler, 1977). The latter was our method of choice, using polystyrene beads, rather than the original colloid gold particles. Glass-bottom 96-well plates were coated with a monolayer of microbeads and then cells expressing different genes were seeded on the bead monolayer and incubated for several (5 to 9) hours and then fixed and examined. Transmitted light images (×10 objective) of adjacent fields recording the entire surface of the well were acquired and "stitched" together to form a continuous montage. This enabled us to view all the tracks throughout the entire well. The images exhibit smooth gray background, with bead-cleared tracks observed as brighter regions. Cells were recognized by the phagocytosed beads usually concentrated around the nucleus. Different cell lines produce characteristically different track shapes, for example, elongated or round, with smooth or rough border, reflecting their migratory behavior (Fig. 8). Computerized analysis of the

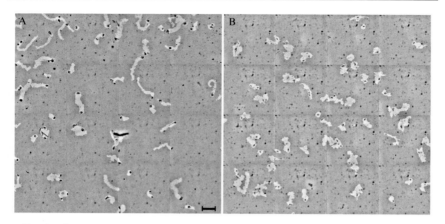

FIG. 8. Phagokinetic tracks of H1299 (nonsmall cell lung carcinoma) and MDA-MB-231 (metastatic breast cancer) cell lines. (A) H1299 cells present thin and elongated tracks, indicating persistent migration, compared to (B) the wide and short tracks of e MDA-MB-231 cells. Scale bar: 200 μm.

track morphology using multiscale segmentation methods allowed extraction of gross track parameters (area, maximum length, ratio between long and short axes of best-fitted ellipsoid), as well as detailed perimeter and border roughness features (detailed description of the screen for migration-inducing genes and track analysis will be published elsewhere).

Protocol for Motility Assay

1. Glass-bottom 96-well plates (Whatmann) are covered with 50 μl fibronectin (Sigma Chem. Co. F-1141; 10 μg/ml in PBS) and incubated overnight at 40°, washed twice with 200 μl PBS using a plate washer (Colombus plus, Tecan, Switzerland), and left with 50 μl PBS in each well.

2. White polystyrene latex 0.34-μm-diameter beads (product number 2–300, batch number 1344, Interfacial Dynamics Corporation/Molecular Probes Microspheres Technologies, Oregon) are sedimented by centrifugation (2 × 5 min at 14,000 rpm) and resuspended in PBS to a final concentration of 0.9×10^{12} beads/ml. The PBS that kept the wells wet is replaced by 70 μl of beads solution, and the plate is incubated at 37° for 2 h. The wells are then washed gently with PBS using the plate washer. After the last wash, 70 μl of PBS is left in each well.

3. PBS is replaced with 50 μl of culture medium. One hundred fifty to 250 cells suspended in a volume of 50 μl are spread gently and evenly over each well area using a multichannel micropipette and incubated to perform their motility at 37°, 5% CO_2 for 5 to 9 h (depending on the cell type).

4. Following incubation, the cells are fixed in 3% paraformaldehyde for 15 min and washed twice with PBS.

5. Fixed plates are imaged by the automated microscope, and the images are analyzed for tracks characterization.

Prospectives

This chapter outlined the development of screening systems and procedures enabling high-throughput analysis of cellular samples (i.e., in the order of thousands of samples per hour), associated with harvesting of high-resolution or "high-content" information. The development of such systems and their wide application requires a dedicated multidisciplinary effort, combining essential biological input, needed for the development of suitable "reporter cells," with innovative know-how of microscope optics and mechanics and advanced image processing. Systems based on such combined expertise are likely to have considerable impact on basic biological research as well as on a variety of biomedical applications. High-resolution automated microscopy can provide valuable information on cutting-edge topics such as functional genomics, enabling the characterization of the cellular action of multiple genes, systems biology, probing the concerted action of multiple biological networks, and more. Novel biomedical applications are also likely to benefit greatly from advanced automated microscopy. This includes drug discovery and development processes, based on primary or secondary microscopy-assisted screens of chemical libraries, or comprehensive analysis of genetic factors affecting drug responsiveness. Naturally, to fully develop the approach described here for large genomic or pharmacological screens, one needs to further develop the automation of the screen, by setting robotic procedures for plate preparation, and processing, and improve the throughput of the microscope. It is also essential to install a fast on-line image acquisition and analysis system capable of quantification of fine changes in multiple cellular features. Given the wide recent interest in such systems it is anticipated that automated microscopy will be developed into one of the important tools in biomedical research.

Acknowledgments

The development of the screening systems was supported by NIGMS, National Institutes of Health Cell Migration Consortium Grant U54 GM64346 and the Israel Science Foundation (to BG) and the Binational Science Foundations (to ZK). Development of the automated system was supported by the Kahn Fund for Systems Biology at the Weizmann Institute of Science. BG holds the E. Neter Chair in Cell and Tumor Biology, and ZK is the Israel Pollak Professor in Biophysics.

References

Abraham, V., Taylor, D., and Haskins, J. (2004). High content screening applied to large scale cell biology. *Trends Biotech.* **22**, 15–22.

Albrecht-Buehler, G. (1977). The phagokinetic tracks of 3T3 cells. *Cell* **11**, 395–404.

Boddy, L., Wilkins, M. F., and Morris, C. W. (2001). Pattern recognition in flow cytometry. *Cytometry* **44**, 195–209.

Boutros, M., Kiger, A. A., Armknecht, S., Kerr, K., Hild, M., Koch, B., Haas, S. A., Consortium, H. F., Paro, R., and Perrimon, N. (2004). Genome-wide RNAi analysis of growth and viability in Drosophila cells. *Science* **303**, 832–835.

Boyden, S. (1962). The chemotactic effect of mixtures of antibody and antigen on poly-morphonuclear leucocytes. *J. Exp. Med.* **115**, 453–466.

Clemons, P. A. (2004). Complex phenotypic assays in high-throughput screening. *Curr. Opin. Chem. Biol.* **8**, 334–338.

Conrad, C., Erfle, H., Warnat, P., Daigle, N., Lorch, T., Ellenberg, J., Pepperkok, R., and Eils, R. (2004). Automatic identification of subcellular phenotypes on human cell arrays. *Genome Res.* **14**, 1130–1136.

Hallab, N., Jacobs, J. J., and Black, J. (2000). Hypersensitivity to metallic biomaterials: A review of leukocyte migration inhibition assays. *Biomaterials* **21**, 1301–1314.

Huh, W. K., Falvo, J. V., Gerke, L. C., Carroll, A. S., Howson, R. W., Weissman, J. S., and O'Shea, E. K. (2003). Global analysis of protein localization in budding yeast. *Nature* **425**, 686–691.

Kam, Z., Jones, M. O., Chen, H., Agard, D., and Sedat, J. (1993). Design and construction of an optimal illumination system for quantitative wide field multidimensional microscopy. *Bioimaging* **1**, 71–81.

Kam, Z., Zamir, E., and Geiger, B. (2001). Probing molecular processes in live cells by quantitative multidimensional microscopy. *Trends Cell Biol.* **11**, 329–334.

Lee, H., Goetzl, E. J., and An, S. (2000). Lysophosphatidic acid and sphingosine 1- phosphate stimulate endothelial cell wound healing. *Am. J. Physiol. Cell Physiol.* **278**, C612–C618.

Lichtenstein, N., Geiger, B., and Kam, Z. (2003). Quantitative analysis of cytoskeletal organization by digital fluorescent microscopy. *Cytometry A* **54**, 8–18.

Liron, Y., Paran, Y., Geiger, B., and Kam, Z. (2006). Laser autofocusing system for high-resolution cell biological imaging. *J. Microscopy* **221**, 145–151.

Mitchison, T. J. (2005). Small-molecule screening and profiling by using automated micro-scopy. *Chembiochem.* **6**, 33–39.

Murphy, R. F. (2005). Cytomics and location proteomics: Automated interpretation of subcellular patterns in fluorescence microscope images. *Cytometry A* **67A**, 1–3.

Perlman, Z. E., Mitchison, T. J., and Mayer, T. U. (2005). High-content screening and profiling of drug activity in an automated centrosome-duplication assay. *Chembiochem.* **6**, 145–151.

Perlman, Z. E., Slack, M. D., Feng, Y., Mitchison, T. J., Wu, L. F., and Altschuler, S. J. (2004). Multidimensional drug profiling by automated microscopy. *Science* **306**, 1194–1198.

Peterson, J. R., and Mitchison, T. J. (2002). Small molecules, big impact: A history of chemical inhibitors and the cytoskeleton. *Chem. Biol.* **9**, 1275–1285.

Sachse, C., Krausz, E., Kronke, A., Hannus, M., Walsh, A., Grabner, A., Ovcharenko, D., Dorris, D., Trudel, C., Sonnichsen, B., and Echeverri, C. J. (2005). High throughput RNA interference strategies for target discovery and validation by using synthetic short interfering RNAs: Functional genomics investigations of biological pathways. *Methods Enzymol.* **392**, 242–277.

Schutze, N. (2004). siRNA technology. *Mol. Cell. Endocrinol.* **213**, 115–119.

Sharon, E., Brandt, A., and Basri, R. (2001). Segmentation and boundary detection using multiscale intensity measurements. *In* "IEEE Comput. Soc. Conf. Comput. Vision Pattern Recognit.," pp. 1469–1476.

Smith, C., and Eisenstein, M. (2005). Automated imaging: Data as far as the eye can see. *Nature Methods* **2**, 547–555.

Stockwell, B. R. (2004). Exploring biology with small organic molecules. *Nature* **432**, 846–854.

Tamura, M., Gu, J., Matsumoto, K., Aota, S., Parsons, R., and Yamada, K. M. (1998). Inhibition of cell migration, spreading, and focal adhesions by tumor suppressor PTEN. *Science* **280**, 1614–1617.

Tanaka, M., Bateman, R., Rauh, D., Vaisberg, E., Ramachandani, S., Zhang, C., Hansen, K. C., Burlingame, A. L., Trautman, J. K., Shokat, K. M., and Adams, C. L. (2005). An unbiased cell morphology-based screen for new, biologically active small molecules. *PLoS Biol.* **3**, e128.

Yarrow, J. C., Feng, Y., Perlman, Z. E., Kirchhausen, T., and Mitchison, T. J. (2003). Phenotypic screening of small molecule libraries by high throughput cell imaging. *Comb. Chem. High Throughput Screen* **6**, 279–286.

Zamir, E., Geiger, B., Cohen, N., Kam, Z., and Katz, B. Z. (2005). Resolving and classifying haematopoietic bone-marrow cell populations by multi-dimensional analysis of flow-cytometry data. *Br. J. Haematol.* **129**, 420–431.

Zamir, E., Katz, B. Z., Aota, S., Yamada, K. M., Geiger, B., and Kam, Z. (1999). Molecular diversity of cell-matrix adhesions. *J. Cell Sci.* **112**(Pt 11), 1655–1669.

Zhou, X., Cao, X., Perlman, Z., and Wong, S. T. (2006). A computerized cellular imaging system for high content analysis in Monastrol suppressor screens. *J. Biomed. Inform.* **39**(2), 115–125.

Zigmond, S. H. (1977). Ability of polymorphonuclear leukocytes to orient in gradients of chemotactic factors. *J. Cell Biol.* **75**, 606–616.

[14]   Adenoviral Sensors for High-Content Cellular Analysis

By Jonathan M. Kendall, Ray Ismail, and Nick Thomas

Abstract

To maximize the potential of high-content cellular analysis for investigating complex cellular signaling pathways and processes, we have generated a library of adenoviral encoded cellular sensors based on protein translocation and reporter gene activation that enable a diverse set of assays to be applied to lead compound profiling in drug discovery and development. Adenoviral vector transduction is an efficient and technically simple system for expression of cellular sensors in diverse cell types, including primary cells. Adenoviral vector-mediated transient expression of cellular sensors, either as fluorescent protein fusions or live cell gene reporters, allows rapid assay development for profiling the activities of

METHODS IN ENZYMOLOGY, VOL. 414   0076-6879/06 $35.00
DOI: 10.1016/S0076-6879(06)14014-8

candidate drugs across multiple cellular systems selected for biological and physiological relevance to the target disease state.

Introduction

There is an emerging realization in drug discovery that pursuing industrialized high-throughput screening (HTS) of massive compound libraries, while generating huge quantities of data, has not significantly improved productivity or generated better drugs. These deficiencies are driving new efforts to improve the drug development process by integrating data from a range of disciplines (Bleicher *et al.*, 2003) into a more unified and knowledge-based process (Bajorath, 2002). This trend is accompanied by a shift in emphasis away from solely the number of data points that can be generated towards a focus on the quality of data obtained (Walters *et al.*, 2003). Recent developments in high-throughput microscopy (Ramm and Thomas, 2003; Mitchison, 2005; Price *et al.*, 2002) allow high-definition intracellular analysis at very high rates, with high-end systems imaging as many as 30,000 wells per day, with 500 cells analyzed per well. Coupled with cellular sensors of appropriate sensitivity and sophistication (Taylor *et al.*, 2001) and matching image analysis software, these high-content screening (HCS) and high-content analysis (HCA) platforms provide the potential to generate detailed and informative functional data at levels of throughput that were previously obtainable only with instrumentation yielding low-complexity data. Ongoing developments in hardware and software are continually improving the power and statistical robustness of cellular imaging, allowing HCA to play an increasingly significant role in decision making toward the early identification of viable and safe drug compounds.

A critical requirement for high-content analysis is the availability of specific and sensitive sensors for key cellular pathways and processes; it is these sensors that are at the heart of producing data-rich cellular images. Many sensors for cellular analysis use genetically encoded elements (Guerrero and Isacoff, 2001; Tsien, 2005; Wouters *et al.*, 2001; Zhang *et al.*, 2002) either as the complete sensor, for example, a green fluorescent protein (GFP)–fusion protein, or as part of a sensor system, for example, a reporter gene generating a fluorescent product.

Cell-based assays used for drug discovery have traditionally used stable cell lines to express genetically encoded sensors (Lundholt *et al.*, 2005; Pagliaro and Praestegaard, 2001). While these engineered cells are ideally suited to primary screens requiring large numbers of cells, the timescales and resources required to generate stable cells generally restrict sensor-expressing cell lines to standard cell types such as HEK293 (Hamdan *et al.*, 2005), CHO (Oosterom *et al.*, 2005), and U2OS (Oakley *et al.*, 2002).

One of the major reasons for the limited diversity of stable cell lines are the restraints imposed by plasmid-based transfection procedures. Chemical transfection procedures using cationic lipid complexes (Dass, 2004), polymers (Dennig, 2002), or peptides (Fischer et al., 2001) often need extensive optimization (Colosimo et al., 2000) and generally yield only low transfection efficiencies in nontransformed and primary cells (Kiefer et al., 2004; Welter et al., 2004; Young et al., 2004). While alternative physical transfection procedures, including novel methods based on electroporation (Distler et al., 2005), are available, these have not been used extensively to generate sensor expressing cell lines.

To overcome the limitations of using stably engineered cells and hence to enable a more diverse set of assays to be applied in lead compound profiling, we have generated a library of adenoviral encoded cellular sensors based on protein translocation and reporter gene activation.

Design and Construction of Adenoviral Sensors for Cellular Assays

Adenoviral vectors have a number of distinct advantages over chemical transfection methods and other viral vectors as a universally applicable cell delivery vehicle for the rapid development of transient cellular assays in different cell types, including primary cells (Table I). Adenoviral vectors are compatible with the transduction of a large range of cell types, including terminally differentiated cells, and have high transduction efficiency, allowing sensor expression to be achieved at a low multiplicity of infection (MOI), minimizing perturbation of cellular processes. Sensor expression from adenoviral vectors does not require integration into the host cell genome, allowing sensor expression without possible modification of chromatin structure or activity arising from random DNA sequence insertion.

Recombinant adenoviral vectors for expression of cellular sensors are engineered (Fig. 1) by replacing the E1 gene of the viral genome (Bett et al., 1993; Hitt et al., 1997) with cDNA encoding the gene of interest fused to a fluorescent protein or with a reporter gene construct. Removal of the E1 gene renders the adenoviral vectors replication deficient; however, they can be propagated by infection of cells such as HEK293 that provide the E1 function by complementation. To further increase the capacity of the vectors for insertion of foreign DNA, the vector system has a further deletion in the nonessential E3 region of the genome (Bett et al., 1994).

Adenoviral vector entry into the cell (transduction) involves a number of interactions between proteins on the capsid coat of the vector and target cell surface (Vellinga et al., 2005). The process starts with absorption of the virus onto target cells through interaction with the coxsackie and adenovirus receptor (CAR) on the cell surface, and entry of the adenoviral vector

TABLE I

COMPARISON OF CHEMICAL AND VIRAL GENE DELIVERY METHODS AND VECTORS

	Chemical	Adenoviral	MMLV (retroviral)	Lentiviral (retroviral)	Baculoviral	Adenoviral advantages
Tropism	Broad	Broad	Broad	Broad	Restricted	Can be used to transduce many different cell types
Transduction efficiency	Low	High	Low	Low	Low	Effective for cell types that are difficult to transfect Low multiplicity of infection required for transduction
Terminally differentiated cells	Variable	Yes	No	Yes	Yes	Enables sensor expression and analysis in physiologically relevant cell types
Expression	Transient/ stable	Transient	Stable	Stable	Transient/ stable	No adverse effects from integration of gene into host cell chromosome

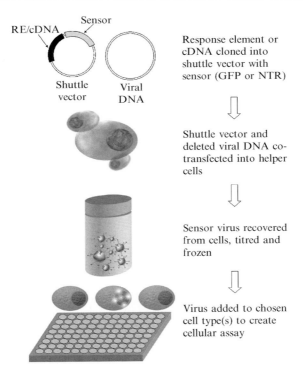

Response element or
cDNA cloned into
shuttle vector with
sensor (GFP or NTR)

Shuttle vector and
deleted viral DNA co-
transfected into helper
cells

Sensor virus recovered
from cells, titred and
frozen

Virus added to chosen
cell type(s) to create
cellular assay

FIG. 1. Engineering adenoviral sensors and assays. A shuttle vector containing a response element (RE) linked to nitroreductase (NTR) for a reporter gene assay or a cDNA sequence linked to GFP for a translocation assay is complemented with modified viral DNA sequences in cotransfection of helper cells in tissue culture. Recombination of sensor and viral DNA, supplemented by additional viral proteins encoded in the helper cells, allows production of replication-deficient adenovirus particles. The sensor encoding adenovirus is recovered from the cell growth media, purified, titered to a defined level of infectivity, and frozen. The addition of adenovirus to assay host cells results in sensor expression without viral replication.

follows via interaction of the capsid penton base with cell surface integrins, $\alpha_v\beta_3$ and $\alpha_v\beta_5$. Viral internalization occurs through clathrin-mediated endocytosis followed by pH-dependent release from endosomes into the cytoplasm. Subsequent transport to the nucleus allows DNA replication and transcription to proceed from an epichromosomal location, resulting in sensor expression.

Transduction by adenoviral vectors is a quick and simple process and does not require any specialized equipment or additional reagents. The adenoviral vector is simply added to cells in culture medium and cells are cultured to allow sufficient expression of the encoded sensor to permit

TABLE II

COMPARISON OF EFFICIENCY OF ADENOVIRAL AND CHEMICAL METHODS FOR
GFP SENSOR DELIVERY

	% transduction/transfection with GFP-glucocorticoid receptor		
	Adenoviral	Fugene 6	Lipofectamine
HeLa	83 ± 11	55 ± 8	51 ± 8
U2OS	81 ± 12	24 ± 11	14 ± 8
HepG2	29 ± 12	10 ± 8	5 ± 4
SW1353	32 ± 32	9 ± 11	7 ± 3

detection and analyses. Expression can be detected within 12 h of transduction, and the majority of GFP translocations and reporter genes can be analyzed 24 h after transduction of target cells.

Adenoviral vectors can efficiently transduce a broad range of cell types and provide higher efficiency delivery than chemical transfection (Table II). They are particularly attractive for studies in primary cells and nondividing cells associated with differentiated tissues such as brain and heart that are refractory to chemical methods. Endogenous levels of CAR and integrins vary between cell types and determine the efficiency of a particular cell line to transduction. However, the amount of virus added to cells can be optimized readily to achieve a desired level of expression, and transduction efficiency is very reproducible, allowing routine use with large quantities of cells, providing an ideal assay development tool.

Adenoviral vectors have been used previously as research tools to transduce cells with GFP translocation sensors (Kajimoto et al., 2001; Kang and Walker, 2005) and reporter gene assays (Hartig et al., 2002; McPhaul et al., 1993). However, despite the wide use of adenovirus for target expression (Darrow et al., 2003; Ogorelkova et al., 2004) for drug screening applications, these vectors have not seen significant use for delivery of sensors for high-content analysis on automated imaging platforms. To increase the potential of HCA in drug activity profiling, we have assembled a library of adenoviral sensors based on cDNA-GFP fusion proteins and a complementary live cell reporter gene assay (Thomas, 2002) based on the bacterial enzyme nitroreductase and a quenched cyanine dye, CytoCy5S.

Employing these sensors alone and in combination (Fig. 2) provides the potential to place reporter gene assays at the terminus of multiple signaling pathways to determine the route of action and specificity of candidate drugs and to deploy GFP translocation sensors at key intervention points on the same pathways to increase the resolution and precision of activity profiling data.

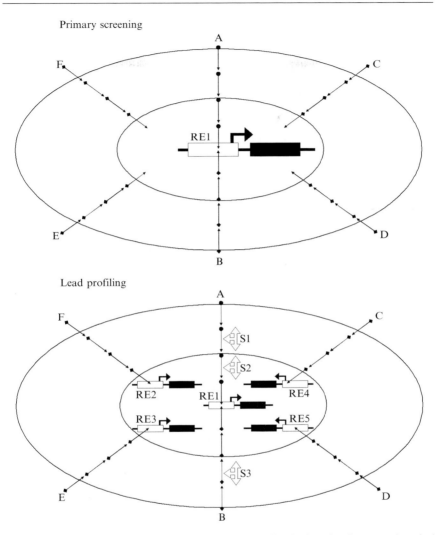

FIG. 2. Application of adenoviral sensors to lead profiling in drug development. A typical primary screening scenario for an assay using a stable reporter gene cell line is illustrated. In the assay, compounds may interact with a number of cell-signaling pathways (A–F). Action of a compound on a target pathway A, linked to the reporter gene via a response element (RE1), will generate an assay "hit." Other compounds may also induce or repress reporter gene activity through alternative pathways (B) impinging on the same response element, producing a false-positive hit. In addition, compounds acting on the target pathway (A) may also interact with other pathways (C, D, E, and F) not connected to the reporter gene. Applying adenoviral sensors to lead profiling allows the activity of compounds arising from primary screening to be characterized in detail. Using a panel of sensors transiently expressed in one or more cell

Validation of Adenoviral Sensors

To validate the performance of adenovirus-encoded sensors, we have compared assay data obtained using transient viral expression with data from the same sensors expressed in transient plasmid-based expression assays and in stable cell lines.

Analysis of the translocation of a rapidly responding translocation sensor, an EGFP fusion to the pleckstrin homology (PH) domain of the delta-1 isoform of phospholipase C, showed comparable assay data (Fig. 3) from a stable CHO cell line and CHO cells transduced with the same fusion protein using adenoviral delivery. Analysis of dose–response curves obtained from image analysis yielded EC_{50} values for ATP stimulation of 6.3 μM for the stably expressed sensor and 13.2 μM for the virally encoded sensor.

Comparison of a reporter gene assay carried out using plasmid transfection and adenoviral transduction showed significantly improved assay statistics using the adenoviral vector (Fig. 4C). Signal to noise (assay range/root sum SD squared) was 7.3 for plasmid assays and 55.3 for adenoviral assays at 100 μM forskolin. Adenoviral transduction allowed nitroreductase (NTR) reporter gene assays to be run in either live or fixed cell format using both HTS and HCS platforms (Fig. 4D) with increased precision, yielding forskolin EC_{50} values (95% confidence limits) of 0.8–4.5 μM for plasmid assays and 0.4–0.8 μM for adenoviral assays.

Application of Adenoviral Sensors to High-Content Analysis

Transient expression of adenoviral sensors allows rapid assay development following a number of key steps. Preliminary literature research should be carried out to determine the biological requirements for the assay and to gather information that is relevant to the drug target and pathway of interest to aid in the choice of one or more sensors. At this stage one or more appropriate cell lines should be selected, taking into account endogenous expression of any key components in the signaling

---

types, the interactions of compounds with a range of signaling pathways may be profiled. An adenoviral nitroreductase (NTR) reporter gene sensor using the same response element used in the primary screen (RE1) may be used to confirm on-target (pathway A) and off-target (pathway B) activity detected in the primary screen. The target pathway may be additionally monitored using adenoviral GFP fusion protein translocation sensors (S1 and S2) upstream of the reporter gene to determine the site of action of the candidate drug at higher resolution. Similarly, further sensors (S3) may be deployed on pathways known to interact with the readout used in primary screening to gain further information on compounds showing off-target activity. Finally, a battery of adenovirus-encoded NTR reporter gene assays using a wide range of response elements (RE2-RE5) may be employed to monitor further off-target activity of hit compounds against other cellular signaling pathways.

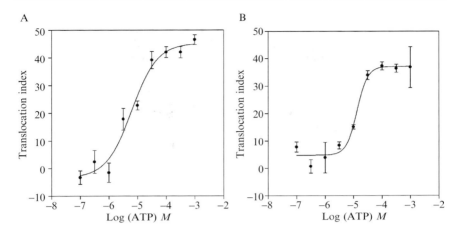

FIG. 3. GFP translocation sensor under stable and transient expression conditions. (A) CHO cells stably expressing a EGFP-phospholipase C delta-1 PH domain (PLCD1-PH) fusion were treated with a range of ATP concentrations to induce translocation of the fusion protein from the plasma membrane to the cytoplasm and the redistribution of the protein quantified by automated image analysis. (B) The same fusion protein was transiently expressed in CHO cells using an adenoviral vector and ATP-induced translocation measured under identical conditions.

pathway(s) under investigation. Any available information on response dynamics, suitable agonists and antagonists for positive controls, and cell culture conditions should be evaluated to design a prototype assay protocol. Once this basic protocol is established, optimization of viral transduction can proceed. At this stage experiments will be required to optimize the MOI (the amount of virus that should be added to each cell). During this period of assay development the aim is to establish the efficiency of transduction, to confirm that no toxic effects occur, and to determine that sensor response to a chosen stimulus can be detected. Once the optimum MOI has been established, further assay development activities addressing cell seeding density, incubation time, replicate numbers, and other assay design factors can be finalized to maximize the assay performance for reproducibility, signal to noise, and other metrics. The protocols and procedures described next are typical of those used in our laboratory for expression and analysis of adenoviral-encoded sensors.

EGFP-Glucocorticoid Receptor (GR) Translocation Sensor

The glucocorticoid receptor, a member of the steroid receptor superfamily, resides predominantly in the cytoplasm and accumulates in the

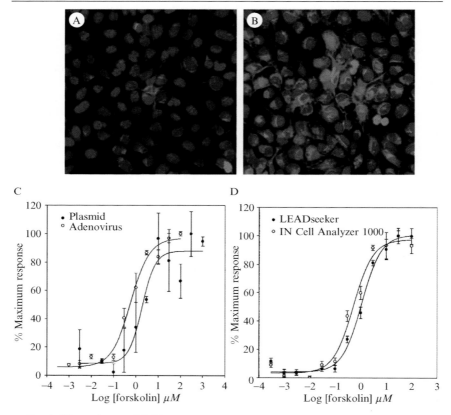

FIG. 4. Nitroreductase (NTR) reporter gene assay. A CRE-NTR construct containing four repeats of the cAMP response element (CRE) upstream of a minimal promoter was used to measure transcriptional activity of cAMP-binding protein (CREB) in transient assays in response to forskolin stimulation. Cells transduced with adenoviral vector encoding the CRE-NTR construct were loaded with CytoCy5S NTR substrate and imaged on IN Cell Analyzer 1000 in the absence (A) and presence (B) of forskolin. (C) Comparison of transient live cell assays performed using plasmid transfection and adenoviral delivery of the CRE-NTR construct showing improved reproducibility with the adenoviral vector. (D) Fixed cell CRE-NTR assay analyzed using macroimaging (LEADseeker) and automated microscopy (IN Cell Analyzer 1000).

nucleus when glucocorticoids such as cortisol bind to the receptor (Kumar and Thompson, 2005). A consequence of the nuclear translocation is the binding of GR to specific DNA elements in the promoter regions of target genes and concomitant regulation of gene transcription rates. The absolute response following gene transcription will depend on the cellular context.

A fusion protein of GR and EGFP provides a sensor to examine the distribution of the receptor within the nucleus and cytoplasm in response to agonists or antagonists of the receptor. The EGFP–GR chimera functions normally in cytoplasmic/nuclear translocation and gene activation and responds to pharmacologically relevant stimuli (Walker *et al.*, 1999). The adenoviral vector containing the gene encoding EGFP fused to the GR enables expression of the transgene in cultured cells, allowing the intracellular translocation of the fusion protein to be monitored on automated epifluorescent microscopes in the presence of an appropriate agonist such as dexamethasone.

Reagent Preparation

Prepare the following reagents. Culture medium; Dulbecco's Modified Eagle's Medium (DMEM), with 10% fetal calf serum (FCS), 2 mM L-glutamine, 100 units/ml penicillin, and 100 μg/ml streptomycin. Culture medium + charcoal-stripped FCS (CS-FCS); DMEM with 2 mM L-glutamine and 10% CS-FCS. Agonist; dexamethasone, prepare 50 μM stock in ethanol and store in suitable aliquots at –20°. Phosphate-buffered saline. Fixative; formalin 10%, dilute to 5% in phosphate-buffered saline (PBS). Hoechst nuclear dye; prepare 10 μM stocks in PBS, store at –20°, and protect from direct light. Prepare assay test compounds by diluting in culture medium + CS-FCS to concentrations twice that of the desired final assay concentration.

Assay Procedure

1. The day before starting the assay, detach logarithmically growing HeLa cells by treatment with trypsin. Adjust the cell count to 2.5×10^4 ml with culture medium to a final volume of 20 ml (i.e., 5×10^5 cells in 20 ml culture medium). Thaw the EGFP-GR adenoviral vector (GE Healthcare) by placing the tube on ice. Add 50 μl (2.5×10^7 ifu) adenovirus to the cell suspension and mix by inverting the tube gently. At this cell density, this procedure will deliver the optimum 50 MOI to the HeLa cells.

2. Dispense 200 μl of the cell suspension per well of a 96-well imaging quality plate. Incubate the cells for 24 h at 37°, 5% CO_2 in a tissue culture incubator.

3. Warm all assay reagents to 37°. Decant the culture medium from the cells and add 100 μl of medium containing CS-FCS. Incubate the cells for 1 h at 37°, 5% CO_2.

4. Add 100 μl of the prepared two-fold dilution stocks of the test compounds.

5. Incubate the plates at 37°, 5% CO_2 for 30 min.

6. Decant the assay medium from the cells. Add 100 μl fixative to each well and incubate for 15 min at room temperature.

7. Decant the fixative and wash the cells in each well with 200 μl PBS.

8. Decant the PBS, add 100 μl of 2.5 μM Hoechst to each well and incubate for 15 min to stain nuclei.

9. Image the cells using automated microscopy (e.g., IN Cell Analyzer 1000 or IN Cell Analyzer 3000, GE Healthcare).

The fixed cell protocol just described provides a convenient assay procedure that does not require complex scheduling where large numbers of test compounds are to be evaluated; however, preliminary experiments may be required to determine the optimal time for fixation following stimulus. Alternatively, the procedure may be carried out using live cell imaging to follow translocation in a dynamic assay. For live cell assays it is advisable to stagger the timing of addition of test compounds to coincide with image acquisition times to ensure that each well is imaged at the same relative time following stimulus. For live cell assays, the Hoechst nuclear stain (5 μM) should be added to the test compounds prior to addition to the cells.

Assay Analysis

Typical image acquisition and analysis procedures for the EGFP-GR sensor using IN Cell Analyzer 1000 and IN Cell Analyzer 3000 are outlined. Image acquisition and procedures on other platforms will vary, but follow the same basic procedure. For image acquisition on a lamp-based imager (e.g., IN Cell Analyzer 1000), use a Q505 long pass dichroic with filters selected for excitation at 360 and 475 nm and emission at 535 nm for both channels. A 10× objective provides sufficient resolution for the assay, but a 20× objective may be employed if preferred. For image acquisition on the line scanning confocal IN Cell Analyzer 3000, EGFP is imaged using 488-nm laser line excitation with emission monitored through the 535-45-nm filter and the Hoechst nuclear stain imaged using 363-nm laser line excitation and a 450-65-nm filter emission filter.

Image analysis is performed using nuclear trafficking analysis software to quantitatively measure the fluorescence intensity in the cytoplasm and the nucleus of each cell. The software provides single cell and population averaged data for the nuclear/cytoplasmic (Nuc/Cyt) ratio of sampled nuclear and cytoplasmic EGFP-GR intensities as a measure of intracellular sensor distribution. Typical data for the EGFP-GR sensor expressed in HeLa cells and imaged on IN Cell Analyzer 1000 are shown in Fig. 5.

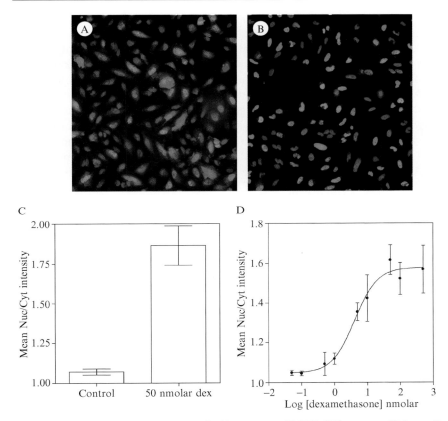

Fig. 5. Adenoviral EGFP-glucocorticoid receptor (EGFP-GR) sensor. HeLa cells expressing the EGFP-GR were imaged on IN Cell Analyzer 1000 after a 30-min incubation in the absence (A) or presence (B) of dexamethasone. Automated image analysis (C) was used to determine the nuclear/cytoplasmic (Nuc/Cyt) distribution of the sensor in 30 replicate control and dexamethasone-treated wells yielding an assay Z factor (Zhang *et al.*, 1999) value of 0.47. Exposure of transduced cells to a range of dexamethasone concentrations (D) yielded an EC_{50} of 4.1 n*M*.

Nuclear Factor of Activated T Cells (NFAT) Nitroreductase Reporter Gene Sensor

Nuclear factor of activated T cell transcription factors (Macian, 2005) are a family of at least four (NFAT1–4) structurally and functionally related proteins expressed in the immune system and in muscle, cardiac, and neuronal cells. In T cells, ligand binding at antigen receptors results in cell activation, leading to cytokine production, which plays a major role in modulating the immune system. A key feature of this type of regulated transcription in T cells is the Ca^{2+} calmodulin-dependent activation of NFAT transcription factors via activation of the phosphatase calcineurin. Activated calcineurin dephosphorylates NFAT proteins, inducing a conformational change that unmasks a nuclear localization sequence inducing NFAT proteins to translocate to the nucleus where they bind to NFAT response elements and modulate gene expression. In the nucleus, NFATs may form synergistic complexes with other transcription factors, including AP-1 and NF-κB, to alter gene transcription. Expression and quantitation of a NTR reporter gene, cloned downstream of a basal promoter and NFAT response elements, provide a measure of the level of activity along the NFAT pathway in application areas such as differentiation, apoptosis, and immunosupression.

Reagent Preparation

Prepare the following reagents. Culture medium; DMEM, with 10% FCS, 2 mM L-glutamine, 100 μg/ml penicillin, and 100 μg/ml streptomycin. Agonists; phorbol myristate acetate (PMA) and ionomycin, prepare stock solutions in ethanol and store in suitable aliquots at –20°. PBS. Fixative; formalin 10%, dilute to 2% in PBS. Hoechst nuclear dye; prepare 25 μM stocks in PBS, store at –20°, and protect from direct light. Prepare assay test compounds by diluting in culture medium + CS-FCS to concentrations twice that of the desired final assay concentration. CytoCy5S solution; reconstitute with DMSO to a concentration of 1 to 5 mM. Dilute in serum-free media to 10 μM immediately before use.

Assay Procedure

1. The day before starting the assay prepare a suspension of trypsinized U2OS cells in culture medium. Thaw the NFAT-NTR adenoviral vector (GE Healthcare) by placing the tube on ice. Add 6 μl of adenovirus/5 $\times$ 10^5 cells and mix by inverting the tube gently to transduce cells at the optimal MOI of 6.

2. Seed cells into a 96-well assay plate at 15,000 cells per well and incubate for 24 h at 37°, 5% CO_2, 95% humidity.

3. Prior to starting the assay, prepare solutions of test compounds at an appropriate concentration in serum-free culture medium. Warm all assay reagents to 37°.

4. After 24 h in culture remove medium from cells and add 90 μl of test compound solution per well. Use 0.1 μg/ml PMA and/or 2 μM ionomycin

as positive controls. Add 90 μl of serum-free culture medium to negative control wells.

5. Incubate cells for 10 min at 37° and then add 10 μl of 10 μM CytoCy5S (GE Healthcare) nitroreductase substrate to each well. Incubate cells for a further 12 to 16 h at 37°, 5% CO_2, 95% humidity.

6. Measure CytoCy5S fluorescence intensity (excitation maximum 647 nm, emission maximum 667 nm) by imaging. For live cell macroimaging (e.g., LEADseeker, GE Healthcare), image the plate without further processing. For live cell imaging by automated microscopy, stain cell nuclei by the addition of 10 μl of 25 μM Hoechst for 30 min.

7. For imaging of fixed cells, decant the assay medium, add 100 μl fixative to each well, and incubate for 30 min at room temperature.

8. Decant the fixative and wash the cells in each well with 200 μl PBS. Decant the PBS and add 100 μl of 2.5 μM Hoechst to each well to stain nuclei.

Assay Analysis

The typical image acquisition and analysis procedures for NTR reporter gene assays using IN Cell Analyzer 1000 are outlined. Image acquisition and procedures on other platforms will vary, but will follow the same basic procedure.

1. Acquire images from each well using the 10× objective and a single tile located at the center of the well to capture data from 200 to 300 cells/ assay. Assays at lower cell density may require more images. Acquisition times are typically 200 to 500 ms for the Hoechst nuclear stain (ex. 360/40 nm, em. 535/50 nm) and 100 to 400 ms for CytoCy5S (ex. 620/60 nm, em. 700/75 nm).

2. Measure CytoCy5S fluorescence in cells using automated Object Intensity image analysis to determine the intensity of the red fluorescent NTR product in cells segmented using the Hoechst nuclear marker.

Data from application of the NFAT-NTR sensor in a multiplexed live cell assay in conjunction with a stably expressed GFP fusion protein are shown in Fig. 6. Transduction of the adenoviral NTR sensor into a cell line reporting cell cycle position via imaging of EGFP intensity and distribution allowed the cell cycle distribution of ionomycin-stimulated gene expression mediated through NFAT to be correlated with cell cycle position on an individual cell basis. Data from this study also demonstrate that delivery of a reporter gene sensor using an adenoviral vector at an optimal MOI does not perturb the cell cycle (Fig. 6D) nor diminish the expression or alter the behavior of a coexpressed GFP fusion protein.

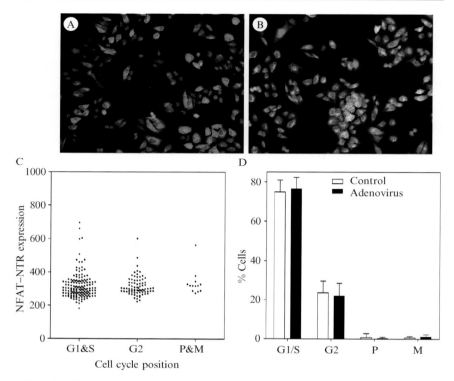

FIG. 6. NFAT-NTR reporter gene sensor in multiplexed assay. U2OS cells stably expressing a cyclin B1-EGFP fusion protein under the control of the cyclin B1 promoter, which dynamically reports cell cycle position (Thomas *et al.*, 2005), were transduced with adenovirus encoding an NFAT-NTR reporter gene construct. NTR expression was imaged in live cells on IN Cell Analyzer 1000 in the absence (A) and presence (B) of 1 μM ionomycin. Automated image analysis was used to determine the cell cycle position and ionomycin-induced NFAT reporter gene expression for each cell (C) and to compare cell cycle distribution in cells in the presence and absence of adenovirus (D).

Conclusions and Future Perspectives

The development of cell-based assays for secondary screening of compounds identified as hits in high-throughput primary screens has traditionally involved using stable cell lines. This approach can severely limit the depth and breadth of interrogation of biological activity of candidate drugs through limitation of the scope and relevance of biological model systems that can be engineered in standard laboratory cell lines. Adenoviral vector transduction provides an efficient and technically simple system for transient expression of assay sensors in diverse backgrounds from transformed to primary cells, providing the potential to profile the biological activity of compounds across a very diverse interrogation matrix.

Biological pathways are often viewed simplistically as a linear chain of cause-and-effect relationships, with the ERK 1 signaling pathway (Cobb *et al.*, 1994) being a typical example. Signal transduction starts at the EGF receptor (EGFR) at the cytoplasmic membrane. Binding of EGF causes receptor activation, transduction of signal to STAT1/3 in the cytoplasm, and internalization of the EGFR. Once activated, STAT1/3 translocate from the cytoplasm to the nucleus where the activated transcription factors promotes gene expression from STAT responsive control elements.

Viewed as an isolated element, this signaling pathway has three levels and at least one intervention point on each level at which virally encoded assays can be engineered to report signal transduction: receptor internalization, STAT translocation, and STAT-induced gene expression. EGFR internalization may be detected by viral expression of EGFR tagged to bind a pH responsive fluorescent dye that increases its fluorescence when the receptor is internalized into acidic vesicles (Adie *et al.*, 2003). STAT translocation may be measured by viral expression of a EGFP–STAT fusion protein, and STAT-induced gene expression may be measured by viral expression of a NTR reporter gene under control of an STAT response element. This basic strategy of inserting sensors at key points in signaling pathways to flag the passage of signal transduction events can be repeated and expanded for many pathways, targeted to distinguish routing of signaling in divergent, convergent, and interacting pathways, and the scope of interrogation diversified by expression of the same sensing networks in a variety of host cells reflecting different genotypes and phenotypes.

In the future we envisage data from multiple sensors expressed in parallel in diverse cellular backgrounds providing a high-definition view of drug activity and specificity, with each adenoviral sensor providing the equivalent of a pixel in an image. When viewed individually they may yield only a small amount of information; as a whole they provide the full picture.

Acknowledgments

The authors thank Sharon Davies and Kathy Lamerton for provision of some of the data used in this chapter.

References

Adie, E. J., Francis, M. J., Davies, J., Smith, L., Marenghi, A., Hather, C., Hadingham, K., Michael, N. P., Milligan, G., and Game, S. (2003). CypHer 5: A generic approach for measuring the activation and trafficking of G protein-coupled receptors in live cells. *Assay Drug Dev. Technol.* **1,** 251–259.

Bajorath, J. (2002). Integration of virtual and high-throughput screening. *Nature Rev. Drug Discov.* **1,** 882–894.

Bett, A. J., Haddara, W., Prevec, L., and Graham, F. L. (1994). An efficient and flexible system for construction of adenovirus vectors with insertions or deletions in early regions 1 and 3. *Proc. Natl. Acad. Sci. USA.* **91,** 8802–8806.

Bett, A. J., Prevec, L., and Graham, F. L. (1993). Packaging capacity and stability of human adenovirus type 5 vectors. *J. Virol.* **67,** 5911–5921.

Bleicher, K. H., Bohm, H. J., Muller, K., and Alanine, A. I. (2003). Hit and lead generation: Beyond high-throughput screening. *Nature Rev. Drug Discov.* **2,** 369–378.

Cobb, M. H., Hepler, J. E., Cheng, M., and Robbins, D. (1994). The mitogen-activated protein kinases, ERK1 and ERK2. *Semin. Cancer Biol.* **5,** 261–268.

Colosimo, A., Goncz, K. K., Holmes, A. R., Kunzelmann, K., Novelli, G., Malone, R. W., Bennett, M. J., and Gruenert, D. C. (2000). Transfer and expression of foreign genes in mammalian cells. *Biotechniques* **29,** 314–324.

Darrow, A. L., Conway, K. A., Vaidya, A. H., Rosenthal, D., Wildey, M. J., Minor, L., Itkin, Z., Kong, Y., Piesvaux, J., Qi, J., Mercken, M., Andrade-Gordon, P., Plata-Salaman, C., and Ilyin, S. E. (2003). Virus-based expression systems facilitate rapid target *in vivo* functionality validation and high-throughput screening. *J. Biomol. Screen.* **8,** 65–71.

Dass, C. R. (2004). Lipoplex-mediated delivery of nucleic acids: Factors affecting *in vivo* transfection. *J. Mol. Med.* **82,** 579–591.

Dennig, J., and Duncan, E. (2002). Gene transfer into eukaryotic cells using activated polyamidoamine dendrimers. *J. Biotechnol.* **90,** 339–347.

Distler, J. H., Jungel, A., Kurowska-Stolarska, M., Michel, B. A., Gay, R. E., Gay, S., and Distler, O. (2005). Nucleofection: A new, highly efficient transfection method for primary human keratinocytes. *Exp. Dermatol.* **14,** 315–320.

Fischer, P. M., Krausz, E., and Lane, D. P. (2001). Cellular delivery of impermeable effector molecules in the form of conjugates with peptides capable of mediating membrane translocation. *Bioconjug. Chem.* **12,** 825–841.

Guerrero, G., and Isacoff, E. Y. (2001). Genetically encoded optical sensors of neuronal activity and cellular function. *Curr. Opin. Neurobiol.* **11,** 601–607.

Hamdan, F. F., Audet, M., Garneau, P., Pelletier, J., and Bouvier, M. (2005). High-throughput screening of G protein-coupled receptor antagonists using a bioluminescence resonance energy transfer 1-based β-arrestin2 recruitment assay. *J. Biomol. Screen.* **10,** 463–475.

Hartig, P. C., Bobseine, K. L., Britt, B. H., Cardon, M. C., Lambright, C. R., Wilson, V. S., and Gray, L. E., Jr. (2002). Development of two androgen receptor assays using adenoviral transduction of MMTV-luc reporter and/or hAR for endocrine screening. *Toxicol. Sci.* **66,** 82–90.

Hitt, M. M., Addison, C. L., and Graham, F. L. (1997). Human adenovirus vectors for gene transfer into mammalian cells. *Adv. Pharmacol.* **40,** 137–206.

Kajimoto, T., Ohmori, S., Shirai, Y., Sakai, N., and Saito, N. (2001). Subtype-specific translocation of the delta subtype of protein kinase C and its activation by tyrosine phosphorylation induced by ceramide in HeLa cells. *Mol. Cell. Biol.* **21,** 1769–1783.

Kang, M., and Walker, J. W. (2005). Protein kinase C delta and epsilon mediate positive inotropy in adult ventricular myocytes. *J. Mol. Cell. Cardiol.* **38,** 753–764.

Kiefer, K., Clement, J., Garidel, P., and Peschka-Suss, R. (2004). Transfection efficiency and cytotoxicity of nonviral gene transfer reagents in human smooth muscle and endothelial cells. *Pharm. Res.* **21,** 1009–1017.

Kumar, R., and Thompson, E. B. (2005). Gene regulation by the glucocorticoid receptor: Structure:function relationship. *J. Steroid Biochem. Mol. Biol.* **94,** 383–394.

Lundholt, B. K., Linde, V., Loechel, F., Pedersen, H. C., Moller, S., Praestegaard, M., Mikkelsen, I., Scudder, K., Bjorn, S. P., Heide, M., Arkhammar, P. O., Terry, R., and Nielsen, S. J. (2005). Identification of Akt pathway inhibitors using redistribution screening on the FLIPR and the IN Cell 3000 Analyzer. *J. Biomol. Screen.* **10,** 20–29.

Macian, F. (2005). NFAT proteins: Key regulators of T-cell development and function. *Nature Rev, Immunol.* **5,** 472–484.

McPhaul, M. J., Deslypere, J. P., Allman, D. R., and Gerard, R. D. (1993). The adenovirus-mediated delivery of a reporter gene permits the assessment of androgen receptor function in genital skin fibroblast cultures: Stimulation of Gs and inhibition of G(o). *J. Biol. Chem.* **268,** 26063–26066.

Mitchison, T. J. (2005). Small-molecule screening and profiling by using automated microscopy. *Chembiochem.* **6,** 33–39.

Oakley, R. H., Hudson, C. C., Cruickshank, R. D., Meyers, D. M., Payne, R. E., Jr., Rhem, S. M., and Loomis, C. R. (2002). The cellular distribution of fluorescently labeled arrestins provides a robust, sensitive, and universal assay for screening G protein-coupled receptors. *Assay Drug Dev. Technol.* **1,** 21–30.

Ogorelkova, M., Elahi, S. M., Gagnon, D., and Massie, B. (2004). DNA delivery to cells in culture: Generation of adenoviral libraries for high-throughput functional screening. *Methods Mol. Biol.* **246,** 15–27.

Oosterom, J., van Doornmalen, E. J., Lobregt, S., Blomenrohr, M., and Zaman, G. J. (2005). High-throughput screening using beta-lactamase reporter-gene technology for identification of low-molecular-weight antagonists of the human gonadotropin releasing hormone receptor. *Assay Drug Dev. Technol.* **3,** 143–154.

Pagliaro, L., and Praestegaard, M. (2001). Transfected cell lines as tools for high throughput screening: a call for standards. *J. Biomol. Screen.* **6,** 133–136.

Price, J. H., Goodacre, A., Hahn, K., Hodgson, L., Hunter, E. A., Krajewski, S., Murphy, R. F., Rabinovich, A., Reed, J. C., and Heynen, M. (2002). Advances in molecular labeling, high throughput imaging and machine intelligence portend powerful functional cellular biochemistry tools. *J. Cell Biochem. Suppl.* **39,** 194–210.

Ramm, P., and Thomas, N. (2003). Image-based screening of signal transduction assays. *Sci. STKE* **177,** PE14.

Taylor, D. L., Woo, E. S., and Giuliano, K. A. (2001). Real-time molecular and cellular analysis: The new frontier of drug discovery. *Curr. Opin. Biotechnol.* **12,** 75–81.

Thomas, N. (2002). Cell based assays: Seeing the light. *Drug Discov. World* 25–31.

Thomas, N., Kenrick, M., Giesler, T., Kiser, G., Tinkler, H., and Stubbs, S. (2005). Characterization and gene expression profiling of a stable cell line expressing a cell cycle GFP sensor. *Cell Cycle* **4,** 191–195.

Tsien, R. Y. (2005). Building and breeding molecules to spy on cells and tumors. *FEBS Lett.* **579,** 927–932.

Vellinga, J., Van der Heijdt, S., and Hoeben, R. C. (2005). The adenovirus capsid: Major progress in minor proteins. *J. Gen. Virol.* **86,** 1581–1588.

Walker, D., Htun, H., and Hager, G. L. (1999). Using inducible vectors to study intracellular trafficking of GFP-tagged steroid/nuclear receptors in living cells. *Methods* **19,** 386–393.

Walters, W. P., and Namchuk, M. (2003). Designing screens: How to make your hits a hit. *Nature Rev. Drug Discov.* **2,** 259–266.

Welter, J. F., Solchaga, L. A., and Stewart, M. C. (2004). High-efficiency nonviral transfection of primary chondrocytes. *Methods Mol. Med.* **100,** 129–146.

Wouters, F. S., Verveer, P. J., and Bastiaens, P. I. (2001). Imaging biochemistry inside cells. *Trends Cell Biol.* **11**, 203–211.

Young, A. T., Moore, R. B., Murray, A. G., Mullen, J. C., and Lakey, J. R. (2004). Assessment of different transfection parameters in efficiency optimization. *Cell Transplant.* **13**, 179–185.

Zhang, J., Campbell, R. E., Ting, A. Y., and Tsien, R. Y. (2002). Creating new fluorescent probes for cell biology. *Nature Rev. Mol. Cell. Biol.* **3**, 906–918.

Zhang, J. H., Chung, T. D., and Oldenburg, K. R. (1999). A simple statistical parameter for use in evaluation and validation of high throughput screening assays. *J. Biomol. Screen.* **4**, 67–73.

[15] Cell-Based Assays Using Primary Endothelial Cells to Study Multiple Steps in Inflammation

By Thomas Mayer, Bernd Jagla, Michael R. Wyler, Peter D. Kelly, Nathalie Aulner, Matthew Beard, Geoffrey Barger, Udo Többen, Deborah H. Smith, Lars Brandén, and James E. Rothman

Abstract

Cell-based assays are powerful tools for drug discovery and provide insight into complex signal transduction pathways in higher eukaryotic cells. Information gleaned from assays that monitor a cellular phenotype can be used to elucidate the details of a single pathway and to establish patterns of cross talk between pathways. By selecting the appropriate cell model, cell-based assays can be used to understand the function of a specific cell type in a complex disease process such as inflammation. We have used human umbilical vein endothelial cells to establish three cell-based, phenotypic assays that query different stages of a major signaling pathway activated in inflammation. One assay analyzes the tumor necrosis factor α (TNFα)-induced translocation of the transcription factor NF-κB from the cytoplasm into the nucleus 20 min after stimulation with TNFα. Two more assays monitor the expression of E-selectin and VCAM-1, 4 and 24 h after stimulation with TNFα. Indirect immunofluorescence and high-throughput automated microscopy were used to analyze cells. Imaging was performed with the IN Cell Analyzer 3000. All assays proved to be highly robust. Z' values between 0.7 and 0.8 make each of the three assays well suited for use in high-throughput screening for drug or probe discovery.

METHODS IN ENZYMOLOGY, VOL. 414
Copyright 2006, Elsevier Inc. All rights reserved.

0076-6879/06 $35.00
DOI: 10.1016/S0076-6879(06)14015-X

Cellular Mechanism of the Inflammatory Response

Chronic inflammatory disease is believed to pose a tremendous medical burden in the developed world, both in terms of patient suffering and in the cost of treatment and loss of worker productivity. The more common inflammatory diseases include rheumatoid arthritis, inflammatory bowel disease, atherosclerosis, chronic obstructive pulmonary disease, and psoriasis. Although each disease has unique aspects regarding the affected tissues and the clinical symptoms, they all share some common biological mechanisms for the establishment and maintenance of the disease state.

Inflammatory disease requires that endothelial cells detect and amplify a proinflammatory signal. This results in the adherence, activation, and ultimately transmigration of lymphocytes at the site of damage. Drugs that block this process would significantly alleviate the symptoms of inflammation.

Some of the major signaling molecules that act on the endothelial cell include the cytokines, tumor necrosis factor α (TNFα), interleukin 1 (IL-1), and CD40L (Aggarwal, 2003; Martin and Wesche, 2002). In each case, the signaling molecule binds to a receptor on the surface of the endothelial cell and activates a cascade of internal events. A key event in the early inflammatory process is the translocation of the transcription factor NF-κB from the cytoplasm to the nucleus of the endothelial cell, an early event (20 min) after receptor stimulation (Ding *et al.*, 1998). NF-κB then activates the transcription of an array of proinflammatory proteins (Senftleben and Karin, 2002). These include proteins that regulate the attachment of lymphocytes to the wall of the blood vessel such as proteins of the selectin family (Karmann *et al.*, 1996). One of these, E-selectin, is clearly detectable on the endothelial cell surface 4 h poststimulation. At later times (24 h poststimulation), lymphocyte recruitment is also mediated by VCAM-1, a cell adhesion molecule, on the endothelial cell surface (Gorczynski *et al.*, 1996). VCAM-1 expression is dependent on activation of the MAP kinase in addition to NF-κB. Disruption of the mechanisms that regulate and limit the inflammatory response results in the development of chronic disease.

This chapter describes three assays that monitor three different phases in the inflammatory process.

Cell-Based Assays Used to Monitor the Effects of Proinflammatory Cytokines

Selection of a Suitable Cell System

Selecting an appropriate cell type is an important consideration in setting up a cell-based assay. Obviously the cell type must be a reliable model system for the underlying biology that the assay is designed to measure, but other considerations are also essential. The cells should be readily available. Their growth characteristics and conditions (media, doubling time, etc.) should be compatible with the desired workflow and enhance assay stability. Cell morphology affects imaging and analysis; large relatively flat cells yield data that are easier to interpret. Also, any special safety risks, such as increased viral risks from human cell types, should be taken into account.

We selected a primary cell type, human umbilical vein endothelial cells (HUVEC), for these assays. These cells are well established as a model system for studying inflammation and behave similarly to cells in the context of a tissue in a human body. Although they are a primary cell type, HUVECs are readily available as well-characterized, pooled preparations. Using pooled cells from multiple individuals reduces the variation that would result from differences in individual phenotypes. Furthermore, HUVECs can be grown for multiple passages *in vitro* (Gimbrone *et al.*, 1974), which is an important consideration when planning cell-based assays that use nonimmortalized cell types. A 100,000 well screen, with a seeding density of 10,000 cells per well, consumes 1×10^9 cells. This number can be grown easily from a few vials (1,000,000 cells per vial) of frozen stocks. We routinely use cells up to passage 10 for pilot experiments and cells between passage 6 and 9 for screens. We selected a growth medium formulation that contains low serum (2%) and defined supplements to provide controlled growth conditions, thus increasing assay robustness by minimizing the impact of lot-to-lot variations of the serum. Because HUVECs are human cells, it is essential to use preparations that have been screened for pathogens.

Selection of the Optimal Probe

Selecting an appropriate probe (assay readout) is another important consideration when setting up cell-based assays. Commonly used probes include immunofluorescent detection of endogenous proteins, green fluorescent protein (GFP)-reporter proteins, and other reporter proteins. Each has advantages and disadvantages.

A major advantage of using GFP-reporter proteins is that these probes are compatible with live cell assays and allow measurements of a single well to be made at multiple time points. Alternatively, the samples can be fixed and analyzed at a later time. Because GFP reporter proteins remain fluorescent after fixation, there is no need for complex immunostaining protocols, which simplifies assay workflow and reduces cost in reagents.

A major disadvantage of using GFP-reporter proteins is that the reporter must be introduced into the cells. This is usually achieved either by using a specially engineered cell line that stably expresses the reporter protein or by transiently introducing an expression vector (either plasmid or viral) encoding the reporter. Another disadvantage is that endogenous proteins are not monitored. GFP-reporter proteins might not faithfully represent the behavior of their endogenous counterparts. This may occur if the GFP moiety affects the proteins function, if the expression levels or stability of the GFP and native proteins are different, or for other reasons. Developing an engineered cell line or transient expression system and validating the reporter protein can be time-consuming, especially if specific expression levels are desired. High-level expression of proteins can deregulate pathways and important cellular functions, further complicating efforts to validate the reporter. Furthermore, immortalized, engineered cell lines are often highly transformed. Immortalized cell lines accumulate mutations over time in culture and may lose the full spectrum of response that is found in the parental cells, again making assay validation essential.

Immunofluorescent detection of endogenous proteins avoids many of the aforementioned disadvantages because endogenous proteins are monitored. Once a suitable antibody has been identified and its availability in sufficient quantities to complete the screen has been ensured, then assay development is usually straightforward. Because this approach minimizes manipulation of the cells, the risk of creating cell systems with new properties is relatively low. Despite these significant advantages, choosing immunofluorescent probes does have some disadvantages. Validating the specificity of primary antibodies for cell imaging is challenging because antibody cross-reactivity cannot be distinguished from the specific signal; indeed, cross-reactivity of antibodies is essentially impossible to detect based on immunomicrograph data alone. Live cell assays are almost impossible to perform using immunofluorescent probes because cells must be permeabilized as a necessary part of the immunostaining procedure. This means that kinetic studies need additional wells for each time point rather than simply monitoring a single well several times during an experiment. The availability of antibodies and batch-to-batch consistency are also factors to consider. These last two can both be more problematic when using polyclonal antibodies compared to monoclonal antibodies.

Sample processing (fixation, permeablization, and staining) for assays that use immunofluorescent probes is also more complicated than for GFP-reporter assays and includes multiple liquid-handling steps. These extra steps add to the cost and labor needed to screen an assay at high throughput.

Based on the relative difficulty of transiently transfecting primary cell types, such as HUVEC, and on our desire to model the behavior of endogenous human endothelial cells as closely as possible, we decided to use immunofluorescent probes in this set of assays. An alternative possibility that we considered was to use a viral system, such as recombinant retro- or adenoviruses, to introduce reporter proteins. However, this would have necessitated additional investment of time and labor in the development and optimization of viral delivery. It would also have added a level of complexity, as, in addition to obtaining specific and constant levels of expression of the target, a homogeneous level of infection among the target cells would need to be ensured. This is especially important for phenotypic cell-based assays (compared to biochemical assays), as the number of individual cells analyzed is limited; typically fewer than several hundred cells are captured in one image. In order to partially alleviate the drawbacks associated with sample processing for immunofluorescent assays, we have developed a standardized protocol (described later) designed to minimize liquid-handling steps compared to conventional staining protocols.

Materials and Methods Common to All Three Assays

We have developed standardized cell culture and postfixing staining procedures for the three assays described in this chapter. This section describes these standardized procedures.

Cell Culture

HUVECs are obtained as frozen stocks (Cambrex). Cells are grown in EGM-2 medium with supplements (Cambrex). To passage cells, the monolayer is washed first with PBS (HyQ HyClone) and then with a trypsin/EDTA solution (0.05% trypsin/0.02% EDTA in PBS). Cells are detached by incubation with 5 ml of trypsin per T175 flask, resuspended in 20 ml of DMEM containing 10% fetal calf serum (Cambrex), and centrifuged for 5 min at 800g (Beckman CS 6KR centrifuge). Cells are resuspended, seeded at dilutions of 1:3, and grown until they are confluent. Experiments are performed with cells kept in culture up to passage 10. Cells are grown at standard conditions (37°, 5% CO_2, and 100% and humidity) in a Nuaire 12 Autoflow Incubator (Nuaire).

Cell Seeding and Stimulation

For phenotypic assays, HUVECs are seeded at 10,000 cells per well in 96-well "ViewPlates" (Perkin Elmer) and grown for 24 h under standard conditions. Proinflammatory signaling is initiated by removing the growth medium and replacing it with 100 μl prewarmed medium containing the desired cytokine, either TNFα (R&D Systems) or IL1 (R&D Systems).

Compound Addition

To test the effects of various compounds on proinflammatory signaling, we seeded cells in view plates, grew them for 16 h, and then replaced the growth medium with 100 μl prewarmed medium containing the desired compound. The cells are preincubated in the presence of compound for a known time: 3 h 40 min for NF-κB translocation assay or 1 h for E-selectin and VCAM-1 assays. Then proinflammatory signaling is initiated by the addition of cytokine, directly to the medium (1 μl of 1 μg/ml TNFα). The incubation is continued for a further known time: 20 min for NF-κB, 4 h for E-selectin, and 24 h for VCAM-1. Then the cells are fixed and stained (described later). All tested compounds are first dissolved as stock solutions in DMSO at a sufficiently high concentration to ensure that the final concentration of DMSO in the assay does not exceed 0.1% (v/v). This concentration of DMSO does not itself affect any of the three assays.

siRNA Transfection

For assays where siRNA is used, cells are seeded at a density of 5000 cells per well in 96-well view plates and grown for 16 h before transfection. The transfection mix is assembled as follows: 100 nM final concentration of siRNA is added in 15 μl Opti-MEM (Invitrogen) and kept for 5 min at room temperature. In a separate tube, 0.5 μl of Oligofectamine (Invitrogen) is added to 2.5 μl of Opti-MEM. Both reactions are combined, mixed gently, and incubated at room temperature for 20 min to allow RNA/lipid complexes to form. Cells are washed once with PBS and 80 μl of Opti-MEM is added. The formed RNA/Lipid complexes are then added to the cells. After 4 h, medium is supplemented with 150 μl EGM-2 media. Cells are incubated for an additional 48 h under standard conditions before being assayed. After 48 h the medium is removed and cells are stimulated with cytokines as described earlier.

Fixing and Staining Solutions

Permeabilizing/blocking solution: 10% (v/v) newborn calf serum (Gibco/Invitrogen) and 0.2% (w/v) Triton X-100 (Sigma) in PBS (MP Biochemicals)

Primary antibody solutions: 2 μg/ml mouse IgG against p65 (Zymed) in PBS containing 10% (v/v) newborn calf serum, 2 μg/ml mouse IgG E-selectin (Chemicon MAB2150) in PBS containing 10% (v/v) newborn calf serum, and 2 μg/ml mouse IgG against VCAM-1 (SantaCruz-13506) in PBS containing 10% (v/v) newborn calf serum

Secondary antibody solution: 2 μg/ml antimouse coupled to Alexa Fluor 647 nm (Molecular Probes/Invitrogen) and 1 μg/ml Hoechst 33342 (Calbiochem) in PBS containing 10% (v/v) newborn calf serum

Fixation and Staining Protocol

Fixation

1. Wash with 200 μl PBS; aspirate.
2. Add 50 μl fix solution (4% paraformaldehyde in PBS).
3. Fix at room temperature for 20 min.
4. Aspirate fix solution and wash with 200 μl PBS; aspirate.
5. Store cells in 100 μl PBS at 4° until ready to stain.

Staining

6. Aspirate PBS from cells.
7. Wash cells with 200 μl PBS; aspirate.
8. Add 50 μl "permeabilizing/blocking solution."
9. Incubate at room temperature for 20 min.
10. Aspirate.
11. Add 50 μl "primary antibody solution" per well.
12. Incubate at room temperature for 20 min.
13. Aspirate, wash once with PBS.
14. Aspirate.
15. Add 50 μl "secondary antibody solution."
16. Incubate at room temperature for 20 min.
17. Aspirate and wash three times with 100 μl PBS; aspirate.
18. Store at 4° in fresh 100 μl PBS and seal until ready to scan.

Image Acquisition

The fixed cells are scanned and analyzed using an IN Cell Analyzer 3000 (GE Healthcare, USA) automated confocal microscope. View plates containing the fixed and immuno-stained cells are sealed and loaded into a Kendro Cytomat Hotel (Kendro, Germany) from which they are transferred automatically to the IN Cell analyzer 3000 by a Mitsubishi MELFA RU-2AJ arm (Mitsubishi, Japan).

The plates are imaged in two channels (blue and red); the respective images are acquired sequentially. Wells are first excited with the 647-nm laser lines, and images are recorded with the 695BP55 emission filter (red channel). The Hoechst dye is then excited with the 365-nm laser line, and images are recorded with the 450BP25 emission filter (blue channel).

A single square field, with dimensions of 0.561 mm^2, is recorded for each well. An average of 200 to 300 cells is imaged in one field. The exposure time is 1.7 ms and the binning factor is 1.

After finishing the run, data files, including images and description files ("run files"), are transferred from the IN Cell computer to a HP DL360 server (Hewlett Packard). Image analyses are performed on PC work stations using IN Cell Analyzer software (GE Healthcare). The modules and the parameters used for the assays mentioned in this chapter are described later. The text files with results of the image analyses are imported into Excel 2003 workbooks to generate graphs. Raw image data, descriptive files, and analysis results are assembled in an Open Microscopy Environment (Goldberg *et al.*, 2005).

Image Analysis

NF-κB Translocation Assay

Image analysis for the NF-κB translocation assay is done using the "nuclear trafficking analysis" module of the IN Cell Analyzer software. An average of 200 to 300 cells per image is analyzed. Individual cells are identified by their stained nuclei (blue channel, Hoechst 33342 staining). The NF-κB transcription factor is identified by immunostaining of its p65 subunit (red channel, Alexa Fluor 647-labeled antibody staining). Identified nuclei are used to define masks that are overlaid onto the red channel images, to define the regions to measure. Cytoplasmic measurements are made by dilating from the nuclear mask by two to three pixels to give a perimeter around the nucleus; measurements taken from this region correspond to cytoplasmic signal. Nuclear measurements are made by eroding the nuclear mask by two pixels and taking measurements. This ensures that sampling of cytoplasmic and nuclear regions is limited to areas within well-defined boundaries so edge-associated artifacts are minimized.

E-Selectin and VCAM-1 Expression Assays

Image analysis for the E-selectin and VCAM-1 expresssion assays is done using the "object intensity module" of the IN Cell Analyzer 3000 software. This module uses two color channels: one for the object "marker" and another for the "signal." First, the algorithm identifies all potential objects in the marker channel (the blue channel showing cell nuclei is used for these assays). Then size or intensity filters are applied and a mask is generated from the qualifying objects. Next, the algorithm establishes a measurement region by dilating the object masks by a specified number of pixels. The masks are then overlaid onto the signal channel (red in these assays) and measurements are taken.

Statistical Analysis

For data presented here, an average value from four wells is used to calculate each data point. Error bars show standard deviations. Z' factors are calculated as described (Zhang *et al.*, 1999), according to the formula:

$$Z' = 1 - [(3 \times \text{SD of the positive controls} + 3 \times \text{SD of the negative controls})/ (\text{mean value from the positive controls-mean mean value from the negative controls})]$$

Assay Data

NF-κB Assay

Assay Principle. Cytokines such as TNFα and interleukin 1 (IL-1) activate a proinflammatory response in endothelial cells. An early event in this response is nuclear translocation of the NF-κB transcription factor. This translocation causes the transcription of proinflammatory genes to be induced (Li and Stark, 2002; Pober, 2002). The nuclear translocation of NF-κB in stimulated primary HUVECs therefore provides a readout that can be used to study the effects of bioactive compounds on the responsiveness of endothelium to proinflammatory stimuli.

NF-κB is a heterodimer, composed of p65 and p50 subunits. Nuclear translocation is monitored by immunolocalization of the p65 subunit of NF-κB in fixed cells and is quantified by measuring the ratio of nuclear to cytoplasmic NF-κB signal (Fig. 1A).

The NF-κB Translocation Assay Produced Data with High Confidence. Figure 1 shows a representative result for HUVECs treated with TNFα (10 ng/ml) for 20 min, then analyzed by immunofluorescence for the translocation of the p65 subunit of NF-κB. Almost 100% of treated cells showed translocation of NF-κB from the cytoplasm to the nucleus under these conditions. Similar results were obtained with IL-1 (data not shown).

Time course experiments showed that NF-κB translocation was maximal 20 min after stimulation by either TNFα or IL-1 (Fig. 2A). Both of these cytokines caused a dose-dependent stimulation of NF-κB translocation in HUVECs. The nuclear-to-cytoplasmic ratio of NF-κB concentration increased up to sixfold after stimulation (Fig. 2B). Using TNFα as an activator, the Z' value of this assay was 0.79.

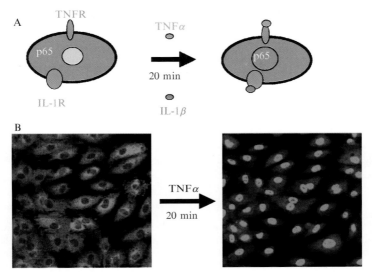

FIG. 1. Nuclear translocation of NF-κB. (A) Translocation of the dimeric transcription factor NF-κB was monitored by immunostaining for the p65 subunit. (B) HUVECs were stimulated with TNFα as described in the text. To detect distribution of the transcription factor, cells were fixed and permeabilized and then incubated with a monoclonal antibody against the p65 subunit of the NF-κB transcription factor and secondary antimouse IgG antibodies coupled to Alexa Fluor 647. Images of nonstimulated (left) and TNFα-stimulated (right) cells.

IL-1 and TNFα Use Different Receptors to Induce NF-κB Translocation

The cytokines IL-1 and TNFα activate different receptors and different upstream arms of the proinflammatory pathway to induce NF-κB translocation (Fig. 3). To prove the specificity and thereby the biological relevance of our assay, we used siRNA technology to demonstrate that the two different cytokines use their natural, physiologically relevant receptors to activate NF-κB. The sequence selection of small double-stranded siRNAs used in these experiments was as described previously (Jagla *et al.*, 2005). Specific knockdown of the receptor for TFNα (TNFR1), by transfecting cells with specific siRNAs 48 h before stimulation, significantly reduced NF-κB translocation in response to TFNα. In contrast, knockdown of the TNFα receptor did not affect IL-1-stimulated NF-κB translocation.

NF-κB Translocation Depends on IκB Kinase (IKK) and Proteasome Activity

We tested the effects of three drugs: curcumin, sulindac sulfide, and RO106–9920. Both curcumin and sulindac target the IKK complex, which

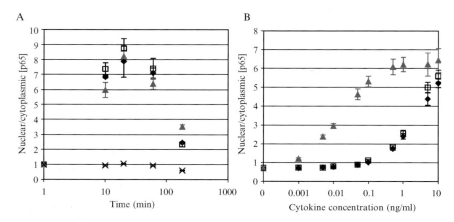

Fig. 2. TNFα and IL-1 induce p65 nuclear translocation. (A) Time course. HUVECs were stimulated with 10 ng/ml TNFα (gray triangles), 20 ng/ml IL-1α (black diamonds), 20 ng/ml IL-1β (open squares), or 1 μg/ml bovine serum albumin (crosses) for the indicated times. Cells were treated as described for Fig. 1B. p65 distribution was calculated as the ratio of nuclear to cytoplasmic staining. (B) Dose response. HUVECs were stimulated (20 min) with the indicated concentrations of TNFα (gray triangles), IL-1α (black diamonds), or IL-1β (open squares). Cells were treated as described for Fig. 1B. p65 distribution was calculated as the ratio of nuclear to cytoplasmic staining.

is a major regulator of the NF-κB pathway (Yamamoto *et al.*, 1999). RO106-9920 blocks an E3-ubiquitin ligase, which has been implicated in translocation of NF-κB (Swinney *et al.*, 2002). All three of these drugs caused a clear dose-dependent inhibition of cytokine-induced NF-κB translocation (Fig. 4). While NF-κB translocation is dependent on IKK kinase activity, as shown by curcumin and sulindac sulfide inhibition data, the process is not dependent on p38 MAP kinase activity. We have tested several p38 inhibitors, but none showed an inhibitory effect on translocation (data not shown).

E-Selectin Assay

Assay Principle. A long-term response to TNFα and other proinflammatory cytokines is the transcription of proinflammatory genes (Li and Stark, 2002; Pober, 2002), such as E-selectin. Cell surface expression of E-selectin following stimulation with cytokines in HUVEC cells provides readout to study the effects of bioactive compounds on the transcriptional activity regulated by proinflammatory stimuli (Fig. 5A).

Figure 5B shows a representative result for HUVECs treated with TNFα (10 ng/ml) for 4 h and then analyzed for expression of E-selectin.

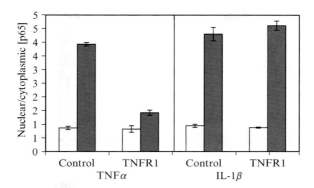

FIG. 3. IL-1 and TNFα use different receptors to induce translocation. HUVECs were seeded and stimulated with 10 or 20 ng/ml IL-1β for 20 min, as described in the text. Where indicated, cells were also treated with siRNA against the TNF receptor 1 (TNFR1) to silence TNFR1 expression in the cells. The distribution of p65 is expressed as a nuclear to cytoplasmic ratio of the fluorescence signal. Silencing of the TNFR1 by siRNA abolishes p65 translocation induced by TNFα but not by IL-1β. Gray bars represent TNFα (left)- or IL-1β (right)-treated cells; white bars represent nontreated control cells.

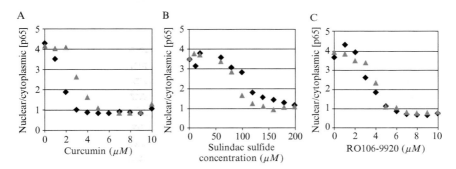

FIG. 4. Effect of compounds on NF-κB translocation. HUVECs were incubated for 1 h (black diamonds) or 4 h (gray triangles) with increasing concentrations of curcumin (A), sulindac sulfide (B), or RO106-9920 (C). Cells were then stimulated for 20 min with TNFα, and distribution was NF-κB measured as described for Fig. 1B.

A dose–response curve (Fig. 6A) indicates that E-selectin can be detected on the surface of approximately 10% of the cells in response to stimulation with as little as 1 pg/ml of TNFα. The response increases with increasing dose up to 50% positive cells at the highest dose tested (100 ng/ml). Time course experiments showed that E-selectin expression was maximal 4 h after stimulation with cytokines (data not shown).

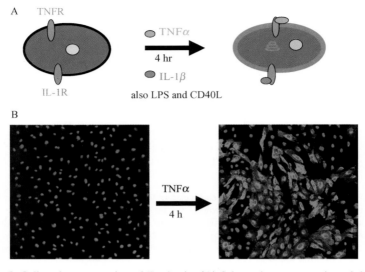

FIG. 5. Cell surface expression of E-selectin. (A) Schematic representation of the assay principle. After 4 h of stimulation with either TNFα or IL-1β, E-selectin accumulates at the cell surface. (B) HUVECs were stimulated with TNFα (10 ng/ml) for 4 h. To detect expression of E-selectin, cells were fixed and then incubated with monoclonal antibody against E-selectin and secondary antimouse IgG antibodies coupled to Alexa Fluor 647. Images of nonstimulated cells (left) and cells stimulated by TNFα for 4 h (right).

E-Selectin Expression Has Multiple Activators

Stimulation with IL-1β (Fig. 6B) results in a similar dose–response curve to that which is observed following stimulation with TNFα (Fig. 6A). A minimal response (5% positive cells) is observed after stimulation with a low cytokine concentration (30 ng/ml). This response increases with dose and plateaus at 50% positive cells following stimulation with 1 μg/ml IL-1β.

CD40L and LPS are also effective inducers of E-selectin expression but they are less potent than either TNFα or IL-1β. The dose–response relationships for CD40L and LPS are shown in Fig. 6C and D, respectively. In both cases, a smaller response is achieved than with cytokine stimulation; a maximum of 8% positive cells for CD40L and of 15% for LPS. Expression of E-selectin is dependent on IKK and proteasome activity, as demonstrated by the inhibitory effect of curcumin and RO106–9920 (Fig. 7).

Statistical analysis revealed a Z' value of 0.77 for activation of the assay with TNFα.

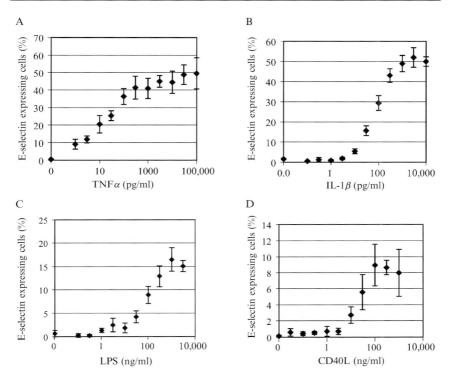

FIG. 6. Expression of E-selectin. (A) Stimulation with TNFα. HUVECs were treated with increasing amounts of TNFα. After 4 h, the cells were fixed and stained as described in Fig. 5B. Images of stained cells were analyzed using the "Object Intensity Module" of the IN Cell Analyzer software. The y axis shows the percentage of cells positive for E-selectin expression on the cell surface. The x axis shows concentration of TNFα. (B) Stimulation with IL-1β. HUVECs were treated with increasing amounts of IL-1β. After 4 h the cells were fixed and stained as described. Cells were analyzed as in A. The x axis shows concentration of IL-1β. (C) Stimulation with LPS. HUVECs were treated with increasing amounts of LPS. Cells were analyzed as in A. The x axis shows concentration of LPS. (D) Stimulation with CD40L. HUVECs were treated with increasing amounts of CD40L. Cells were analyzed as in A. The x axis shows concentration of CD40L.

VCAM-1 Assay

Assay Principle. Proinflammatory cytokines induce the transcription of proinflammatory genes (Li and Stark, 2002; Pober, 2002). One such protein, induced at late time points, is the vascular cell adhesion molecule-1, VCAM-1. The cell surface expression of VCAM-1 upon stimulation with cytokines in HUVECs therefore provides a model system that can be used

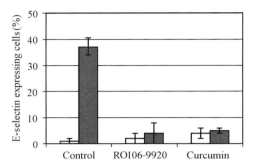

FIG. 7. E-Selectin expression is dependent on IKK and proteasome. HUVECs were treated with 10 ng/ml of TNFα and either 10 μM curcumin or 10 μM RO106-9920. After 4 h, the cells were analyzed as described in Fig. 6A.

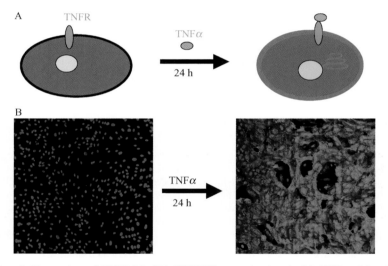

FIG. 8. Expression of VCAM-1. (A) HUVECs are stimulated with TNFα. After 24 h, VCAM-1 accumulates at the cell surface. (B) HUVECs were stimulated with TNFα for 24 h. To detect expression of VCAM-1, cells were fixed and incubated with a monoclonal antibody against VCAM-1 and secondary antimouse IgG antibodies coupled to Alexa Fluor 647. Images of nonstimulated cells (left) and cells stimulated with TNFα (right).

to study the effects of bioactive compounds on the responsiveness of endothelium to proinflammatory stimuli at late time points (Fig. 8A).

HUVECs were stimulated with TNFα for 24 h, and expression of VCAM-1 was monitored. Fixed cells were incubated with a monoclonal antibody against VCAM-1 and secondary antimouse IgG antibodies

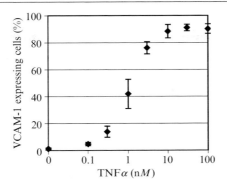

FIG. 9. TNFα induces VCAM-1 expression in a dose–response manner. HUVECs were stimulated for 24 h with the indicated concentrations of TNFα. After 4 h cells were fixed and stained as described. Images of stained cells were analyzed using the "object intensity module" of the IN Cell Analyzer. The percentage of cells showing expression of VCAM-1 was measured.

coupled to Alexa Fluor 647 (Fig. 8B). The left side of Fig. 8 shows images of nonstimulated cells, and the right side of Fig. 8 shows images of cells stimulated with TNFα.

The induction of VCAM-1 with TNFα was a robust and highly reproducible result, with a Z' value calculated at 0.8.

A dose-response curve, shown in Fig. 9, indicates that TNFα induces VCAM-1 expression in a dose-dependent manner with an EC_{50} of 2 ng/ml. The response plateaus at 5 ng TNFα with over 80% of the cells expressing VCAM-1.

The VCAM-1 Assay Can be Used to Screen for Bioactive Compounds

The expression of VCAM-1 is dependent on IKK and NF-κB translocation as tested by IKK inhibitors (data not shown). VCAM-1 expression is strongly dependent on active p38 MAP kinase activity, as tested with the compound SB202190 (a p38 MAP kinase inhibitor). SB202190 blocked VCAM-1 expression with an IC_{50} of approximately 2 μM (Fig. 10). Taken together, these results also validate our assay, as VCAM-1 induction is known to be dependent on both IKK and p38 MAP kinase pathways.

Concluding Remarks and Future Perspectives

We have shown that by using immunofluorescent probes in combination with high-speed automated microscopy and automated image analysis, one can develop robust assays in nontransformed human cells. As shown

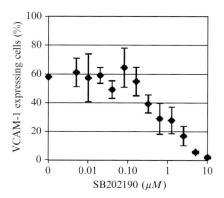

FIG. 10. p38 MAPK inhibitor SB202190 inhibits VCAM-1 expression after TNFα stimulation. HUVECs were incubated with TNFα (10 ng/ml) and increasing amounts of the MAP kinase inhibitor SB202190. After 24 h cells were fixed and stained as described. Images of stained cells were analyzed using the "object intensity module" of the IN Cell Analyzer. The y axis shows the percentage of cells positive for VCAM-1 expression.

earlier, antibodies can be applied to monitor nuclear translocation (NF-κB assay) or changes in the expression profile of target proteins (VCAM-1 and E-selectin assay).

Posttranslational modifications are often indicators of the activity status of a protein. Antibodies specific for these modifications, including Ser/Thr phosphorylation, Tyr phosphorylation, and Lys acetylation, are available for a growing number of proteins and are widely used to study signaling in mammalian cells. The challenge is to develop assays based on these antibodies that are robust enough for automated screening. Once these assays are developed they can directly monitor the activity of a target protein. Such assays will allow us to identify new small molecule-based drugs and research probes that modulate the targets activity.

Acknowledgments

This work was supported by MLSCN Grant 1U54 H6003914-01 to J.E.R and NIH Grants 1RO3 MH076344-01, 1RO3 MH076509-01, and 1RO3 MH076343-01 to support assay development to T.M. We thank Dr. Ouerfelli (MSKCC) for providing the siRNA. We thank Dr. Martin Wiedmann for his suggestions.

References

Aggarwal, B. B. (2003). Signalling pathways of the TNF superfamily: A double-edged sword. *Nature Rev. Immunol.* **3,** 745–756.

Ding, G. J., Fischer, P. A., Boltz, R. C., Schmidt, J. A., Colaianne, J. J., Gough, A., Rubin, R. A., and Miller, D. K. (1998). Characterization and quantitation of NF-kappaB nuclear translocation induced by interleukin-1 and tumor necrosis factor-alpha: Development and use of a high capacity fluorescence cytometric system. *J. Biol. Chem.* **273**, 28897–28905.

Gimbrone, M. A., Jr., Cotran, R. S., and Folkman, J. (1974). Human vascular endothelial cells in culture: Growth and DNA synthesis. *J. Cell Biol.* **60**, 673–684.

Goldberg, I. G., Allan, C., Burel, J. M., Creager, D., Falconi, A., Hochheiser, H., Johnston, J., Mellen, J., Sorger, P. K., and Swedlow, J. R. (2005). The Open Microscopy Environment (OME) Data Model and XML file: Open tools for informatics and quantitative analysis in biological imaging. *Genome Biol.* **6**, R47.

Gorczynski, R. M., Chung, S., Hoang, Y., Sullivan, B., and Chen, Z. (1996). Altered patterns of migration of cytokine-producing T lymphocytes in skin-grafted naive or immune mice following *in vivo* administration of anti-VCAM-1 or -ICAM-1. *Immunology* **87**, 573–580.

Jagla, B., Aulner, N., Kelly, P. D., Song, D., Volchuk, A., Zatorski, A., Shum, D., Mayer, T., De Angelis, D. A., Ouerfelli, O., Rutishauser, U., and Rothman, J. E. (2005). Sequence characteristics of functional siRNAs. *Rna* **11**, 864–872.

Karmann, K., Min, W., Fanslow, W. C., and Pober, J. S. (1996). Activation and homologous desensitization of human endothelial cells by CD40 ligand, tumor necrosis factor, and interleukin 1. *J. Exp. Med.* **184**, 173–182.

Li, X., and Stark, G. R. (2002). NFkappaB-dependent signaling pathways. *Exp. Hematol.* **30**, 285–296.

Martin, M. U., and Wesche, H. (2002). Summary and comparison of the signaling mechanisms of the Toll/interleukin-1 receptor family. *Biochim. Biophys. Acta* **1592**, 265–280.

Pober, J. S. (2002). Endothelial activation: intracellular signaling pathways. *Arthritis Res.* **4** (Suppl. 3), S109–S116.

Senftleben, U., and Karin, M. (2002). The IKK/NF-kappaB pathway. *Crit. Care Med* **30**, S18–S26.

Swinney, D. C., Xu, Y. Z., Scarafia, L. E., Lee, I., Mak, A. Y., Gan, Q. F., Ramesha, C. S., Mulkins, M. A., Dunn, J., So, O. Y., Biegel, T., Dinh, M., Volkel, P., Barnett, J., Dalrymple, S. A., Lee, S., and Huber, M. (2002). A small molecule ubiquitination inhibitor blocks NF-kappa B-dependent cytokine expression in cells and rats. *J. Biol. Chem.* **277**, 23573–23581.

Yamamoto, Y., Yin, M. J., Lin, K. M., and Gaynor, R. B. (1999). Sulindac inhibits activation of the NF-kappaB pathway. *J. Biol. Chem.* **274**, 27307–27314.

Zhang, J. H., Chung, T. D., and Oldenburg, K. R. (1999). A simple statistical parameter for use in evaluation and validation of high throughput screening assays. *J. Biomol. Screen* **4**, 67–73.

[16] Development and Implementation of Multiplexed Cell-Based Imaging Assays

By BONNIE J. HOWELL, SEUNGTAEK LEE, and
LAURA SEPP-LORENZINO

Abstract

Fluorescence microscopy, image analysis, and automated screening technologies are some of the most powerful tools enabling cell biologists to investigate complex signaling pathways and compound or siRNA effects on cellular function in individual cells. Researchers can now use multiple fluorescent probes to quantify effects on intracellular molecular events, measure phenotypic changes, and provide contextual information about cellular pathways not discernible by traditional single-parameter, end point experiments. This chapter focuses on fluorescent labeling techniques and methods for designing image-based assays, multiplexed readouts, and image analysis routines. Case studies are presented describing the use of cell-based imaging assays for monitoring cell proliferation, cell cycle stage, and apoptosis.

Introduction

One of the most powerful tools enabling cell biologists to monitor physiological processes and interrogate protein function at the cellular and molecular level is the fluorescence microscope. This tool, coupled with sophisticated image analysis and automated screening technologies, provides a unique opportunity to investigate complex signaling pathways and effects of compounds or other agents on cellular function in individual cells (Mitchison, 2005). Through the use of multiple fluorescent probes, researchers can now quantify intracellular molecular events, measure phenotypic changes, and provide contextual information about cellular pathways not discernible by traditional single-parameter, end point experiments. Furthermore, this multiparametric approach can filter out false positives and false negatives quite effortlessly during screening campaigns, resulting in higher-quality leads. This chapter focuses on fluorescent labeling techniques and methods for image-based assay design, including multiplexed readouts and image analysis routines for high-content screening. Case studies are presented describing the use of cell-based imaging assays for monitoring cell cycle stage and apoptosis.

METHODS IN ENZYMOLOGY, VOL. 414
0076-6879/06 $35.00
DOI: 10.1016/S0076-6879(06)14016-1

General Considerations

Several labeling approaches exist for visualizing intracellular components and physiological processes in cells, including incubation of cells with membrane-permeable chemical fluorophores, microinjection of fluorescently labeled proteins, transfection with DNA constructs expressing fluorescent proteins, for example, green fluorescent protein (GFP), and immunostaining with antigen-specific antibodies. The desired labeling approach is often dictated by the biology, hardware accessibility, reagent availability, throughput, and cost. For example, if temporal and spatial information is desired, then time-lapse fluorescence microscopy using nondestructive, cell-permeable chemical fluorophores, fluorescently labeled proteins, or fluorescent fusion proteins is the preferred approach. If kinetic information is not necessary and/or if multiplexing is desired, then immunofluorescence or a combination of labeling approaches is preferred. Advantages and limitations of fluorescent labeling approaches are discussed later and reviewed in Table I.

Fluorescent dye technology is a rapidly evolving approach to labeling cells. Chemical dyes, together with appropriate imaging technologies, allow characterization of cellular architecture and detection of cellular events in the absence of cell perturbation. Increased utilization of chemical fluorophores is driven, in part, by improved brightness and photostability of dyes, wider selection and improved filters to separate colors, low cost, and flexibility to use across cell lines. Shortcomings include possible compound interference and limited choice in wavelength selection. Overall, fluorescent dyes offer the benefit of multiplexing with other probes, do not require sophisticated hardware capabilities, and are widely available through several vendors.

A second and more conventional approach to studying cellular processes and protein behavior is indirect immunofluorescence. Traditionally, this method has been performed on slides with complex, laborious protocols. However, with today's advanced liquid-handling technologies, immunofluorescence assays have become a popular tool for high-throughput screening, target validation, and lead optimization. Advantages of fixed-cell immunofluorescence studies include stable fluorescent signal for convenient imaging, flexibility in cell line choice, and large availability of antigen-specific antibodies. However, the greatest advantage is the ability to multiplex probes and proteins to produce multiparametric, high-content information in one simple assay. Historically, multiplexing has been limited to three or four probes; however, advances in multispectral imaging equipment and quantum dot technologies offer wide spectral resolution and new promises in this area. Immunofluorescence approaches have the drawbacks of possible fixation and permeabilization artifacts, lack of kinetic information, and laborious assay development.

TABLE I
CELLULAR MARKERS FOR USE WITH AUTOMATED FLUORESCENCE MICROSCOPY

Technology	Biology	Application
DAPI, Hoechst 33342, DRAQ5	Nucleic acid stain/marker; DAPI can be used for fixed cell assays, whereas Hoechst and DRAQ5 can be used for both live and fixed cell assays	Nuclear size and intensity, cell cycle profile, endoreduplication, apoptosis, cytotoxity, morphology
Propidium iodide	Only taken up by dead cells or cells with compromised cell membrane. Binds DNA	Cell viability
Calcein AM	Nonfluorescent membrane permeant cellular marker cleaved by intracellular esterase to fluorescent form in living cells	Cell viability
Ethidium homodimer	Cell-impermeant marker; only taken up by dead cells or cells with compromised cell membrane. Binds DNA	Cell toxicity
7-AAD	Nucleic acid stain taken up by dead or apoptotic cells	Cell cycle and apoptosis analysis
Cyclin B1-GFP, cyclin B1 antibody	Cyclin B1 marker	Identifies G1/S, G2, mitotic cells
PhosphoHistone H3 antibody	G2/mitosis marker	Identifies G2 and mitotic cells
MitoTracker	Mitochondrial membrane potential sensing dyes	Measures disruption of active mitochondria for apoptosis analysis
Annexin V	Binds to phosphatidyl serine	Early stage apoptotic marker
Cleaved caspase 3 antibody	Binds to active caspase 3	Midstage apoptotic marker
Cleaved PARP antibody	Binds to active PARP	Detects late-stage apoptosis
TUNEL	Detects DNA strain breaks	Detects late-stage apoptosis

The field of cellular imaging has exploded even more within the last decade with recent advances in molecular cloning, optical imaging, and the discovery of fluorescent proteins. Fluorescent proteins, for example, GFP, provide a nondestructive method for studying dynamic processes *in vivo* and offer several advantages over conventional immunofluorescence. These labeling technologies allow quantification of cellular events and protein behavior in real time, are often less expensive, and provide flexibility in cell line choice. Additionally, the fluorescence from fluorescent proteins, for example, GFP, remains after fixation of cells with paraformaldehyde and therefore this approach can also be combined with conventional immunofluorescence or membrane-permeable dyes. Disadvantages include lengthy assay development periods for establishing transfection conditions or making stable cell lines, need for specialized equipment for kinetic studies, and concern that the fluorescent tag may alter protein folding or function.

Because assay design and image quality lay the foundation for distinguishing between on-target vs off-target effects, careful consideration should be given when choosing cell lines, reagents, image acquisition technologies, and image analysis routines. Lee and Howell (2006) provide detailed discussion regarding assay design and choice of hardware/software. This chapter describes basic protocols for visualizing cell cycle and apoptosis events using automated fluorescence microscopy and outlines caveats to bear in mind when choosing between different fluorescent labeling approaches.

Monitoring Cell Cycle Progression

One of the most fundamental processes occurring in eukaryotic cells is the cell division process in which cells duplicate their contents and distribute them equally to two daughter cells. The cell cycle is regulated both temporally and spatially and is divided into four discrete phases: G1, S, G2, and mitosis (M). Historically, flow cytometry has been the gold standard for measuring DNA content and determining cell cycle stage. With this technology, cells are first incubated with a fluorescent DNA-binding dye and then passed through a detector one cell at a time. Because G2/M cells display exactly twice the DNA content of G1 cells, and S cells have an intermediate DNA content, flow cytometry can be used to generate distribution plots of cell number and DNA content and quickly access fractions of cells in different phases of the cell cycle. This approach can be applied to automated fluorescence microscopy (Fig. 1); however, microscopy and image analysis offer additional measurements compared to flow cytometry. With the same DNA stain, information such as nuclear size and DNA fragmentation (measuring irregular staining and shape of the nucleus) can be obtained along with the cell cycle profile.

An alternative approach to using DNA-binding dyes is quantification of protein expression and/or posttranslational modifications of cell cycle-regulated proteins. One example is cyclin proteins, as these undergo a cycle of synthesis and degradation during cell cycle progression. Here we show the use of cyclin B1 to distinguish among G2, M, and G1/S phases of

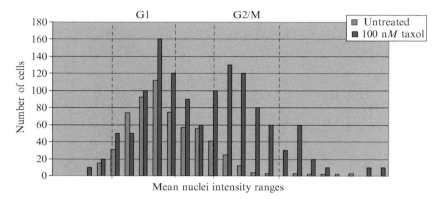

Fig. 1. Cell cycle analysis by quantification of nuclear stain. HCT116 cells were left untreated or treated with 100 nM paclitaxel for 24 h, fixed with paraformaldehyde, and stained with DAPI. Nuclear intensity was quantified using the object intensity algorithm from GE Healthcare and cells were binned according to the intensities and cell numbers. Paclitaxel treatment (red bars) resulted in increased G2/M population compared to untreated cells (blue bars).

the cell cycle. Using cell cycle analysis algorithms from GE Healthcare, one can quantify the localization and intensity of cyclin B1 in individual cells and classify cells into distinct phases of the cell cycle (http://www.mdyn. com/aptrix/upp01077.nsf/Content/ProductsTree; Fig. 2). This approach is multiplexed easily with other probes to interrogate cell cycle dependence of key processes or pathways and can be run either as an end point, immunofluorescence study using antibodies to cyclin B1, or as a real-time kinetic study using a fluorescent fusion protein, for example, GFP.

Posttranslational protein modifications also have a major role in driving cell cycle regulation, as well as many other biological processes, and therefore serve as excellent markers of cellular events. For cell cycle analysis of G2/M cells, we chose antibodies that recognize phosphorylated histone H3. Histones are among the numerous DNA-binding proteins that control DNA condensation/decondensation during cell cycle progression, and phosphorylation of histone H3 has been tightly correlated with chromosome condensation during mitosis (Prigent and Dimitrov, 2003). Therefore, antibodies to phosphorylated histone H3 can be used to identify cells in G2 and M phases of the cell cycle, and this can be multiplexed with other probes for correlating cell cycle dependency with drug efficacy or toxicity. The following sections describe various methods for identifying cell cycle stage using fluorescence microscopy.

A

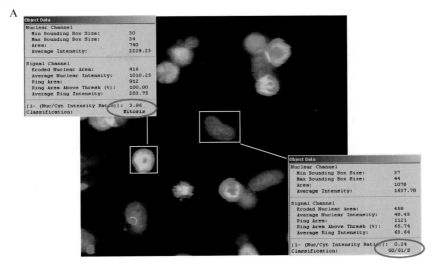

B

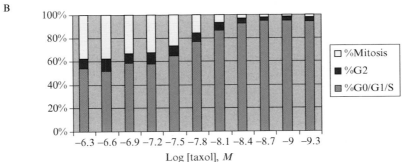

FIG. 2. Cell cycle analysis using cyclin B1. U2OS cells expressing cyclin-B1-GFP were treated with paclitaxel at various concentrations, and cell cycle stage was determined using the cell cycle phase module by GE Healthcare. This algorithm classifies individual cells based on intensity and localization of cyclin B (A) and can then be used to generate a histogram of cell cycle response for the entire well (B).

Protocols for Cell Cycle Analysis

Cell Culture

Seed cells onto 96-well microtiter plates in 100 μl media and incubate at least 12 h, preferably 24 h, to allow them to adhere. For drug treatments, add 100 μl of a 2× stock of the preferred drug, or add directly from dimethyl sulfoxide stock. For inducing mitotic arrest, paclitaxel (Taxol) can be used as a positive control at a final concentration of 0.1 to 1 μM. For cells with a doubling time <25 h, cells should be incubated with drug for ~18–24 h to see maximal mitotic arrest.

Determining Cell Cycle Distribution by Quantification of Nuclei Intensity

1. For live cell assays, stain nuclei by adding 1 to 10 μg/ml Hoechst 33342 (final concentration; Invitrogen) and incubate 30 to 60 min before imaging.
2. For fixed cell assays, remove media, fix for 15 min with 2% paraformaldehyde/phosphate-buffered saline (PBS), rinse once with PBS, and then add 100 μl PBS containing 1 to 5 μM DAPI (Invitrogen) or 1 to 10 μg/ml Hoechst 33342 (Invitrogen). Incubate 15 min and read on a fluorescence microscope.
3. Quantify nuclear intensity and display as histogram of binned intensities

Note: Some high-content screening (HCS) companies, for example, Cellomics (http://www.cellomics.com), offer analysis modules that specifically bin cells into one of the four cell cycle stages based on DNA staining intensity (similar to flow cytometry applications).

Cyclin-B1-GFP Live Cell Assay. The cyclin-B1-GFP assay described here was developed by GE Healthcare (http://www.mdyn.com/aptrix/upp01077.nsf/Content/ProductsTree). For this assay, the N-terminal region of cyclin B1 containing the localization and destruction domain is fused to GFP under control of the cyclin B1 promoter and stably expressed in U2OS cells. The assay can be performed as a kinetic or an end point assay, and the cells can be imaged while still alive or after paraformaldehyde fixation.

Cells are grown per manufacturer's recommendations, seeded in 96-well microtiter plates at 8000 cells per well in 100 μl of growth medium, and incubated for 24 h at 37°, 5% CO_2/air prior to drug treatment. Phenol red-free media should be used to reduce background autofluorescence.

1. Treat cells with 2× drug in 100 μl phenol red-free media.
2. For kinetic studies, incubate for desired period, add live-cell nuclear stain (e.g., Hoechst 33342 or DRAQ5), and image at defined time points. Be sure platform is equipped with appropriate environmental controls.
3. For end-point studies, incubate for desired period, remove media, and add 4% paraformaldehye (final concentration) in PBS. Incubate at room temperature for 30 min, remove fixative, rinse twice with PBS, and add 100 μl nuclear stain (e.g., Hoechst 33342, DRAQ5, or DAPI) in PBS.

4. The cell cycle stage can be quantified using the cell cycle phase module from GE Healthcare or custom algorithms that quantify fluorescence intensity and localization of cyclin B1-GFP.
5. For the GE Healthcare software module, results are reported as percentage of cells in G0/G1/S, G2, prophase, and mitosis.

Cyclin B1 Immunofluorescence Assay. Cell lines are cultured according to vendor recommended growth conditions.

1. Plate cells at appropriate seeding density (e.g., 8000–10,000 cells/ well in 96-well microtiter plate), incubate overnight to allow adherence, and drug treat as described earlier.
2. Incubate for desired time periods.
3. Remove media by gentle aspiration, leaving ∼15 μl in the well so as not to disturb the bottom layer of cells.
4. Add 100 μl fresh 4% paraformaldehyde in PBS and incubate 15 min (perform this operation in a hood).
5. Remove paraformaldehyde, leaving ∼15 μl.
6. Rinse three times with 200 μl PBS, leaving ∼15 μl after each rinse.
7. Permeabilize, block, and immunostain cells in TBS supplemented with 0.1% (v/v) Triton X-100, 5% bovine serum albumin (BSA), and 1 to 2 μg/ml FITC-conjugated anticyclin B1 antibodies (Santa Cruz) for 90 min at 37°, 4 h at room temperature, or overnight at 4° (high backgrounds can be reduced by adding 0.1% Triton X-100 to the rinse step and lengthening its duration).
8. Remove staining solution and add 150 μl PBS supplemented with DNA stain, such as 1 to 10 μg/ml Hoechst 33342 (Invitrogen) or 1–5 μM DAPI (Invitrogen).
9. Image on a fluorescence microscope.
10. The cell cycle stage can be quantified using the cell cycle phase module from GE Healthcare, as described earlier.

Phosphohistone H3 Immunfluorescence Assay for Detecting Late G2 and Mitotic Populations. Cells are cultured according to vendor-recommended conditions.

1. Plate cells at appropriate seeding density (e.g., 8000–10,000 cells/well in 96-well microtiter plate), incubate overnight to allow adherence, and drug treat as described earlier. Incubate for desired time periods.
2. Remove media by gentle aspiration, leaving ∼15 μl in the well so as not to disturb the bottom layer of cells.
3. Add 100 μl fresh 4% paraformaldehyde in PBS and incubate 15 min (perform this operation in a hood).

4. Remove paraformaldehyde, leaving ~15 μl.
5. Rinse three times with 200 μl PBS, leaving ~15 μl after each rinse.
6. Permeabilize, block, and immunostain cells in TBS supplemented with 0.1% (v/v) Triton X-100, 5% BSA, and 1:50–1:100 dilution of fluorophores-conjugated antiphosphohistone H3 antibodies (Cell Signaling) for 60 min at 37°, 2 h at room temperature, or overnight (high backgrounds can be reduced by adding 0.1% Triton X-100 to the rinse step and lengthening its duration).
7. Remove staining solution and add 150 μl PBS supplemented with DNA stain, such as 1 to 10 μg/ml Hoechst 33342 (Invitrogen) or 1 to 5 μM DAPI (Invitrogen).
8. Image on a fluorescence microscope.
9. The cell cycle stage can be quantified by thresholding for cells with phosphoHistone H3 positive signal colocalizing with nuclear stain. The mitotic index is reported as number of thresholded objects divided by number of total objects × 100.

Visualizing Apoptosis by Fluorescence Microscopy

Apoptosis is a well-regulated form of cell death initiated by a wide range of physiological and pathological conditions (Vermes *et al.*, 1995). It is unique from other forms of cell death, such as necrosis, autophagy, and mitotic catastrophe, and can be characterized by a sequence of biochemical and morphological events beginning with cell surface changes and caspase activation and ending with DNA fragmentation and cell death/rupture. The time interval between reception of an apoptotic signal and execution of downstream signaling events may vary between cell lines and between apoptotic stimuli, as well as the duration of the stage (window for detection). Therefore, it is essential to designs experiments that carefully monitor apoptotic stage of interest.

Several tools exist for detecting apoptosis by fluorescence microscopy. For early apoptosis detection, researchers can monitor loss of cell membrane integrity using dyes that are impermeant to intact membranes but selectively pass through the plasma membrane of apoptotic cells (Darzynkiewicz *et al.*, 1997). Second, dyes that detect changes in mitochondrial membrane potential can also be used (Darzynkiewicz *et al.*, 1997). Often, annexin V conjugates or antibodies to phosphatidyl-serine (PS) are used to detect the apoptosis-initiated translocation of PS from the inner surface of the cell membrane to the external surface (Vermes *et al.*, 1995). Finally, antibodies to activated caspases can be used, as described in detail later, or fluorescent inhibitors to caspase, which are nonfluorescent until they enter cells and bind activated caspases (Villa *et al.*, 1997). Many of these reagents are used to detect early to midstage apoptotic events; however, late-stage apoptosis can be visualized using antibodies to cleaved PARP or TUNEL labeling to detect DNA strand breaks (Villa *et al.*, 1997). The following sections describe approaches for apoptosis induction using antibodies to cleaved caspase 3 or cleaved PARP and the TUNEL method for detecting DNA breaks.

Apoptosis Detection Using Antibodies to Cleaved Caspase 3

Cells are cultured according to vendor-recommended conditions.

1. Plate cells at appropriate seeding density (e.g., 8000–10,000 cells/well in a 96-well microtiter plate), incubate overnight to allow adherence, drug treat, and incubate for desired time period.
2. Remove media by gentle aspiration, leaving ~15 μl in the well so as not to disturb the bottom layer of cells.
3. Add 100 μl fresh 4% paraformaldehyde in PBS and incubate 15 min (perform this operation in a hood).
4. Fix cells for 10 min at room temperature in 4% paraformaldehyde/PBS solution.
5. Remove formaldehyde, leaving a small amount (~20 μl).
6. Rinse three times with PBS (150 μl per rinse). Leave a small volume between rinse steps when using a plate washer (~20 μl).
7. Block with 100 μl of 2% FBS in PBST (PBS supplemented with 0.1% Triton X-100) for 30 min at room temperature. Rock gently (optional).
8. Remove blocking solution.
9. Add 1:100 dilution of FITC-conjugated rabbit anticaspase-3 antibody (Cell Signaling Technologies) in PBST + 2% FBS.
10. Incubate overnight at 4° in dark.
11. Remove staining solution and add 200 μl PBST per well. Repeat twice, leaving 10-min incubations between rinses.
12. Add 100 μl DAPI or 1–10 μg/ml Hoechst 33342 in PBS.
13. Incubate at room temperature for 15 to 30 min.
14. Image plates.

Note: To minimize cell loss, we suggest spinning plates at 300g for 5 min between wash steps and before imaging.

Apoptosis Detection Using Antibodies to Cleaved PARP

Cells are cultured according to vendor-recommended conditions.

1. Plate cells at appropriate seeding density (8000–10,000 cells/well in a 96-well microtiter plate), incubate overnight to allow adherence, drug treat, and incubate for desired time period.
2. Remove media by gentle aspiration, leaving ~15 μl in the well so as not to disturb the bottom layer of cells.
3. Add 100 μl fresh 4% paraformaldehyde in PBS and incubate 15 min (perform this operation in a hood).

4. Fix cells for 10 min at room temperature in 4% paraformaldehyde/ PBS solution.
5. Remove formaldehyde, leaving a small amount ($\sim$20 μl).
6. Rinse three times with PBS (150 μl per rinse). We use a plate washer for this and leave a small volume in the wells between rinse steps ($\sim$20 μl).
7. Block in 100 μl of 1% BSA/TBST (TBS supplemented with 0.1% Triton X-100) for 30 min at room temperature.
8. Remove blocking solution and rinse once with 150 μl PBS, leaving 20 μl.
9. Add 100 μl of staining solution (1–10 μg/ml DAPI or Hoechst 33342, 1–5 μg/ml FITC-conjugated PARP antibody in 1% BSA/TBST). The antibody is from BD Biosciences.
10. Incubate for 90 min at room temperature.
11. Wash plate twice with PBST (PBS supplemented with 0.1% Triton X-100), leaving a small volume between rinses.
12. Add 100 μl PBS and image plates.

Note: To minimize cell loss, we suggest spinning plates at 300g for 5 min between wash steps and before imaging.

TUNEL Method for Apoptosis Detection

This assay uses reagents obtained from the Invitrogen Apo-BrdU TUNEL labeling kit, and the protocol has been modified for multiplexing with cell cycle probes and for preventing loss of apoptotic cells during high-content screening.

Reagent Preparation

DNA labeling mix (50 μl per sample): 5 μl TdT reaction buffer (green cap), 0.38 μl TdT enzyme (yellow cap), 4 μl Brd-UTP (violet cap), and 40.63 μl of distilled water. Keep on ice when using or at 4° for up to 24 h.

Anti-BrdU staining mix (50 μl per sample): 2.5 μl Alexa Fluor 488 anti-BrdU (orange cap), 47.5 μl rinsing buffer (red cap), and 1 to 10 μg/ml DAPI or Hoechst 33342. Keep on ice when using or at 4° for up to 24 h. Protect from light.

Protocol. Cells are cultured according to vendor-recommended conditions.

1. Plate cells at appropriate seeding density (e.g., 8000–10,000 cells/ well in a 96-well microtiter plate), incubate overnight to allow adherence, drug treat, and incubate for desired time period.

2. Remove media by gentle aspiration, leaving ~15 μl in the well so as not to disturb the bottom layer of cells.
3. Add 100 μl fresh 4% paraformaldehyde in PBS and incubate 15 min (perform this operation in a hood).
4. Fix cells for 10 min at room temperature in 4% paraformaldehyde/PBS solution.
5. Remove formaldehyde, leaving a small amount (~20 μl).
6. Rinse three times with PBS (150 μl per rinse) or with wash buffer provided in kit. We use a plate washer for this and leave a small volume in the wells between rinse steps (~20 μl).
7. Spin plate(s) at 300g for 10 min, remove buffer, leaving 20 μl, and add 50 μl of DNA labeling reaction per well.
8. Incubate 1 hl at 37° (shake every 15 min or place on rocker if available).
9. Add 150 μl rinsing buffer (red cap) from kit or PBS.
10. Centrifuge plates at 300g for 10 min, remove buffer, leaving 20 μl, and add 100 μl anti-BrdU staining solution.
11. Incubate 30 min at room temperature in the dark.
12. Add 150 μl rinse buffer (from kit) or PBS, spin at 300 g for 10 min, and remove buffer, leaving a small amount. Repeat.
13. Resuspend in 125 μl PBS and image plate.

An example image is shown in Fig. 3.

Detecting Morphological Changes Associated with Apoptosis

An alternative approach to detecting apoptotic cells other than using an apoptosis-specific marker is to quantify the morphological hallmarks associated with late-stage apoptosis, such as DNA fragmentation. To apply this approach, cells are first stained with a nuclear marker (e.g., DAPI, Hoechst, DRAQ5) and then "granularity" or "spot detection" algorithms are used to quantify the punctate appearance of nuclei. The following methods are used for visualizing apoptosis in cells by fluorescence microscopy.

1. Cells are cultured according to vendor specifications, plated in 96-well microtiter plates, drug treated, and incubated for desired time periods.
2. Nuclear dye, for example, 1 to 5 μM DAPI, Hoechst, or DRAQ5 in PBS, is added to wells and is incubated for 15 to 60 min before imaging.
3. Spot detecting algorithms are used to identify fragmented nuclei and to calculate apoptotic index. Examples are seen in Fig. 4.

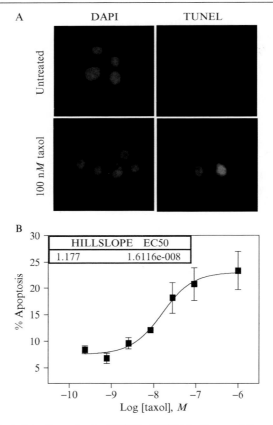

FIG. 3. Apoptosis detection using by TUNEL. HCT116 cells were left untreated or treated with paclitaxel for 48 h, fixed with paraformaldehyde, and labeled for DNA strand breaks via the TUNEL method. Paclitaxel treatment results in positive TUNEL staining (A) and an increase in the total number of apoptotic cells (B).

Multiplexing Cell Cycle and Apoptosis Assays

Many fluorescent probes and HCS assays are multiplexed easily. For example, nuclear stains are commonly used to identify individual cells in HCS assays; however, these stains can also be used to classify cell cycle stage, quantify apoptotic events, detect endoreduplication, and measure cytotoxicity. An example can be seen in Fig. 4 for cell cycle/apoptosis analysis in which the nuclear stain is also used to detect fragmented nuclei. A second approach is to combine antibodies, as described later in the

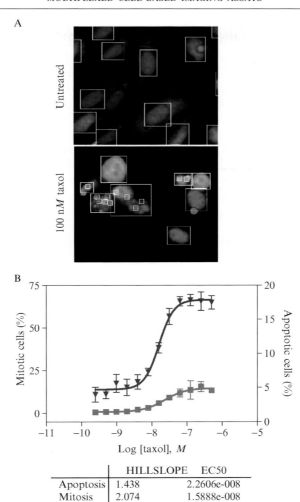

	HILLSLOPE	EC50
Apoptosis	1.438	2.2606e-008
Mitosis	2.074	1.5888e-008

Fig. 4. Apoptosis detection using morphological quantification. U2OS cells expressing cyclin-B1-GFP were treated with paclitaxel at various concentrations and cell cycle stage and apoptosis effects. Paclitaxel treatment resulted in increased DNA fragmentation (A), a morphological hallmark characteristic of late stage apoptosis, which was quantified using the granularity algorithm by GE Healthcare (B). Cell cycle quantification was performed using the cell cycle phase module (B), as described in Fig. 2.

phosphoHistone H3/TUNEL assay for cell cycle and apoptosis detection. Immunofluorescence markers are multiplexed easily but care should be taken to use antibodies from different species or primary-conjugated antibodies to avoid cross-reactivity. A major challenge to this approach is ensuring that the fixative used adequately preserves the antigen epitope and maintains antibody recognition. Advances in antibody-labeling technologies, such as the zenon-labeling approach from Invitrogen (http://www. Invitrogen.com), offer scientists a unique opportunity to use primary antibodies derived from the same species, rapidly label them with fluorophores of their choice, and avoid the use of secondary detection reagents. This allows versatility in assay design and minimizes cross-reactivity.

Multiplexing of PhosphoHistone H3/TUNEL

This assay uses reagents obtained from the Invitrogen Apo-BrdU TUNEL labeling kit, and the protocol has been modified for multiplexing with cell cycle probes and for preventing loss of apoptotic cells during high-content screening.

Reagent Preparation

> DNA labeling mix (50 μl per sample): 5 μl TdT reaction buffer (green cap), 0.38 μl TdT enzyme (yellow cap), 4 μl Brd-UTP (violet cap), and 40.63 μl of distilled water. Keep on ice when using or at 4° for up to 24 h
>
> Anti-BrdU staining mix (50 μl per sample): 2.5 μl Alexa Fluor 488 anti-BrdU (orange cap), 47.5 μl rinsing buffer (red cap), and 1–10 μg/ml DAPI or Hoechst 33342. Keep on ice when using or at 4° for up to 24 h. Protect from light.

Protocol. Cells are cultured according to vendor-recommended conditions.

1. Plate cells at appropriate seeding density (e.g., 8000–10,000 cells/well in a 96-well microtiter plate), incubate overnight to allow adherence, drug treat, and incubate for desired time period.
2. Remove media by gentle aspiration, leaving ∼15 μl in the well so as not to disturb the bottom layer of cells.
3. Add 100 μl fresh 4% paraformaldehyde in PBS and incubate 15 min (perform this operation in a hood).
4. Fix cells for 10 min at room temperature in 4% paraformaldehyde/PBS solution.
5. Remove formaldehyde, leaving a small amount (∼20 μl).

6. Rinse three times with PBS (150 μl per rinse) or with wash buffer provided in kit. We use a plate washer for this and leave a small volume in the wells between rinse steps ($\sim$20 μl).
7. Spin plate(s) at 300g for 10 min, remove buffer, leaving 20 μl, and add 50 μl of DNA labeling reaction per well.
8. Incubate 1 h at 37° (shake every 15 min or place on rocker if available).
9. Add 150 μl rinsing buffer (red cap) from kit or PBS.
10. Centrifuge plates at 300g for 10 min, remove buffer, leaving 20 μl, and add 100 μl anti-BrdU staining solution supplemented with 1:50 to 1:100 dilution of fluorophore-conjugated antiphosphohistone H3 antibodies (Cell Signaling) and DNA stain (e.g., 1–10 μg/ml Hoechst 33342 or 1–5 μM DAPI).
11. Incubate 30 min at room temperature in the dark.
12. Add 150 μl rinse buffer (from kit) or PBS, spin at 300g for 10 min, and remove buffer, leaving small amount. Repeat.
13. Resuspend in 125 μl PBS and image plate.

Conclusions

Multiple approaches exist for fluorescent labeling of intracellular components in cells, including immunostaining with antibodies, using dyes or stains specific to cellular components or events, and fluorescent proteins for live-cell imaging. Whatever the need, there is a plethora of choices in fluorescent probes, imaging platforms, automation, and image analysis modules that offer researchers unique opportunities to interrogate complex signaling pathways, measure phenotypic changes, and exploit mechanistic effects of agents on cellular function in individual cells. For successful screening campaigns, care should be given when choosing cellular probes (e.g., dyes, fluorescent proteins, and antigen-specific, noncross-reactive antibodies), fixatives, buffers, imaging hardware, analysis software, and visualization tools. Nevertheless, with these resources in hand, automated fluorescence microscopy and high-content screening have the potential to reduce target and lead identification time and accelerate the drug discovery process.

Acknowledgments

The authors acknowledge Mike Cerra, Brandi Dickinson, Frank Traynor, and Shawn Gauby for their assistance with tissue culture and immunostaining.

References

Darzynkiewicz, Z., Juan, G., Li, X., Gorczyca, W., Murakami, T., and Traganos, F. (1997). Cytometry in cell necrobiology: Analysis of apoptosis and accidental cell death (necrosis). *Cytometry.* **27,** 1–20.

Lee, S., and Howell, B. J. (2006). High-content screening: Emerging hardware and software technologies. *Methods Enzymol.* **414**(this volume).

Mitchison, T. J. (2005). Small-molecule screening and profiling by using automated microscopy. *Chembiochem.* **6,** 33–39.

Prigent, C., and Dimitrov, S. (2003). Phosphorylation of serine 10 in histone H3, what for? *J. Cell Sci.* **116,** 3677–3685.

Vermes, I., Haanen, C., Steffens-Nakken, H., and Reutellingsperger, C. (1995). A novel assay for apoptosis flow cytometric detection of phosphatidylserine expression on early apoptotic cells using fluorescein labelled annexin V. *J. Immunol. Methods.* **184,** 39–51.

Villa, P., Kaufmann, S. H., and Earnshaw, W. C. (1997). Caspases and caspase inhibitors. *Trends Biochem. Sci.* **22,** 388–393.

[17] High-Throughput Screening for Modulators of Stem Cell Differentiation

By Paul J. Bushway and Mark Mercola

Abstract

Realizing the potential of stem cell biology requires the modulation of self-renewal and differentiation, both of which are incompletely understood. This chapter describes methods for the design, development, and implementation of cell-based screens of small molecules, genes and expressed proteins for modulation of stem and progenitor cell fate. These include the engineering of embryonic and other stem cells with gene promoter–reporter protein constructs and their application in automated screening. We discuss considerations of promoter reporter selection, assay development and implementation, and image acquisition, analysis, and data handling. Such black-box screens are useful for the identification of probes of developmental processes and should provide tools that will identify druggable targets for biochemical assays.

Introduction

Stem cells combine the potential for self-renewal with the ability to give rise to one or more differentiated cell types, and the ability to manipulate these cells offers unprecedented opportunities to increase our understanding of developmental biology and to provide a renewable source of cells to

METHODS IN ENZYMOLOGY, VOL. 414   0076-6879/06 $35.00
Copyright 2006, Elsevier Inc. All rights reserved.   DOI: 10.1016/S0076-6879(06)14017-3

address crippling degenerative diseases by replacing lost or damaged tissue. Recent advances have identified stem cells in nearly all tissue compartments of the body, and, increasingly, it is becoming possible to isolate the stem cells, maintain them in culture, and, in certain instances, expand their numbers. Despite advances in basic stem cell biology, it remains quite challenging to stimulate them to differentiate efficiently into the cell types needed for clinical application, either in culture or after transplantation into damaged tissue. The reasons for inefficient differentiation include often extreme difficulty in obtaining large numbers of appropriate types of stem cells, poor survival in culture, and, upon transplantation, incomplete understanding of the natural molecules that promote differentiation. To promote efficient differentiation requires specific knowledge of the molecules and signaling pathways that are relevant to the given type of stem cell. Moreover, in those instances in which multiple cell types develop from a single class of stem cell, the molecules and signaling pathways must also be matched to the lineage of the particular differentiated cell type that is desired. Thus, before the biotechnological and medical potential of stem cells can be realized, it will be essential to have in hand molecules that can influence their survival, renewal, and differentiation.

The incomplete knowledge of natural modulators has prompted the development of cell-based assays to screen chemical and genome libraries for small molecules, genes, and proteins capable of controlling stem and progenitor cell biology. This chapter focuses on the development of cell-based assays and considers the advantages of this approach, as well as the limitations that their current design imposes on the assay development and implementation process.

Assay Design

General Considerations of Cell-Based Assay Technology Applied to Stem and Progenitor Cell Biology

Cell-based assays offer the ability to perform unbiased screens of libraries for molecules that modulate a wide range of biological phenomena (see other chapters in this volume). In the stem cell arena, screens have been implemented to identify small molecules that stimulate differentiation along multiple lineages (Ding et al., 2003; Wei et al., 2004; Wu et al., 2004). The readouts of these assays can be based on detailed analysis of microscopic images collected, capable of detecting subcellular events, such as localization of a protein (high-content screening, HCS) (e.g., Granas et al., 2005; Morelock et al., 2005; Starkuviene et al., 2004; Thomsen et al., 2005; Werner et al., 2006), which provides advantages over detection of a particular cellular constituent using a plate reader-based assay (conventional high-throughput screening, HTS) because of the potential to reveal information about subcellular processes and detect relatively rare differentiation events against a background of nonresponding cells. This section focuses on the development of assays to measure the activity of individual marker gene promoters as a readout of differentiation because of their ubiquitous application in stem and developmental biology and because the fundamentals are the same as for detection of structural or biochemical features, which present more content.

Analysis of the wells is often best achieved through image acquisition by automated microscopy because this method affords the most precise means of reporting data on the basis of the number of cells that are affected (e.g., differentiate into a target cell type). This is particularly important when relatively few cells in a well might be expected to be competent to respond to the inducer, such as when cells in the well are heterogeneous [as in differentiating embryonic stem cells (ESCs) or primary cell material] or when the response is expected to vary stochastically by activity and the primary hit is modestly effective at the test dose or if toxicity or other confounding activities coexist. Detailed discussion of the HCS image analysis methods suitable for stem cell biology can be found in other chapters in this volume.

Promoter–Reporter Constructs

Some of the first stem cell-based screens were implemented using antibody binding to visualize differentiation (Ding *et al.*, 2003), and this continues to be a common approach for cell-based assays. Immunological detection of differentiation markers, however, has several obvious drawbacks for scale-up, including cost and availability of immunological and detection reagents and incorporation of wash steps that increase the risk of cell detachment.

The use of gene promoters to direct expression of reporter proteins is routine in developmental biology, and fluorescent proteins, such as green fluorescent protein (GFP) and dsRED variants, offer a straightforward detection method (Hadjantonakis *et al.*, 2002; Zhang *et al.*, 2002). If expressed as a fusion protein to a localization domain, a transcription factor or membrane-association domain for instance, the fluorescent protein can be used in real time to detect regulated translocation to spatially constrained compartments within cells. Fluoresent proteins are also engineered into real-time biosensors of numerous cellular-signaling processes, including detection of metabolites or protein–protein interactions indicative of flux through a signaling pathway (e.g., see Miyawaki *et al.*, 1997; Nalbant *et al.*, 2004;

Tallini *et al.*, 2006; Zaccolo and Pozzan, 2002; reviewed in Tavare *et al.*, 2001; Zaccolo and Pozzan, 2000; Zhang *et al.*, 2002). Alternatives to fluorescent proteins include enzymes that catalyze production of a fluorophore [e.g., β-lactamase directed cleavage of LiveBLAzer fluorescence resonance energy transfer (FRET) substrate (Invitrogen)], although these systems traditionally have lacked the ability to visualize subcellular localization of reporter proteins. Alternative fluorescent technologies that permit subcellular localization use covalent modification of the reporter protein with a small fluorescent probe, such as HaloTag (Promega).

A potential pitfall of cell-based, differentiation assays might be encountered when multiple steps in the differentiation program separate the readout (e.g., expression of a reporter protein at a particular stage of differentiation) from compound addition. Even a potently active compound might not be detected if culture conditions are not conducive to sustaining differentiation through successive steps until the readout is manifest. However, if a compound is active at a late stage of differentiation, the culture conditions might give rise to too few cells that are competent to respond. For these reasons, the readout of differentiation should follow compound addition as closely as possible, preferably within 24 to 48 h. This will also have the advantage of minimizing media changes with attendant reagent savings.

Lentiviral Technology to Create Stable Reporter Lines

Placing reporter proteins under the control of well-characterized differentiation-specific gene promoters requires that the transcriptional elements be located within a contiguous stretch of DNA suitable for introduction into cells. A number of relatively short, <10-kb promoter fragments have been described as capable of directing transgene expression to stem, progenitor, or terminally differentiated cell types, including the genes Oct4 for embryonic stem cells (Yang *et al.*, 2005), αMHC (Subramaniam *et al.*, 1991; Takahashi *et al.*, 2003) or MLC2V for cardiomyocytes (Yan *et al.*, 2003), PDX-1 for pancreatic tissue (Melloul *et al.*, 2002), and insulin for pancreatic β cells (Odagiri *et al.*, 1996). VSV-G pseudotyped HIV lentiviral vectors are well suited for transducing promoter–reporter cassettes incorporating such fragments. VSV-G pseudocoated vectors are generated by three plasmid (vector, packaging, and VSV-G plasmids) calcium phosphate cotransfection of 293T cells, yielding serum-free, viral supernatants with titers of 10^8 infectious viral particles/ml after concentration (as in Reiser, 2000) that can be stored at $-80°$. For infections, concentrated viral supernatant containing polybrene is added to stem cells, including murine or human ESCs, in suspension at a multiplicity of infection (MOI) of 10 to 100. The mixture is then plated as for routine culture.

Infection efficiency of ESCs under these conditions often exceeds 50%. The infection efficiency can be considerably higher for many other cell lines, but can be considerably lower for nondividing primary cells.

Targeted Introduction of Reporters into Marker Gene Loci by Homologous Recombination

The spatiotemporal patterns of gene expression that characterize many developmentally interesting genes require transcriptional elements that are distributed over large regions of DNA and thus these gene promoters exceed the carrying capacity of plasmid, HIV lentiviral, or common DNA viral transduction systems.

Homologous recombination in transgenic mice is routinely used to integrate reporter cDNAs under the control of marker DNA loci. Because targeted recombination is performed in mouse ESCs, ESCs harboring fluorescent proteins in reporter loci represent a potential source of cells for HCS and HTS.

Homologous recombination in other cell types is more difficult to achieve for a variety of reasons, such as poor transfection efficiency. An alternative is to target integration into reporter loci contained on bacterial artificial chromosomes (BACs) or P1-derived artificial chromosomes (PACs) because of their ability to contain large regions of genomic DNA and their stability. Commercially available libraries of BACs harboring upward of 150 kb of cargo DNA can be maintained and targeted by homologous recombination within *Escherichia coli*. Recombination-mediated genetic engineering of BACs, known as recombineering, is a powerful method for fast and efficient manipulation of large genomic regions for subsequent cell culture experiments. Protocols are available on the Recombineering Web site maintained by the NCI (http://www.recombineering.ncifcrf.gov). Recombineered BAC or PAC DNA is often prepared using a commercial kit (such as Qiagen) or using CsCl density sedimentation. Transfection into ESCs is often performed using a protocol involving lipid: DNA complex reagents (Bauchwitz and Costantini, 1998), but nonetheless is often inefficient and requires optimization (Montigny *et al.*, 2003). An efficient method to overcome inefficient transfection by retrofitting a drug selection cassette has been developed by Wang *et al.* (2001).

Validation of Cell Lines

To confirm promoter–reporter fidelity, engineered cell lines must be tested for correct expression of the reporter transgenes by colocalization of transgene expression with expression of the endogenous gene by *in situ* hybridization or immunohistochemistry. In the case of ESCs, differentiation of certain derivatives can be achieved in differentiating aggregates of cells maintained generally in suspension, termed embryoid bodies (EBs), (Robertson, 1987). Within 3 to 4 days, 65% of the cells express brachyury, a mesendodermal marker, by FACS analysis (unpublished results), by which time EBs can be either maintained floating or plated onto gelatin-coated cell culture plastic dishes or coverslips. In either case, beating cardiomyocytes arise by day 8. Because EBs are highly heterogeneous, immunohistochemistry or *in situ* hybridization is essential to verify coincident reporter protein expression with the spatiotemporal pattern of endogenous gene expression. For lineages that form poorly in EBs, such as later stages of the endocrine pancreas, mouse ESC reporter lines can be evaluated rigorously in chimeric mouse embryos obtained by injection of murine ESCs into preimplantation blastocysts that are then introduced into the uterus of pseudopregnant females (Robertson, 1987). At appropriate stages, founder chimeric embryos should be examined for ESC contribution and marker expression in relevant tissues to verify that fluorescent protein expression coincides with the spatiotemporal pattern of endogenous gene expression. The evaluation of reporter expression in human ESCs, for which the production of interspecies chimeras might be limited by poor contribution, as well as by law and material transfer agreements, verifying the fidelity of reporter expression, presents considerable challenges and is an argument in favor of targeting reporter proteins to human gene promoters in BACs to maximize the likelihood of faithful expression.

Assay Development

Confidence in the ability of an assay to resolve hits from background is based on its signal-to-noise ratio and the overall dynamic range. In the stem cell arena, it might be desirable to use cells that are not clonal and might even be primary cell isolates. The signal-to-noise ratios and dynamic range values in assays with these cells are likely to be more problematic than with established, clonal cell lines, resulting in an increased likelihood of false outcomes.

In general, high-throughput assays are considered robust if the coefficient of variation is less than 10% across trials and the signal-to-noise ratio and dynamic range results in a Z' value of >0.5. Z' is calculated according to the following formula:

$$Z' = 1 - (3 \times \text{sd}[S_0] + 3 \times \text{sd}[S_{max}])/\text{ABS}(\text{mean}[S_{max}] - \text{mean}[S_0]),$$

where sd is standard deviation and S_0 and S_{max} are values obtained for negative [S_0] and positive (S_{max}) controls. An image-based HCS assay might very well be amenable to screening at lower Z' values [e.g., 0.3 (Granas *et al.*, 2005)] for several reasons, not least of which is the ability to visually confirm meaningful hits from image data. Evaluating an assay during development requires appropriate positive and negative controls. Physiologically relevant controls might not exist for a number of reasons, however, such as novelty of the assay for which no hit has yet been described. In such a case, it is imperative that the investigator optimize the assay using compounds or reagents that realistically approximate what would be expected from a scientifically interesting hit. Additionally, a pilot screening of about 10,000 compounds can be performed in the hope of blindly identifying an assay "hit" that can serve as a positive control. This is an undisciplined approach and is a risky and costly endeavor for a number of reasons.

- There is no expectation of the assay's physiological dynamic range.
- Scale-up and automation of the assay suffers from improper validation.
- No clear metric exists to validate success or failure of the screen on completion.
- Data extraction, quantification, and statistical inferences are nebulous.
- Elevated false positive and negative outcomes drive up confirmation time and cost.

Stem cell assays that use promoter–reporter readouts are often evaluated using image analysis algorithms because the number of responding cells might be a minor fraction of the total number of cells in the well. This would be true if the cell population were heterogeneous, such as differentiating ESCs, for which a small number of responders might be a meaningful positive outcome because only a small fraction of the cells would be expected to be competent to differentiate along one particular lineage. In this context, applying an algorithm that measures the signal generated by a few cells against background nonresponders might lead to a poor Z' value. Approaches that segment the images into single cells and evaluate the number of differentiating cells might circumvent this problem. Alternatively, a metric in which a given well is deemed "on" or "off" is determined relative to a threshold for activity could be imposed and validated statistically across many replicates. Such a strategy is worth using in a rapid and simple repeat of the primary screen when discrimination of hits by the primary screen is suboptimal.

Methods

Handling and Growth of Murine ESCs prior to Screening

The typical bench-top mESC differentiation assay calls for the growth on mouse embryonic fibroblasts (MEFs) on gelatin and subsequent differentiation within EBs. Each component is problematic in a high-content screening approach. The use of poly-D-lysine-coated tissue culture plastic was used to rapidly filter out MEFs during assay ramp-up. MEFs adhere tightly to poly-D-lysine-coated tissue culture plastic, whereas mESCs loosely tether to the surface. After 2 days in culture, the ESCs are then lifted from the culture dish with cold media, leaving MEFs behind. Two to three passages in this manner will typically clear the culture of all contaminating MEFs. A high level of residual MEFs reduces well-to-well reproducibility for several reasons, including that they might secrete factors that affect mESC differentiation.

Choice of matrix material influences numerous assay parameters, including maintaining the cells in a spread, two-dimensional environment more suitable for image analysis than EBs and the important consideration that certain matrix constituents will positively or negatively influence differentiation. Not surprisingly, evaluation of matrix components has been shown to optimize liver cell differentiation (Flaim *et al.*, 2005). For mESC cardiogenesis assays, we found that a combination of poly-D-lysine and gelatin (see later) was best at promoting mESC survival and adherence, while maintaining differentiation potency, during the assay.

In HC/HT formats it is not entirely unexpected to see deviations from the biology observed at the bench. One example of this was the longer assay development time observed when scaling the assay to a 384-well format. Differentiation in EBs and in larger well plate formats typically results in cardiomyocyte differentiation by 8 days. This time line reproducibly shifted to 10 to 11 days when converting the assay to a 384-well plate format. One possible explanation might be the change in cell number to media volume and alterations in media change procedure, which would alter the concentration of secreted factors that accumulate in conditioned medium.

Imaging and Processing a Stem Cell Differentiation Assay

Other chapters in this volume provide extensive discussion of the imaging processing of particular high-content screening assays. The expenditure of effort up front to increase the ability of the algorithm to recognize features of the cells that reflect the desired biological response will return

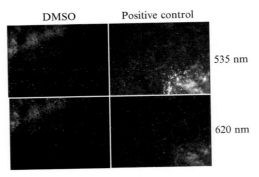

FIG. 1. Comparison of a "strongly" positive focus as compared to a negative outcome in an ESC-based cardiogenesis assay. Each panel consists of nine-image, single-well mosaics of images. Images were acquired at 460 nm (nuclear, not shown), 535 nm (specific channel, upper panels), and 620 nm (nonspecific channel, lower channels); emissions were gathered on the INCell 1000 with a 10× objective and 4 × 4 binning. The focus of eGFP+ cardiomyocytes in this case coincides with increased autofluorescence, probably reflecting cell death in cells that had grown into a multilayered structure. From an imaging perspective, this dictates image capture of as much of the well as possible so as to maximize the likelihood of capturing a positive event. The appearance of a high-intensity 535-nm signal is localized to the lower right quadrant of the positive control well.

the investment by improving the hit identification. This sections discusses a few points relevant to acquisition and processing of images for cell differentiation assays. The number of images acquired per well depends on the incidence of responding cells. Figure 1 demonstrates the appearance of a "strong" positive focus of differentiation as compared to a negative outcome in promoter–reporter ESC cardiogenesis assay in which relatively few cells are expected to respond to the inducing stimulus. Nine image single-well mosaics of both 535-nm (specific channel, eGFP) and 620-nm (nonspecific channel, autofluorescence) emissions were gathered on the GE INCell 1000 with a 10× objective and 4 × 4 binning. From an imaging perspective, this dictates image capture of as much of the well as possible so to maximize the likelihood of capturing a positive event. To achieve this end, one can capture many images with a higher power objective or fewer with a low-power objective. The requirement for spatial resolution will ultimately drive the decision of objective and binning. Promoter–reporter assays that measure gene activity produce a simple yes/no response; thus, lower resolution is preferable due to speed and memory efficiency.

In imaging a HC/HT screening campaign, processing speed, computational power, and memory constraints should be considered for overall process efficiency. A typical image captured with a 10× objective, 12-bit depth, with 4 × 4 binning will approximate 170 kB in standard TIFF image format. Table I summarizes the approximate data load one might expect as image quality increases. Process time and storage requirement will increase proportionately.

It will be essential to include operations on an image used to correct systemic problems due to the image capture process. Examples of image processing include correction of the illumination field (commonly referred to as flat-field correction), removal of corrupt pixels, noise suppression, and so on. The need for processing is exemplified in Fig. 2, which shows the (x,y) pixel coordinate map with intensity on the z axis of a blank well from a 384-well plate. Most of the variance is due to imperfections in the illuminating field with occasional punctuate artifacts.

Quantification of the images involves the calculation of segmented pixels in an image (usually an area encompassing the region of interest) dictated by user preferences. The summed time of these two processes can rapidly saturate the computational power of even the most powerful workstations when dealing with large volumes of high-resolution images. As shown earlier, from the perspective of cost effectiveness, it is preferable to use only the minimum resolution required to achieve the computational objective confidently. However, it might be desirable to capture high-resolution images if archival for subsequent data mining for other purposes is anticipated. Using the example of the cardiomyocyte differentiation assay, three colors and nine fields were captured per well with a 10× objective on the GE INCell1000 with 12-bit images binned 4 × 4. The data load of this process on a single 384-well plate is approximately 3.7 GB. This is the lowest resolution that can be achieved on the INCell1000 as of February 2006. This resolution exceeds that which is necessary to resolve "positive" events defined by promoter–reporter activity. Furthermore, simplicity in processing and quantification is also desirable to enhance

TABLE I
APPROXIMATE DATA LOAD

Objective	12-bit image binning	Number of images	Image file size in MB	Three-color imaging	Image file size/384-well plate in MB
10×	1 × 1	1	2.832	8.496	3262
10×	2 × 2	1	0.708	2.124	815.6
10×	4 × 4	1	0.177	0.531	203.9
20×	1 × 1	1	11.32	33.98	13049
20×	2 × 2	1	2.832	8.496	3262
20×	4 × 4	1	0.708	2.124	815.6

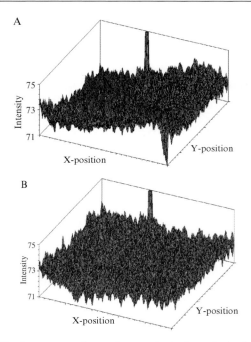

FIG. 2. The need for image correction and processing is shown in the (x, y) pixel coordinate maps with intensity on the z axis of a blank well from a 384-well plate. Most of the variance in an uncorrected image (A) is due to imperfections in the illuminating field with occasional punctuate artifacts. The correction applied here (B) is referred to commonly as flat-field correction. Variance in the illumination field is a normal component of CCD-based image capture and is corrected easily using calibration images gathered after each imaging run.

efficiency. In the ESC-based cardiogenesis assay, the emphasis was on calculation and summarizing events; in the assay described here, 1 well = 1 sample = 1 test. To process images rapidly in as simple a manner possible, an image mask was generated by calculating the global average of each image field applying and accepting all pixels above this threshold plus an empirically defined differential (percentage of global average). After application of a size exclusion filter, the result is the image processing scheme depicted in Fig. 3.

Notes on Data Handling

The success of a HC/HT screening assay depends on the handling of quantified data generated from the images. A general assumption can be made of a population of random samples; given enough samples, the population will

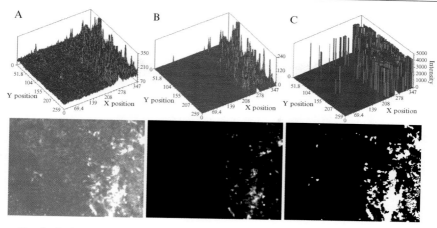

FIG. 3. An image-processing scheme of a ESC-based cardiomyocyte assay. (A) The image and three-dimensional area plot represents the image as it is captured from the microscope. (B) The same image after application of the process for removing nonspecific signal from the image calculation using general statistics from the entire image. (C) The image and plot represent the generation of a two bit-mask, which can be applied to segment images collected in all wavelengths (e.g., 460, 535, 620, 680 nm) to identify features present within the mask. One can apply many methods to achieve a similar result; however, all image-processing algorithms must be validated on appropriate test cases to ensure fidelity to the desired outcome.

have a normal distribution. When zeros and extreme outliers are removed and the population is normalized across all well plates, one should expect a near-normal distribution. In the mESC cardiomyocyte example, the desired result was to obtain a small number of hits that biased mESC lineage toward a cardiomyocyte outcome. In the primary screen, the majority of compounds would be expected to have little or no effect, with a small population of wells harboring artifactual fluorescence due to fluorescent compounds or toxicity. This can be summarized numerically in two categories: zeros (background level) and numbers of extraordinary scale relative to the population at large. Figure 4 depicts this graphically, with the intensity value of the wells on the x axis in quantiles and the frequency of observations on the y axis. Figure 4A shows that zeros are present in the population of data points at the greatest abundance, skewing the normality of the data population. Figure 4B shows curve normality after filtering out all zeros and artifacts. The box plot (Fig. 4D) depicts near and far outliers distributed greater than $1.5\times$ the value of the interquartile region (IRQ, which equals the distance between

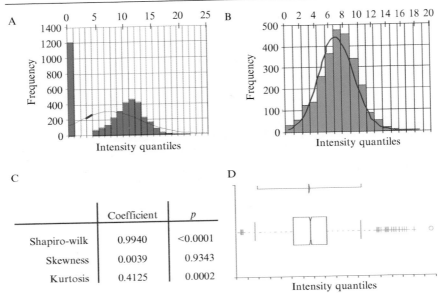

FIG. 4. Graphic summary of raw and filtered data obtained during the screening process. (A) Using the ESC-based cardiomyocyte differentiation assay as an example, samples can be parsed numerically into two categories: zeros and numbers of extraordinary scale relative to the population at large. The frequency distribution curve shows the intensity value of the observation after normalization, in quantiles, is represented on the x axis and the number of observations falling in that quantile is represented on the y axis. Zeros are present in the population of data points at the greatest abundance (reporter OFF), skewing the normality of the data population. (B) The frequency distribution curve shows normality after filtering out all zeros and artifacts. (C) General statistics for the distribution indicating a slightly peaked curve with kurtosis at 0.4 and only slight right tail skew at 0.004. (D) The box plot depicts the intensity distribution of compounds with the boundaries of the box reflecting the 25 and 75th percentile and the internal bar set at the 50th percentile. Blue tick marks represent 2.5 and 97.5th percentiles. Red + and O symbols reflect near and far outliers, respectively, lying greater than 1.5 IQRs above the upper quartile. IQR is the interquartile region, the region between the 25 and the 75th percentiles. Summarizing data in such a way makes it possible to view the assay results in a top-down view and highlights any important concerns with the data population that need to be addressed.

25 and 75th percentiles) above the upper quartile with red + and O symbols, respectively. Using general statistics to summarize data makes it possible to view the assay results in a top-down view and highlights any concerns with the data population that should be addressed.

Conclusion

Illustrations were provided to describe the approaches taken to develop assays for stem cell differentiation. The major advantages of black-box, cell-based screens are that the compounds identified are selected based on biological function and can serve as powerful probes of basic mechanisms of cell phenomena, such as stem cell differentiation. A consideration with such screens is achieving sufficient dynamic range and a signal-to-noise ratio with minimal plate-to-plate and day-to-day variation to discern hits. This can be achieved with judicious selection of promoter–reporter constructs and image analysis techniques. As data handling and storage capabilities and complementary algorithms for data mining improve, it will be practical to archive image data for subsequent analysis long after the initial screening campaign. We envision that this will enhance probe discovery in the future.

Appendix

As an illustration, a protocol is provided for an ESC cardiogenesis screen, which would need to be adapted to suit differentiation along alternate lineage or for other cell types, so is provided as an illustration.

Growth of ESCs

Cells are a murine embryonic stem cell line carrying the αMHC promoter (REF) fused to a fluorescent eGFP reporter (REF). Cells are recovered from a fresh thaw by coculture with MEFs and 10% fetal calf serum (FCS)-enriched DMEM fortified with penicillin and strepfamycin, non-essential amino acids (NEAA), Na-pyruvate, β-mercaptoethanol, and leukemia inhibitory factor (LIF). On third passage cells are converted to poly-D-lysine-coated plates (0.01%; 300 kDa) and weaned to screening growth media with identical formulation of recovery media with the exception of 7% FCS. Cells are passaged twice prior to seeding 384- or 1536-well plates.

Cell Plating

The 384- and 1536-well plates are prepared with a matrix of poly-D-lysine and gelatin at 0.005 and 0.05%, respectively. The matrix is left on the plates overnight at room temperature and removed and washed once with water. Residual water is aspirated from the plates, which are dried at 42° overnight. Seeding densities determined for 384- and 1536-well plate formats are 750 to 1000 and 175 to 225 cells per well, respectively. Twofold shifts from these approximations can result in failure of the assay due to lack of sufficient cells or excessive spontaneous differentiation. Cells are resuspended in a half-well volume of growth medium without LIF and applied to the plates and left to grow for 2 full days.

Assay Execution

On days 2 and 4, compound fortified medium is applied to the wells. In brief, the compound is prepared in CRG8 growth media at $2\times$ final concentration and seeding volume equivalent added. Residual compound fortified media is then diluted twofold prior to second addition on day 4. Compounds are removed on day 6, and fresh growth medium is reapplied to cultures every other day. Ideally, each plate should contain positive and negative controls.

Assay Termination

On approximately day 10 ($\pm$ 1 day) the assay is terminated by evacuation of media from wells and the addition of 4% paraformaldehyde for 45 min. Plates are then washed twice in $1\times$ phosphate-buffered saline(PBS) and then stained in 1 μg/ml of 4′,6′-diamidine-2′-phenylindole dihydrochloride (DAP1) in $1\times$ PBS for 30 min. DAPI is removed from the wells with one wash of $1\times$ PBS and preserved in a solution of 50% glycerol and water.

Quantification

Images gathered from the GE INCell 1000 are flat-field corrected and quantified using an image subtraction technique utilizing the image global average and a global average based differential to remove the nonspecific 535-nm fluorescent signal. This process is executed using the GE software package Developer Toolbox v1.5 (formerly known as IRI). An image bit mask produced from the 535-nm channel is passed through a size-exclusion filter of approximately 3px and applied to both 535- and 620-nm channels. Average and integrated densities in addition to area measurement are provided as output to data tables.

Data Handling

The data set is normalized across plates. We generally normalize using medians and application of simple linear scalers within each plate. A common alternative is to rely on control samples in each plate. Data are then power transformed according to the data set for visual statistical viewing of data. Data are binned and filtered according to user criteria (rank, stdev, morphology, correlation across gathered excitations, frequency, etc.).

Acknowledgments

The authors acknowledge the help of Suzette Farber-Katz, Maria Barcova, Pam Itkin-Ansari, and Fred Levine for developing the MIN6 insulin promoter screen and Joaquim Teixeira, Tomo Takahashi, and Richard T. Lee (Brigham and Women's Hospital, Boston) for developing the αMHC screen. This work would not have been possible without the many helpful discussions of Fred Levine, Jeff Price, Joaquim Teixeira, Pam Itkin-Ansari, Maria Barcova, Rosa Guzzo, and Steve Vasile at the BIMR. The research was supported by grants from the Juvenile Diabetes Foundation and the NIH (HL059502, HL071913 and DK068715) to MM.

References

Bauchwitz, R., and Costantini, F. (1998). YAC transgenesis: A study of conditions to protect YAC DNA from breakage and a protocol for transfection. *Biochim. Biophys. Acta* **1401,** 21–37.

Ding, S., Wu, T. Y., Brinker, A., Peters, E. C., Hur, W., Gray, N. S., and Schultz, P. G. (2003). Synthetic small molecules that control stem cell fate. *Proc. Natl. Acad. Sci. USA* **100,** 7632–7637.

Flaim, C. J., Chien, S., and Bhatia, S. N. (2005). An extracellular matrix microarray for probing cellular differentiation. *Nature Methods* **2,** 119–125.

Granas, C., Lundholt, B. K., Heydorn, A., Linde, V., Pedersen, H. C., Krog-Jensen, C., Rosenkilde, M. M., and Pagliaro, L. (2005). High content screening for G protein-coupled receptors using cell-based protein translocation assays. *Comb. Chem. High Throughput Screen* **8,** 301–309.

Hadjantonakis, A. K., Macmaster, S., and Nagy, A. (2002). Embryonic stem cells and mice expressing different GFP variants for multiple non-invasive reporter usage within a single animal. *BMC Biotechnol.* **2,** 11.

Melloul, D., Marshak, S., and Cerasi, E. (2002). Regulation of pdx-1 gene expression. *Diabetes* **51**(Suppl. 3), S320–S325.

Miyawaki, A., Llopis, J., Heim, R., McCaffery, J. M., Adams, J. A., Ikura, M., and Tsien, R. Y. (1997). Fluorescent indicators for Ca^{2+} based on green fluorescent proteins and calmodulin. *Nature* **388,** 882–887.

Montigny, W. J., Phelps, S. F., Illenye, S., and Heintz, N. H. (2003). Parameters influencing high-efficiency transfection of bacterial artificial chromosomes into cultured mammalian cells. *Biotechniques* **35,** 796–807.

Morelock, M. M., Hunter, E. A., Moran, T. J., Heynen, S., Laris, C., Thieleking, M., Akong, M., Mikic, I., Callaway, S., DeLeon, R. P., Goodacre, A., Zacharias, D., and Price, J. H. (2005). Statistics of assay validation in high throughput cell imaging of nuclear factor kappaB nuclear translocation. *Assay Drug Dev. Technol.* **3,** 483–499.

Nalbant, P., Hodgson, L., Kraynov, V., Toutchkine, A., and Hahn, K. M. (2004). Activation of endogenous Cdc42 visualized in living cells. *Science* **305,** 1615–1619.

Odagiri, H., Wang, J., and German, M. S. (1996). Function of the human insulin promoter in primary cultured islet cells. *J. Biol. Chem.* **271,** 1909–1915.

Reiser, J. (2000). Production and concentration of pseudotyped HIV-1-based gene transfer vectors. *Gene Ther.* **7,** 910–913.

Robertson, E. J. (1987). "Teratomacarcinomas and Embryonic Stem Cells: A Practical Approach." IRL Press, Oxford.

Starkuviene, V., Liebel, U., Simpson, J. C., Erfle, H., Poustka, A., Wiemann, S., and Pepperkok, R. (2004). High-content screening microscopy identifies novel proteins with a putative role in secretory membrane traffic. *Genome Res.* **14,** 1948–1956.

Subramaniam, A., Jones, W. K., Gulick, J., Wert, S., Neumann, J., and Robbins, J. (1991). Tissue-specific regulation of the alpha-myosin heavy chain gene promoter in transgenic mice. *J. Biol. Chem.* **266,** 24613–24620.

Takahashi, T., Lord, B., Schulze, P. C., Fryer, R. M., Sarang, S. S., Gullans, S. R., and Lee, R. T. (2003). Ascorbic acid enhances differentiation of embryonic stem cells into cardiac myocytes. *Circulation* **107,** 1912–1916.

Tallini, Y. N., Ohkura, M., Choi, B. R., Ji, G., Imoto, K., Doran, R., Lee, J., Plan, P., Wilson, J., Xin, H. B., Sanbe, A., Gulick, J., Mathai, J., Robbins, J., Salama, G., Nakai, J., and Kotlikoff, M. I. (2006). Imaging cellular signals in the heart *in vivo*: Cardiac expression of the high-signal Ca^{2+} indicator GCaMP2. *Proc. Natl. Acad. Sci. USA* **103,** 4753–4758.

Tavare, J. M., Fletcher, L. M., and Welsh, G. I. (2001). Using green fluorescent protein to study intracellular signalling. *J. Endocrinol.* **170,** 297–306.

Thomsen, W., Frazer, J., and Unett, D. (2005). Functional assays for screening GPCR targets. *Curr. Opin. Biotechnol.* **16,** 655–665.

Wang, Z., Engler, P., Longacre, A., and Storb, U. (2001). An efficient method for high-fidelity BAC/PAC retrofitting with a selectable marker for mammalian cell transfection. *Genome Res.* **11,** 137–142.

Wei, Z. L., Petukhov, P. A., Bizik, F., Teixeira, J. C., Mercola, M., Volpe, E. A., Glazer, R. I., Willson, T. M., and Kozikowski, A. P. (2004). Isoxazolyl-serine-based agonists of peroxisome proliferator-activated receptor: Design, synthesis, and effects on cardiomyocyte differentiation. *J. Am. Chem. Soc.* **126,** 16714–16715.

Werner, T., Liebisch, G., Grandl, M., and Schmitz, G. (2006). Evaluation of a high-content screening fluorescence-based assay analyzing the pharmacological modulation of lipid homeostasis in human macrophages. *Cytometry A* **69,** 200–202.

Wu, X., Ding, S., Ding, Q., Gray, N. S., and Schultz, P. G. (2004). Small molecules that induce cardiomyogenesis in embryonic stem cells. *J. Am. Chem. Soc.* **126,** 1590–1591.

Yan, X., Price, R. L., Nakayama, M., Ito, K., Schuldt, A. J., Manning, W. J., Sanbe, A., Borg, T. K., Robbins, J., and Lorell, B. H. (2003). Ventricular-specific expression of angiotensin II type 2 receptors causes dilated cardiomyopathy and heart failure in transgenic mice. *Am. J. Physiol. Heart Circ. Physiol.* **285,** H2179–H2187.

Yang, H. M., Do, H. J., Oh, J. H., Kim, J. H., Choi, S. Y., Cha, K. Y., and Chung, H. M. (2005). Characterization of putative cis-regulatory elements that control the transcriptional activity of the human Oct4 promoter. *J. Cell Biochem.* **96,** 821–830.

Zaccolo, M., and Pozzan, T. (2000). Imaging signal transduction in living cells with GFP-based probes. *IUBMB Life* **49,** 375–379.

Zaccolo, M., and Pozzan, T. (2002). Discrete microdomains with high concentration of cAMP in stimulated rat neonatal cardiac myocytes. *Science* **295,** 1711–1715.

Zhang, J., Campbell, R. E., Ting, A. Y., and Tsien, R. Y. (2002). Creating new fluorescent probes for cell biology. *Nature Rev. Mol. Cell. Biol.* **3,** 906–918.

[18]   High-Content Kinetic Calcium Imaging in
        Drug-Sensitive and Drug-Resistant
             Human Breast Cancer Cells

By Maria A. DeBernardi and Gary Brooker

Abstract

Intracellular calcium $(Ca^{2+})_i$ is involved in the regulation of a variety of biological functions in cancer cells, including growth inhibition, tumor invasiveness, and drug resistance. To gain insight into the possible role played by Ca^{2+} in the development of drug resistance in breast cancer, we performed a comparative high-content analysis of the intracellular Ca^{2+} dynamics in drug-sensitive human breast cancer MCF-7 cells and five drug-resistant, MCF-7-derived clonal cell lines. Fura-2 single cell ratiometric fluorescence microscopy was used to monitor real-time quantitative changes in cytosolic-free Ca^{2+} concentration ($[Ca^{2+}]_i$) upon addition of phosphoinositol-coupled receptor agonists. While the magnitude and the onset kinetics of the $[Ca^{2+}]_i$ rise were similar in drug-sensitive and drug-resistant cell lines, the decay kinetics of the $[Ca^{2+}]_i$ increase was found to be consistently slower in drug-resistant than drug-sensitive cells. Such a delay in reestablishing homeostatic $[Ca^{2+}]_i$ persisted in the absence of extracellular Ca^{2+} and was independent of the expression or function of specific drug efflux pumps associated with drug resistance. Moreover, intracellular Ca^{2+} pools releasable by phosphoinositol-coupled receptor agonists or thapsigargin appeared to be differentially shared in drug-sensitive and drug-resistant cells. In light of the clinical relevance that drug resistance has in the treatment of cancer, the molecular and biochemical relationship between alterations in Ca^{2+} dynamics and drug resistance demands to be further investigated and tested in a wider array of cell types. Automated microscopy will help greatly in this pursuit by facilitating both sample imaging and data analysis, thus allowing high-content as well as high-throughput screening of large sample sets. A protocol for studying $[Ca^{2+}]_i$ kinetics with a commercially available automated imaging platform is described.

Introduction

Human breast cancer cell lines (bcc) selected for resistance to antineoplastic agents are a valuable research tool for studying the mechanisms associated with the development of drug resistance, the major cause for

METHODS IN ENZYMOLOGY, VOL. 414
0076-6879/06 $35.00
DOI: 10.1016/S0076-6879(06)14018-5

chemotherapy failure in breast cancer patients. *In vitro*, drug-resistant (DR) bcc exhibit numerous differential biochemical features when compared to drug-sensitive (DS) cells. These include overexpression of various membrane drug efflux pumps (e.g., P-glycoprotein [P-gp], multidrug resistance-associated protein [MRP], and breast cancer resistance protein [BCRP], Barrand *et al.*, 1997; Doyle *et al.*, 1998), reduced susceptibility to apoptosis (Ogretmen and Safa, 1996), altered enzyme activity (Herman *et al.*, 2006), and gene expression (Kudoh *et al.*, 2000; Rahbar and Fenselau, 2005). Ca^{2+} is a ubiquitous intracellular molecule involved in the control of vital cellular processes and, thus, its spatial and temporal dynamics are finely controlled (Berridge *et al.*, 2003). Little is known, however, about the potential role played by Ca^{2+} in breast cancer drug resistance. Higher resting intracellular Ca^{2+} levels ($[Ca^{2+}]_i$) and deficient intracellular Ca^{2+} pools have been described in two separate lines of human breast cancer MCF-7 cells resistant to doxorubicin (Chen *et al.*, 2002; Mestdagh *et al.*, 1994). Interestingly, the epidermal growth factor receptor, whose activation leads to increased $[Ca^{2+}]_i$, is overexpressed in adriamycin-resistant MCF-7 cells but returns to the expression level exhibited by DS parental cells when chemosensitivity is restored (Dickstein *et al.*, 1993). Finally, drug resistance can be reversed by compounds that, albeit through different mechanisms, modulate Ca^{2+} dynamics independently. Verapamil (VER, a well-known voltage-operated Ca^{2+} channel and Ca^{2+} influx blocker) and cyclosporin A (an inhibitor of calcineurin, a Ca^{2+} calmodulin-dependent protein phosphatase shown to potentiate Ca^{2+} responses; Bandyopadhyay *et al.*, 2000; LoRusso *et al.*, 1997) are widely studied chemosensitizers in tumors overexpressing P-gp (Barrand *et al.*, 1997; Hall *et al.*, 1999; Twentyman *et al.*, 1990). Probenecid (an organic anion transport blocker shown to either depress or increase Ca^{2+} responses; DiVirgilio *et al.*, 1990; Packham *et al.*, 1996) and genistein (a protein tyrosine kinase inhibitor reported to prevent capacitative Ca^{2+} entry; Foster and Conigrave, 1999) are examples of compounds that modulate MRP pumps (Gollapudi *et al.*, 1997; Versantvoort *et al.*, 1993).

This chapter describes a protocol to identify potential differences in Ca^{2+} dynamics between DS and DR bcc. We chose single-cell fluorescence microscopy-based high-content analysis (HCA) as a tool for our investigation. This technology offers significant advantages over both fluorimetric Ca^{2+} measurements and whole-well, end-point plate reader-based high-throughput Ca^{2+} assays (e.g., FLIPR), as heterogeneous single-cell responses can be appreciated, real-time Ca^{2+} kinetics (rather than just the peak response) and spatially localized Ca^{2+} signals can be evaluated, and simultaneous analysis of other biological responses can be performed (DeBernardi and Brooker, 1996). Automated microscopy instrumentation allows the processing of high volumes of samples, thus moving HCA into a higher throughput mode (high-content screening, HCS). Here, we describe in detail a procedure for HCA of Ca^{2+} kinetics in DS and DR bcc and provide a protocol to apply this HCA to a commercially available, automated imaging station for HCS purposes.

High-Content Analysis of Ca^{2+} Dynamics by Fluorescence Microscopy

Human Breast Cancer Cell Lines

Three different wild-type MCF-7 cell lines (wt, DS bcc) are tested and are from American Type Culture Collection (ATCC, Manassas, VA), the Lombardi Cancer Center, Georgetown University (GU), Washington, DC, and Dr. Susan Bates, NCI, NIH (Bethesda, MD). Five drug-resistant MCF-7-derived cell lines are used: MCF-7/ADR cells [resistant to adriamycin (Vickers et al., 1988) and overexpressing P-gp (Fairchild et al., 1987)]; MCF-7/VP cells (resistant to VP-16 etoposide and overexpressing MRP; Schneider et al., 1994); MCF-7/C4 (resistant to camptothecin and not overexpressing P-gp or MRP; Fujimori et al., 1996); MCF-7/Melph cells (resistant to melphalan and expressing neither P-gp nor MRP; Moscow et al., 1993); and MCF-7/Ad-Vp cells [resistant to adriamycin and VER and expressing BCRP (Doyle et al., 1998) but not P-gp or MRP (Lee et al., 1997)]. Cells are grown at 37° in a humidified atmosphere of 95% air/5% CO_2, refed twice a week, and passaged once a week (not to exceed 80% confluence). DR cells are grown in the presence of the appropriate drug (in μM) : MCF-7/ADR–doxorubicin, 2; MCF-7/VP–VP-16 etoposide, 4; MCF-7/Melph–melphalan, 6; MCF-7/Ad-Vp 3000–doxorubicin, 5; and VER, 10. Pairs of DS and DR clones are plated and imaged the same day; we compared MCF-7/ADR and MCF-7/Ad-Vp cells with their parental DS lines from GU and NIH, respectively, and the other DR clones with MCF-7 cells from ATCC.

Media and Reagents

1. IMEM medium, 10% fetal calf serum, 2 mM L-glutamine, 100 U/ml penicillin, 100 μg/ml streptomycin, and 50 μg/ml gentamycin. Used for all cell lines except MCF-7/C4.
2. RPMI-1640 with 5% fetal calf serum, 1 mM nonessential amino acids, 0.1 mM sodium pyruvate, 2 mM L-glutamine, 100 U/ml penicillin, 100 μg/ml streptomycin, and 50 μg/ml gentamycin. Used only for MCF-7/C4 cells.
3. H/H buffer (Ham's F-10 medium supplemented with 20 mM Na-HEPES, pH 7.4)

4. Fura 2-AM (1 mM stock in anhydrous DMSO; store at $-20°$)
5. Fura-2 standards: 5 μM Fura-2 pentaK (10 mM stock in H$_2$O) in 10 mM Na-HEPES, pH 7.4, with either 1 mM CaCl$_2$ ("high" standard) or 1 mM EGTA tetrasodium salt ("low" standard).

Ca^{2+} Imaging

[Ca^{2+}]$_i$ is measured by Fura-2 fluorescence ratio imaging. Fura-2 is a dual-excitation, single-emission Ca^{2+} indicator that, by virtue of its ratiometric properties, allows quantitative Ca^{2+} measurements that are independent of artifacts such as uneven cell loading and intracellular dye distribution, wavelength and intensity of the excitation light, dye leakage, and photobleaching.

Cell Growth and Labeling

1. Plate cells (20,000 cell/ml) on 25-mm-round, thickness No.1 glass coverslips (Fisher Scientific) in six-well plates and grow for 48 to 72 h (60 to 70% confluence). To avoid artifacts in [Ca^{2+}]$_i$ measurements, grow DR cells in drug-free medium during this time.
2. Label cells with 5 μM Fura-2 AM (1 ml/well) in growth medium (45 min/37°).
3. Wash cells twice with H/H buffer (1 ml/well) and let sit for 15 min at room temperature to allow complete dye deesterification.
4. Transfer coverslips to stainless-steel chambers (Attofluor Coverlips Holder, Molecular Probes), fitting the microscope stage holder (H/H buffer volume = 950 μl/chamber).
5. Manually add compounds (20× in H/H buffer) to the chambers in 50-μl volumes to ensure fast diffusion over the cells being imaged.

Fluorescence Microscopy

Cells are imaged at room temperature using an Attofluor RatioVision fluorescence microscopy system (Atto Instruments, Rockville, MD) equipped with a Zeiss Axiovert 135 microscope, a F-Fluar 40×, 1.3 NA, oil-immersion objective, and an ICCD camera (high gain used for MCF-7/ADR and Ad/Vp cells). Fura-2 dual excitation ratio imaging is performed with emission at >510 nm and excitation at 340 nm (maximal signal with high [Ca^{2+}]$_i$) and 380 nm (maximal signal with low [Ca^{2+}]$_i$); the 340/380-nm excitation ratio increases as a function of the [Ca^{2+}]$_i$. Accurate day-to-day comparison of [Ca^{2+}]$_i$ is ensured by calibrating the instrument daily. A two-point calibration is done *in vitro* using 5 μM Fura-2 pentaK standards ("high" standard: maximal emission at 340 nm; "low" standard: maximal emission at 380 nm). Ratio values are converted by the software to $[Ca^{2+}]_i$ nM according to the equation described by Grynkiewicz et al. (1985), with a K_d of 225 nM.

Data Capture and Statistical Analysis

To identify single cells, regions of interests (ROIs) are placed manually and $[Ca^{2+}]_i$ is recorded over time (2-s interval) from each ROI. Thirty to 99 cells per microscopic field/coverslip are imaged simultaneously. Single-cell responses can be averaged to yield $[Ca^{2+}]_i$ population means ($\pm$ SEM) and plotted against time. Statistical analyses use the two-tailed, unpaired t test with a 95% confidence interval for differences between means (StatMate, GraphPad Software, San Diego, CA).

Ca^{2+} Elevating Agents

The status of the intracellular Ca^{2+} pool and Ca^{2+} mobilization from the endoplasmic reticulum (ER) are studied by either activating ER inositol 1,4,5- trisphosphate (IP_3) receptors [ATP and UTP (50 μM), bradykinin (BK, 10 μM), α-thrombin (THR, 5 μM) are phosphoinositol (PI)-coupled receptor agonists, which activate IP_3 receptors and Ca^{2+} release] or bypassing the PI pathway using thapsigargin (TG, 1 μM), which poisons ER Ca^{2+}-ATPase pumps, thus preventing the reuptake into the stores of Ca^{2+} passively diffused out into the cytosol (Tharstrup et al., 1990).

Measurement of Resting Ca^{2+} State and Stimulated $[Ca^{2+}]_i$ Response

Resting $[Ca^{2+}]_i$ is recorded for $\sim$60 s before agonist addition and ranges between 40 and 70 nM with no statistically significant differences detected among the three DS MCF-7 cell lines and between DS and DR cells, although DR cells exhibit a wider range of variability (data not shown). MCF-7/ADR and MCF-7/Ad-Vp cells show a lower fluorescence signal than the other cell lines, likely reflecting a high degree of dye extrusion by the drug efflux pumps (Homolya et al., 1993); however, pump inhibitors are not used during dye loading to avoid affecting the native Ca^{2+} responses of the cells. The number of cells responsive to a given agonist varies among all the cell lines with <30% of the cells (both DS and DR) responding to THR and BK, whereas >95% of the cells (both DS and DR) consistently respond to ATP and UTP (Dixon et al., 1997). Onset kinetics are comparable among all the DS and DR cell lines and rather fast [average onset $t_{1/2}$ (time to reach 50% of the $[Ca^{2+}]_i$ peak) is $\sim$10 s, Fig. 1]. The magnitude of the

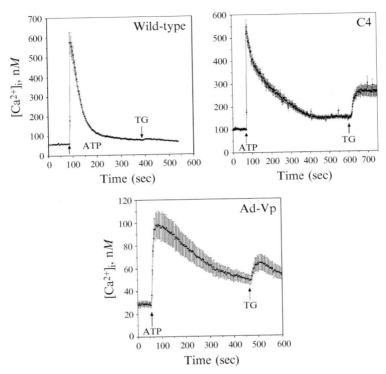

Fig. 1. HCA of intracellular Ca^{2+} kinetics in wild-type MCF-7 cells and two representative MCF-7-derived drug-resistant cell lines. ATP (50 μM) and then TG (1 μM) were added as indicated by arrows. Ca^{2+} responses from 30 to 99 individual cells were monitored simultaneously; data are presented as population means ± SEM. of $[Ca^{2+}]_i$ in nM values. In the experiments shown, ATP-induced $[Ca^{2+}]_i$ decay $t_{1/2}$ (calculated as described in the text) were: wild-type cells, 34 s; C4 cells, 84 s; Ad-Vp cells, 164 s. Ca^{2+} traces from each cell line are representative of at least 10 experiments.

response (fold increase = peak $[Ca^{2+}]_i$ /basal $[Ca^{2+}]_i$ varies among DS and DR cells depending on the agonist (Table I). All three MCF-7 wt cell lines respond to any given agonist with similar magnitude and, thus, data shown represent pooled responses. With selected agonists, quantitative differences in peak $[Ca^{2+}]_i$ are observed between DS and DR cells, which, however, are found not to be statistically significant (Table I). MCF-7/ADR cells pretreated with VER (25 μM, 2 h) show a significantly enhanced dye loading and magnitude of the ATP-evoked Ca^{2+} (fold increase: MCF-7/ADR no VER: 1.5 ± 0.2, $n = 5$; with VER: 3.9 ± 0.7, $n = 9$, $P < 0.05$).

TABLE I
Fold $[Ca^{2+}]_i$ Increase Evoked by Various Agonists in Human Breast Cancer Cell Lines

	ATP	UTP	BK	THR	TG
Drug-sensitive cells					
MCF-7 wt	7.9 ± 1.8[a] (n = 67)	6.0 ± 1.3 (n = 9)	4.1 ± 0.8 (n = 16)	2.8 ± 0.1 (n = 3)	2.8 ± 0.5 (n = 21)
MDA-MB-231	16.5 ± 4.8 (n = 10)	n.d.[b]	6.5 ± 1.8 (n = 3)	13.5 ± 5.1 (n = 3)	3.5 ± 0.9 (n = 3)
Drug-resistant cells					
MCF-7/ADR	2.9 ± 0.7 (n = 25)[c]	2.2 ± 0.3 (n = 3)[d]	1.9 ± 0.5 (n = 3)[e]	2.8 ± 1.1 (n = 3)	2.7 ± 0.7 (n = 10)
MCF-7/AdVp	4.5 ± 1.2 (n = 12)	3.7 ± 0.8 (n = 3)	3.5 ± 1.0 (n = 7)	n.d	2.4 ± 0.7 (n = 11)
MCF-7/VP	4.4 ± 1.2 (n = 11)	5.2 ± 1.9 (n = 3)	3.3 ± 1.7 (n = 3)	n.d.	1.5 ± 0.1 (n = 6)[f]
MCF-7/C4	5.9 ± 0.8 (n = 12)	n.d.	3.0 ± 0.6 (n = 3)	3.3 ± 0.3 (n = 3)	3.4 ± 0.8 (n = 7)
MCF-7/Melph	16.0 ± 6.1 (n = 17)[g]	3.7 ± 0.3 (n = 5)	n.d.	n.d.	n.d.

[a] Mean ± SEM.
[b] Not determined.
[c] P = 0.0976 versus ATP response in DS MCF-7 cells.
[d] P = 0.1341 versus UTP response in DS MCF-7 cells.
[e] P = 0.2030 versus BK response in DS MCF-7 cells.
[f] P = 0.1834 versus TG response in DS MCF-7 cells.
[g] P = 0.0841 versus ATP response in DS MCF-7 cells.

All cell lines responded to TG when applied as the first drug, with a comparable $\sim$3-fold $[Ca^{2+}]_i$ increase (Table I). As expected from the mode of action of TG, onset Ca^{2+} kinetics ($t_{1/2} \sim$40 s) are slower than those evoked by PI-coupled receptor agonists but are similar in DS and DR bcc. Interestingly, in four DR clones, but not in MCF-7 wt or MCF-7/Vp cells (the only DR clone with receptor-mediated $[Ca^{2+}]_i$ decay similar to wt cells), TG increased $[Ca^{2+}]_i$ when applied after ATP (Fig 1). Also, selected DR clones are able to respond to ATP after TG induced a robust $[Ca^{2+}]_i$ increase (authors' personal communication). Thus, the DS and DR bcc under study exhibit similar Ca^{2+} homeostatic levels and comparably filled IP_3- and TG-sensitive intracellular stores that release Ca^{2+} (via active IP_3 release or constitutive leakage) with seemingly equal efficiency and kinetics. However, TG- and IP_3-releasable pools seem to be shared differentially in DS and DR bcc and differential sensitivity to TG of the ER Ca^{2+}-ATPase pumps may exist.

Decay Kinetics of Agonist-Stimulated $[Ca^{2+}]_i$ Response

The decay kinetics of $[Ca^{2+}]_i$ increased by all PI-coupled receptor agonists are markedly slower in four out of five DR MCF-7 clones than in DS bcc (Figs. 1 and 2). Also, single-cell HCA of Ca^{2+} kinetics revealed greater variation in cell behavior among DR than DS cells, as reflected by larger data error bars (Fig. 1). To quantify Ca^{2+} decay, we used the $[Ca^{2+}]_i$ decay $t_{1/2}$, calculated as the time (seconds) at which $[Ca^{2+}]_i$ reaches 50% of the peak during the decay phase, *minus* the time at which the $[Ca^{2+}]_i$ peak is achieved. Compared to DS cells, (1) MCF-7/ADR and MCF-7/Ad-Vp cells showed the longest decay $t_{1/2}$, (2) MCF-7/C4 and MCF-7/Melph cells exhibited a decay $t_{1/2}$ about twice as long, and (3) MCF-7/VP cells showed a decay $t_{1/2}$, which, for all the agonists tested, was consistently, although not significantly, longer (Figs. 1 and 2). Importantly, even a 24-h treatment with VER (25 μM) did not accelerate ATP-evoked Ca^{2+} decay $t_{1/2}$ in MCF-7/ADR (MCF-7 wt: 46 $\pm$ 6.3 s, $n = 9$; MCF-7/ADR: 132 $\pm$ 20 s, $n = 9$; $P = 0.0008$). Chelation of extracellular Ca^{2+} by EGTA did not significantly affect the magnitude of ATP responses (indicating that the primary cause for the $[Ca^{2+}]_i$ increase is ion mobilization from internal stores) but accelerated the $[Ca^{2+}]_i$ decay in both DS and DR cells; nevertheless, decay kinetics in DR cells remained significantly longer than DS cells (data not shown). In both DS and DR bcc, $[Ca^{2+}]_i$ increased by TG was slower to return to baseline than the receptor-mediated $[Ca^{2+}]_i$ rise. In particular, two DR clones (Ad-Vp and VP cells) showed a significantly slower recovery than DS cells (authors' personal communication).

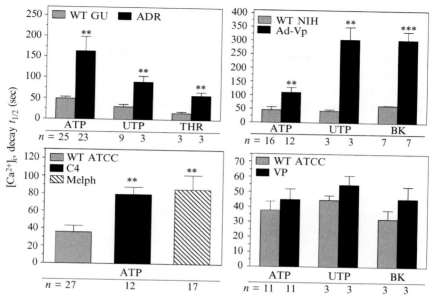

Fig. 2. Summary of $[Ca^{2+}]_i$ decay $t_{1/2}$ in wild-type and drug-resistant MCF-7 cells following exposure to PI-coupled receptor agonists. The source of the various DS MCF-7 lines is indicated. The decay $t_{1/2}$ of the $[Ca^{2+}]_i$ increase evoked by each agonist was calculated as described in the text. Data are means ± SEM. n = number of experiments. **$P<0.01$, ***$P< 0.001$ indicate decay $t_{1/2}$ significantly different from wild-type cells. Thirty to 99 individual cells were imaged per experiment.

Conclusions from HCA and Rationale for High-Content Screening

Single-cell, real-time HCA of intracellular Ca^{2+} dynamics identified slower $[Ca^{2+}]_i$ decay kinetics in DR bcc when compared to DS bcc. This phenomenon does not correlate with the functional expression of known drug efflux pumps, occurs after IP_3-dependent and -independent Ca^{2+} increase, and thus localizes downstream from specific Ca^{2+}-signaling pathways, suggesting that DR bcc are impaired in their ability to readily remove Ca^{2+} accumulated in the cytosol and restore Ca^{2+} homeostasis.

Ca^{2+} transients ("on" reaction) are biphasic, consisting of a rapid-onset phase (release from intracellular stores) and a slower compensatory influx (capacitative entry; Berridge *et al.*, 2003) that amplifies the initial Ca^{2+} signal and replenishes the stores. A recovery phase ("off "reaction) then follows when cells rectify the increased $[Ca^{2+}]_i$ via Ca^{2+} binding to cytoplasmic proteins (both buffer and effector proteins), Ca^{2+} efflux (through plasmalemmal Ca^{2+}-ATPase pumps and Na^+Ca^{2+} exchangers), and Ca^{2+} sequestration into the stores (through ER Ca^{2+}-ATPase pumps and mitochondrial Ca^{2+} transporters). Thus, any of these "off" mechanisms could be impaired in the DR bcc tested in this study.

While sporadic reports imply a potential association between some of these mechanisms and drug resistance in cancer (e.g., McAlroy et al., 1999; Padar et al., 2004; Villa and Doglia, 2004; Zhou et al., 2006), no systematic studies are available where DS and DR bcc are interrogated for possibly causal links between any of these mechanisms and drug resistance. In view of the clinical relevance that drug resistance has in breast cancer therapy, further investigations are needed to (1) identify the mechanism(s) responsible for the slower decay, (2) study the specificity and occurrence of anomalous Ca^{2+} dynamics in a wider array of DR cancer cells, and (3) establish a potential relationship (cause/effect) between Ca^{2+} and drug resistance. Such studies call for higher cell imaging and data analysis throughput than HCA can provide. HCS of Ca^{2+} dynamics and automated microscopy will make these studies feasible.

High-Content Screening

In our laboratory, automated fluorescence microscopy is performed on a BD Pathway Bioimager (BD Bioscience, Rockville, MD), which displays advantageous features that facilitate the HCS studies we propose as well as other HCA kinetic and end point applications (Chan et al., 2005; DeBernardi et al., 2006). Relevant to our project, this system uses mercury arc lamps as light sources, which, combined with 340- to 380-nm excitation filters and a UV transmission-optimized 20× objective, 0.75 NA, provide optimal excitation for Fura-2 and, compared to laser-based systems, minimizes photobleaching and phototoxicity greatly (a major concern when imaging live cells). A high-sensitivity CCD camera, with adjustable gain, affords detection of suboptimal emission signals (such as those exhibited by DR bcc overexpressing drug efflux pumps). The imaging station has environmental control (5% CO_2 and 37° ensure physiological conditions for live cells) and an on-stage fluidic head (with disposable tips) for automated compound addition compatible with 96-well plates (high-density plates provide high sample throughput). Because the stage is stationary and the objective moves underneath the 96-well plate, imaging takes place while compounds are added and, thus, changes in early-onset Ca^{2+} kinetics can be detected. The software (AttoVision) performs an automated autofocus routine, segmentation [single cells are identified as ROIs (virtually unlimited number)], image capture, drug addition, ratio calculations, and real-time data display. To increase sample throughput, multiple wells within a

96-well plate can be interleaved and processed at the same time [the x–y mechanical resolution (100 nm) allows the objective to reposition accurately on previously imaged fields] with user-settable time spacing between imaging and fluidic steps. Further improvement in throughput can be achieved by performing cell segmentation and data analysis off-line (Reanalysis Option), after the entire plate is processed. To perform HCS on bcc, we are in the process of adapting our HCA Ca^{2+} kinetics on the Pathway. Here we describe the modifications to the HCA protocol, the advantages offered by the automated data analysis, and provide sample data obtained with rat C6 glioma cells, which we use routinely for Ca^{2+} kinetics studies.

Cell Plating, Labeling, and Imaging Setup

1. Plate cells in 96-well plastic plates (Greiner Bio-one, μClear black wall plates) at a density of 5000 to 10,000 cell/well (100 μl/well) and grow cells 24 to 48 h prior to imaging.

2. Label cells as described for HCA (loading: 60 μl/well dye solution; washing: 100 μl/well H/H buffer; imaging: 80 μl/well H/H buffer).

3. Prepare compounds ($5\times$ in H/H buffer) and place in specific wells of a 96-well conical-bottom, polypropylene plate; automated delivery of 20 μl/well volumes at a rate of 40 μl/s ensures even diffusion over the cells with no need for mechanical mixing.

4. To calibrate the instrument, place the two Ca^{2+} standards (100 μl/well) in adjacent wells of a Greiner plate. Open the Dye Setup dialog box (Calibration Tab) and capture >510-nm emissions after excitation at 340 (numerator) or 380 (denominator) nm of the high or low standards, respectively, and set exposure times to yield maximal signal with either standard (usually, denominator exposure time is twice that of the numerator); the software uses the captured ratio and intensity values to generate a calibration curve and interpolate ratio values into $[Ca^{2+}]_i$ (Grynkiewicz et al., 1985).

5. Set up and save a Fura-2 Ca^{2+} ratio-imaging "template macro" and apply it to all the wells to be processed—"compound macro"; this macro is indeed an automated assay protocol listing all the various steps to be performed in each well.

- Autofocus setup course and fine; define the best algorithm (e.g., Vollath F4), the maximal z span that the objective must travel above/below starting z position, and the optimal z interval to ensure that focus on the cells is attained. Test at least a couple of wells on each side of the 96-well plate to determine best settings. These parameters are plate and cell type specific.

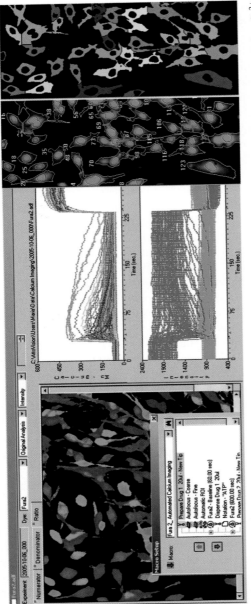

FIG. 3. HCS of Ca²⁺ kinetics with automated fluorescence microscopy. Screen shot from the Pathway Bioimager showing Ca²⁺ changes over time (seconds) in Fura-2-labeled C6 cells exposed to ATPs (50 μM) first and then to the Ca²⁺ ionophore, ionomycin (1 μM). The upper graph shows color-coded, single-cell Ca²⁺ responses (one trace = one cell; 106 cells being imaged) and the lower graph shows emission signals after 340-nm (green traces) and 380-nm (red traces) excitation. On the left side of the graphs, a pseudo-colored 340/380-nm ratio image of the cells highlights heterogeneity in the ATP responses (ratiometric color model scale for [Ca²⁺]ᵢ level: purple/ blue = low; green/yellow = high). Overlaid on this image is a screen shot capture of the Macro Setup dialog box showing the automated steps carried out in each well of the plate. The two panels on the right side of the graphs depict the automated segmentation process: using the dual channel exclude center option, single cells are identified as ROIs, numbered, and a segmentation mask is generated where the colors assigned to individual ROIs match those of the corresponding Ca²⁺ traces in the graph.

• Auto-ROI setup: use the dual channel/exclude center shape (which places ROIs on the cytoplasm excluding the nucleus) to segment single cells; reject dead or unhealthy cells by size (limiting object pixel min/max) and/or brightness (lowering upper threshold intensity). Use shading and watershed options to correct for uneven field illumination and improve segmentation if cells are too clustered.

• Imaging setting: in light path setup, define filter configuration: numerator A lamp = 340 and denominator B lamp = 380/ND, epifluorescence dichroic = 400DCLP, emission filter = 435 nm LP; in dye setup, define the exposure time (seconds) and camera gain for both 340- and 380-nm excitation (note that if working in calibrated mode, these parameters are set during the calibration procedure and cannot be changed for the calibration to remain valid) as well as the numerator and denominator threshold for ratio capture; in probe cycle, define sampling frequency (2 s) and duration of kinetic imaging (baseline and after compound addition).

• Fluid handling setup (prepare and dispense steps), define the volume (20 μl), rate of delivery (40 μl/s), origin, and destination well of each compound to be added to the cells.

6. Set up a treatment plate map indicating the agonist (and its concentration) added to each well to be processed and associate it with the macro. Assign the macro to the assay launch box. For daily use, launch the macro from here and adjust parameters such as exposure times, camera gain, and ROI settings to compensate (if necessary) for variation in cell labeling, morphology, and density. Create new treatment plate maps as needed.

Data Display, Analysis, and Report

1. Single cell kinetic responses from each ROI are analyzed on the fly (unless a reanalysis mode is chosen) and displayed in graphs showing changes over time in both the emission signals (after 340 and 380 nm excitation) and the 340/380-nm ratio (or $[Ca^{2+}]_i$, nM in calibrated mode).

2. Numbered ROIs in the images and their corresponding ratio traces in the graph are color coded and interactive; thus, "clicking" on a trace highlights the cell of origin and vice versa, allowing the identification of individual cell responses within a heterogeneous population (Fig. 3).

3. As the wells are processed, corresponding average traces are plotted on the plate map for immediate visualization. Each well is assigned an ADF file (proprietary binary format) containing measured numeric data (kinetic data points) and images saved during the experiment (TIFF files). Also, an experiment summary (text file) is generated that contains

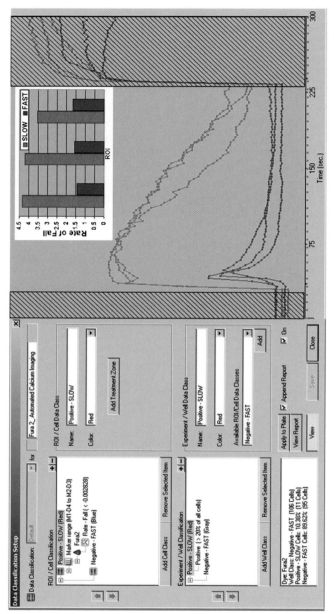

FIG. 4. HCS Ca^{2+} decay ROI data analysis. Screen shot from the Pathway Bioimager showing the ROI classification process, based on the ATP-increased Ca^{2+} rate-of-fall data attribute, applied to the experiment displayed in Fig. 3 (for clarity purpose, only six representative cell traces are shown). The data classification setup dialog box shows the treatment zone range (corresponding to the area between the two hatched regions on the graph) and the rate-of-fall setting. Ca^{2+} traces with decay slower than the preset value are classified as slow (red), whereas traces with faster decay are classified as fast (blue). An ROI Summary.txt file is generated with single cell rate-of-fall values. These values (negative integers) can then be transformed within Microsoft Excel or BD Image Data Explorer into positive integers and plotted.

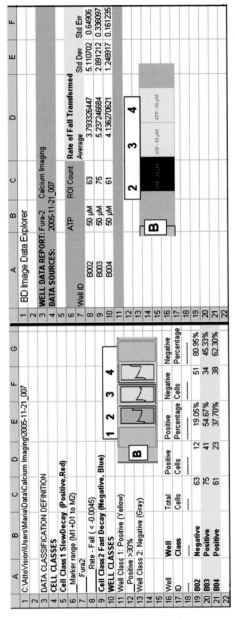

Fig. 5. HCS Ca^{2+} decay well data analysis and report. Data classification based on ATP-increased Ca^{2+} rate of fall is applied to a plate (three representative wells, B 2, 3, and 4 are shown). (Left) A Report.txt file is generated within AttoVision summarizing both ROI (percentage of positive and negative cells in each well) and well classification data (wells are considered positive and color coded in yellow if more than 30% of the ROIs exhibit a decay slower than the preset value; negative wells are displayed in gray). (Right) ROI summary and treatment plate map files are imported into the IDE program, which generates a plate report listing the average ± SD and SEM of the rate-of-fall values in each well analyzed and a plate heat map (B2, negative, fast decay = black; B3–4, positive, slower decay = yellow).

relevant assay parameter information about each experiment and can be used for documentation and data storage/mining.

4. Experimental data are binned into user-defined classes by using a hierarchical data classification structure wherein ROIs and wells are classified as positive or negative (and color coded accordingly) based on specific data attributes (amplitude [max/min/average], rate-of-rise [onset, slope of fluorescence intensity increase over time] or, as in our case, rate-of-fall [decay, slope of fluorescence intensity decrease over time]) over the entire data range or within specific treatment zones ($TZ_{1,2,3,n}$) (e.g., start to first compound addition [baseline], from compound addition to peak response, or, as in our case, from peak response back to baseline]). A preset data classification can be associated with the kinetic macro prior to running the experiment and all the ROIs and wells are then classified (and color coded accordingly) on the fly (Fig. 4).

5. For each experiment, an ROI summary (with single cell kinetic data and classification parameters) and a plate report (with percentage positive/negative wells) are automatically generated as text files and can be imported into traditional spreadsheet or graphing programs.

6. In our HCS, the ROI summary and treatment plate map files are imported into BD Image Data Explorer (IDE), a Microsoft Excel-based program that performs single cell-based data analysis and generates reports and charts (e.g., dose–response curves; EC_{50}, Z' and S/N calculations; histograms, scatter plots). Within IDE, new arithmetic parameters are created to, for example, normalize raw amplitude values (by diving peak ratio/basal ratio or peak $[Ca^{2+}]_i$/basal $[Ca^{2+}]_i$) and transform rate-of-fall raw values (negative integers) into positive integers (more convenient to analyze and plot). Wells positive for the desired data attribute (in our case, a Ca^{2+} rate of fall less than a preset threshold value from drug-sensitive cells) are visually identifiable in a heat map, a pseudo-colored overview of the plate where both scale and color are user adjustable.

7. The final IDE report will show the rate-of-fall values (or the values of any chosen data attribute) per well (ROI average $\pm$SD and SEM); user-generated graphs can be saved with this report, printed, and entered into hard-copy, record-keeping books (Fig. 5).

Acknowledgments

The authors are grateful to Drs. Bates, Moscow, Pommier, and Schneider for the precious gift of the drug-resistant MCF-7 clones. This work was supported in part by Grant ROI HL 28940 from NIH.

References

Bandyopadhyay, A., Shin, D. W., and Kim, D. H. (2000). Regulation of ATP-induced calcium release in COS-7 cells by calcineurin. *Biochem. J.* **348,** 173–181.

Barrand, M. A., Bagrij, T., and Neo, S. Y. (1997). Multidrug resistance-associated protein: A protein distinct from P-glycoprotein involved in cytotoxic drug expulsion. *Gen. Pharmacol.* **28,** 639–645.

Berridge, M. J., Bootman, M. D., and Roderick, H. L. (2003). Calcium signaling: Dynamics, homeostasis and remodelling. *Nature Rev. Mol. Cell Biol.* **47,** 517–529.

Chan, G. K., Richards, G. R., Peters, M., and Simpson, P. B. (2005). High content kinetic assay of neuronal signaling implemented on BD Pathway HT. *ASSAY Drug Dev. Technol.* **3,** 623–636.

Chen, J. S. K., Agarwal, N., and Mehta, K. (2002). Multidrug-resistant MCF-7 breast cancer cells contain deficient intracellular calcium pools. *Breast Cancer Res. Treat.* **71,** 237–247.

DeBernardi, M. A., and Brooker, G. (1996). Single cell Ca^{2+}/cAMP cross-talk monitored by simultaneous Ca^{2+}/cAMP fluorescence ratio imaging. *Proc. Natl. Acad. Sci. USA* **93,** 4577–4582.

DeBernardi, M. A., Hewitt, S. M., and Kriete, A. (2006). Automated confocal imaging and high-content screening for Cytomics. *In* "Handbook of Biological Confocal Microscopy" (J. B. Pawley, ed.), pp. 818–828. Springer Science, New York.

Dickstein, B., Valverius, E. M., Wosikowsky, K., Saceda, M., Pearson, J. W., Martin, M. B., and Bates, S. E. (1993). Increased epidermal growth factor receptor in an estrogen-responsive adriamycin-resistant MCF-7 cell line. *J. Cell Physiol.* **157,** 110–118.

DiVirgilio, F., Steinberg, T., and Silverstein, S. C. (1990). Inhibition of Fura-2 sequestration and secretion with organic anion transport blockers. *Cell Calcium* **11,** 57–62.

Dixon, C. J., Bowler, W. B., Fleetwood, P., Ginty, A. F., Gallagher, J. A., and Carron, J. A. (1997). Extracellular nucleotides stimulate proliferation in MCF-7 breast cancer cells via P2-purinoceptors. *Br. J. Cancer* **75,** 34–39.

Doyle, L. A., Yang, W., Abruzzo, L. V., Krogmann, T., Gao, Y., Rishi, A. K., and Ross, D. D. (1998). A multidrug resistance transporter from human MCF-7 breast cancer cells. *Proc. Natl. Acad. Sci. USA* **26,** 15665–15670.

Fairchild, C. R., Ivy, S. P., Kao-Shan, C. S., Whang-Peng, J., Rosen, N., Israel, M. A., Melera, P. W., Cowan, K. H., and Goldsmith, M. E. (1987). Isolation of amplified and overexpressed DNA sequences from adriamycin-resistant human breast cancer cells. *Cancer Res.* **47,** 5141–5148.

Foster, F. M., and Conigrave, A. D. (1999). Genistein inhibits lysosomal enzyme release by suppressing Ca^{2+} influx in HL-60 granylocytes. *Cell Calcium* **25,** 69–76.

Fujimori, A., Gupta, M., Hoki, Y., and Pommier, Y. (1996). Acquired camptothecin resistance in human breast cancer MCF-7/C4 cells with normal topoisomerase I and elevated DNA repair. *Mol. Pharmacol.* **50,** 1472–1478.

Gollapudi, S., Kim, C. H., Tran, B. N., Sangha, S., and Gupta, S. (1997). Probenecid reverses multidrug resistance in multidrug resistance-associated protein-overexpresssing HL60/AR and H69/AR cells but not in P-glycoprotein-overexpressing HL60/Tax and P388/ADR cells. *Cancer Chemother. Pharmacol.* **40,** 150–158.

Grynkiewicz, G., Poenie, M., and Tsien, R. Y. (1985). A new generation of Ca^{2+} indicators with greatly improved fluorescence properties. *J. Biol. Chem.* **260,** 3440–3450.

Hall, J. G., Cory, A. H., and Cory, J. H. (1999). Lack of competition of substrates for P-glycoprotein in MCF-7 cells overexpressing MDR1. *Adv. Enzyme Regul.* **39,** 113–128.

Herman, J. F., Mangala, L. S., and Mehta, K. (2006). Implications of increased tissue transglutaminase (TG2) expression in drug-resistant breast cancer (MCF-7) cells. *Oncogene* Jan 30 Epub ahead of print.

Homolya, L., Hollo, Z., Germann, U., Pastan, I., Gottesmann, M. M., and Sarkadi, B. (1993). Fluorescent cellular indicators are extruded by the multidrug resistance protein. *J. Biol. Chem.* **268,** 21493–21496.

Kudoh, K., Ramanna, M., Ravatn, R., Elkahloun, A. G., Bittner, M. L., Meltzer, P. S., Trent, J. M., Dalton, W. S., and Chin, K-V. (2000). Monitoring the expression profiles of doxorubicin-induced and doxorubicin-resistant cancer cells by cDNA microarray. *Cancer Res.* **60,** 4161–4166.

Lee, J. S., Scala, S., Matsumoto, Y., Dickstein, B., Robey, R., Zhan, Z., Altenberg, G., and Bates, S. E. (1997). Reduced drug accumulation and multidrug resistance in human breast cancer cells without associated P-glycoprotein or MRP overexpression. *J. Cell Biochem.* **65,** 513–526.

Lo Russo, A., Passaquin, A. C., Cox, C., and Ruegg, U. T. (1997). Cyclosporin A potentiates receptor-activated $[Ca^{2+}]_i$ increase. *J. Recept. Signal Transduct. Res.* **17,** 149–161.

McAlroy, H. L., Bovell, D. L., Plumb, J. A., Thompson, P., and Wilson, S. M. (1999). Drug extrusion,[125] I-efflux and the control of intracellular Ca^{2+} in drug-resistant ovarian epithelial cells. *Exp. Physiol.* **84,** 285–297.

Mestdagh, N., Vandewalle, B., Hornez, L., and Henichart, J. P. (1994). Comparative study of intracellular calcium and adenosine $3',5'$-cyclic monophosphate levels in human breast carcinoma cells sensitive and resistant to adriamycin: Contribution to reversion of chemoresistance. *Biochem. Pharmacol.* **48,** 709–716.

Moscow, J. A., Swanson, C. A., and Cowan, K. H. (1993). Decreased melphalan accumulation in a human breast cancer cell line selected for resistance to melphalan. *Br. J. Cancer* **68,** 732–737.

Ogretmen, B., and Safa, A. R. (1996). Down-regulation of apoptosis-related bcl-2 but not bcl-xL or bax protein in multidrug-resistant MCF-7/Adr breast cancer cells. *Int. J. Cancer* **67,** 608–614.

Packham, M. A., Rand, M. L., Perry, D. W., Ruben, D. H., and Kinlough-Rathbone, R. L. (1996). Probenecid inhibits platelet responses to aggregating agents *in vitro* and has a synergistic inhibitory effect with penicillin G. *Thromb. Haemost.* **76,** 239–244.

Padar, S., van Breemen, C., Thomas, D. W., Uchizono, J. A., Livesey, J. C., and Rahimian, R. (2004). Differential regulation of calcium homeostasis in adenocarcinoma cell line A549 and its taxol-resistant subclone. *Br. J. Pharmacol.* **142,** 305–316.

Rahbar, A. M., and Fenselau, C. (2005). Unbiased examination of changes in plasma membrane proteins in drug resistant cancer cells. *J. Proteome Res.* **4,** 2148–2153.

Schneider, E., Horton, J. K., Yang, C. H., Nakagawa, M., and Cowan, K. H. (1994). Multidrug resistance-associated protein gene overexpression and reduced drug sensitivity to topoisomerase II in breast carcinoma MCF-7 cells selected for etoposide resistance. *Cancer Res.* **54,** 152–158.

Tharstrup, O., Cullen, P. J., Drobak, B. K., Hanley, M. R., and Dawson, A. P. (1990). Thapsigargin, a tumor promoter, discharges intracellular Ca^{2+} stores by specific inhibition of endoplasmic Ca^{2+}-ATPases. *Proc. Natl. Acad. Sci. USA* **87,** 2466–2470.

Twentyman, P. R., Reeve, J. G., Koch, G., and Wright, K. A. (1990). Chemosensitization by verapamil and cyclosporin A in mouse tumour cells expressing different levels of P-glycoprotein and CP22 (sorcin). *Br. J. Cancer* **62,** 89–95.

Versantvoort, C. H., Schuurhuis, G. J., Pinedo, H. M., Eekman, C. A., Kuiper, C. M., Lankelma, J., and Broxterman, H. J. (1993). Genistein modulates the decreased drug accumulation in non-P-glycoprotein mediated multidrug resistant tumour cells. *Br. J. Cancer* **68,** 939–946.

Vickers, P. J., Dickson, R. B., Shoemaker, R., and Cowan, K. H. (1988). A multidrug-resistant MCF-7 breast cancer cell line which exhibits cross-resistance to antiestrogens and hormone-independent tumor growth *in vivo. Mol. Endocrinol.* **2,** 886–892.

Villa, A. M., and Doglia, S. M. (2004). Mitochondria in tumor cells studied by laser scanning confocal microscopy. *J. Biomed. Opt.* **9,** 385–394.

Zhou, Y., Xu, Y., Qi, J., Xiao, Y, Yang, C., Zhu, Z., and Xiong, D. (2006). Sorcin, an important gene associated with multidrug-resistance in human leukemia cells. *Leuk. Res.* **30,** 469–476.

[19] Measurement and Analysis of Calcium Signaling in Heterogeneous Cell Cultures

By GILLIAN R. RICHARDS, ANDREW D. JACK,
AMY PLATTS, and PETER B. SIMPSON

Abstract

High-content imaging platforms capable of studying kinetic responses at a single-cell level have elevated kinetic recording techniques from labor-intensive low-throughput experiments to potential high-throughput screening assays. We have applied this technology to the investigation of heterogeneous cell cultures derived from primary neural tissue. The neuronal cultures mature into a coupled network and display spontaneous oscillations in intracellular calcium, which can be modified by the addition of pharmacological agents. We have developed algorithms to perform Fourier analysis and quantify both the degree of synchronization and the effects of modulators on the oscillations. Functional and phenotypic experiments can be combined using this approach. We have used post-hoc immunolabeling to identify subpopulations of cells in cocultures and to dissect the calcium responses of these cells from the population response. The combination of these techniques represents a powerful tool for drug discovery.

Introduction

High-content screening has become established as a valuable tool for drug discovery, allowing rapid quantification of end point parameters such as neurite outgrowth (Simpson *et al.*, 2001), cell motility (Richards *et al.*, 2004), and intracellular translocations (Ding *et al.*, 1998) that were previously not amenable to high-throughput screening. While these fixed end

METHODS IN ENZYMOLOGY, VOL. 414   0076-6879/06 $35.00
                        DOI: 10.1016/S0076-6879(06)14019-7

point assays have proved extremely useful over a number of years, it is becoming more apparent that being able to quantify the kinetics of cellular events is important for elucidating drug targets and molecular mechanisms (Simpson, 2005; Simpson and Wafford, 2006). High-throughput imaging of rapid and/or transient cellular events is now possible with the development of imaging systems such as the GE INCell 3000, BD Pathway HT, Cellomics KineticScan, and EvoTec Opera (for a review, see Zemanova et al., 2003).

The BD Pathway HT Bioimager offers confocal imaging capability, flexible combinations of excitation and emission filters, and integrated liquid handling. The imager allows high-resolution single-cell imaging of many individual cells, or "regions of interest" (ROI), within each well across a microtiter plate. We have used this system to implement a variety of kinetic assays (Chan et al., 2006).

The ability to track individual cell responses within a population permits the use of more complex cell cultures in screening, such as stem cell-derived populations (Richards et al., 2004) and primary cell cultures (Chan et al., 2006). In central nervous system drug discovery, the use of cultures of this nature has the advantage of providing an *in vitro* model that may resemble tissue or *in vivo* responses more accurately than traditional recombinant cell lines. Combining this ability with kinetic imaging techniques represents a powerful tool for drug discovery whereby functional and phenotypic responses can be studied simultaneously. We have established a number of techniques for analyzing high-content kinetic calcium signaling responses from entire heterogeneous cell cultures and subpopulations within them.

Characterization of Calcium Signaling in Rat Cortical Cultures

Overview

We have established methodology that enables E17 embryonic rat cortical cultures to be used as an *in vitro* model of neuronal development and network formation. Calcium imaging studies indicate that these cultures display spontaneous calcium signals and form a complex synaptically coupled network over time. Between 1 and 3 days after plating very few, if any, cells show evidence of calcium flux. At 5 to 8 days of culture, the cells begin to exhibit asynchronous intracellular calcium transients. From 9 to 14 days, the cell population showed synchronized oscillations of intracellular calcium with modest amplitude (see Fig. 1). The synchronous oscillations observed in mature cortical cultures can be modulated pharmacologically, producing changes to the frequency and amplitude of the waveform (see Fig. 2). To enable this system to be utilized as a tool within drug discovery, we have developed a method for performing Fourier analysis of imaging-derived data sets to rapidly quantify the degree of synchronization of a culture and the characteristics of synchronized oscillations. Fourier analysis is a commonly used mathematical tool and can be performed by a variety of commercially available software, such as MATLAB (The MathWorks Inc., Natick, MA; see Uhlen, 2004) and Statistica (StatSoft Inc., Tulsa, OK).

Calcium signaling data are acquired using automatic ROI identification defined in such a way that each ROI represents a single cell or small cluster of cells. Series of images are converted to numerical data indicating the variation of intensity with time for each ROI. Numerical data are imported into Statistica 6.1 for multistep analysis consisting of detrending, splitting, and Fourier analysis. For each ROI, the median intensity value

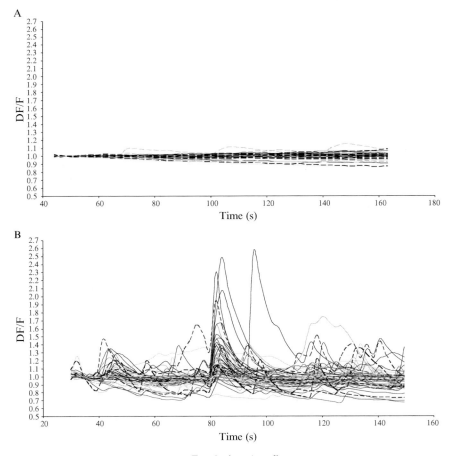

Fig. 1. (*continued*)

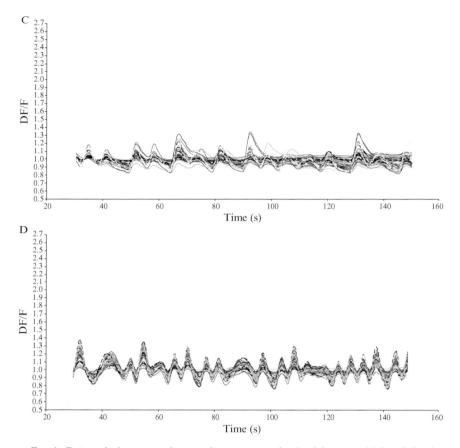

FIG. 1. Rat cortical neuron primary cultures were maintained for up to 14 days following plating, and spontaneous calcium fluxes were recorded using the BD Pathway HT imaging system. Activity develops with time in culture, from quiescence to asynchronous to synchronized oscillation. Example traces are shown from 2 (A), 7 (B), 9 (C), and 14 (D) days in culture. Data are expressed as the fold change from basal fluorescence (DF/F).

is subtracted from each data point in order to center each ROI series around zero. Figure 3A shows raw data obtained, with the average fluorescence intensity varying from cell to cell, and Fig. 3B illustrates normalization of data by subtracting the median and rescaling. As photobleaching or leakage of the calcium indicator can occur, resulting in a gradual decrease in intensity, the option to calculate and remove this linear trend from data is also included. Where appropriate, data can be split into regions on the

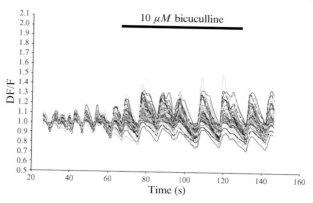

FIG. 2. Synchronous oscillations in intracellular calcium can be modulated. In this example the addition of 10 μM bicuculline increased the amplitude and decreased the frequency of the oscillations.

basis of time, for example, "basal" and "drug addition" (see Fig. 3A and B; drug addition is indicated by the black bar). Single-series Fourier analysis is performed to identify and quantify the underlying cyclical components of the calcium signaling (for further information, see Bloomfield, 2000) and produces a plot of the spectral density against frequency (see Fig. 3C). Fourier analysis is performed on each ROI series individually, and the results can be displayed as an average of all the ROIs within a well, as in Fig. 3C, or as individual plots, as in Fig. 4.

The spectral density provides a method to determine the frequency, which contributes most to the overall periodic behavior of calcium signaling observed within the cortical culture. Although the plots do not provide a direct measure of amplitude, they do provide an indication of the relative strength of the different frequencies that form the oscillation. Spectra plotted in Fig. 3C thus reflect and quantify the drug-induced decrease in frequency and increase in amplitude observed in Fig. 3A and 3B. Statistical comparisons of both the frequency at which the peak spectral density occurs and the magnitude of the peak spectral density can therefore be performed using this analysis. Downstream signaling pathways can be influenced by information encoded in both the frequency and the amplitude of calcium oscillations (Berridge, 1997), and the analysis methods described permit quantification of these complex aspects of cell signaling behavior.

We have used the techniques described here to determine a measure of the degree of synchronization of rat cortical cultures (see Fig. 4) and to analyze the effects of acute (see Fig. 3) and of chronic additions of drugs

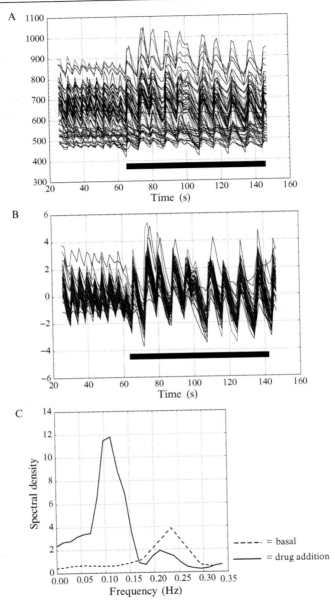

Fig. 3. Series of images are analyzed to produce numerical intensity values for each ROI, which can be plotted against time (A; black bar indicates addition of 10 μM bicuculline). The individual ROIs vary in intensity levels depending on dye loading and cellular

to these cells. Using this analysis technique permits rapid quantification of alterations to the oscillatory properties of the cultures and is beneficial for performing statistical comparison of complex cell signaling. This

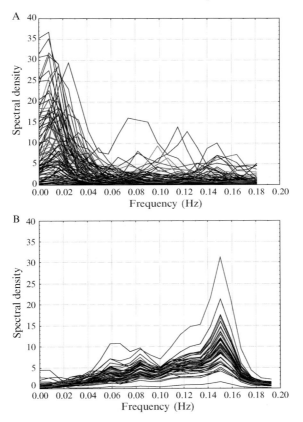

FIG. 4. Spectral analysis of spontaneous calcium signaling can be applied to determine the degree of synchronization of a cortical culture. Each line on the spectra corresponds to an individual ROI. In early cultures with asynchronous signaling, individual spectra are disparate (A), whereas in later cultures where synchronous oscillations are established, individual spectra have similar profiles (B).

characteristics. Before any further analysis, each data series is normalized by subtracting its median value from each data point to center each series around zero (B; black bar indicates addition of a drug). Data are divided in "basal" and "drug addition" regions, and then single-series Fourier analysis is performed. The average spectral density for all the ROIs is plotted against frequency for each region (C).

approach is not only applicable to kinetic measurements of calcium signaling, but equally could be applied to any cell culture system or assay where cyclical changes in signal intensity are observed, for example, NF-κB translocation (Nelson *et al.*, 2002, 2004).

Assay Protocol

1. Obtain rat E17 embryonic cortices according to local regulations governing the use of laboratory animals in experimentation. Dissect cortices under sterile Hanks balanced salt solution (HBSS; Invitrogen, Paisley, UK), removing the meningeal layer and refractory bulb. Chop cortices in a fresh culture dish with bowspring scissors and transfer to a 1:10 trypsin solution at 37° for 30 min. Neutralize the digest with Neurobasal media supplemented with penicillin, streptomycin, fungizone, L-glutamine, and 10% fetal calf serum (FCS) (all from Invitrogen). Pellet the cells by centrifugation, remove supernatant, and triturate the pellet in 1 ml Neurobasal medium. Add 19 ml of Neurobasal media and then filter the solution with a 70-μm cell strainer. Plate cells in poly-D-lysine-coated 96-well plates (Biocoat, BD Biosciences, Oxford, UK) at 50,000 cells/well in 100 μl/well of Neurobasal medium supplemented with 1:50 B27 (Invitrogen). Change the medium following overnight incubation at 37°/5% CO_2. Maintain cultures at 37°/5% CO_2 with 50% medium changes every 3 to 4 days. Note that it may be best to avoid using wells at the edges of the plate for experiments, as edge effects may be particularly prominent in long-term cultures.

2. Wash cells with 100 μl/well physiological salt solution (KHB: 118 mM NaCl, 4.7 mM KCl, 4.2 mM NaHCO$_3$, 1.2 mM KH$_2$PO$_4$, 1.3 mM CaCl$_2$, 1.2 mM MgSO$_4$, and 10 mM HEPES; all from Sigma-Aldrich) and then incubate with 100 μl/well 5 μM Fluo-3-AM/ 0.004% pluronic acid (TefLabs, Austin, TX) for 1 h at room temperature in the dark. Following incubation, wash cells three times in KHB, leaving a residual volume of 100 μl/well.

3. Transfer the cell plate to a microplate-based high-content imaging system with the capability of recording kinetic fluorescent responses from individual cells. Data presented here were acquired using a BD Pathway HT bioimager. For experiments where compound additions are required, prepare solutions in KHB at 10× [final] and add 11 μl per well during recording.

4. Acquire a series of images from each well using appropriate filter settings and exposure times to record the kinetic activity of the culture. Typically, images are acquired every 0.5 to 1 s with an exposure of 0.3 to

0.8 s for a period of 2 to 5 min. For experiments where compounds are applied, ensure that a sufficient period of basal activity is recorded prior to the addition to allow analysis of compound-evoked changes.

5. Analyze the images to produce intensity values for each cell or ROI and produce an output file suitable for further analysis.

6. Center data about zero by calculating the median fluorescence for each ROI and subtracting this median from each time point of the ROI. Then rescale data by calculating the median of the *absolute values* of the centered ROI and dividing each time point by this new median value (because data have been centered, the absolute value is necessary, otherwise the median would be approximately zero). This centering and scaling procedure will adjust for the different signal strength of each ROI. If the ROIs have been measured under different conditions, then split data into different variables according to time at which each treatment condition began and ended. Each of these new variables is now analyzed separately.

7. Centered and scaled data are analyzed using the default single series Fourier analysis in Statistica. This default analysis also subtracts the series average and removes any linear trend before calculating the Fourier decomposition. This is important, for example, if photobleaching has produced a general decline in fluorescent values with time. For each individual ROI, the spectral density at each frequency is generated by the Fourier analysis. The frequencies reported by Statistica assume that the observations come from time 1, 2, 3 ... so in order to interpret the frequencies correctly, the reported frequencies are divided by the average interval between images in original data (note that it is vital that the images are taken at equally spaced intervals).

8. If the average spectrum is required for a set of ROIs (and/or for a particular treatment condition), then a simple average of the spectral density for each frequency is taken across the ROIs. Once the final densities have been calculated, then the frequency that has the highest density can be calculated and used to make comparisons across different conditions.

Subpopulation Analysis of Calcium Signaling in Cocultures

Overview

In recent years many researchers have focused their attention on stem cell therapy, with both transplantation and activation of endogenous populations of progenitors considered to be potential strategies (for reviews, see Mitchell *et al.*, 2004; Zhu *et al.*, 2005). For either approach, the ability of precursors to divide, migrate, differentiate, and integrate in an appropriate

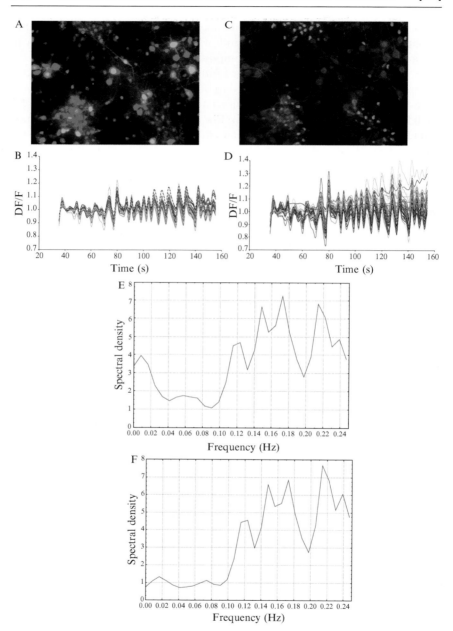

FIG. 5. Cocultures of rat cortical cells and human neural precursors are loaded with Fluo-3, and calcium imaging of spontaneous signaling is recorded from the entire population (A).

manner will be critical for replacing damaged tissue and correcting functional deficits.

We have previously developed high-content screening methodologies suitable for quantifying migration and differentiation in neural cell populations (Richards *et al.*, 2004; Simpson *et al.*, 2001), which were suitable for fixed end point analysis. The availability of the cortical cultures as a model system and the ability to perform high-content combined kinetic and phenotypic experiments have enabled us to develop methodology to study the integration of human neural precursors into an established neural culture in a manner amenable to higher throughput screening.

Human neural precursor cells are seeded onto established rat cortical cultures produced as described earlier. Cells are then cocultured for the desired period of time and then subject to calcium imaging, as described earlier, and data are recorded from the entire population (see Fig. 5A and B). Following kinetic experiments, cell plates are fixed and labeled with a primary antibody that specifically recognizes human nuclei and then a fluorescent secondary antibody. The cell plates are reimaged (see Fig. 5C) and new ROIs are defined that represent only those cells positive for human nuclear labeling. Original kinetic calcium signaling data can be readily reanalyzed to reflect the responses of this subpopulation alone (see Fig. 5D). Spectral analysis can then be performed to quantify the similarities or differences between the populations of cells (see Fig. 5E and F). The diversity of fluorescent secondary antibodies and the flexibility of excitation and emission filters in an imager such as the BD Pathway HT mean that this methodology can be adapted to include a number of different primary antibodies and yield still further information from the same experiment. We have typically included antibodies to neural progenitors, glia, and neurons to facilitate investigation of both functional responses and cell fate.

Assay Protocol

1. Obtain rat cortical neuron cultures as described earlier.
2. Culture human neural precursor cells and obtain a single cell

---

Series of images are analyzed to produce numerical intensity values for each region of interest (ROI), which are plotted against time (B). Cell plates are fixed and labeled with human nuclei-specific antibodies (red), a neuronal marker (green), and a nuclear label (blue) (C). Identified human nuclei are used to redefine a new set of ROIs, which are used to reanalyze the kinetic data file. Data for the subpopulation of human cells can thus be plotted (D). Spectral analysis can be performed on the two sets of ROIs to compare to the results (E and F).

suspension, as described previously (Richards *et al.*, 2004).

3. Seed the precursor cells onto the cortical cultures at a density of 5000 cells/well. Typically, cells are seeded onto cortical networks that are either 1 or 7 days old and cultured for a further 2 weeks.

4. Coculture the cells for the required period of time at 37°/5% CO_2 with 50% medium changes every 3 to 4 days.

5. Wash cells with 100 μl/well physiological salt solution (KHB: 118 mM NaCl, 4.7 mM KCl, 4.2 mM NaHCO$_3$, 1.2 mM KH$_2$PO$_4$, 1.3 mM CaCl$_2$, 1.2 mM MgSO$_4$, and 10 mM HEPES; all from Sigma-Aldrich) and then incubate with 100 μl/well 5 μM Fluo-3-AM/ 0.004% pluronic acid (TefLabs) for 1 h at room temperature in the dark. Following incubation, wash cells three times in KHB, leaving a residual volume of 100 μl/well.

6. Transfer the cell plate to a microplate-based high-content imaging system with the capability of recording kinetic fluorescent responses from individual cells. Data presented here are acquired using a BD Pathway HT bioimager.

7. Acquire a series of images from each well using appropriate filter settings and exposure times to record the kinetic activity of the culture. Typically, images are acquired every 0.5 to 1 s with an exposure of 0.3 to 0.8 s for a period of 2 min. Analyze the images to produce intensity values for each cell or ROI and produce an output file suitable for plotting.

8. Remove the cell plate from the imager and wash gently in phosphate-buffered saline, pH 7.4, (PBS). Fix the cells with 4% paraformaldehyde for 10 min, wash three times with PBS, and add blocking buffer (PBS/0.1% Triton-X /5% normal goat serum; all from Sigma-Aldrich) for 1 h at room temperature. Wash cell plates three times with PBS and add mouse antihuman/nuclei antibody 1:100 and rabbit anti-MAP2 (an early neuronal marker) 1:200 in blocking buffer (both from Chemicon, Temecula, CA) and incubate overnight at 4°. Wash cell plates three times with PBS and add staining solution containing 1:500 Alexa-594 conjugated antimouse secondary antibody, 1:500 Alexa-488 conjugated antirabbit secondary antibody, and 5 μM Hoechst 33342 in PBS (both from Invitrogen, Paisley, UK) for 1 h at room temperature. Wash the cell plate three times with PBS and then return it to the high-content imager.

9. For each well, acquire an image using appropriate filter settings to indicate the location of the Alexa-594-labeled nuclei. Use this image to redefine a new set of ROIs and reanalyze previously acquired kinetic data. Plot the intensity values for this subpopulation of cells and determine whether the human neural precursor cells display similar calcium signaling to the underlying cortical culture. Additional images of all nuclei and MAP2-positive neurons can also be acquired to identify whether any of the human neural precursors have adopted a neuronal phenotype.

10. Spectral analysis can be performed, as described earlier, to permit statistical analysis of the signaling properties of the two populations of cells.

Acknowledgments

We are grateful to Mathew Leveridge for cell culture and David Pellat for provision of Excel-based data plotting macros.

References

Berridge, M. J. (1997). The AM and FM of calcium signalling. *Nature* **386,** 759–760.

Bloomfield, P. (2000). "Fourier Analysis of Time Series: An Introduction." Wiley, New York.

Chan, G. K. Y., Richards, G. R., Peters, M., and Simpson, P. B. (2006). High content kinetic assays of neuronal signalling implemented on BD Pathway HT. *Assay Drug Dev. Technol.* **4,** 143–152.

Ding, G. F. J., Fischer, P. A., Boltz, R. C., Schmidt, J. A., Colaianne, J. J., Gough, A., Rubin, R. A., and Miller, D. K. (1998). Characterization and quantitation of NF-κB nuclear translocation induced by interleukin-1 and tumor necrosis factor-α. *J. Biol. Chem.* **273,** 28897–28905.

Mitchell, B. D., Emsley, J. G., Magavi, S. S., Arlotta, P., and Macklis, J. D. (2004). Constitutive and induced neurogenesis in the adult mammalian brain: Manipulation of endogenous precursors toward CNS repair. *Dev. Neurosci.* **26,** 101–117.

Nelson, D. E., Ihekwaba, A. E., Elliott, M., Johnson, J. R., Gibney, C. A., Foreman, B. E., Nelson, G., See, V., Horton, C. A., Spiller, D. G., Edwards, S. W., McDowell, H. P., Unitt, J. F., Sullivan, E., Grimley, R., Benson, N., Broomhead, D., Kell, D. B., and White, M. R. (2004). Oscillations in NF-κB signaling control the dynamics of gene expression. *Science* **306,** 704–708.

Nelson, G., Paraoan, L., Spiller, D. G., Wilde, G. J., Browne, M. A., Djali, P. K., Unitt, J. F., Sullivan, E., Floettmann, E., and White, M. R. (2002). Multi-parameter analysis of the kinetics of NF-κB signalling and transcription in single living cells. *J. Cell Sci.* **115,** 1137–1148.

Richards, G. R., Millard, R. M., Leveridge, M., Kerby, J., and Simpson, P. B. (2004). Quantitative assays of chemotaxis and chemokinesis for human neural cells. *Assay Drug Dev. Technol.* **2,** 465–472.

Simpson, P. B. (2005). Getting a handle on neuronal behaviour in culture. *Eur. Pharm. Rev.* **10,** 56–63.

Simpson, P. B., Bacha, J. I., Palfreyman, E.L, Woollacott, A. J., McKernan, R. M., and Kerby, J. (2001). Retinoic acid evoked-differentiation of neuroblastoma cells predominates over growth factor stimulation: An automated image capture and quantitation approach to neuritogenesis. *Anal. Biochem.* **15,** 163–169.

Simpson, P. B., and Wafford, K. A. (2006). New directions in kinetic high information content assays. Submitted for publication.

Uhlen, P. (2004). Spectral analysis of calcium oscillations. *Sci. STKE* **258,** 115.

Zemanová, L., Schenk, A., Valler, M. J., Nienhaus, G. U., and Heilker, R. (2003). Confocal optics microscopy for biochemical and cellular high-throughput screening. *Drug Disc. Today* **8,** 1085–1093.

Zhu, J., Wu, X., and Zhang, H. L. (2005). Adult neural stem cell therapy: Expansion *in vitro,* tracking in vivo and clinical transplantation. *Curr. Drug. Targets* **6,** 97–110.

[20] Multiplex Analysis of Inflammatory Signaling Pathways Using a High-Content Imaging System

By MALENE BERTELSEN

Abstract

This chapter describes a robust high-content cellular screening assay to simultaneously analyze the spatiotemporal activation of three different kinase-associated signaling pathways involving NF-κB, JNK, and p38, all of which are closely implicated in proliferative and proinflammatory responses. Signal transduction is dependent on the translocation of NF-κB p65 and phosphorylated c-Jun and p38 from the cytosol to the nucleus, and fluorescent immunolabeling was used to monitor changes in their cellular distribution. Cellular screening, data acquisition, and data interpretation were conducted on the ArrayScan HCS Reader (Cellomics Inc., Pittsburgh, PA). Assay adaptation to various cellular systems is feasible when sufficient separation of the nuclear and cytosolic compartment can be achieved and if cell adhesion properties permit proper attachment to the culture plates. Substitution of NF-κB p65 and phosphorylated forms of c-Jun and p38 as targets to analyze other translocating components is possible and is limited primarily by antibody specificity and the risk of fluorescent bleed-through between emission channels. Because assay validity is particularly confounded by inadequate spectral separation of the detection dyes in multicolor labeling assays, means of eliminating or counterbalancing staining artifacts are illustrated. Also, protocol parameter settings important for imaging and image processing are described, including object identification, image exposure settings, separation of cytosolic and nuclear regions, number of cells sufficient for analysis, and the use of gating thresholds critical for cell sorting and subpopulation analysis. This assay is a useful tool to investigate the interplay between signaling pathways and the mode of action, potency, and selectivity of compound inhibition of specific target molecules in a cellular context.

Introduction

The automation of high-content biology, defined as investigations of the biological activity of biomolecules and their temporal and spatial organization in intact cells, has resulted in some of the most information-rich screening platforms available. One example is the fluorescence-based

METHODS IN ENZYMOLOGY, VOL. 414
0076-6879/06 $35.00
DOI: 10.1016/S0076-6879(06)14020-3

ArrayScan HCS reader (Cellomics Inc., Pittsburgh, PA), which is an automated high-quality image acquisition and interactive data analysis system with the capacity to extract complex sets of data from single cells or cell populations (Giuliano *et al.*, 2003). It represents a valuable and efficient tool to analyze changes in cell cycle, gene expression, cell motility, receptor internalization, and trafficking of intracellular components between various cellular compartments, while also providing information about morphological changes, off-target effects, and cellular toxicity (Gasparri *et al.*, 2004; Ghosh *et al.*, 2004; Giuliano, 2003). Because the ArrayScan HCS reader permits concurrent detection of four different fluorophores, several subcellular events can be monitored simultaneously. This chapter describes a multiplex assay to quantify the interleukin (IL)-1α-mediated spatiotemporal activation of NF-κB p65 and the stress-response mitogen-activated protein kinases p38 and c-Jun N-terminal kinase (JNK). All three signaling components play central roles in the promotion of cellular proliferation and differentiation, as well as in the induction and exacerbation of inflammatory responses (Craig *et al.*, 2000; Wada and Penninger, 2004). Not surprisingly, NF-κB p65, p38, and JNK associated signaling cascades have received considerable interest as sources of potential targets in the search for antitumor and anti-inflammatory drugs. Extensive cross talk among p38, JNK, and NF-κB signaling pathways does, however, complicate the design of selective inhibitors. For example, p38 has been reported to exert both stimulating and inhibitory effects on NF-κB-dependent gene transcription via NF-κB p65 phosphorylation, and through negative regulation of transforming growth factor-β activated kinase 1 (TAK1) stimulated IκB degradation (Cheung *et al.*, 2003; Schmeck *et al.*, 2004). Studies have also demonstrated p38-mediated blockage of JNK signaling, and c-Jun has been implicated in the transcription of NF-κB-driven genes (Chen *et al.*, 2000; Xiao *et al.*, 2004). As a tool to further investigate signaling interplay and dissect positive and negative feedback loops, we have developed a multiplex assay on the ArrayScan HCS reader where translocation of cytosolic NF-κB p65 and phosphorylated forms of c-Jun (phospho-c-Jun) and p38 (phospho-p38) to the nucleus is utilized as a measure of signal transduction (Bertelsen and Sanfridson, 2005). The optimization and validation of this multiplex assay are described thoroughly in this chapter and may serve as a template for establishing screening platforms analyzing different combinations of translocating signaling components.

Assay Procedures

The protocol used for immunofluorescence labeling of NF-κB p65, phospho-c-Jun, and phospho-p38 prior to high-content analysis closely resembles standard immunocytochemistry protocols. Accordingly, primary

(1:100–1:500 dilution) and secondary (1:250–1:1000 dilution) antibodies were titrated in order to achieve a robust window between basal and IL-1α-induced phospho-c-Jun, phospho-p38, and NF-κB p65 distribution patterns in the human cervical epithelioid carcinoma-derived HeLa cell line. Because the capacity of the applied image processing algorithm to separate individual cells for analysis is reduced significantly in wells with near confluent cells, optimization of initial cell seeding density was required. A significant IL-1α-induced effect on target translocation was achieved using 9000 cells/well while maintaining a relatively high cell count per optical field, which is essential for limiting plate scan time. The following protocol represents the optimized method for evaluation of IL-1α-stimulated NF-κB p65, phospho-c-Jun, and phospho-p38 nuclear accumulation in HeLa cells.

Immunofluorescence Labeling

1. Culture HeLa cells (American Type Culture Collection, Manassas, VA) in minimum essential medium eagle (MEME) containing 1% sodium pyruvate, 1% nonessential amino acids, and 10% fetal calf serum at 37° in humidified air/CO_2 (95%/5%) and seed log-phase cells in black-walled optically clear-bottomed 96-well Packard ViewPlates (PerkinElmer, Boston, MA) at 9×10^3 cells/well (100 μl) 18 h prior to the experiment.
2. Incubate cells in 100 μl 10 ng/ml IL-1α [Sigma, St. Louis, MO; dissolved in sterile-filtered phosphate-buffered saline (PBS) containing 0.1% bovine serum albumin (BSA)] in serum-free MEME for 20 min. To verify staining selectivity, preincubate cells with pharmacological inhibitors [dissolved in dimethyl sulfoxide (DMSO)] or vehicle (maximum 0.5% DMSO) for 1 h prior to IL-1α addition.
3. Fix cells at room temperature in 100 μl prewarmed (37°) 4% paraformaldehyde for 10 to 15 min and wash with 3×200 μl PBS by flicking off the wash buffer and gently tapping the plate on a pad of tissues.
4. Permeabilize the cells in 100 μl PBS containing 0.2% Triton X-100 for 5 min and rinse once with 200 μl PBS.
5. Block nonspecific binding sites by incubation in 100 μl PBS containing 2% BSA for 1 h prior to a 2-h incubation at room temperature with a 50-μl primary antibody mixture containing 0.45 μg/ml rabbit polyclonal antibody against p38 phosphorylated at Thr180/Tyr182 (Cell Signaling Technology, Beverly, MA), 0.5 μg/ml mouse monoclonal antibody against c-Jun phosphorylated at Ser63, and 1 μg/ml goat polyclonal anti-NF-κB p65 (Santa Cruz Biotechnology, Santa Cruz, CA) in PBS supplemented with 0.01% Tween 20 and 0.2% BSA. To verify staining specificity, immunoneutralize primary

antibodies in PBS with excess blocking peptide (8- to 10-fold over the antibody) for 3 h at 4° before immunolabeling.

6. Wash cells in 200 μl PBS containing 0.05% Tween 20 for 5 min on a shaker device and rinse twice in 200 μl PBS before a 1-h incubation at room temperature with 50 μl of secondary antibodies [Alexa Fluor 488 donkey antimouse (1/250), Alexa Fluor 555 donkey antirabbit (1/200) Alexa Fluor 680 donkey antigoat (1/100) (Molecular Probes, Leiden, The Netherlands)] in PBS supplemented with 0.01% Tween 20, 0.2% BSA, and 0.625 μg/ml Hoechst 33342 (Molecular Probes, Leiden, The Netherlands) to fluorescently label cell nuclei. Centrifuge (2000g, 5 min, 4°) Hoechst 33342 and Alexa-conjugated antibody stock solutions prior to use to pellet aggregates.

7. Wash the cells in 200 μl PBS containing 0.05% Tween 20 for 5 min on a shaker device and rinse twice in 200 μl PBS. Leave the last wash on the cells and seal the plate with sealing tape. Plates can be stored at 4° in the dark for several weeks or scanned immediately for high-content analysis.

High-Content Analysis

High-content analysis can be performed using a range of different bioapplications available on the ArrayScan HCS reader. They differ significantly in the extent of user-defined input parameters and the number of assay output parameters describing cell- and well-related biological features. While some require minimal user interaction and are characterized by modest flexibility, others are tailored for a multiplex approach, analyzing target expression and distribution between several cellular compartments. The latter type of analysis demands a considerably higher level of user experience in data interpretation and assay optimization. In the current protocol, high-content analysis is performed using the Molecular Translocation algorithm. This bioapplication is ideal for both multiplexing and screening by allowing a high degree of assay optimization flexibility while limiting output parameters to averages of changes in cellular target distribution. The principle behind quantification of target nuclear translocation is outlined in Fig. 1. In order to determine nuclear protein distribution in the three target channels, a nuclear region mask is created based on Hoechst DNA staining (channel 1). By expanding the nuclear region mask, while remaining within cell boundaries, a concentric ring is generated and used as an approximation of the cytosolic compartment. To ascertain that the cytosolic region mask does not exceed the cell perimeter and is separated appropriately from the nuclear region, it is recommended to perform initial experiments using 20 or 40× magnification. Moreover, this helps

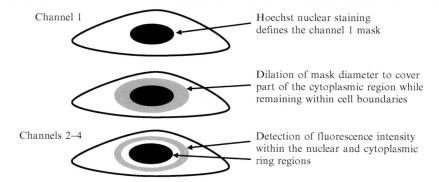

Channel 1 — Hoechst nuclear staining defines the channel 1 mask

Dilation of mask diameter to cover part of the cytoplasmic region while remaining within cell boundaries

Channels 2–4 — Detection of fluorescence intensity within the nuclear and cytoplasmic ring regions

FIG. 1. Schematic of the Molecular Translocation algorithm. Based on Hoechst nuclear staining in channel 1, a nuclear mask is created, which is dilated to create a concentric ring covering part of the cytosolic compartment. The distribution of fluorescently labeled targets between nuclear and cytosolic regions is detected in channel 2–4 to measure changes in target translocation.

verify that high-content analysis output features accurately describe target distribution, target expression, and cell morphology. For screening purposes the assay may subsequently be adopted to a 5 or $10\times$ objective lens to limit plate scan times. In the current study, using a $10\times$ objective lens, one pixel separates the cytosolic ring from the nuclear boundary while one pixel defines the cytosolic ring width. Because cytosolic measurements are based on only a fraction of the entire cytosolic compartment, it is essential that the target is distributed homogeneously throughout the cytosol. Cytosolic and nuclear staining intensities in the target channels (channel 2–4) are normalized to total nuclear region and cytosolic ring area, allowing for the quantification of protein translocation between the nucleus and the cytosol in all three reporter channels. The difference between mean nuclear and cytosolic target staining intensities is denoted Mean-NucCytDiffCh$_n$. Using the Cellomics vHCS View software, all data can, upon scan completion, be exported to an Excel spreadsheet and expressed as means $\pm$ SEM or transferred to Spotfire for further data visualization.

Image Acquisition and Object Identification

Images are captured, using a 12-bit high-resolution CCD camera, and automatically transferred to a Microsoft SQL database from where they can be retrieved easily for analysis. In the current study, images are acquired from multiple optical fields within each well in four independent channels using an Omega XF93 filter set with excitation/emission wavelengths of $485 \pm 10/515 \pm 15$ nm (channel 2, phospho-c-Jun), $555 \pm 15/600 \pm 15$ nm (channel 3, phospho-p38), and $650 \pm 25/725 \pm 30$ nm (channel 4, NF-κB p65). To ensure even field illumination, an illumination correction factor is applied, generated from measuring the illumination pattern for the filter dichroic using the illumination correction wizard. In the current protocol, Hoechst nuclear staining in channel 1 serves to identify cells as well as to optimize the focus plane. Because image overexposure is likely to prevent accurate focusing, the optimal scan exposure time in channel 1 is calculated from autoexposing three to four representative wells using a 24% upper limit for image staining saturation [Auto Exposure Options, Peak Target (percentile method)]. If Hoechst and other dye-labeled antibody aggregates have not been completely removed from the staining solution, bright specks will appear on the acquired images. Consequently, calculations are performed most accurately by excluding the upper 0.5% of the gray level image histogram (Auto Exposure settings, Skip Fraction) in both the object channel (channel 1) and the target channels (channel 2–4).

In cells characterized by a low cell to nucleus diameter ratio where the cytosolic ring domain is likely to exceed the cell perimeter, it is particularly relevant to apply fluorescence intensity thresholds to exclude background pixels from the analysis. Elongated cells represent a typical example where the cytosolic compartment in discrete areas may be too narrow to entirely omit background pixels from cytosolic ring measurements unless fluorescence intensity thresholds are specified. Object pixel identification is either performed by manually specifying an intensity threshold value (Fixed Threshold Method) or by employing an autothreshold technique based on iterative analysis of the gray level image histogram (IsoData Method). Importantly, however, if image staining is dim and object pixel intensities are approaching background intensity values, less stringent threshold measures may be required. In the current study the computed threshold values were lowered by 20% (phospho-p38; IsoData Threshold Value $= -0.20$) and 40% (NF-κB p65 and phospho-c-Jun; IsoData Threshold Value $= -0.40$) to include all pixels within the cytosolic compartment in untreated and IL-1α-stimulated cells in the analysis. To prevent erosion of the nuclear region mask and risk of inadequate separation of the nuclear and cytosolic domains, calculated Hoechst staining threshold values were reduced by 40% (IsoData Threshold Value $= -0.40$).

Optimization of Fluorophore Labels to Minimize Channel Bleed Through

Bleed through between reporter channels caused by insufficient spectral separation may interfere with target detection, hence representing a potentially confounding factor in multicolor labeling assays. We have evaluated the extent of bleed through associated with secondary antibodies conjugated to Alexa Fluor 488 (channel 2), Alexa Fluor 555 (channel 3), and Alexa Fluor 680 (channel 4) dyes characterized by absorption and fluorescence emission spectra displayed in Fig. 2A. The bandwidth of the excitation and emission window of the Omega XF93 filter set is depicted in Fig. 2B, together with the extent of light transmission. Moreover, the capacity of the dichroic beam splitter to separate emitted fluorescence from scattered excitation light is indicated.

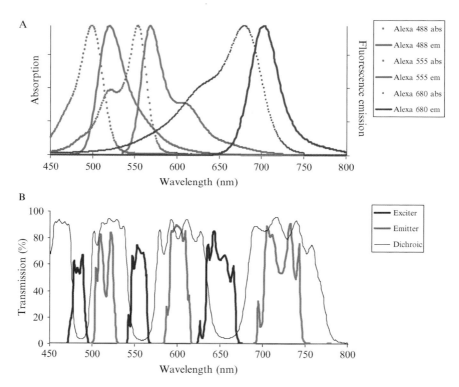

FIG. 2. Complete spectral separation of fluorophore absorption and fluorescence emission spectra with the XF93 triple-band filter set is not achievable. Absorption (stippled line, abs) and fluorescence emission (full line, em) spectra are illustrated for Alexa Fluor 488 (green), Alexa Fluor 555 (red), and Alexa Fluor 680 (purple) dyes (A). Optical transmission curves for individual XF93 filter set components are shown in blue (excitation filter), red (emission filter), and black (dichroic beam splitter) (B).

When comparing Alexa fluorophore absorbance/emission spectra with XF93 excitation/emission ranges it became clear that higher wavelength fluorophores could potentially interfere with measurements in lower wavelength reporter channels. For example, Alexa 555 and Alexa 680 are partially excited in the 485 ± 10- and 555 ± 15-nm excitation windows, respectively (Fig. 2). Although only a fraction of the resulting fluorescence emission is collected in the corresponding 515 ± 15- and 600 ± 15-nm emission windows, we found that channel bleed through had to be taken into account. The extent of bleed through was estimated using fixed target channel exposure times (phospho-c-Jun: 1.1 s; phospho-p38: 1.3 s; NF-κB p65: 5.0 s). These settings were obtained in the software interactive window by autoexposing IL-1α-stimulated cells using a maximum 25 to 30% saturation of target staining intensity [Auto Exposure Options, Peak Target (percentile method)]. If using Alexa Fluor 555 as the detection dye for phospho-c-Jun for analysis in channel 3 (Fig. 3B), significant bleed through was observed in channel 2 (Fig. 3A). Considering that bleed through from phospho-c-Jun staining was of the same magnitude as phospho-p38 MeanNucCytDiff values in channel 2 (Fig. 3A), this would clearly perturb measurements of cellular phospho-p38 distribution in a multiplexed assay. However, because the Alexa Fluor 555-derived fluorescence signal was weaker for phospho-p38 compared to phospho-c-Jun, bleed-through interference could be resolved. If interchanging the Alexa Fluor dyes associated with the secondary antibodies in order to detect phospho-c-Jun in channel 2 (Fig. 3C) and phospho-p38 in channel 3 (Fig. 3D), channel bleed through was negligible.

Because NF-κB p65 is very abundant, the less bright Alexa 680-conjugated fluorophore was chosen to track NF-κB p65 spatial responses. However, bleed through was evident in channel 3, which could only partly be compensated for by reducing the concentration of the Alexa Fluor 680 dye. For example, phospho-p38 MeanNucCytDiff values in IL-1α-stimulated cells were increased by 25 to 30% as a result of Alexa Fluor 680 dye bleed through (data not shown). This issue was overcome because a linear relationship exists between Alexa Fluor 680 staining in channels 3 and 4, provided the image is not overexposed. Thus, if using fixed exposure times in both channels, a bleed-through correction curve can be constructed by measuring the level of staining filtering through to channel 3 in cells labeled exclusively with increasing concentrations of anti-NF-κB p65 antibodies and stained with Alexa Fluor 680-conjugated antibodies (Fig. 3E, $R = 0.989$). When subsequently measuring NF-κB p65-dependent staining in the multiplex assay using fixed exposure times in both channels, it is possible to manually estimate the extent of bleed through collected in channel 3 and hence calculate true phospho-p38 staining.

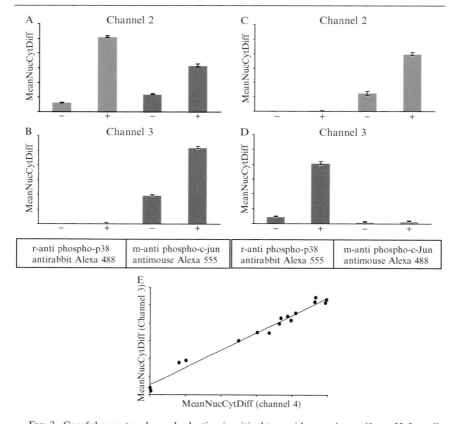

Fig. 3. Careful reporter channel selection is critical to avoid screening artifacts. HeLa cells were incubated with vehicle (−) or 10 ng/ml IL-1α (+) for 20 min prior to fixation, permeabilization, and incubation with either rabbit (r) antiphospho-p38 or mouse (m) antiphospho-c-Jun antibodies. Cells were immunolabeled with antirabbit Alexa Fluor 488 antibodies (green bars, A, B), antimouse Alexa Fluor 555 antibodies (red bars, A, B), antirabbit Alexa Fluor 555 antibodies (red bars, C, D), or antimouse Alexa Fluor 488 antibodies (green bars, C, D) prior to imaging on the ArrayScan HCS reader. The Molecular Translocation algorithm was used to calculate MeanNucCytDiff values from images collected in channel 2 (excitation: 485 ± 10 nm) (A, C) and channel 3 (excitation: 555 ± 15 nm) (B, D). In cells incubated exclusively with increasing concentrations of antigoat NF-κB p65 antibodies and labeled with Alexa Fluor 680-conjugated antigoat antibodies, the relationship between target staining in channel 4 and bleed-through staining in channel 3 is depicted (E) (adapted from Bertelsen and Sanfridson, 2005).

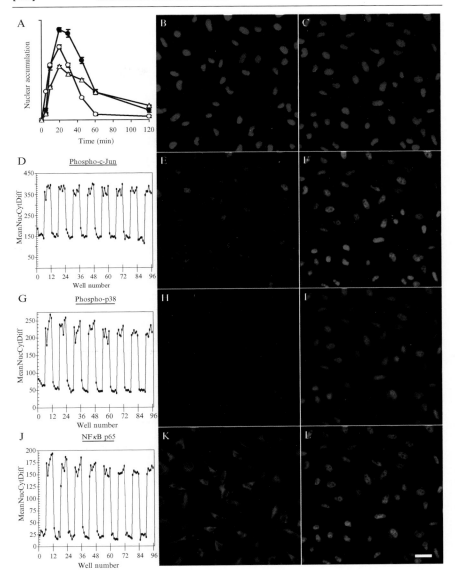

FIG. 4. IL-1α stimulates nuclear accumulation of phospho-c-Jun, phospho-p38, and NF-κB p65. HeLa cells were incubated with vehicle (B, E, H, K) or 10 ng/ml IL-1α (C, F, I, L) for 20 min prior to fixation, permeabilization, Hoechst nuclear staining (B, C), and immunolabeling with antibodies targeting phospho-c-Jun (E, F), phospho-p38 (H, I), and NF-κB p65 (K, L). Cells were imaged on the ArrayScan HCS reader using a 20× objective lens with each column reflecting images collected from the respective fluorescent channels using the same optical

Imaging Intracellular Target Distribution

The temporal sequence of IL-1α-mediated NF-κB p65, phospho-c-Jun, and phospho-p38 redistribution demonstrates a peak in target nuclear accumulation following a 20-min IL-1α exposure (Fig. 4A). This time point was subsequently used to ensure the most robust screening assay for analyzing changes in target distribution. Representative images are shown of Hoechst-stained nuclei in vehicle (Fig. 4B) and IL-1α (Fig. 4C) exposed cells captured using a 20× objective lens. Basal phospho-c-Jun levels were significant and mainly confined to the nuclear compartment (Fig. 4E), which may be associated with a putative role of phospho-c-Jun in cell cycle progression in exponentially growing cells (Bakiri *et al.*, 2000). IL-1α treatment further induced nuclear phospho-c-Jun accumulation (Fig. 4F), increasing MeanNucCytDiff measurements 2.5-fold (Fig. 4D), which was calculated from a 96-well plate where columns 1–6 were incubated with vehicle and columns 7 to 12 were stimulated with 10 ng/ml IL-1α. Untreated cells displayed a uniform phospho-p38 staining pattern just above background signal (Fig. 4H). The capacity of IL-1α to promote phospho-p38 nuclear translocation (Fig. 4I) was illustrated by a 4.7-fold increase in MeanNucCytDiff values (Fig. 4G). Finally, a typical NF-κB p65 distribution pattern was observed (Handa *et al.*, 2004), with the vast majority of the transcription factor present throughout the cytosolic compartment before stimulation (Fig. 4K), followed by a 16.5-fold increase in NFκB p65 nuclear translocation upon IL-1α treatment (Fig. 4J and 4L).

The abundance and cellular distribution pattern of all three targets were also analyzed in the lung epithelial cell line NCI-H292. Although NCI-H292 cells have a tendency to grow in multilayered clusters, which prohibits proper focusing, this issue was of limited extent if using an initial cell seeding density of 10,000 cells/well. A comparable response to IL-1α for all three proteins was observed in HeLa and NCI-H292 cells (data not shown), demonstrating the utility of this assay in other cellular systems.

Evaluation of Assay Reliability

To determine the suitability of the assay as a screening tool, the Z' factor was calculated for all three targets from means of measurements and

---

field. The temporal sequence of 10 ng/ml IL-1α-stimulated phospho-c-Jun ($\bullet$), phospho-p38 ($\bigcirc$), and NF-κB p65 ($\triangle$) nuclear accumulation is depicted (A). MeanNucCytDiff measurements were calculated using the Molecular Translocation algorithm from a 96-well plate where columns 1–6 were incubated with vehicle and columns 7–12 were incubated with IL-1α for 20 min. Cellular distribution of phospho-c-Jun (D), phospho-p38 (G), and NF-κB p65 (J) is depicted. Scale bar: 40 μm (adapted from Bertelsen and Sanfridson, 2005).

standard deviation from measurements on vehicle and IL-1α-treated cells (Fig. 4D, 4G, and 4J) according to the method of Zhang et al. (1999). In the current study, a minimum of 200 cells/well were imaged using a 10× objective lens, resulting in a scanning time of approximately 45 min to complete a full 96-well plate scan. This protocol produced Z' values of 0.69 (phospho-c-Jun), 0.59 (phospho-p38), and 0.65 (NF-κB p65) if omitting outer wells often associated with lower cell densities and altered target translocation kinetics (inclusion of outer wells reduced the Z' factors by 0.1–0.2). Increasing the number of cells included for analysis to 400 cells/ well merely prolonged plate scan times without enhancing assay robustness significantly. A recent study addressing microplate edge effects in cell-based assays documented that thermal gradients during cell adhesion are most pronounced in the outer wells of the plate, resulting in uneven cell distribution, possibly affecting cell responsiveness (Lundholt et al., 2003). A brief preincubation period at ambient conditions prior to placement in the CO_2 incubator, as suggested by the authors, may therefore lower whole plate data variation and further improve assay reliability as a result of a more even cell distribution profile.

In order to verify the specificity of target recognition, nuclear target translocation was analyzed following pharmacological inhibition of JNK, p38, and IκB kinase (IKK2/IKKβ) activity. IKK2 constitutes part of the IKK complex responsible for phosphorylating the inhibitory subunit IκB, thereby targeting it for degradation, which results in the unmasking of the NF-κB nuclear localization signal. Cellular redistribution of phospho-c-Jun, phospho-p38, and NF-κB p65 was inhibited in a dose-dependent manner by SP600125 ($IC_{50} = 6.8$ μM), SB203580 ($IC_{50} = 1.4$ μM), and a specific IKK2 inhibitor ($IC_{50} = 0.4$ μM), respectively (Bertelsen and Sanfridson, 2005). Further evidence supporting staining specificity was provided by antibody immunoneutralization where blocking peptides against antibodies targeting phospho-p38 and NF-κB p65 eliminated IL-1α-mediated target nuclear accumulation (data not shown). A commercially available antiphospho-c-Jun antibody-blocking peptide could not be obtained. However, a similar IL-1α redistribution response was observed using an antibody raised against a different synthetic phospho-peptide in rabbit, thus supporting monoclonal antiphospho-c-Jun antibody specificity.

Output Parameters Related to Cell Health

Several output features inherent to the Molecular Translocation algorithm are relevant for monitoring general cell health. These include average cell count/well and measures of nuclear morphology, such as nuclear size (NucArea), shape (NucShapeP2A and NucShapeLWR), and fluorescence intensity (NucTotalIntenCh1 and NucAvgIntenCh1). Reductions in average nuclear fluorescence intensity, indicative of DNA degradation, together with nuclear condensation and fragmentation, are all indicators of cellular toxicity. Measurements of nuclear fragmentation are not offered by the Molecular Translocation algorithm; however, images can be reanalyzed using the Cell Health Profiling algorithm where Hoechst staining variability within each nucleus is calculated as a measure of nuclear integrity (MeanObject VarIntenCh1). In the current study, no apparent indications of inhibitor or IL-1α-related cell toxicity were observed. However, to obtain a more thorough cell health profile, additional indicators of cellular toxicity have to be evaluated. These may include changes in mitochondrial transmembrane potential, release of cytochrome c from mitochondria, caspase activation, F-actin reorganization, and increased cell permeability (Rudner *et al.*, 2001; van Engeland *et al.*, 1997). Examples of an ArrayScan-based multiplex approach to analyze several of these parameters have been described and demonstrated to aid understanding of the cellular mechanism of compound toxicity (Abraham *et al.*, 2004; Lovborg *et al.*, 2004).

The output features of the Molecular Translocation algorithm have been optimally tailored to report on average changes in cellular target distribution within the entire cell population. In contrast, nuclear morphology measurements are provided on an individual cell basis and not as a well average. To obtain a clear notion of cell health status, additional data analysis steps are therefore required, which may be a disadvantage in large compound screens. Alternatively, high-content analysis can be performed using the more flexible and information-rich Compartmental Analysis algorithm, which provides average well-level nuclear intensity and size measurements. Protocol settings relating to object identification and image acquisition are interchangeable between the Molecular Translocation and the Compartmental Analysis algorithms. In addition to reporting on differences in average pixel intensity between nuclear and cytosolic mask regions (CircRingAvgIntenDiff), this bioapplication also provides average target intensity measurements within either the cytosolic (RingAvgInten) or the nuclear (CircAvgInten) region. These parameters may be useful in order to elucidate if decreased target nuclear accumulation is, in part, a result of reduced cellular target expression.

Subpopulation Analysis

Because images and data are collected for each individual cell, it is possible to select or reject cells or cell populations based on morphological features or the presence of a particular marker. Initially, cells are elected for further analysis based on object identification (nuclei in the current protocol) and selection settings in channel 1. Cells are rejected if failing to meet object identification isodata thresholds requirements (described in the Image acquisition and object identification section) or if not within a defined range for nuclear intensity, area, and shape. Subpopulation analysis is subsequently performed based on intensity measurements (AvgIntenCh$_n$, TotalIntenCh$_n$) in target channels 2–6. For example, by adjusting object selection intensity criteria in the FITC channel it is possible to monitor changes in target expression or distribution in cells transfected with a GFP-labeled marker while gating out nontransfected cells from the analysis. Alternatively, it may be of interest to analyze cellular readouts in a subpopulation not associated with phosphorylation or expression of a particular fluorescently labeled protein. This may be appropriate if analyzing cells transfected with small interfering RNA (siRNA). Because transfection efficiency is generally less than 100%, only cells with a strong siRNA uptake and significant target knockdown are relevant to include in the analysis. Based on immunolabeling of the siRNA target, subpopulation analysis is performed by lowering the maximum value of the object selection parameters; AvgIntenCh$_n$ and TotalIntenCh$_n$ in the interactive window. The object selection parameters are determined most conveniently by moving the cursor over an acquired image in the interactive window and monitoring pixel intensity, characterized by a 12-bit (0–4095) dynamic range, at the bottom of the screen.

Several of the bioapplications, including the Molecular Translocation algorithm, have the capacity to compare cellular responses to a control population typically derived from nonstimulated cells. Upper and lower thresholds defining specific features of the control population are either manually specified or computed from Reference Wells, which also provide the benefit of limiting run-to-run variation. Homogeneous control cell populations are characterized by normal physiological distribution histograms where approximately 95% of the cells fall within 2 (correction coefficient, CC) standard deviations of the mean. Distribution histograms describing different cellular features are visualized easily using the Cell Detail function in the Cellomics vHCS View software. In the current study, vehicle-treated cells were used as reference wells and lower and upper thresholds for Mean-NucCytDiff values were defined using a correction coefficient of 2 in all target channels. In 86% of the cell population, IL-1α-stimulated phospho-p38 and NF-κB p65 nuclear accumulation was significantly higher compared to control (%HighCircRingAvgInten DiffCh$_n$), whereas only 70% of the cells displayed statistically higher phospho-c-Jun nuclear levels. If calculating the percentage of cells characterized by significantly induced phospho-c-Jun translocation, as well as significant nuclear accumulation of phospho-p38, NF-κB p65, or both (%HighCombinedAvgIntenDiffCh$_n$Ch$_n$), it was evident that phospho-c-Jun redistribution was always accompanied by both phospho-p38 and NF-κB p65 translocation.

Concluding Remarks

In summary, the multiplex assay described in this chapter is an example of an information-rich screening tool used to analyze the spatial and temporal distribution of signaling components in intact cells as a measure of biological activation. In contrast to biochemical model systems, this assay offers information about compound-associated cell toxicity and biological potency where the complex effects of signaling interplay are taken into account. Adapting the assay to other cell types has been demonstrated successfully, but requires that protein and phosphorylation levels of p38, c-Jun, and NF-κB p65 are analyzed carefully to reveal possible staining artifacts due to emission channel bleed through. The assay can be modified to track cellular redistribution of different combinations of signaling components, provided that antibody specificity is evaluated thoroughly. Establishing assays on the ArrayScan reader to analyze cellular responses in suspension cells or semiadherent cells represents a particular challenge and is beyond the scope of this chapter. However, the use of binding matrixes, such as collagen, poly-L-lysine, fibronectin, vitronectin, and laminin, is likely to help facilitate cell attachment to the culture plates. It is, however, critical to perform appropriate control experiments to ascertain the reliability of data, as cell–matrix interactions may induce cell activation and initiate detrimental cellular effects (Tang *et al.*, 1998).

In conclusion, we feel that this type of multiplex assay can be modified and expanded easily to generate useful cellular information, which would benefit a wide variety of drug discovery and signal elucidation projects.

Acknowledgments

I am grateful to Dr. Annika Sanfridson for helpful discussions and critical review of the manuscript. I also thank Dr. Britt-Marie Kihlberg for critical reading and comments.

References

Abraham, V. C., Taylor, D. L., and Haskins, J. R. (2004). High content screening applied to large-scale cell biology. *Trends Biotechnol.* **22,** 15–22.

Bakiri, L., Lallemand, D., Bossy-Wetzel, E., and Yaniv, M. (2000). Cell cycle-dependent variations in c-Jun and JunB phosphorylation: A role in the control of cyclin D1 expression. *EMBO J.* **19,** 2056–2068.

Bertelsen, M., and Sanfridson, A. (2005). Inflammatory pathway analysis using a high content screening platform. *Assay Drug Dev. Technol.* **3,** 261–271.

Chen, G., Hitomi, M., Han, J., and Stacey, D. W. (2000). The p38 pathway provides negative feedback for Ras proliferative signaling. *J. Biol. Chem.* **275,** 38973–38980.

Chefung, P. C., Campbell, D. G., Nebreda, A. R., and Cohen, P. (2003). Feedback control of the protein kinase TAK1 by SAPK2a/p38alpha. *EMBO J.* **22,** 5793–5805.

Craig, R., Larkin, A., Mingo, A. M., Thuerauf, D. J., Andrews, C., McDonough, P. M., and Glembotski, C. C. (2000). p38 MAPK and NF-κ B collaborate to induce interleukin-6 gene expression and release: Evidence for a cytoprotective autocrine signaling pathway in a cardiac myocyte model system. *J. Biol. Chem.* **275**, 23814–23824.

Gasparri, F., Mariani, M., Sola, F., and Galvani, A. (2004). Quantification of the proliferation index of human dermal fibroblast cultures with the ArrayScan high-content screening reader. *J. Biomol. Screen.* **9**, 232–243.

Ghosh, R. N., Grove, L., and Lapets, O. (2004). A quantitative cell-based high-content screening assay for the epidermal growth factor receptor-specific activation of mitogen-activated protein kinase. *Assay Drug Dev. Technol.* **2**, 473–481.

Giuliano, K. A. (2003). High-content profiling of drug-drug interactions: Cellular targets involved in the modulation of microtubule drug action by the antifungal ketoconazole. *J. Biomol. Screen.* **8**, 125–135.

Giuliano, K. A., Haskins, J. R., and Taylor, D. L. (2003). Advances in high content screening for drug discovery. *Assay Drug Dev. Technol.* **1**, 565–577.

Handa, O., Naito, Y., Takagi, T., Shimozawa, M., Kokura, S., Yoshida, N., Matsui, H., Cepinskas, G., Kvietys, P. R., and Yoshikawa, T. (2004). Tumor necrosis factorα-induced cytokine-induced neutrophil chemoattractant-1 (CINC-1) production by rat gastric epithelial cells: Role of reactive oxygen species and nuclear factor-κB. *J. Pharmacol. Exp. Ther.* **309**, 670–676.

Lovborg, H., Nygren, P., and Larsson, R. (2004). Multiparametric evaluation of apoptosis: Effects of standard cytotoxic agents and the cyanoguanidine CHS 828. *Mol. Cancer Ther.* **3**, 521–526.

Lundholt, B. K., Scudder, K. M., and Pagliaro, L. (2003). A simple technique for reducing edge effect in cell-based assays. *J. Biomol. Screen.* **8**, 566–570.

Rudner, J., Lepple-Wienhues, A., Budach, W., Berschauer, J., Friedrich, B., Wesselborg, S., Schulze-Osthoff, K., and Belka, C. (2001). Wild-type, mitochondrial and ER-restricted Bcl-2 inhibit DNA damage-induced apoptosis but do not affect death receptor-induced apoptosis. *J. Cell Sci.* **114**, 4161–4172.

Schmeck, B., Zahlten, J., Moog, K., van Laak, V., Huber, S., Hocke, A. C., Opitz, B., Hoffmann, E., Kracht, M., Zerrahn, J., Hammerschmidt, S., Rosseau, S., Suttorp, N., and Hippenstiel, S. (2004). Streptococcus pneumoniae induced p38 MAPK dependent phosphorylation of RelA at the interleukin-8 promotor. *J. Biol. Chem.* **279**, 53241–53247.

Tang, M. J., Hu, J. J., Lin, H. H., Chiu, W. T., and Jiang, S. T. (1998). Collagen gel overlay induces apoptosis of polarized cells in cultures: Disoriented cell death. *Am. J. Physiol.* **275**, C921–C931.

van Engeland, M., Kuijpers, H. J., Ramaekers, F. C., Reutelingsperger, C. P., and Schutte, B. (1997). Plasma membrane alterations and cytoskeletal changes in apoptosis. *Exp. Cell Res.* **235**, 421–430.

Wada, T., and Penninger, J. M. (2004). Mitogen-activated protein kinases in apoptosis regulation. *Oncogene* **23**, 2838–2849.

Xiao, W., Hodge, D. R., Wang, L., Yang, X., Zhang, X., and Farrar, W. L. (2004). NF-κB activates IL-6 expression through cooperation with c-Jun and IL6-AP1 site, but is independent of its IL6-NFkappaB regulatory site in autocrine human multiple myeloma cells. *Cancer Biol. Ther.* **3**, 1007–1017.

Zhang, J. H., Chung, T. D., and Oldenburg, K. R. (1999). A simple statistical parameter for use in evaluation and validation of high throughput screening assays. *J. Biomol. Screen.* **4**, 67–73.

[21] Generation and Characterization of a Stable MK2-EGFP Cell Line and Subsequent Development of a High-Content Imaging Assay on the Cellomics ArrayScan Platform to Screen for p38 Mitogen-Activated Protein Kinase Inhibitors

By RHONDA GATES WILLIAMS, RAMANI KANDASAMY, DEBRA NICKISCHER, OSCAR J. TRASK, JR., CARMEN LAETHEM, PATRICIA A. JOHNSTON, and PAUL A. JOHNSTON

Abstract

This chapter describes the generation and characterization of a stable MK2-EGFP expressing HeLa cell line and the subsequent development of a high-content imaging assay on the Cellomics ArrayScan platform to screen for p38 MAPK inhibitors. Mitogen-activated protein kinase activating protein kinase-2 (MK2) is a substrate of p38 MAPK kinase, and p38-induced phosphorylation of MK-2 induces a nucleus to cytoplasm translocation (Engel *et al.*, 1998; Neininger *et al.*, 2001; Zu *et al.*, 1995). Through a process of heterologous expression of a MK2-EGFP fusion protein in HeLa cells using retroviral infection, antibiotic selection, and flow sorting, we were able to isolate a cell line in which the MK2-EGFP translocation response could be robustly quantified on the Cellomics ArrayScan platform using the nuclear translocation algorithm. A series of assay development experiments using the A4-MK2-EGFP-HeLa cell line are described to optimize the assay with respect to cell seeding density, length of anisomycin stimulation, dimethyl sulfoxide tolerance, assay signal window, and reproducibility. The resulting MK2-EGFP translocation assay is compatible with high-throughput screening and was shown to be capable of identifying p38 inhibitors. The MK2-EGF translocation response is susceptible to other classes of inhibitors, including nonselective kinase inhibitors, kinase inhibitors that inhibit upstream kinases in the p38 MAPK signaling pathway, and kinases involved in cross talk between different modules (ERKs, JNKs, and p38s) of the MAPK signaling pathways. An example of mining "high-content" image-based multiparameter data to extract additional information on the effects of compound treatment of cells is presented.

Introduction

There has been a growing trend in drug discovery toward the implementation of cell-based assays where the target is screened in a more physiological

METHODS IN ENZYMOLOGY, VOL. 414   0076-6879/06 $35.00
         DOI: 10.1016/S0076-6879(06)14021-5

context than in biochemical assays of isolated targets (Johnston and Johnston, 2002). Fluorescence microscopy, confocal or wide field, is one of the most powerful tools that cell biologists have used to interrogate the amount and spatial temporal location of biomolecules that comprise the molecular mechanisms of the cell (Giuliano et al., 1997; Mitchison, 2005). High-content screening (HCS) platforms automate the capture and analysis of fluorescent images of thousands of cells in the wells of microtiter plates with a throughput and capacity that has made fluorescence microscopy and image analysis compatible with drug discovery (Almholt et al., 2004; Lundholt et al., 2005; Oakley et al., 2002; Ramm et al., 2003). The ArrayScan was one of the first HCS platforms introduced to the drug discovery market by Cellomics, Inc. (Pittsburgh, PA) in 1997, and there are now more than a dozen models of HCS imagers on the market. By selection of the appropriate probes (antibodies, fluorescent protein fusion partners, biosensors, and stains), fluorescence microscopy can be applied to many drug target classes. HCS assays may be configured to simultaneously detect multiple target readouts (multiplexing) and can provide information on cellular morphology, population distributions, and subcellular localizations and relationships. Image-based assays provide multiparameter quantitative and qualitative information beyond the single parameter target data typical of most other assay formats, and thus are termed "high content" (DeBiasio et al., 1987; Giuliano et al., 1997; Mitchison, 2005).

The introduction of green fluorescent protein (GFP), and a variety of its spectral variants, into heterologous organisms has allowed cell biologists to monitor the dynamics of GFP-fusion proteins in living cells, thus addressing both temporal and spatial aspects (Giuliano and Taylor, 1998; Roessel and Brand, 2004; Tavare et al., 2001). GFP and its variants have been used to study gene expression profiling, protein trafficking, protein translocation, lipid metabolism, protein–protein interactions, second messenger cascades (Ca^{2+}, cAMP, cGMP, and $InsP_3$), protein phosphorylation, proteolysis, intracellular ion concentrations, and intracellular pH (Giuliano and Taylor, 1998; Roessel and Brand, 2002; Tavare et al., 2001). The combination of heterologous GFP-fusion protein expression with automated HCS imaging platforms has been utilized to screen a number of drug target classes: G-protein coupled receptors (GPCRs), kinase inhibitors, nuclear export inhibitors, signaling pathway inhibitors, and cell cycle targets (Almholt et al., 2004; Lundholt et al., 2005; Oakley et al., 2002; Thomas and Goodyer, 2003). Automated imaging platforms are being deployed in many phases of the drug discovery process: target identification/target validation, primary screening and lead generation, hit characterization, lead optimization, toxicology, bio-marker development, diagnostic histopathology, and other clinical applications. The purpose of this chapter is to describe the generation and characterization of a stable mitogen-activated protein kinase-activated protein kinase-2 (MK2) enhanced-GFP cell line (MK2-EGFP) and the subsequent development of a high-content imaging assay on the Cellomics ArrayScan automated imaging platform to screen for p38 MAPK inhibitors.

Definition of the Cell Model

Mitogen-activated protein (MAP) kinases are members of the signaling cascades for diverse extracellular stimuli that regulate fundamental cellular processes. Four distinct MAP kinase families have been described: extracellular signal-regulated kinases (ERKs), c-jun N-terminal (JNK) or stress-activated protein kinases (SAPK), ERK5/big MAP kinase 1 (BMK1), and the p38 group of protein kinases (Cowan and Storey, 2003; Garrington and Johnson, 1999; Ono and Han, 2000). p38 (reactivating kinases, RKs or p40) kinases are known to mediate stress responses and are activated by heat shock, ultraviolet light, bacterial lipopolysaccharide (LPS), or the proinflammatory cytokines interleukin (IL)-1β or tumor necrosis factor (TNF)-α (Cowan and Storey, 2003; Garrington and Johnson, 1999; Ono and Han, 2000). Activation of the p38 pathway results in phosphorylation of downstream kinases, transcription and initiation factors that affect cell division, apoptosis, invasiveness of cultured cells, and the inflammatory response (Cowan and Storey, 2003; Garrington and Johnson, 1999; Ono and Han, 2000). In addition to p38α (CSBP, MPK2, RK, Mxi2), there are three p38 homologues: p38β, p38γ (ERK6, SAPK3), and p38δ (SAPK4) (Ono and Han, 2000). p38α and p38β are expressed ubiquitously, p38γ is expressed predominantly in skeletal muscle, and p38δ is enriched in lung, kidney, testis, pancreas, and small intestine (Ono and Han, 2000). While p38 MAP kinase phosphorylates a number of transcription factors, including ATF2, CHOP/GADD153, and MEF2, it also activates many downstream protein kinases, such as the MAP kinase-activated protein kinases MK2 and MK3, the MAP-interacting kinases MNK1 and MNK2, and the p38-regulated/activated protein kinases PRAK and MSK1 (Cowan and Storey, 2003; Garrington and Johnson, 1999; Ono and Han, 2000). Selective inhibition of p38α-MAPK would be expected to inhibit production of TNF-α and IL-1β, as well as other proinflammatory mediators such as IL-6 and COX-2, thereby reducing the inflammation and/or joint destruction associated with rheumatoid arthritis (English and Cobb, 2002; Fabbro et al., 2002; Noble et al., 2004; Regan et al., 2002). The hypothesis that p38 may be a potential drug target is a popular one, as indicated by the large number of patent applications (>48) that have been submitted by 15 pharmaceutical companies describing small molecule modulators of this pathway (English and Cobb, 2002; Regan et al., 2002).

MK2 is a 370 amino acid Ser/Thr kinase with a proline-rich N terminus, a highly conserved catalytic domain, and a C-terminal region containing an autoinhibitory A-helix (AH), a leucine-rich nuclear export signal (NES), and a functional nuclear localization signal (NLS) (Engel *et al.*, 1998; Neininger *et al.*, 2001; Zu *et al.*, 1995). The proline-rich N terminus of MK2 may interact with proteins that contain SH3 domains. In MK2 there are two regulatory phosphorylation sites: threonine 205 (T205), which is located at the activation loop of the kinase between subdomains VII and VIII, and threonine 317 (T317), which is adjacent to the autoinhibitory AH domain (Engel *et al.*, 1998; Neininger *et al.*, 2001; Zu *et al.*, 1995). Phosphorylation of T205 contributes to activation of the kinase by changing the conformation of the activation loop within the catalytic domain. Based on a number of deletion mutants and GFP fusion studies, a model has been proposed for the role of phosphorylation of T317 in MK2 activation and translocation from the nucleus to the cytoplasm (Engel *et al.*, 1998; Neininger *et al.*, 2001; Zu *et al.*, 1995). The NLS is accessible in the inactive and active form of the kinase, whereas the NES is only functional in the phosphorylated and activated form of the enzyme. It has been proposed that the inactive form of MK2 exists in a closed conformation that resides exclusively in the nucleus. Upon activation of the MAP kinase pathways by stress, p38-mediated phosphorylation of MK2 T317 produces a conformational switch to a more open form that activates the kinase by reducing the interaction between the AH and catalytic domains and demasks the overlapping NES. The unmasked NES of MK2 becomes accessible to exportin or an exportin-binding adaptor, and the activated kinase is exported rapidly from the nucleus. MK2 activation and its nuclear export are therefore coupled by this conformational opening. In the active state, both the NES and the NLS are accessible, but nuclear export appears to be more effective that its import, leading to a steady state where most of the continuously shuttling protein is cytoplasmic (Engel *et al.*, 1998; Neininger *et al.*, 2001; Zu *et al.*, 1995). MK2 is essential for LPS-induced TNFα biosynthesis (Kotlyarov *et al.*, 1999) and targets the AU-rich 3'-untranslated regions of the proinflammatory cytokines TNFα and IL-6, thereby regulating their biosynthesis (Neininger *et al.*, 2002).

We decided to generate a stable MK2-EGFP expressing cell line to take advantage of the p38-induced nucleus to cytoplasm translocation that could be quantified on an HCS imaging platform as a cell-based model for identifying inhibitors of p38.

Generation and Characterization of MK2-EGFP HeLa Cell Line

A MK2 clone obtained from colleagues at Lilly research laboratories (Indianapolis, IN) contains the cDNA described in GenBank #X75346. This cDNA sequence was originally amplified to contain the coding region from AA 32 to the end of the coding region of the GenBank entry.

An N-terminal truncated form of MK2 missing the first 31 amino acids was expressed in *Escherichia coli* and was utilized successfully as a p38 substrate for in gel kinase assays (data not shown). Because the regulatory p38 phosphorylation sites are contained in this clone and it is the C-terminal region that contains the T317 site, AH, NES, and NLS motifs (Engel *et al.*, 1998; Neininger *et al.*, 2001; Zu *et al.*, 1995), we decided to proceed with this truncated insert rather than generate a full-length clone. The pGEX/MK2 plasmid is digested with *Bam*H1 and *Xho*I, run on a 1% agarose gel, and the ~1.1-kb MK2 cDNA fragment was subcloned into the *Bgl*II/*Sal*I cut pLEGFPC1 retroviral vector. The pLEGFPC1/MK2 was transfected into Phoenix Amphotropic packaging cells, and retrovirus harvested 48 h posttransfection is used to infect HeLa (CCL-2, ATCC, Rockville, MD) cells.

Single-cell clones were isolated from a population of MK2-EGFP retrovirus-infected HeLa cells that were placed under selection with G418 48 h postinfection and sorted using the Beckman-Coulter Elite flow cytometer sorter equipped with an autoclone single-cell deposition device (Hialeah, FL). EGFP fluorescence was excited with a 15-mW 488-nm Ag laser, and emission was collected though 550-nm DCLP and 525-nm band-pass filter sets. Three distinct populations of cells based on cell size and EGFP fluorescent intensity, high, medium, or low, were isolated using the cell-sorting procedure. A number of cell clones were expanded in culture for 2 weeks under G418 selection conditions and characterized by FACS analysis (Fig. 1, Table I). Six separate flow-sorted clonal cell lines were compared for EGFP expression levels, fluorescent intensity, cell viability, and the robustness of their MK2-EGFP translocation responses on the ArrayScan (see later). For 9 of the 10 cell lines analyzed by FACS (Table I), the number of cells positive for EGFP expression ranges between 56 and 80%. For those same cell lines, the mean fluorescent intensity (MFI) of the EGFP signal was more varied, ranging between 70 and 266 MFI. The cell clone designated "A4" was 78% positive for EGFP expression with an MFI of 105 and was selected for further assay development (see later).

Fluorescence microscopy was used to analyze the subcellular distribution of MK2-EGFP in HeLa cells infected with the MK2-EGFP retrovirus (Fig. 2). The C2 stable population of HeLa-MK2-EGFP cells was seeded in four-well chamber slides, treated with ±100 n*M* anisomycin for 30 min, fixed with 3.7% formaldehyde containing Hoechst dye for 10 min, and examined on a Nikon Eclipse TE300 (Japan) fluorescence microscope equipped with a 300-W Hg lamp and GFP filter set (Chroma, Battlebrook, VT) (Fig. 2). Images were captured on a Micropublisher (Qimaging, Burnaby, BC, Canada) camera. In untreated cells, MK2-EGFP is predominantly located in the nucleus of the cells where it appears to colocalize with the nuclear Hoechst stain (Fig. 2).

Treatment with anisomycin induces the translocation of the majority of the MK2-EGFP into the cytoplasm (Fig. 2).

Cellomics ArrayScan Automated Imaging Platform

The largest category of HCS systems in the market place, and by far the largest installed base, is wide-field fixed cell imaging platforms. The ArrayScan platform marketed by Cellomics is one of the most widely

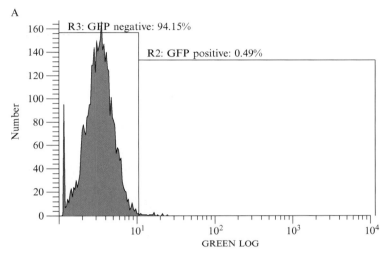

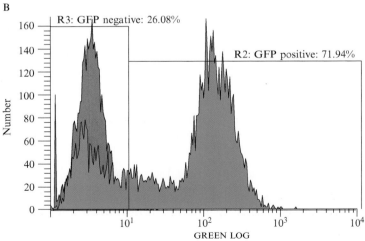

FIG. 1. (*continued*)

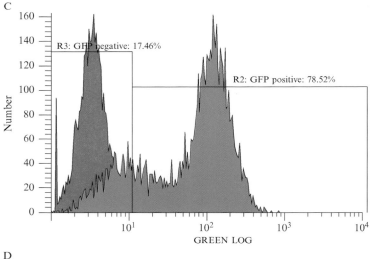

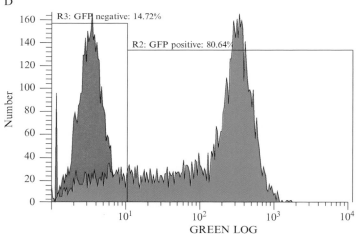

FIG. 1. FACS analysis of parental HeLa cells and MK2-EGFP-HeLa cell lines. MK2-EGFP retrovirus-infected HeLa cells were sorted using the Beckman-Coulter Elite flow cytometer sorter equipped with an autoclone single-cell deposition device (Hialeah, FL). EGFP fluorescence was excited with a 15-mW 488-nm Ag laser, and emission was collected though 550-nm DCLP and 525-nm BP filter sets. Three distinct populations of cells based on cell size and EGFP fluorescent intensity, high, medium or low, were isolated using the cell-sorting procedure. Several cell clones were expanded under G418 selection conditions for further FACS analysis. (A) Parental HeLa Cells ~0% EGFP positive with 0 MFI, (B) C1 stable infection with 71.94% EGFP positive and 123 MFI, (C) A4 clone 78.5% EGFP positive with 105 MFI, and (D) C5 clone 80.64% EGFP positive with 266 MFI.

TABLE I
FACS ANALYSIS OF FLOW SORTER-DERIVED MK2-EGFP CELL LINES

Clone	% GFP positive	MFI
C2 infected	74.64	136.51
C2-AO2 pool	78.46	101.79
C5	80.64	266.03
A4	78.52	105.54
F2	74.94	92.22
D7	74.6	136.09
D10	68.56	70.30
A7	63.92	175.11
C6	56.24	124.37
F5	18.66	19.90
Parental	0.68	NA

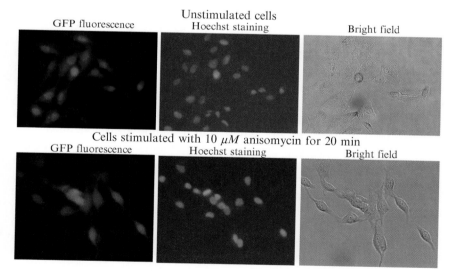

FIG. 2. Fluorescence and bright-field images of MK2-EGFP HeLa cells. HeLa-MK2-GFP infected cells were seeded in four-well chamber slides, treated with ±100 nM anisomycin for 30 min, fixed with 3.7% formaldehyde containing Hoechst dye for 10 min, and examined on a Nikon Eclipse TE300 (Japan) fluorescence microscope equipped with a 300-W Hg lamp and GFP filter set (Chroma, Battlebrook, VT). Images were captured on a Micropublisher (Qimaging, Burnaby, BC, Canada) camera.

deployed HCS systems. The studies described in this chapter are performed on an ArrayScan II that had the software upgraded to a 3.1 version. The ArrayScan 3.1 houses a Zeiss Axiovert S100 inverted microscope outfitted with 5×/0.25 NA, 10×/0.3 NA, and 20×/0.4 NA Zeiss objectives. Illumination is provided by a Xe/Hg arc lamp source (EXFO, Quebec, Canada), and fluorescence is detected by a 12-bit high sensitivity −20° cooled CCD camera (Photometrics Quantix). The ArrayScan 3.1 provides the capability of imaging multiwavelength fluorescence by acquiring wavelength channels sequentially in which each fluorophore is separately excited and detected on the chip of a monochromatic CCD camera. Channel selection is accomplished using a fast excitation filter wheel combined with a multiband emission filter, although single band emission filters can be used to improve selectivity. The system comes with filter sets designed for the common fluorescent probes and can distinguish up to four labels in a single preparation with minimal cross talk between channels. The ArrayScan 3.1 was a wide-field imaging system that illuminates a "large" area of the specimen and directly images that area all at once. It uses an image-based autofocus system that images a fluorescent label in cells, typically fluorescently stained nuclei, but any feature of interest could be used, and an algorithm measures the relative sharpness of the image. The ArrayScan 3.1 was integrated with a Zymark Twister Robot and 80 plate stacker for fixed end point assays.

The cytoplasm-to-nuclear translocation algorithm developed by Cellomics may be used to quantify the relative distribution of a fluorescently tagged target between two cellular compartments, namely the cytoplasm and the nucleus (Giuliano et al., 1997). Labeling with a nucleic acid dye such as Hoechst 33342, DAPI, or DRAQ5 identifies the nuclear region, and this signal is used to focus the instrument and to define a nuclear mask. The mask is eroded to reduce cytoplasmic contamination within the nuclear area, and the final reduced mask is used to quantify the amount of target channel fluorescence within the nucleus. The nuclear mask is then dilated to cover as much of the cytoplasmic region as possible without going outside the cell boundary. Removal of the original nuclear region from this dilated mask creates a ring mask that covers the cytoplasmic region outside the nuclear envelope. The "Cytonuc" difference measurement is calculated as the difference of the average nuclear intensity minus the average cytoplasmic ring intensity on a per cell basis or may be reported as an overall well average value (Giuliano et al., 1997).

Characterization of MK2-EGFP Clones via Imaging

The Cellomics ArrayScan 3.1 was used to analyze the subcellular distribution of MK2-EGFP in HeLa cells infected with the MK2-EGFP retrovirus.

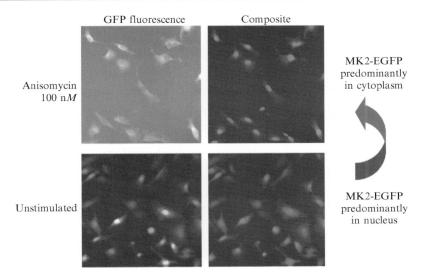

FIG. 3. Fluorescence images of MK2-EGFP HeLa cells captured on the ArrayScan. HeLa-MK2-EGFP A4 cells (5×10^3) were seeded into each of the 96-wells of Packard View plates in EMEM + 10% FBS and incubated overnight at 37° and 5% CO_2. Cells were treated $\pm$ 100 nM of anisomycin for 25 min and fixed in 3.7% formaldehyde + 2 μg/ml Hoechst dye, and fluorescent images were collected on the ArrayScan 3.1.

HeLa-MK2-EGFP A4 cells (5×10^3) per well were seeded in EMEM + 10% fetal bovine serum (FBS) and incubated overnight at 37° and 5% CO_2. Cells were either treated with 100 nM of the protein synthesis inhibitor anisomycin for 25 min or were left untreated and fixed in 3.7% formaldehyde +2 μg/ml Hoechst dye, and fluorescent images were collected on the ArrayScan 3.1 (Fig. 3). Consistent with the fluorescent images in Fig. 2, in untreated cells, MK2-EGFP is located predominantly in the nucleus and appears to colocalize with the Hoechst stain (Fig. 3). Treatment with anisomycin induces the translocation of the majority of the MK2-EGFP into the cytoplasm (Fig. 3). To compare the translocation responses of the different stable cell populations and flow-sorting derived clones (Fig. 1, Table I), anisomycin-induced translocation dose–response curves were run in triplicate (Fig. 4). Using the nuclear translocation algorithm to analyze images captured on the ArrayScan 3.1, the A4 clone and the C2 retroviral-infected population produced the most robust dose-dependent assay signal window (Fig. 4). The A4 clone was selected for further assay development.

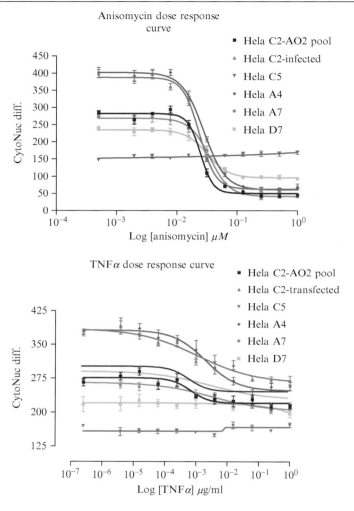

Fig. 4. Comparison of the anisomycin-induced nucleus to cytoplasm MK2-EGFP translocation responses in different flow-sorted cell populations quantified by the nuclear translocation algorithm from images captured on the ArrayScan. To compare the translocation responses of the different stable cell populations and flow-sorting-derived autoclones (Fig. 1, Table I), anisomycin-induced translocation dose–response curves were run in triplicate. HeLa-MK2-EGFP cells (5×10^3) from the indicated cell lines were seeded into each of the 96 wells of Packard View plates in EMEM + 10% FBS and incubated overnight at 37° and 5% CO_2. Cells were treated with the indicated doses of anisomycin for 25 min and fixed in 3.7% formaldehyde + 2 μg/ml Hoechst dye, and fluorescent images were collected on the ArrayScan 3.1. The nuclear translocation algorithm was used to analyze the images captured on the ArrayScan 3.1 and quantify the anisomycin-induced translocation response.

Characterization of the p38 MAPK Signaling Pathway in HeLa-MK2-EGFP Cells

To confirm that the p38 MAPK signaling pathway in the HeLa-MK2-EGFP-A4 clone is intact and similar to wild-type HeLa cells, both cell populations were treated with 100 nM anisomycin for 30 min. Cells were then solubilized in SDS sample buffer, their total cellular proteins were separated on 10% SDS-PAGE gels, transferred to nitrocellulose, and the resulting blots were probed with specific antibodies for total and phosphorylated p38 and MK-2 (Fig. 5). The total p38 signal detected on Western blots appears identical in both HeLa-MK2-EGFP-A4 population and wild-type HeLa cells and is unaffected by anisomycin treatment. In contrast, although the A4 clone and wild-type HeLa cells both exhibit a weak endogenous total MK2 signal at the appropriate molecular mass of ~45.5 kDa, the A4 cells also express a much stronger immunoreactive total MK2 signal at ~75 kDa, corresponding to 45 kDa MK2 plus the additional 30 kDa contributed by the EGFP fusion partner. Neither the endogenous MK2

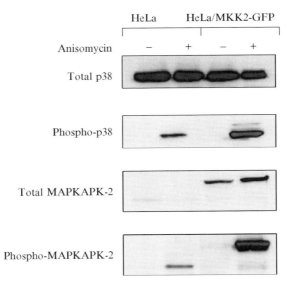

FIG. 5. Comparison of total and phosphorylated p38 and MK2 expression in parental HeLa and MK2-EGFP-HeLa A4 cells by Western blotting analysis. Parental HeLa cells and cells from the HeLa-MK2-EGFP-A4 clone were treated ± 100 nM anisomycin for 30 min. Cells were then solubilized in SDS sample buffer, their total cellular proteins were separated on 10% SDS-PAGE gels, transferred to nitrocellulose, and the resulting blots were probed with specific antibodies for total and phosphorylated p38 and MK-2.

nor the MK2-EGFP total protein signals are affected by anisomycin treatment. Activation of the p38 MAPK signaling pathway by anisomycin treatment dramatically increases the phosphorylation signals for p38 and MK2 in both cell populations relative to untreated controls. As expected from the total MK2 signals, the phospho-MK2 signal is stronger in the A4 clone relative to wild-type HeLa cells, and the majority of the immunoreactive signal migrates at the higher molecular mass of ~75 kDa. Interestingly, there is a stronger phospho-p38 signal apparent in the A4 clone relative to wild-type HeLa. In the A4 clone, overexpression of the MK2-EGFP fusion protein has no apparent effect on the qualitative pattern of the p38 and MK2 phosphorylation responses to activation by anisomycin, despite an overall increase in signal strength (Fig. 5).

Assay Development on the ArrayScan 3.1 Imaging Platform

Treatment of MK2-EGFP HeLa cells with 100 ng/ml anisomycin produces a significantly larger MK2-EGF translocation response than treatment with 50 ng/ml TNF-α when quantified on the ArrayScan 3.1 (Fig. 6A and B). However, the time course of MK2-EGFP export from the nucleus appears similar for both stimuli (Fig. 6A). The amount of MK2-EGFP that translocates from the nucleus to the cytoplasm appears to increase in a roughly linear fashion for 20 to 25 min post stimulation and then remains stable for as long as 90 min (Fig. 6A). A 25-min period of stimulation with anisomycin (100 ng/ml) was selected as the standard treatment to induce maximal MK2-EGFP translocation. The observed time course for MK2-EGFP translocation is consistent with published data (Engel et al., 1998.). Seeding densities between 0.5 and 16×10^3 A4 cells per well produce fairly comparable MK2-EGFP translocation responses (Fig. 6B) for both stimuli. A seeding density of 5×10^3 cells per well was selected for the remaining assay development effort. Because compound-screening libraries are typically solubilized in dimethyl sulfoxide (DMSO), we evaluated the DMSO tolerance of the MK2-EGFP translocation response in A4 cells that were treated for 25 min with media, 100 ng/ml anisomycin, or 50 ng/ml TNF-α at the indicated DMSO concentrations (Fig. 6C). The MK2-EGFP translocation response appears unaffected at DMSO concentrations $\leq 0.65\%$. However, at DMSO concentrations $\geq 1.25\%$, irrespective of the stimulus, the Cytonuc difference was increased. At DMSO concentrations $\geq 1.25\%$, the majority of the A4 cell population assumes a rounded morphology rather than the well-attached, flat morphology more typical of this clone (Figs. 2 and 3). The rounded A4 cells have a much smaller cytoplasmic area than normal, and the Cytonuc difference algorithm therefore has difficulty segmenting the cytoplasm and nuclear areas to make the difference calculation.

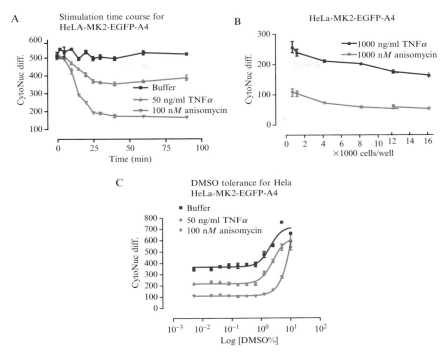

FIG. 6. MK2-EGFP translocation assay development. (A) Stimulation time course. HeLa-MK2-EGFP cells (5×10^3) from the HeLa-MK2-EGFP-A4 clone were seeded into each of the 96 wells of Packard View plates in EMEM + 10% FBS and incubated overnight at 37° and 5% CO_2. Cells were treated ± 100 ng/ml anisomycin or ± 50 ng/ml TNFα for the indicated times and fixed in 3.7% formaldehyde + 2 μg/ml Hoechst dye, and fluorescent images were collected on the ArrayScan 3.1. (B) Cell-seeding density. The indicated numbers of cells from the HeLa-MK2-EGFP-A4 clone were seeded into each of the 96 wells of Packard View plates in EMEM + 10% FBS and incubated overnight at 37° and 5% CO_2. Cells were treated ± 100 ng/ml anisomycin or ± 50 ng/ml TNFα for 25 min and fixed in 3.7% formaldehyde + 2 μg/ml Hoechst dye, and fluorescent images were collected on the ArrayScan 3.1. (C) DMSO tolerance. HeLa-MK2-EGFP cells (5×10^3) from the HeLa-MK2-EGFP-A4 clone were seeded into each of the 96 wells of Packard View plates in EMEM + 10% FBS and incubated overnight at 37° and 5% CO_2. Cells were treated ± 100 ng/ml anisomycin or ± 50 ng/mL TNFα containing the indicated concentrations of DMSO for 25 min and fixed in 3.7% formaldehyde + 2 μg/ml Hoechst dye, and fluorescent images were collected on the ArrayScan 3.1. The nuclear translocation algorithm was used to analyze the images captured on the ArrayScan 3.1 and quantify the anisomycin and TNFα-induced translocation response.

Standard Operating Procedure for the MK2-EGFP
    Translocation Assay

1. Sub-confluent HeLa-MK2-EGFP-A4 monolayers (70–80%) are detached from tissue culture flasks with trypsin-versene and centrifuged for 5 min at 800g.

2. Cells are resuspended in complete media (EMEM + 10% FBS + 2 mM L-glutamine + penicillin/streptomycin + 800 μg/ml G418) to a cell density to 5×10^4 cells/ml and seeded into 96-well Packard View plates using the Multidrop (Titertek, Huntsville, AL) at 100 μl/well (5000 cells/well).

3. Plates are incubated at 37°, 5% CO_2 and 95% humidity for 18 to 24 h.

4. Twenty-five microliters of compounds and controls are transferred to the wells of assay plates using a 96-well head on the Multimek (Beckman, Fullerton, CA), and plates are incubated at 37°, 5% CO_2, and 95% humidity for 12 min.

5. Twenty-five microliters of anisomycin (100 ng/ml final concentration) is transferred to the assay wells using a 96-well head on the Multimek, and plates are incubated at 37°, 5% CO_2, and 95% humidity for 25 min.

6. Plates are fixed by the addition of 150 μl of (37°) formaldehyde (3.7% final) + 2 μg/μl Hoechst 33342 stain using the Multidrop (Titertek) and incubated at room temperature in a fume hood for 12 min.

7. The fixation solution is aspirated, 100 μl of phosphate-buffered saline (PBS) is added to wash the monolayers, the PBS is aspirated, another 100 μl of PBS is added, and plates are sealed. This aspiration and wash process is performed on a MAP C2 96-well plate handler (Titertek).

8. Images are acquired on the ArrayScan 3.1 platform, and MK2-EGFP translocation is analyzed using the cytoplasm-to-nucleus translocation algorithm.

MK2-EGFP Translocation Assay Reproducibility and Signal
    Widow Evaluation

To evaluate the reproducibility of the MK2-EGF translocation response in A4 cells, EC_{50} values for anisomycin are determined in three independent experiments run on separate days (Fig. 7A). The EC_{50} for anisomycin-induced MK2-EGFP translocation ranged from 27 to 32 nM and produced, on average, an EC_{50} of 29.54 ± 2.75 nm, indicating that the translocation response was very reproducible.

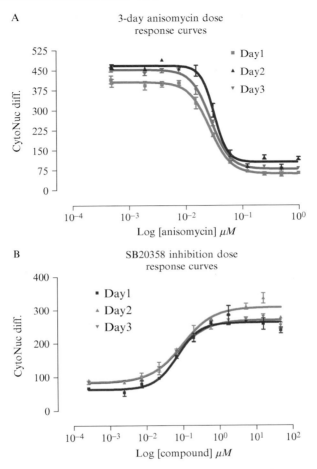

FIG. 7. Three-day MK2-EGFP translocation activation and inhibition curves. (A) Three-day EC$_{50}$ curves. HeLa-MK2-EGFP-A4 cells were seeded at 5 × 10^3 cells per well in Packard View plates and on the following day were treated with the indicated doses of anisomycin for 25 min and fixed in 3.7% formaldehyde + 2 μg/ml Hoechst dye, and fluorescent images were collected on the ArrayScan 3.1. The nuclear translocation algorithm was used to analyze the images and quantify the anisomycin-induced translocation response. Data are presented from three independent experiments, each performed in triplicate wells and run on separate days. (B) Three-day IC$_{50}$ curves. HeLa-MK2-EGFP-A4 cells were seeded at 5 × 10^3 cells per well in Packard View plates and on the next day were pretreated with the indicated doses of SB203580 for 12 min, followed by the addition of 100 ng/ml anisomycin (final) and a further incubation for 25 min. Plates were fixed in 3.7% formaldehyde + 2 μg/ml Hoechst dye, and fluorescent images were collected on the ArrayScan 3.1. The nuclear translocation algorithm was used to analyze the images and quantify the anisomycin-induced translocation response. Data are presented from three independent experiments, each performed in triplicate wells and run on separate days.

To examine the reproducibility of inhibition of the translocation response, three independent experiments are run on three separate days. On the day after seeding, cells are pretreated with the indicated doses of the p38 inhibitor SB203580 for 12 min, followed by the addition of 100 ng/ml anisomycin (final) and a further incubation for 25 min (Fig. 7B). The IC_{50} for SB203580 inhibition of anisomycin-induced MK2-EGFP translocation ranged from 72 to 99 nM and produced, on average, an IC_{50} of 82.12 $\pm$ 15.21 nm, indicating that the inhibition was reproducible.

Finally, the robustness of the MK2-EGF translocation response is quantified in full plate assays used to determine the Z factor (Zhang et al., 1999). Three 96-well plates are treated under the following conditions: media alone (blue squares), 100 ng/ml anisomycin (red triangles), and p38 inhibitor + 100 ng/ml anisomycin (green circles) (Fig. 8). A comparison of either the media control plate or p38 inhibitor plate data to anisomycin plate data produced a Z factor (Zhang et al., 1999) of 0.61 and an average assay signal window of 3.6-fold, indicating a good reproducible assay compatible with HTS.

p38 Inhibitor Data

Because p38 is an attractive target for anti-inflammatory and anticancer therapies, a number of pharmaceutical companies have developed p38 inhibitor compounds (English and Cobb, 2002; Fabbro et al., 2002; Noble et al., 2004; Regan et al., 2002). In addition to five "selective" p38 inhibitors, we included a number of other kinase inhibitors to evaluate the selectivity of the MK2-EGF translocation response (Table II). The SP 600125 compound is a JNK-1/2 inhibitor, UO126 and PD 98059 are MEK-1/2 inhibitors, wortmannin is a PI3-kinase inhibitor, and staurosporin is a nonselective kinase inhibitor. All five p38 inhibitors produced IC_{50} values for inhibition of anisomycin-induced MK2-EGFP translocation <100 nM. Neither the PI3-kinase inhibitor wortmannin nor the JNK-1/2 inhibitor SP 600125 inhibited anisomycin-induced MK2-EGFP translocation. UO126 and PD 98059, the MEK-1/2 inhibitors, produced IC_{50} values of 33 and 39 μM respectively, while the nonselective kinase inhibitor staurosporin produced an IC_{50} of 0.75 μM (Fig. 7B, Table II).

Secondary Analysis Parameters

In addition to the target readout for nucleus to cytoplasm translocation, the algorithm provides a range of image-derived features at the individual cell level or at the well averaged level (Table III). These image-derived features provide information on cell morphology, cytotoxicity, and potential fluorescent artifacts produced by fluorescent compounds. The ArrayScan

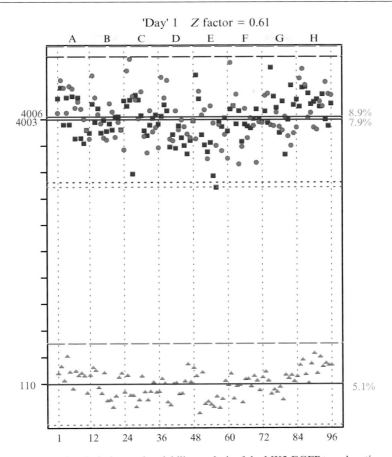

FIG. 8. Assay signal window and variability analysis of the MK2-EGFP translocation assay. HeLa-MK2-EGFP cells (5×10^3) from the HeLa-MK2-EGFP-A4 clone were seeded into each of the 96 wells of Packard View plates in EMEM + 10% FBS and incubated overnight at 37° and 5% CO_2. One full 96-well plate was treated for 25 min under the following conditions: media alone (blue squares), 100 ng/ml anisomycin (red triangles), and p38 inhibitor + 100 ng/ml anisomycin (green circles). Plates were fixed in 3.7% formaldehyde + 2 μg/ml Hoechst dye, and fluorescent images were collected on the ArrayScan 3.1. The nuclear translocation algorithm was used to analyze the images and quantify the anisomycin-induced translocation response. The Z factor was calculated according to the method of Zhang *et al.* (1999).

software provides data visualization tools that allow the user to analyze these multiparameter data and the images from which they are derived (Fig. 9). During the course of the p38 inhibitor studies described earlier (Table II, Fig. 7B), we observed that the two highest concentrations of the

TABLE II
MK2-EGFP KINASE INHIBITOR IC$_{50}$ COMPARISON

| Compound | Anisomycin | |
	% Inhibition at 50 μM	IC$_{50}$ μM
SP600125	29.5	>50
VX-745	146	0.045
Merch p38 inhib	−30.5	0.007
RWJ 68354	122	0.087
SB242235	96.5	0.028
SB203580	64	0.053
PD98059	58.5	39.000
U0126 # 9903	64	33.000
Wortmannin	33.5	>50
Staurosporin	180.5	0.755

TABLE III
CELL AND WELL FEATURES DERIVED BY THE NUCLEAR TRANSLOCATION IMAGE
ANALYSIS ALGORITHM

Cell features	Well features
Nuc Area	Valid Object Count
Nuc Perimeter/Area	Selected Object Count
Nuc Area / Bounding Box	% Selected Object Count
Nuc Length / Width	Mean Nuc Area
Nuc Intensity (Nuc Ch)	Mean Nuc Intensity
Nuc Total Intensity (Nuc Ch)	Mean Nuc-Cyto Intensity Diff
Nuc Area (Target Ch)	Avg Cell Density/Field
Nuc Intensity (Target Ch)	Mean Nuc Intensity(Target Ch)
Nuc Total Intensity (Target Ch)	Mean Cyto Ring Intensity(Target Ch)
Cyto Ring Area (Target Ch)	Number of Valid Fields
Cyto Intensity (Target Ch)	Z-position
Cyto Total Intensity (Target Ch)	CV & StdDev of features
Nuc-Cyto Intensity Diff (Target Ch)	
Object Intensity / Area Ch2 – Ch6	
Object Total Intensity Ch2 – Ch6	

Merck p38 inhibitor compound, 50 and 16.6 μM, were producing obvious outliers in several of these parameters: the valid object counts, the average cell density per field, and the number of valid fields (Fig. 9A). A visual inspection of the images captured in these wells revealed that there were fewer cells in wells treated with 50 and 16.6 μM of the Merck inhibitor compared to the 5.5 μM dose or images from control wells (Fig. 9B). It is

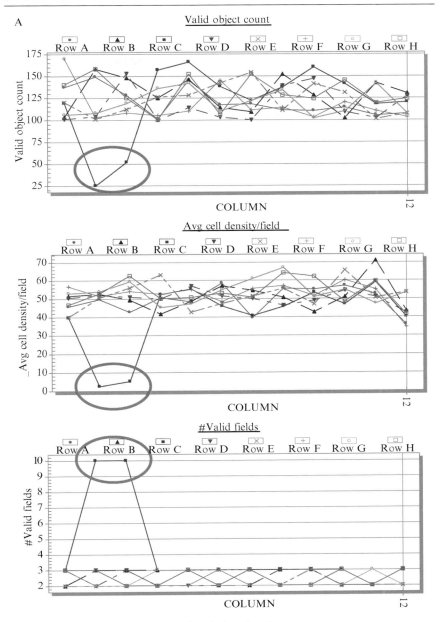

A

Valid object count

Avg cell density/field

#Valid fields

FIG. 9. (*continued*)

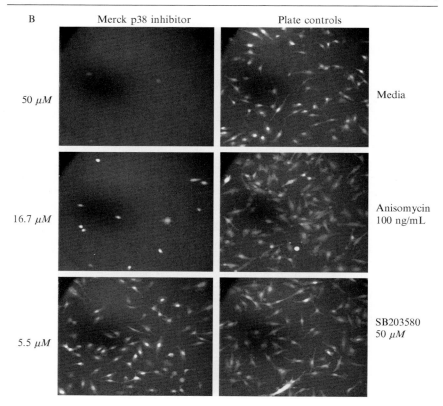

FIG. 9. Secondary analysis of multiparameter image-derived data. The nuclear translocation algorithm provides a range of image-derived features at the individual cell level or at the well averaged level (Table III). The ArrayScan software provides data visualization tools that allow the user to analyze these multiparameter data and the images from which they have been derived. (A) Multiparameter data views. Data views for the valid object counts, the average cell density per field, and the number of valid fields are presented with obvious outliers; data are circled in red. (B) Image views. Representative images captured in wells treated with 50 and 16.6 μM of the p38 Merck inhibitor compared to the 5.5 μM dose, together with images from plate control wells.

likely that the Merck inhibitor compound, at 50 and 16.6 μM, was either cytotoxic or reduced the adherence of the HeLa cells significantly.

Discussion

This chapter described the generation and characterization of a stable MK2-EGFP expressing HeLa cell line and the subsequent development of a high-content imaging assay on the Cellomics ArrayScan platform to screen for p38 MAPK inhibitors. The assay took advantage of the well-substantiated hypothesis that mitogen-activated protein kinase-activating protein kinase-2 is a substrate of p38 MAPK kinase and that p38-induced phosphorylation of MK-2 induces a nucleus-to-cytoplasm translocation (Engel et al., 1998; Neininger et al., 2001; Zu et al., 1995).

Through a process of heterologous expression of a MK2-EGFP fusion protein in HeLa cells using retroviral infection, antibiotic selection, and flow sorting, we were able to isolate a MK2-EGFP-HeLa cell line in which the MK2-EGFP translocation response could be robustly quantified on the Cellomics ArrayScan HCS platform (Figs. 1–4, Table I). Several clonal populations were isolated from the original retrovirus-infected population by flow sorting and selection, and their GFP expression and intensity levels were measured by FACS analysis (Table I) and compared to their anisomycin-induced MK2-EGFP translocation responses (Figs. 3 and 4). Five of the six cell lines produced a dose-dependent anisomycin-induced MK2-EGFP translocation response (Fig. 4). Only the highest-expressing cell line, the C5 clone, which was 80% positive for EGFP expression with an MFI of 266, failed to exhibit an anisomycin-induced MK2-EGFP translocation response (Fig. 4). The A4 clone of MK2-EGFP-HeLa cells, which was 78% positive for EGFP expression with an MFI of 105 (Table I), was selected for further assay development because it exhibited the most robust assay signal window for anisomycin-induced MK2-EGFP translocation (Fig. 4). Similar responses were observed when TNFα was used as the stimulus to activate the p38 MAPK pathway (data not shown). The characterization and selection of the appropriate clone or cell line were the most critical steps of the assay development process. Overexpression of the fusion protein in the A4-MK2-EGFP-HeLa cell population had no effect on the qualitative pattern of the p38 and MK2 phosphorylation responses to p38 MAPK activation induced by anisomycin, although there was an apparent overall increase in phospho-p38 and phospho-MK2 signal strength relative to wild-type HeLa cells (Fig. 5).

The A4-MK2-EGFP-HeLa cell line was used to develop an assay for identifying p38 inhibitors that would be compatible with HTS on the Cellomics ArrayScan imager. A series of experiments were conducted to optimize the assay with respect to cell seeding density, length of anisomycin stimulation, DMSO tolerance, assay signal window, and reproducibility (Figs. 6–8). The resulting MK2-EGFP translocation assay was compatible with HTS and was shown to be an effective cell-based assay to identify p38 inhibitors (Fig. 7B, Table II). p38 has been a major target for drug discovery by the pharmaceutical industry, as indicated by the numerous patent applications and small molecule inhibitors that have been developed (English and Cobb, 2002; Fabbro et al., 2002; Noble et al., 2004; Regan *et al.*, 2002). Many of these p38 inhibitors have exhibited significant efficacy in cellular assays and animal disease models and several have progressed into human clinical trails for the treatment of inflammation and cancer (English and Cobb, 2002; Fabbro *et al.*, 2002; Noble *et al.*, 2004; Regan *et al.*, 2002). For example, SB203580 and BIRB 796 and their analogs have been shown to be potent and selective inhibitors of p38 MAPK (English and Cobb, 2002; Fabbro *et al.*, 2002; Noble *et al.*, 2004; Regan *et al.*, 2002). SB203580 and BIRB 796 produced IC_{50} values of 60 and 18 nM, respectively, in a THP-1 TNFα production cell model (Regan *et al.*, 2002). In the MK2-EGFP translocation assay described here, SB203580 produced an IC_{50} of 53 nM, and the other published p38 inhibitors produced IC_{50} values below 100 nM. In a similar MK2-GFP redistribution assay developed in a BHK-1 cell background and quantified on the In cell analyzer 3000 imaging platform, SB203580 produced an IC_{50} of 3.4 μM for anisomycin-induced redistribution (Almholt *et al.*, 2004), ~60-fold less sensitive than the assay described here. The BHK-1 MK2 redistribution assay was utilized for an HTS of 183,375 compounds that identified two main classes of inhibitors: direct inhibitors of the MK2 nuclear export process and inhibitors of the upstream p38 MAPK pathway (Almholt *et al.*, 2004). Hits from the primary screen were categorized via a number of secondary assays and compounds were identified that both structurally and functionally resembled p38 kinase inhibitors (Almholt *et al.*, 2004). In the MK2-EGFP translocation assay described here, neither the PI3-kinase inhibitor wortmannin nor the JNK-1/2 inhibitor SP 600125 inhibited anisomycin-induced MK2-EGFP translocation, indicating some degree of selectivity (Table II). However the MK2-EGFP translocation assay was susceptible to inhibition by the nonselective kinase inhibitor staurosporin, which produced an IC_{50} of 0.75 μM. The MEK-1/2 inhibitors UO126 and PD 98059, produced IC_{50} values of 33 and 39 μM, respectively, perhaps indicating that ERK1/2 may be capable of phosphorylating MK2 as well as MK1, or perhaps reflecting the cross talk between the different modules of MAPK signaling pathways (Cowan and Storey 2003; Garrington and Johnson, 1999). In addition to p38 inhibitors, the MK2-EGF translocation response may therefore be susceptible to other classes of inhibitors: nonselective kinase inhibitors, kinase inhibitors that inhibit upstream kinases in the p38 MAPK signaling pathway, kinases involved in the cross talk between different modules (ERKs, JNKs, and p38s) of the MAPK signaling pathways, and inhibitors of the MK2 nuclear export process. Trask *et al.* (2006) describe the conversion of this assay to a 384-well format and optimization of the assay for the screening of a 32K kinase-biased library to identify p38 MAPK inhibitors.

Some 518 protein kinases encoded in the human genome share a catalytic domain, the ATP-binding site, conserved in sequence and structure (English and Cobb, 2002; Noble *et al.*, 2004). The vast majority of known kinase inhibitors was found to be competitive with ATP, and thus are believed to interact within the ATP-binding site (English and Cobb, 2002; Noble *et al.*, 2004). The conservation of the ATP-binding site within the kinase family and the large number of cellular proteins that bind and/or utilize ATP, together with intracellular concentrations of ATP reported in the millimolar range, raise significant concerns about inhibitor potency in cellular assays, kinase selectivity, and adverse effects (English and Cobb, 2002; Noble *et al.*, 2004). In addition to the direct target readout that can be obtained from the images, the image analysis algorithm also measures and reports multiple features and image-based parameters that provide information on cell morphology, cytotoxicity, and potential interference by fluorescent compounds (Giuliano *et al.*, 1997). Thus, by mining these "high-content" data it is possible to extract additional information on the effects of compound treatment of cells (Fig. 9). For example, we observed that the two highest concentrations of the Merck p38 inhibitor compound were producing obvious outliers in several parameters relative to the other wells on the plate (Fig. 9A). A visual inspection of the images revealed that there were fewer cells in the fields of view captured from wells treated with 50 and 16.6 μM of the Merck inhibitor compared to the 5.5 μM dose or images from control wells (Fig. 9B). The Merck p38 inhibitor compound produced an IC$_{50}$ of 7 nM in the MK2-EGFP translocation assay, but at 50 and 16.6 μM doses was either cytotoxic or significantly reduced the adherence of the HeLa cells. While the acute effects on cytotoxicity or cell morphology observed in our MK2-EGFP HeLa cell-based model may not be predictive of *in vivo* toxicity, it is important to note that a number of the p38 inhibitors, including the Merck p38 inhibitor, have been withdrawn from clinical trials because of adverse toxicity profiles (English and Cobb, 2002; Fabbro *et al.*, 2002; Noble *et al.*, 2004; Regan *et al.*, 2002). It will be interesting to see whether the ability of HCS to measure and discriminate between the on-target and off-target effects of lead compounds will provide cell-based models that will improve the conversion rate of drug candidates to successful drugs.

References

Almholt, D. L., Loechel, F., Nielsen, S. J., Krog-Jensen, C., Terry, R., Bjorn, S. P., Pedersen, H. C., Praestegaard, M., Moller, S., Heide, M., Pagliaro, L., Mason, A. J., Butcher, S., and Dahl, S. W. (2004). Nuclear export inhibitors and kinase inhibitors identified using a MAPK-activated protein kinase 2 redistribution screen. *Assay Drug Dev. Technol.* **2,** 7–20.

Cowan, K. J., and Storey, K. B. (2003). Mitogen-activated protein kinases: New signaling pathways functioning in cellular responses to environmental stress. *J. Exp. Biol.* **206,** 1107–1115.

DeBiasio, R., Bright, G. R., Ernst, L. A., Waggoner, A. S., and Taylor, D. L. (1987). Five-parameter fluorescence imaging: Wound healing of living Swiss 3T3 cells. *J. Cell Biol.* **105,** 1613–1622.

Engel, K., Kotlyarov, A., and Gaestel, M. (1998). Leptomycin B-sensitive nuclear export of MAPKAP kinase 2 is regulated by phosphorylation. *EMBO J.* **17,** 3363–3371.

English, J. M., and Cobb, M. H. (2002). Pharmacological inhibitors of MAPK pathways. *Trends Pharmacol. Sci.* **23,** 40–45.

Fabbro, D., Ruetz, S., Buchdunger, E., Cowan-Jacob, S. W., Fendrich, G., Liebetanz, J., Mestan, J., O'Reilly, T., Traxler, P., Chaudhuri, B., Fretz, H., Zimmermann, J., Meyer, T., Caravatti, G., Furet, P., and Manley, P. W. (2002). Protein kinases as targets for anticancer agents: From inhibitors to useful drugs. *Pharmacol. Ther.* **93,** 79–98.

Garrington, T. P., and Johnson, G. L. (1999). Organization and regulation of mitogen activated protein kinase signaling pathways. *Curr. Opin. Cell Biol.* **11,** 211–218.

Giuliano, K. A., DeBiasio, R. L., Dunlay, R. T., Gough, A., Volosky, J. M., Zock, J., Pavlakis, G. N., and Taylor, D. L. (1997). High-content screening: A new approach to easing key bottlenecks in the drug discovery process. *J. Biomol. Screen.* **2,** 249–259.

Giuliano, K. A., and Taylor, D. L. (1998). Fluorescent-protein biosensors: New tools in drug discovery. *TIBTech.* **16,** 135–140.

Johnston, P. A., and Johnston, P. A. (2002). Cellular platforms for HTS: Three case studies. *Drug Discov. Today* **7,** 353–363.

Kotlyarov, A., Neininger, A., Schubert, C., Eckert, R., Birchmeier, C., Volk, H. D., and Gaestel, M. (1999). MAPKAP kinase 2 is essential for LPS-induced TNF-alpha biosynthesis. *Nature Cell Biol.* **1,** 94–97.

Lundholt, B. K., Linde, V., Loechel, F., Pedersen, H. C., Moller, S., Praestegaard, M., Mikkelsen, I., Scudder, K., Bjorn, S. P., Heide, M., Arkhammar, P. O., Terry, R., and Nielsen, S. J. (2005). Identification of Akt pathway inhibitors using redistribution screening on the FLIPR and the IN Cell 3000 analyzer. *J. Biomol. Screen.* **10,** 20–29.

Mitchison, T. J. (2005). Small-molecule screening and profiling by using automated microscopy. *Chembiochem.* **6,** 33–39.

Neininger, A., Kontoyiannis, D., Kotlyarov, A., Winzen, R., Eckert, R., Volk, H. D., Holtmann, H., Kollias, G., and Gaestel, M. (2002). MK2 targets AU-rich elements and regulates biosynthesis of tumor necrosis factor and interleukin-6 independently at different post-transcriptional levels. *J. Biol. Chem.* **277,** 3065–3068.

Neininger, A., Thielemann, H., and Gaestel, M. (2001). FRET-based detection of different conformations of MK2. *EMBO Rep.* **2,** 703–708.

Noble, M. E. M., Endicott, J. A., and Johnson, L. N. (2004). Protein kinase inhibitors: Insights into drug design and structure. *Science* **303,** 1800–1805.

Ono, K., and Han, J. (2000). The p38 signal transduction pathway, activation and function. *Cell. Signal.* **12,** 1–13.

Ramm, P., Alexandrov, Y., Cholewinski, A., Cybuch, Y., Nadon, R., and Soltys, B. J. (2003). Automated screening of neurite outgrowth. *J. Biomol. Screen.* **8,** 7–18.

Regan, J., Breitfelder, S., Cirillo, P., Gilmore, T., Graham, A. G., Hickey, E., Klaus, B., Madwed, J., Moriak, M., Moss, N., Pargellis, C., Pav, S., Proto, A., Swinamer, A., Tong, L., and Torcellini, C. (2002). Pyrazole urea-based inhibitors of p38 MAP kinase: From lead compound to clinical candidate. *J. Med. Chem.* **45,** 2994–3008.

Roessel, P. V., and Brand, A. H. (2001). Imaging into the future: Visualizing gene expression and protein interactions with fluorescent proteins. *Nature Cell Biol.* **4,** E15–E20.

Tavare, J. M., Fletcher, L. M., and Welsh, G. I. (2001). Using green fluorescent protein to study intracellular signaling. *J. Endocrinol.* **170,** 297–306.

Thomas, N., and Goodyer, D. (2003). Stealth sensors: Real-time monitoring of the cell cycle. *Targets* **2,** 26–33.

Trask, O. J., Jr., Baker, A., Williams, R. G., Kickischer, D., Kandasamy, R., Laethem, C., Johnston, P. A., and Johnston, P. A. (2006). Assay development and case history of a 32K-biased library high-content MK2-EGFP translocation screen to identify p38 MAPK inhibitors on the ArrayScan 3.1 imaging platform. *Methods Enzymol.* **414** (this volume).

Zhang, J. H., Chung, T. D., and Oldenburg, K. R. (1999). A simple statistical parameter for use in evaluation and validation of high throughput screening assays. *J. Biomol. Screen.* **4,** 67–73.

Zu, Y. L., Ai, Y., and Huang, C. K. (1995). Characterization of an autoinhibitory domain in human mitogen-activated protein kinase-activated protein kinase 2. *J. Biol. Chem.* **270,** 202–206.

[22] Development and Implementation of Three Mitogen-Activated Protein Kinase (MAPK) Signaling Pathway Imaging Assays to Provide MAPK Module Selectivity Profiling for Kinase Inhibitors: MK2-EGFP Translocation, c-Jun, and ERK Activation

By DEBRA NICKISCHER, CARMEN LAETHEM, OSCAR J. TRASK, JR., RHONDA GATES WILLIAMS, RAMANI KANDASAMY, PATRICIA A. JOHNSTON, and PAUL A. JOHNSTON

Abstract

This chapter describes the development and implementation of three independent imaging assays for the major mitogen-activated protein kinase (MAPK) signaling modules: p38, JNK, and ERK. There are more than 500 protein kinases encoded in the human genome that share an ATP-binding site and catalytic domain conserved in both sequence and structure. The majority of kinase inhibitors have been found to be competitive with ATP, raising concerns regarding kinase selectivity and potency in an environment of millimolar intracellular concentrations of ATP, as well as the potential for off-target effects via the many other cellular proteins that bind and/or utilize ATP. The apparent redundancy of the kinase isoforms and functions in the MAPK signaling modules present additional challenges for kinase

METHODS IN ENZYMOLOGY, VOL. 414   0076-6879/06 $35.00
Copyright 2006, Elsevier Inc. All rights reserved.   DOI: 10.1016/S0076-6879(06)14022-7

inhibitor selectivity and potency. Imaging assays provide a method to address many of these concerns. Cellular imaging approaches facilitate analysis of the targets expressed in the context of their endogenous substrates and scaffolding proteins and in a complex environment for which subcellular localization, cross talk between pathways, phosphatase regulatory control, and intracellular ATP concentrations are relevant to the functions of the kinase. The assays described herein provide a strategy to profile kinase inhibitors for MAPK pathway selectivity while simultaneously providing information on cell morphology or toxicity. Results suggest that the MAPK pathways are indeed susceptible to nonselective kinase inhibitors such as staurosporin and inhibitors that inhibit upstream MAPK Kinase Kinases (MKKKs) and MAPK Kinases (MKKs) in the MAPK signaling pathway, especially those involved in cross talk between the pathways. However, selective MAPK inhibitors were identified that exhibited pathway selectivity as evidenced by significantly lower IC_{50} values for their respective p38, JNK, or ERK signaling pathway assays.

Introduction

There are more than 500 protein kinases encoded in the human genome that share an ATP-binding site catalytic domain that is conserved in both sequence and structure (English and Cobb, 2002; Noble et al., 2004). Not surprisingly, the majority of kinase inhibitors identified to date interact within the ATP-binding site and have been found to be competitive with ATP (English and Cobb, 2002; Noble et al., 2004). The degree of conservation of the ATP-binding site raises concerns over the ability to build selective inhibitors (English and Cobb, 2002; Noble et al., 2004). Another significant hurdle for ATP-competitive kinase inhibitors is their ability to exhibit potent activity in cellular environments where concentrations of ATP are reportedly in the millimolar range (English and Cobb, 2002; Noble et al., 2004). In addition to kinases, numerous other cellular proteins bind and/or utilize ATP, each providing the potential for off-target binding of ATP-competitive kinase inhibitors leading to adverse effects (English and Cobb, 2002; Noble et al., 2004).

The emergence of automated cell-based imaging assays has provided a powerful tool to interrogate the inhibition of kinases and signaling pathways in the cellular context (Almholt et al., 2004; Giuliano et al., 1997; Mitchison, 2005; Trask et al., 2006; Williams et al., 2006). Imaging assays may be configured to measure multiple kinase targets or signaling pathway readouts, thereby providing a selectivity profile (Almholt et al., 2004; Giuliano et al., 1997; Mitchison, 2005) Image analysis algorithms also provide multiple features and image-based parameters that provide information on cell morphology and cytotoxicity (Giuliano *et al.*, 1997). Thus, by mining "high-content" data it is possible to extract additional information on the effects of compound treatment of cells beyond the single target readout typical of most other assay formats (Giuliano *et al.*, 1997; Mitchison, 2005).

Williams *et al.* (2006) and Trask *et al.* (2006) described the development and implementation of a mitogen-activated protein kinase-activated protein kinase-2 (MK2) translocation assay to screen for inhibitors of the p38 mitogen-activated protein kinase (MAPK) signaling pathway. This chapter describes the development of two additional imaging assays to identify inhibitors of extracellular signal-regulated kinases (ERK) and stress-activated protein (c-jun N-terminal, JNK) mitogen-activated signaling pathways. Furthermore, it describes the implementation of these three independent MAPK signal transduction pathway assays to assess the hits from a p38 inhibitor MK2-EGFP translocation screen (Trask *et al.*, 2006) and to profile compounds arising from two lead optimization efforts focused on the construction of selective p38 and JNK inhibitors.

Definition of the Cell Model

Mitogen-activated protein kinases are members of the signaling cascades for diverse extracellular stimuli that regulate fundamental cellular processes, including embryogenesis, differentiation, mitosis, apoptosis, movement, and gene expression (Cowan and Storey, 2003; English and Cobb, 2002; Garrington and Johnson, 1999; Johnson and Lapadat, 2002). Four distinct MAP kinase families have been described: extracellular signal-regulated kinases (ERKs), c-jun N-terminal (JNK) or stress-activated protein kinases (SAPK), ERK5/big MAP kinase 1 (BMK1), and the p38 group of protein kinases (Cowan and Storey, 2003; Garrington and Johnson, 1999; Johnson and Lapadat, 2002; Ono and Han, 2000). MAP kinases phosphorylate other protein kinases, phospholipases, transcription factors, and cytoskeleton proteins (Cowan and Storey, 2003; Garrington and Johnson, 1999; Johnson and Lapadat, 2002). The core unit of a MAPK pathway is a three-member protein kinase cascade (Cowan and Storey, 2003; Garrington and Johnson, 1999; Johnson and Lapadat, 2002). MAP kinases are regulated by phosphorylation and serve as substrates for MAPK kinases (MKKs). The third component of the phospho-relay system are the MKKKs that phosphorylate and activate specific MKKs (Cowan and Storey, 2003; Garrington and Johnson, 1999; Johnson and Lapadat, 2002). Specificity of MAPK responses is achieved by the activation of distinct MKKK-MKK-MAPK signaling modules in response to different stimuli (Cowan and Storey, 2003; Garrington and Johnson, 1999; Johnson and Lapadat, 2002). The importance of MAPK signaling pathways as potential drug targets is indicated by the large number of patent applications that have been submitted by numerous pharmaceutical companies describing small molecule modulators of these pathways (English and Cobb, 2002; Fabbro *et al.*, 2002; Noble *et al.*, 2004; Regan *et al.*, 2002). A number of small molecule inhibitors of the MAPK signaling modules have exhibited efficacy in animal disease models, and several have advanced into human clinical trials for cancer and inflammation therapies (English and Cobb, 2002; Fabbro *et al.*, 2002; Noble *et al.*, 2004; Regan *et al.*, 2002).

The p38 (reactivating kinases, RKs, or p40) kinase module is known to mediate stress responses and is activated by heat shock, ultraviolet light, bacterial lipopolysaccharide, or the proinflammatory cytokines interleukin (IL)-1β or tumor necrosis factor (TNF)-α (Cowan and Storey, 2003; Garrington and Johnson, 1999; Ono and Han, 2000). Activation of the p38 pathway results in the phosphorylation of downstream kinases, transcription and initiation factors that affect cell division, apoptosis, and invasiveness of cultured cells and the inflammatory response (Cowan and Storey, 2003; Garrington and Johnson, 1999; Ono and Han, 2000). In addition to p38a (CSBP, MPK2, RK, Mxi2), there are three p38 homologues: p38β, p38γ (ERK6, SAPK3), and p38δ (SAPK4) (Ono and Han, 2000). p38α and p38β are expressed ubiquitously, whereas p38γ is expressed predominantly in skeletal muscle, and p38δ is enriched in lung, kidney, testis, pancreas, and small intestine (Ono and Han, 2000). Upstream kinases acting on p38 include MKK3 and MKK6, which in turn are activated by MLKs and ASK1 (MKKKs).

The ERK MAPK module responds primarily to growth factors and mitogens by stimulating transcriptional responses involved in cell division, migration, and survival (Cowan and Storey, 2003; Garrington and Johnson, 1999; Johnson and Lapadat, 2002). ERK1 and ERK2 are widely expressed and are activated upstream by MEK1 and MEK2, which are in turn activated by Raf (Cowan and Storey, 2003; Garrington and Johnson, 1999; Johnson and Lapadat, 2002).

The JNK MAPK module responds to a wide variety of stress signals, including heat shock, osmotic stress, proinflammatory cytokines, ischemia, and ultraviolet exposure (Cowan and Storey, 2003; Garrington and Johnson, 1999; Johnson and Lapadat, 2002). JNK1 and JNK2 are widely expressed in many tissues, whereas JNK3 is brain specific (Cowan and Storey, 2003; Garrington and Johnson, 1999; Johnson and Lapadat, 2002). JNK1 and JNK2 are activated upstream by MKK4 and MKK7, which are in turn activated by a variety of MKKKs: MEKKs 1–4, ASK1, and MLKs (Cowan and Storey, 2003; Garrington and Johnson, 1999; Johnson and Lapadat, 2002).

Thus, the MAPK signaling pathways are cascades composed of at least three kinases: MKKK-MKK-MAPK. This signaling cascade is exquisitely sensitive to regulatory input at multiple levels involving several mechanisms (Cowan and Storey, 2003; Garrington and Johnson, 1999; Johnson and Lapadat, 2002). On the basis of sequence homology and function, 12 members of the MAPK family have been identified together with 7 MKKs and 14 MKKKs (Garrington and Johnson, 1999; Johnson and Lapadat, 2002). The MKKKs are diverse in structure and have a variety of regulatory motifs not found in MKKs or MAPKs: pleckstrin homology domains, proline-rich sequences for binding Src-homlogy domains, leucine-zipper dimerization sequences, and binding sites for GTP-binding proteins (Garrington and Johnson, 1999). The diversity of regulatory domains in different MKKKs may provide the flexibility that allows the MAPK modules to respond to diverse stimuli. The MKK family has the fewest members in the MAPK module and has the highest specificity for their MAPK substrates (Garrington and Johnson, 1999). However, there is a considerable degree of cross talk between the MAPK modules that may be involved in modulating or fine-tuning the response to stimuli (Cowan and Storey, 2003; Garrington and Johnson, 1999). For example, each MKK can be phosphorylated by multiple MKKKs, and several MAPK substrates, such as Elk1 and MK3, can be phosphorylated by all three MAPK signaling pathways (Cowan and Storey, 2003; Garrington and Johnson, 1999).

Given the apparent redundancy within MAPK signaling modules and cross talk between pathways, how then may specificity be achieved? One level of specificity may be provided at the level of cell/tissue specific and/or developmental expression patterns of the different isoforms that comprise the MAPK modules (Cowan and Storey, 2003; Garrington and Johnson, 1999; Johnson and Lapadat, 2002). Kinase activity can also be regulated by subcellular location, thus specificity and regulation of MAPK pathways may therefore be provided by scaffolding or anchoring proteins that bring together specific kinases for the selective activation, sequestration, and localization of signaling complexes (Cowan and Storey, 2003; Garrington and Johnson, 1999; Johnson and Lapadat, 2002). Because MAPKs are regulated by phosphorylation, it therefore follows that phosphatases will be a key element of their control, and a family of dual specificity (Ser/Thr/Tyr) MAPK phosphatases (MKP1, MKP2, and MKP3) has been implicated (Cowan and Storey, 2003). Imaging assays provide a means to investigate kinase targets in the intracellular milieu where these regulatory mechanisms are intact.

Cellomics ArrayScan Automated Imaging Platform

The ArrayScan platform marketed by Cellomics (Fisher Scientific, Hampton, NH) is one of the most widely deployed high-content screening (HCS) systems. The studies described in this chapter and in Williams *et al.* (2006) and Trask *et al.* (2006) are performed on an ArrayScan II that has the software upgraded to a 3.1 version. The ArrayScan 3.1 houses a Zeiss Axiovert S100 inverted microscope outfitted with 5×/0.25 NA, 10×/0.3NA, and 20×/0.4 NA Zeiss objectives. Illumination is provided by a Xe/Hg arc lamp source (EXFO, Quebec, Canada), and fluorescence is detected by a 12-bit high sensitivity −20°-cooled CCD camera (Photometrics Quantix). The ArrayScan 3.1 provides the capability of imaging multiple fluorescent probes by acquiring wavelength channels sequentially in which each fluorophore is excited separately and then collected through single- or multiband-pass filter sets and detected on the chip of a monochromatic CCD camera. Channel selection is accomplished using a fast excitation filter wheel combined with single or multiband emission filter. The system comes with filter sets designed for the common fluorescent probes and can distinguish up to four labels in a single preparation with minimal cross talk between channels. The ArrayScan 3.1 is a wide-field imaging system that illuminates a "large" area of the specimen and directly images that area all at once. It uses an image-based autofocus system that images a fluorescent label in cells, typically fluorescently stained nuclei, but any feature of interest could be used, and an algorithm measures the relative sharpness of the image. It should be noted that any fluorescent particle could potentially be focused upon. The ArrayScan 3.1 was integrated with a Zymark Twister Robot and 80 plate stacker for fixed end point assays.

The cytoplasm-to-nuclear translocation algorithm developed by Cellomics may be used to quantify the relative distribution of a fluorescently tagged target between two cellular compartments, namely the cytoplasm and the nucleus (Giuliano *et al.*, 1997). Labeling with a nucleic acid dye such as Hoechst 33342, DAPI, DRAQ5, or other fluorophore identifies the nuclear region, and this signal is used to focus the instrument and to define a nuclear mask. The mask is eroded to reduce cytoplasmic contamination within the nuclear area, and the final reduced mask is used to quantify the amount of target channel fluorescence within the nucleus. The nuclear mask is then dilated to cover as much of the cytoplasmic region as possible without going outside the cell boundary. Removal of the original nuclear region from this dilated mask creates a ring mask that covers the cytoplasmic region outside the nuclear envelope. The "Cytonuc" difference measurement is calculated as the difference of the average nuclear intensity minus the average cytoplasmic ring intensity on a per cell basis or may be reported as an overall well average value (Giuliano *et al.*, 1997).

Development of the JNK MAPK Signaling Pathway Assay

To complement the p38 MAPK assay described elsewhere (Trask *et al.*, 2006; Williams *et al.*, 2006), we set out to develop a c-Jun activation assay leveraging a similar HCS approach. The c-Jun activation assay was developed in the HeLa adenocarcinoma cervical cell line (ATCC-CCL2) using a commercially available screening kit (Hitkit) purchased from Cellomics. To confirm that the SAPK signaling pathway is intact in the HeLa cell line, cells are either left untreated or are treated with 100 nM of the protein synthesis inhibitor anisomycin for 30 min. Cells are then solubilized in SDS sample buffer, their total cellular proteins are separated on 10% SDS-PAGE gels, transferred to nitrocellulose, and the resulting blots are probed with specific antibodies for the total and phosphorylated forms of JNK1, JNK2, and c-Jun (Fig. 1A). The total JNK1, JNK2, and c-Jun signals detected on the Western blots appear unaffected by anisomycin treatment. However, activation of the JNK/SAPK signaling pathway by anisomycin treatment dramatically increases the phosphorylation signals for JNK1, JNK2, and c-Jun relative to untreated controls (Fig. 1A).

The c-Jun activation kit from Cellomics utilizes a combination of a primary mouse monoclonal antibody to phospho-c-Jun (Ser 63) and an Alexa 488-conjugated goat anti-mouse IgG secondary antibody to measure the amount of phospho-c-Jun in cells by indirect immunofluorescence. HeLa cells (5×10^3) per well were seeded in EMEM + 10% fetal bovine serum (FBS) and 2 mM L-glutamine and were incubated overnight at 37° and 5% CO_2. Medium, with or without 100 nM anisomycin, was added and the plates were incubated at 37° and 5% CO_2 for 25 min. The cells were fixed with formaldehyde, permeabilized, incubated with the primary mouse antibody against phoshorylated c-Jun, washed, and then incubated with the goat anti-mouse secondary antibody conjugated with Alexa Fluor 488 containing Hoechst dye. The Cellomics ArrayScan 3.1 is used to capture images of the phospho-c-Jun in fields of view of untreated and anisomycin-treated HeLa cells (Fig. 1B). The level of phospho-c-Jun fluorescent signal is significantly brighter in anisomycin-treated cells than in untreated cells, and the majority of the staining of both populations appears localized within the nucleus. Labeling nuclei with Hoechst 33342 identifies the nuclear region, and this signal is used to focus the instrument and in defining a nuclear mask. The cytoplasm-to-nuclear translocation algorithm measures the relative distribution of the target, in this case c-Jun, between two cellular compartments, the cytoplasm and the nucleus. Bit maps of individual cells, from both nuclear and c-Jun channels for untreated and treated cells, are also shown (Fig. 1B). As determined by the algorithm, the nuclear masks are indicated in blue and the cytoplasmic ring masks are indicated in green. While the

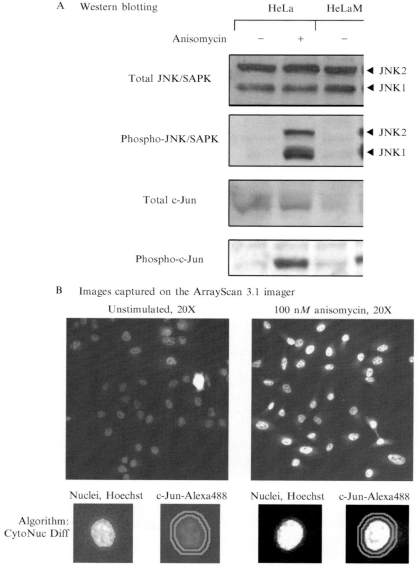

FIG. 1. Anisomycin-induced SAPK pathway activation: Western blotting and images captured on the ArrayScan 3.1 imager. (A) HeLa cells either were left untreated or were treated with 100 nM of the protein synthesis inhibitor anisomycin for 30 min. Cells were then solubilized in SDS sample buffer, their total cellular proteins were separated on 10% SDS-PAGE gels, and they were transferred to nitrocellulose, and the resulting blots were probed

nuclear staining appears equivalent in both cells, the phospho-c-Jun fluorescent signal is significantly brighter in the nuclear region of the anisomycin-treated cell than in the untreated cell. The "Cytonuc" difference (average nuclear intensity – average cytoplasmic ring intensity) for c-Jun staining will therefore be significantly higher in anisomycin-treated cells.

A series of experiments was conducted to optimize the c-Jun activation assay and to provide a robust and reproducible readout of the JNK/SAPK pathway activation (Figs. 2 and 3). To evaluate the c-Jun activation responses of HeLa cells, several stimuli were examined in dose–response experiments: IL-1β, platelet-derived growth factor, phorbol ester (PMA), and anisomycin (Fig. 2A). Even though anisomycin, IL-1, and PMA all produce dose-dependent c-Jun activation responses, anisomycin was selected for further assay development, as it exhibits the largest signal window. Cell density effects were examined as follows: 2.5, 5.0, and 10.0×10^3 HeLa cells per well are seeded in EMEM + 10% FBS and incubated overnight at $37°$ and 5% CO_2. Medium, with or without 100 nM anisomycin, was added and the plates were incubated at $37°$ and 5% CO_2 for 25 min. The ArrayScan imager and cytoplasm-to-nuclear translocation algorithm are able to adequately quantify the c-Jun activation response at all three seeding densities (Fig. 2B). Data presented are the means from at least 250 valid objects (cells), or 10 fields, whichever came first. A seeding density of 5.0×10^3 cells per well was selected for further assay development to reduce both the cell culture demands for the assay and the number of fields required to obtain a valid object count greater than 250 while maintaining good physical separation between cells. To identify the optimal stimulation time, HeLa cells were plated at 5.0×10^3 cells per well; after overnight culture, medium containing the indicated concentrations of anisomycin was added and plates were incubated for the indicated time points prior to fixation and staining (Fig. 2C). Although there is a measurable response to anisomycin

with specific antibodies for the total and phosphorylated forms of JNK1, JNK2, and c-Jun. (B) HeLa cells (5×10^3) per well were seeded in EMEM + 10% FBS and incubated overnight at $37°$ and 5% CO_2. Medium, with or without 100 nM anisomycin, was added and the plates were incubated at $37°$ and 5% CO_2 for 25 min. The cells were fixed with formaldehyde, permeabilized, incubated with the primary mouse antibody against phosphorylated c-Jun, washed, and then incubated with the goat antimouse secondary antibody conjugated with Alexa Fluor 488 and containing Hoechst dye. The Cellomics ArrayScan 3.1 was used to capture images of the phospho-c-Jun in fields of view of untreated and anisomycin-treated HeLa cells. The cytoplasm-to-nuclear translocation algorithm measures the relative distribution of the target, in this case c-Jun, between two cellular compartments, the cytoplasm and the nucleus. Bit maps of individual cells, from both nuclear and c-Jun channels for untreated and treated cells, are also shown; nuclear masks are indicated in blue and the cytoplasmic ring masks are indicated in green.

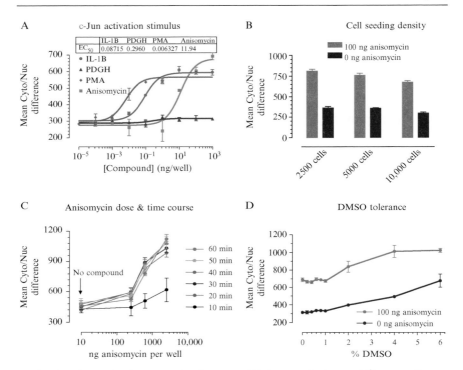

FIG. 2. c-Jun activation assay development. (A) HeLa cells (5.0×10^3) per well were seeded in EMEM + 10% FBS and incubated overnight at $37°$ and 5% CO_2. The c-Jun activation dose responses of Hela cells treated with the indicated doses of interleukin-1β (IL-1), platelet-derived growth factor (PDGH), or phorbol ester (PMA) were compared to anisomycin-induced responses quantified by the cytoplasm-to-nuclear translocation algorithm from images captured on the ArrayScan. (B) HeLa cells (2.5, 5.0, and 10.0×10^3) per well were seeded in EMEM + 10% FBS and incubated overnight at $37°$ and 5% CO_2. Medium, with or without 100 nM anisomycin, was added and the plates were incubated at $37°$ and 5% CO_2 for 25 mins. The c-Jun activation responses were quantified by the cytoplasm-to-nuclear translocation algorithm from images captured on the ArrayScan. (C) HeLa cells (5.0×10^3) per well were plated and, after overnight culture, medium containing the indicated concentrations of anisomycin was added and plates were incubated for the indicated time points prior to fixation and staining. The c-Jun activation responses were quantified by the cytoplasm-to-nuclear translocation algorithm from images captured on the ArrayScan. (D) HeLa cells (5.0×10^3) per well were plated and, after overnight culture, were treated for 30 min with medium or 100 ng/ml anisomycin at the indicated DMSO concentrations. The c-Jun activation responses were quantified by the cytoplasm-to-nuclear translocation algorithm from images captured on the ArrayScan.

treatment after 10 min of incubation, at least 20 min is required for a maximal response. The response appears stable between 20 and 60 min, and a 30-min incubation period was selected for further assay development. Finally, compound screening libraries are typically solubilized in dimethyl sulfoxide (DMSO), and we wanted to evaluate the DMSO tolerance of the c-Jun activation response in HeLa cells. HeLa cells were plated at 5.0×10^3 cells per well and after overnight culture, were treated for 30 min with media or 100ng/mL anisomycin at the indicated DMSO concentrations (Fig. 2D). The c-Jun activation response appears unaffected at DMSO concentrations $\leq 1.0\%$. However, at DMSO concentrations $>1.0\%$, the Cytonuc difference increases with the DMSO concentration, independently of the treatment conditions. We therefore decided that for the c-Jun activation assay, compounds would be diluted such that DMSO concentrations would not exceed 1.0%. Having completed these assay development experiments enabled us to finalize a c-Jun activation protocol.

c-Jun Activation Protocol

Materials

> HeLa cells (ATCC)
> Formaldehyde (Sigma F1268)
> EMEM (BioWhittaker) containing 10% FBS (JRH Biosciences), 2 mM L-glutamine (BioWhittaker), and 10 mM HEPES (Bio-Whittaker)
> Packard 96-well ViewPlates
> c-Jun Activation Hit Kit (Cellomics)

Buffers and Working Solutions

> 1× wash buffer: Add 20 ml of 10× wash buffer (from Hit Kit) to 160 ml with deionized water, adjust pH to 7.2, and bring to a final volume of 200 ml. May be stored for several days at 4°.
> Fixation solution: In a fume hood, add 2.2 ml 37% formaldehyde to 19.8 ml 1× wash buffer. Prepare fresh for each assay and warm to 37° before use.
> Permeabilization buffer: Add 4 ml of 10× permeabilization buffer (Hit Kit) to 30 ml deionized water, adjust pH to 7.2, and bring to a final volume of 40 ml. May be stored for several days at 4°.
> Detergent buffer: Add 8 ml of 10× detergent buffer (from Hit Kit) to 60 ml of deionized water, adjust pH to 7.2, and bring to a final volume of 80 ml. May be stored for several days at 4°.
> Primary antibody solution: Add 27.5 µl of c-Jun primary antibody (from Hit Kit) to 5.5 ml 1× wash buffer. Prepare fresh daily.

Staining solution: Add 55 μl of secondary antibody (from Hit Kit) and 2.75 μl Hoechst Dye (from Hit Kit) to 5.5 ml 1× wash buffer. Prepare fresh daily.

Protocol

1. Culture HeLa cells to approximately 80% confluency in EMEM plus 2 mM glutamine and 10% FBS. (Split 1:5 every 3 to 4 days.)
2. Harvest cells with 0.05% trypsin-EDTA, adjust the cell number to 5×10^4 cells/ml in media, and plate 100 μl per well (5000 cells per well).
3. Incubate plates overnight at 37° and 5% CO_2.
4. Add 20 μl of medium containing 7× compounds and controls in 7% DMSO.
5. Add 20 μl of medium containing 5 μg/ml anisomycin.
6. Incubate 30 min at 37° and 5% CO_2.
7. Aspirate culture medium and add 200 μl of prewarmed fixation solution.
8. Incubate in fume hood for 10 min at room temperature (all remaining steps are performed at room temperature).
9. Aspirate fixation solution and add 200 μl of 1× wash buffer.
10. Aspirate wash buffer and add 200 μl of permeabilization buffer and incubate 90 s.
11. Aspirate permeabilization buffer and add 200 μl of 1× wash buffer.
12. Aspirate wash buffer, add 50 μl of primary antibody solution, and incubate 1 h.
13. Aspirate primary antibody solution, add 200 μl detergent buffer, and incubate 5 min.
14. Aspirate detergent buffer and add 200 μl of 1× wash buffer; perform procedure twice.
15. Add 50 μl of staining solution and incubate 1 h.
16. Aspirate staining solution, add 200 μl of detergent buffer, and incubate 5 min.
17. Aspirate detergent buffer and add 200 μl of 1× wash buffer; perform procedure twice.
18. Add 200 μl of wash buffer, seal plates, and read on the ArrayScan.

c-Jun Activation Signal Window and Reproducibility

To evaluate the reproducibility of the c-Jun activation response in HeLa cells, cells were plated at 5.0×10^3 cells per well and on the following day are treated with the indicated doses of anisomycin in three independent experiments run on separate days (Fig. 3A). The EC_{50} for anisomycin-induced c-Jun activation ranged from 52 to 114 nM and produced on average an EC_{50} of 87 ± 32 nm, indicating a reproducible assay. To evaluate the ability of the c-Jun activation response in HeLa cells to identify SAPK pathway inhibitors, cells were plated and on the next day the indicated doses of the JNK 1/2 inhibitor SP 600125 were added simultaneously with 100 ng/ml anisomycin (final) and cells were incubated for 30 min (Fig. 3B). In three independent experiments, the IC_{50} for the JNK 1/2 inhibitor SP 600125 ranged from 3.0 to 5.7 μM and produced on average an IC_{50} of 4.77 ± 1.51 μM. To further evaluate the robustness and reproducibility of the c-Jun activation response in HeLa cells, two full 96-well plates were treated under the following conditions, one plate per condition: medium alone (blue squares) and 100 ng/ml anisomycin (green circles) (Fig. 3C). In addition to the population scatter plot, plate heat map views of the two plates are also presented (Fig. 3D). The average Cytonuc difference for the medium control plate was 356, and the average Cytonuc difference for the anisomycin-treated control plate was 1580, producing an average assay signal to background window of 4.4-fold. A comparison of medium control plate variability data to anisomycin-treated plate variability data produced a Z factor of 0.77, indicating that the assay would definitely be compatible with HTS (Zhang et al., 1999).

Development of the ERK MAPK Signaling Pathway Assay

To complement our p38 and JNK MAPK signaling pathway HCS assays, an ERK activation assay was developed in the HeLa adenocarcinoma cervical cell line (ATCC-CCL2) using antiphospho-p44/42 MAPK (ERK1/2) antibodies. To confirm that the ERK signaling pathway is intact in the HeLa cell line, cells were left untreated or were treated with either 10 ng/ml oncostatin M (OSM) or 50 ng/ml PMA for 6 min. Cells were then solubilized in SDS sample buffer, their cellular proteins separated on 10% SDS-PAGE gels, transferred to nitrocellulose, and the resulting blot was probed with specific antibodies for the total and phosphorylated forms of ERK1 and ERK2 (Fig. 4A). The total ERK1 and ERK2 signals detected on the Western blots appear unaffected by OSM and PMA treatment. However, activation of the ERK signaling pathway by both OSM and PMA treatment dramatically increases the phosphorylation signals for ERK1 and ERK2 relative to untreated controls (Fig. 4A).

The ERK activation imaging assay utilizes a combination of a primary polyclonal rabbit anti-phospho-p44/42 MAPK (Thr202/Try204) (ERK1/2) antibody purchased from Cell Signaling Technologies and an Alexa 488-conjugated goat antirabbit IgG secondary antibody from Molecular Probes to measure the amount of phospho-ERK1/2 in cells by indirect immunofluorescence (Fig. 4B). HeLa cells (5×10^3) per well were seeded in DMEM

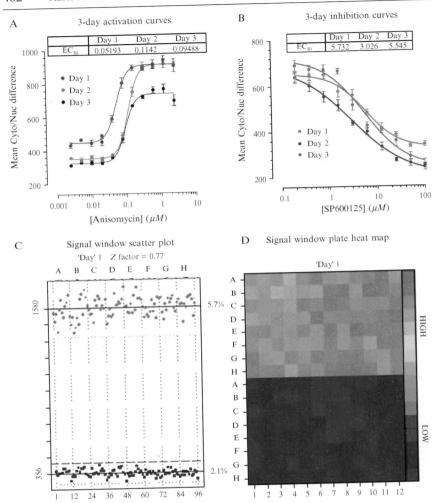

Fig. 3. c-Jun activation reproducibility and variability. (A) HeLa cells (5.0×10^3) per well were plated and on the following day were treated with the indicated doses of anisomycin in three independent experiments run on separate days. (B) HeLa cells (5.0×10^3) per well were plated and, on the next day, the indicated doses of the JNK1/2 inhibitor SP 600125 were added simultaneously with 100 ng/ml anisomycin (final) and cells were incubated for 30 min. Three independent inhibition experiments were run on separate days. (C) HeLa cells (5.0×10^3) per well were plated and, on the next day, two full 96-well plates were treated under the following conditions, one plate per condition: media alone (blue squares) and 100 ng/ml anisomycin (green circles). The Z factor was calculated according to the method of Zhang *et al.* (1999). (D) A plate heat map view of the two plates is also presented. The c-Jun activation responses were quantified by the cytoplasm-to-nuclear translocation algorithm from images captured on the ArrayScan.

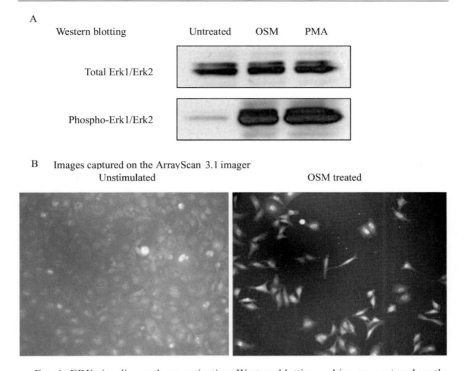

FIG. 4. ERK signaling pathway activation: Western blotting and images captured on the ArrayScan 3.1 imager. (A) HeLa cells were left untreated or were treated with either 10 ng/ml oncostatin M (OSM) or 50 ng/ml phorbol ester (PMA) for 6.0 min. Cells were then solubilized in SDS sample buffer, their total cellular proteins were separated on 10% SDS-PAGE gels, and they were transferred to nitrocellulose, and the resulting blots were probed with specific antibodies for the total and phosphorylated forms of ERK1 and ERK2 . (B) HeLa cells (5 × 10^3) per well were seeded in DMEM + 0.5% FBS and incubated overnight at 37° and 5% CO_2. Medium, with or without 10 ng/ml OSM, was added, and the plates were incubated at 37° and 5% CO_2 for 10 min. Cells were fixed with formaldehyde, permeabilized, incubated with the primary polyclonal rabbit antiphospho-p44/42 MAPK antibody, washed, and then incubated with the goat antirabbit secondary antibody conjugated with Alexa Fluor 488 and containing Hoechst dye. The Cellomics ArrayScan 3.1 was used to capture images of the phospho-ERK1/2 in fields of view of untreated and OSM-treated HeLa cells.

+ 0.5% FBS and were incubated overnight at 37° and 5% CO_2. Medium, with or without 10 ng/ml OSM, was added and the plates were incubated at 37° and 5% CO_2 for 10 min. The cells were fixed with formaldehyde, permeabilized, incubated with the primary polyclonal rabbit anti-phospho-p44/42 MAPK antibody, washed, and then incubated with the goat anti-rabbit secondary antibody conjugated with Alexa Fluor 488 and containing Hoechst dye. The Cellomics ArrayScan 3.1 was used to capture images of the phospho-ERK1/2 in fields of view of untreated and OSM-treated HeLa cells (Fig. 4B). In untreated HeLa cells, the phospho-ERK1/2 fluorescent signal is relatively dim and appears to be localized predominantly in the cytoplasm. In contrast, the level of phospho-ERK1/2 fluorescent signal is significantly brighter in OSM-treated cells, and the majority of the staining appears localized within the nucleus. The cytoplasm-to-nuclear translocation algorithm measures the relative distribution of the target, in this case phospho-ERK1/2, between two cellular compartments, the cytoplasm and the nucleus. The "Cytonuc" difference (average nuclear intensity - average cytoplasmic ring intensity) for phospho-ERK1/2 staining will therefore be significantly higher in OSM-treated cells.

A series of assay development experiments was conducted to optimize the ERK activation assay and provide a robust and reproducible readout of the ERK pathway activation (Figs. 5 and 6). Because the ERK MAPK module responds to a variety of growth factors and mitogens, a number of different stimuli were tested in the ERK1/2 activation assay: EGF, TNFa, PMA, OSM, and TGFα (Fig. 5A). EGF, PMA, OSM, and TGFα all induced robust dose-dependent ERK1/2 activation responses as quantified by the cytoplasm-to-nuclear translocation algorithm from images captured on the ArrayScan (Fig. 5A). TNFα treatment, however, only produced a weak activation of phospho-Erk1/2 when measured on the ArrayScan. OSM and PMA are the stimuli selected for further assay development. To evaluate the effects of seeding density, HeLa cells were seeded at a variety of densities between 2 and 10×10^3 cells per well in DMEM + 0.5% FBS and are incubated overnight at 37° and 5% CO_2. Medium, with or without the indicated doses of PMA, was added and the plates were incubated at 37° and 5% CO_2 for 30 min (Fig. 5B). The ArrayScan imager and cytoplasm-to-nuclear translocation algorithm are able to adequately quantify the ERK activation response at all six seeding densities (Fig. 5B). Data presented are the means from at least 250 valid objects (cells), or 10 fields, whichever came first. A seeding density of 5.0×10^3 cells per well was selected for further assay development to reduce both the cell culture demands for the assay and the number of fields required to obtain a valid object count ≥ 250. To identify the optimal stimulation time, HeLa cells were plated 5.0×10^3 cells per well and, after overnight culture, medium containing either 10 ng/ml OSM or 50 ng/ml PMA was added and plates were incubated for the indicated time points prior to fixation and staining (Fig. 5C). The ERK activation induced by PMA or OSM treatment is relatively rapid and reaches a maximal response at 5 and 10 min, respectively (Fig. 5C). With OSM treatment the maximal activation of ERK1/2 appears stable between 10 and 15 min post-treatment and then declines to pre-stimulation levels in a roughly linear fashion over the next 45 min. For

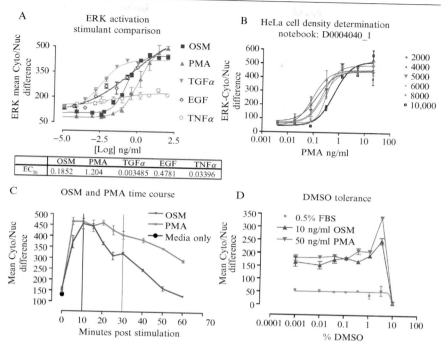

FIG. 5. ERK activation assay development. (A) HeLa cells (5.0×10^3) per well were plated and, after overnight culture, were treated with the indicated doses of EGF, TNFα, PMA, OSM, or TGFα. (B) HeLa cells were seeded at a variety of densities between 2 and 10×10^3 cells per well in DMEM + 0.5% FBS and incubated overnight at 37° and 5% CO_2. Medium, with or without the indicated doses of PMA, was added, and the plates were incubated at 37° and 5% CO_2 for 30 min. (C) HeLa cells (5.0×10^3) per well were plated and, after overnight culture, medium containing either 10 ng/ml OSM or 50 ng/ml PMA was added and plates were incubated for the indicated time points prior to fixation and staining. (D) HeLa cells (5.0×10^3) per well were plated and, after overnight culture, were treated with either 10 ng/ml OSM or 50 ng/ml PMA for 10 min at the indicated DMSO concentrations. Cells were fixed with formaldehyde, permeabilized, incubated with the primary polyclonal rabbit anti-phospho-p44/42 MAPK antibody, washed, and then incubated with the goat anti-rabbit secondary antibody conjugated with Alexa Fluor 488 and containing Hoechst dye. The Cellomics ArrayScan 3.1 was used to capture images of the phospho-ERK1/2, and ERK activation responses were quantified by the cytoplasm-to-nuclear translocation algorithm.

PMA treatment, the maximal activation of ERK1/2 appears stable between 5 and 20 min posttreatment and then declines to ~50% of prestimulation levels in a roughly linear fashion over the next 40 min. A treatment period of 10 min was selected for further assay development with both OSM and PMA. Finally, because compound libraries are typically solubilized in DMSO, we wanted to evaluate the DMSO tolerance of the ERK activation response. HeLa cells were plated 5.0×10^3 cells per well and, after overnight culture, were treated with either 10 ng/ml OSM or 50 ng/ml PMA for 10 min at the indicated DMSO concentrations (Fig. 5D). The ERK activation response appears unaffected at DMSO concentrations $\leq 1.0\%$. At 5% DMSO there is an apparent increase in the Cytonuc difference for both OSM and PMA, but at 10% DMSO there are so few cells left in the wells that no values can be measured. We therefore decided that for the ERK activation assay, compounds would be diluted such that DMSO concentrations would not exceed 0.5%. Based on the results of these assay development experiments we were able to compose an ERK activation protocol.

ERK1/2 Activation Protocol

Materials

> HeLa cells (ATCC)
> Formaldehyde (Sigma F1268)
> DMEM (BioWhittaker) containing 10% FBS (JRH Biosciences), 2 mM L-glutamine (BioWhittaker), 10,000 units penicillin/streptomycin (BioWhittaker), and 10 mM HEPES (BioWhittaker)
> Dulbecco's PBS Mg^{2+} and Ca^{2+} free (BioWhittaker)
> Packard 96-well View Plates
> Phospho-p44/42 map kinase (Thr202/Try204) antibody (Cell Signaling)
> Alexa 488 goat anti-rabbit IgG (Molecular Probes)
> Hoechst 33342 (Sigma)
> Triton X-100 (Roche)
> Tween-20 (Roche)
> Phorbol Ester (PMA) (Sigma)
> Oncostatin M (OSM) (R&D)

Buffers and Working Solutions

> Wash buffer: Dulbecco's PBS Mg^{2+} and Ca^{2+} free
> Permeabilization buffer: PBS/Triton X-100, 0.5%
> Blocking Buffer: PBS/Tween 20, 0.1%
> Formaldehyde, 3.7%: Add 1.2 ml 37% formaldehyde to 10.8 ml wash buffer. Keep warmed to 37° until ready for use. (Prepare formaldehyde solution in a fume hood.)
> Primary antibody solution: 1 μg/ml (final) phospho-p44/42 map kinase antibody in PBS
> Secondary antibody solution: 10 μg/ml (final) Alexa 488 goat anti-rabbit IgG and 2 μg/ml (final) Hoechst 33342 dye in PBS

Protocol

1. Harvest HeLa cells and seed 5000 cells per well into 96-well plate(s) in medium.
2. Culture cells at $37°$, 5% CO_2 overnight.
3. Add 25 μl of compound (inhibitor) to the wells and incubate for 15 min at $37°$, 5% CO_2.
4. Add 25 μl of 10 ng/ml OSM activation stimulus to the wells and incubate for 10 min at $37°$, 5% CO_2.
5. Aspirate medium and add 100 μl of prewarmed fixation solution (3.7% formalin) and incubate for 10 min at room temperature.
6. Wash once with 100 μl wash buffer, aspirate wash buffer, add 100 μl of permeabilization buffer, and incubate for 90 s.
7. Aspirate permeabilization buffer and add 100 μl wash buffer.
8. Aspirate buffer and add 50 μl of primary antibody; 1 μg/ml (final) phospho-p44/42 MAP kinase (Thr202/Try204) antibody, and incubate for 1 h at room temperature.
9. Aspirate primary antibody, wash once with blocking buffer, and incubate for 15 min at room temperature.
10. Aspirate and wash once with 100 μl PBS and aspirate.
11. Add 50 μl secondary antibody and stain; 10 μg/ml (final) Alexa 488 goat anti-rabbit IgG and 2 μg/ml (final) Hoechst 33342 dye in PBS and incubate for 1 h at room temperature in the dark.
12. Aspirate and wash once with 100 μl blocking buffer and incubate for 10 min at room temperature.
13. Aspirate buffer and wash once with 100 μl wash buffer.
14. Aspirate buffer, add 100 μl of wash buffer, and seal plate.
15. Read plate on the ArrayScan.

ERK1/2 Activation Signal Window and Reproducibility

To evaluate the reproducibility of the ERK activation response in HeLa cells, cells were plated and on the following day are treated with the indicated doses of OSM or PMA in three independent experiments run on separate days (Fig. 6A and B). The EC_{50} for OSM-induced ERK activation ranges from 0.1 to 0.22 ng/ml and produces on average an EC_{50} of 0.169 ± 0.062 ng/ml, indicating a reproducible assay (Fig. 6A). The EC_{50} for PMA-induced ERK activation ranges from 0.81 to 0.95 ng/ml and produces, on average, an EC_{50} of 0.884 ± 0.068 ng/ml, indicating a reproducible assay (Fig. 6B). To evaluate the ability of the ERK activation response in HeLa cells to identify ERK pathway inhibitors, cells were plated and on the next day the indicated doses of the MEK1/2 inhibitor U0126 were added 15 min prior to the addition of 10 ng/ml OSM and cells were incubated for an additional 10 min (Fig. 6C). In three independent experiments, the IC_{50} for the MEK1/2 inhibitor U0126 ranged from 0.489 to 0.711 μM and produced, on average, an IC_{50} of 0.616 $\pm$ 0.115 μM. To further evaluate the robustness and reproducibility of the ERK activation response in HeLa cells, three full 96-well plates were treated under the following conditions, one plate per condition: medium alone (green circles), 10 ng/ml OSM (red triangles), and 10 ng/ml OSM + 10 μM staurosporin (blue squares)(Fig. 6D). The average Cytonuc difference for the media control plate was 71, and the average Cytonuc difference for the anisomycin-treated control plate was 177, producing an average assay signal-to-background window of 2.49-fold. A comparison of medium control plate variability data to anisomycin-treated plate variability data produced a Z factor of 0.54, indicating that the assay would be compatible with HTS (Zhang et al., 1999).

MAPK Pathway Inhibitor Test Cassette

To evaluate the three independent MAPK signaling pathway imaging assays that we had developed and to determine their ability to discriminate selectivity profiles across the MAPK family, we assembled a test cassette of known MAPK pathway inhibitors (Table I). In addition to five "selective" p38 inhibitors, we included the JNK-1/2 inhibitor SP 600125, the MEK-1/2 inhibitors UO126 and PD 98059, the PI3-kinase inhibitor wortmannin, and staurosporin as a nonselective kinase inhibitor. We also selected two activation stimuli for each MAPK pathway assay: anisomycin and TNFα for the p38 pathway, anisomycin and IL-1β for the JNK pathway, and OSM and PMA for the ERK pathway (Table I).

All five p38 inhibitors produced IC_{50} values for inhibition of anisomycin- and TNFα-induced MK2-EGFP translocation <100 nM. UO126 and PD 98059, the MEK-1/2 inhibitors, produced IC_{50} values in the 20 to 39 μM range for both activating stimuli, whereas the nonselective kinase inhibitor staurosporin produced submicromolar IC_{50} values for both stimuli (Table I). Neither the PI3-kinase inhibitor wortmannin nor the JNK-1/2 inhibitor SP 600125 inhibited anisomycin- or TNFα-induced MK2-EGFP translocation. Overall, the MK2-EGFP translocation assay appears exquisitely sensitive to p38 inhibitors relative to other kinase inhibitors.

In the c-Jun activation assay, the JNK-1/2 inhibitor SP 600125 and staurosporin produced IC_{50} values of 8.6 and 1.4 μM, respectively, with IL-1ß as the stimulus , and IC_{50} values of 14 and 7.2 μM, respectively, with anisomycin as the stimulus (Table I). Three of the five "selective" p38 inhibitors, the Merck p38 inhibitor, RWJ 68354, and SB203580, produced IC_{50} values between 20 and 40 μM with anisomycin as the activating stimulus, whereas only the Merck

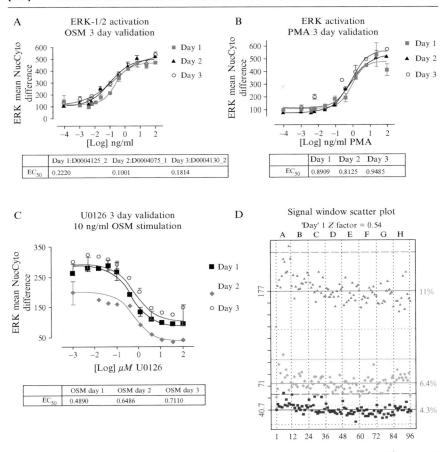

FIG. 6. ERK activation reproducibility and variability. (A) HeLa cells (5.0×10^3) per well were plated and, after overnight culture, were treated with the indicated doses of OSM in three independent experiments run on separate days. (B) HeLa cells (5.0×10^3) per well were plated and, after overnight culture, were treated with the indicated doses of PMA in three independent experiments run on separate days. (C) HeLa cells (5.0×10^3) per well were plated and on the next day the indicated doses of the MEK1/2 inhibitor U0126 were added 15 min prior to the addition of 10 ng/ml OSM and cells were incubated for an additional 10 min. Three independent experiments were run on separate days. (D) HeLa cells (5.0×10^3) per well were plated and, after overnight culture, three full 96-well plates were treated under the following conditions, one plate per condition: medium alone (green circles), 10 ng/ml OSM (red triangles), and 10 ng/ml OSM + 10 μM staurosporin (blue squares). The Z factor was calculated according to the method of Zhang *et al.* (1999). Cells were fixed with formaldehyde, permeabilized, incubated with the primary polyclonal rabbit anti-phospho-p44/42 MAPK antibody, washed, and then incubated with the goat anti-rabbit secondary antibody conjugated with Alexa Fluor 488 and containing Hoechst dye. The Cellomics ArrayScan 3.1 was used to capture images of the phospho-ERK1/2, and the ERK activation responses were quantified by the cytoplasm-to-nuclear translocation algorithm.

TABLE I

PROFILING OF A SELECTED KINASE INHIBITOR TEST CASSETTE IN DISTINCT MAPK SIGNALING PATHWAY IMAGING ASSAYS[a]

Test cassette of commercially available kinase inhibitors

| | MK2-GFP translocation | | | | cJun activation | | | | ERK activation | | | |
| | Anisomycin | | TNFα | | Anisomycin | | IL-1β | | OSM | | PMA | |
Compound	% Inhib at 50 μM	IC$_{50}$ μM	% Inhib at 50 μM	IC$_{50}$ μM	% Inhib at 50 μM	IC$_{50}$ μM	% Inhib at 50 μM	IC$_{50}$ μM	% Inhib at 50 μM	IC$_{50}$ μM	% Inhib at 50 μM	IC$_{50}$ μM
SP600125	29.5	>50	36.5	>50	95.5	14	107	8.6	15	>50	-52	>50
VX-745	146	0.045	119	0.050	22.5	>50	-10	>50	0	>50	-36	>50
Merck p38 inhib	-30.5	0.007	-99.5	0.011	214	19	235.5	21	117	10.95	117.5	12.5
RWJ 68354	122	0.087	97	0.074	105	20.5	48	38.5	-0.5	>50	12	>50
SB242235	96.5	0.028	81	0.103	41	>50	12	>50	-7.5	>50	5	>50
SB203580	64	0.053	16	0.220	59.5	41	-10	>50	37.5	>50	75.5	26
PD98059	58.5	39.000	78	24.500	26.5	>50	-8.5	>50	44	>50	-0.5	>50
U0126 # 9903	64	33.000	67.5	28.000	26.5	>50	26	44	96.5	1.275	94	6.95
Wortmannin	33.5	>50	38.5	>50	-57.5	>50	-123.5	>50	-25.5	>50	-73	>50
Staurosporin	180.5	0.755	282	0.505	173	7.2	196.5	1.4	85	0.003	83	0.003

[a] Data represent the mean of duplicate independent determinations.

p38 inhibitor and RWJ 68354 inhibited c-Jun activation by IL-1ß. Two of the five "selective" p38 inhibitors, VX-745 and SB242235, did not inhibit c-Jun activation by either stimulus. Neither of the MEK-1/2 inhibitors UO126 and PD 98059 nor the PI3-kinase inhibitor wortmannin inhibited c-Jun activation by anisomycin, although U0126 inhibited IL-1ß-induced activation (Table I). Clearly the c-Jun activation assay was inhibited by the JNK-1/2 kinase inhibitor; however, it was also susceptible to three of the five p38 inhibitors, albeit at >100-fold less sensitivity that the MK2-EGFP translocation p38 MAPK pathway assay.

The MEK-1/2 inhibitor U0126 produced IC_{50} values in the 1 to 7 μM range for both ERK activation stimuli, whereas the nonselective kinase inhibitor staurosporin produced nanomolar IC_{50} values for both stimuli, and the Merck p38 inhibitor produced IC_{50} values in the 11 to 12 μM range for both stimuli (Table II). SB203580 produced an IC_{50} of 26 μM with PMA as the activating stimulus, but failed to inhibit OSM-induced ERK activation. None of the other compounds inhibited ERK activation by either stimulus: the MEK-1/2 inhibitor PD 98059, three of the "selective" p38 inhibitors, VX-745, RWJ 68354, and SB242235, the PI3-kinase inhibitor wortmannin, and the JNK-1/2 inhibitor SP 600125 (Table I). The ERK activation assay was the least susceptible to inhibition by compounds that are reportedly selective for the other MAPK signaling pathways (Table I).

Staurosporin is a potent inhibitor of numerous kinases *in vitro* that is active in many cell-based assays. The ability of staurosporin to inhibit all three MAPK signaling pathways assays (Table I) is likely due to its ability to bind to the ATP-binding sites of multiple kinases in these pathways. The Merck p38 inhibitor also produced IC_{50} values in all three assays, but the ability to inhibit the JNK and ERK signaling pathways (Table I) was likely due to the apparent cytotoxicity, or significantly reduced adherence of the HeLa cells, observed at the two highest concentrations of the compound (Williams *et al.*, 2006). Images revealed that there were much fewer cells in the fields of view captured from wells treated with 50 and 16.6 μM of the Merck p38 inhibitor compared to the 5.5 μM dose, which produced obvious outliers in several cell number and morphology parameters derived by the image analysis algorithm (Williams *et al.*, 2006). The high-content nature of image-derived data therefore provides additional information that may be a consequence of target inhibition or may be due to off-target effects of the compound. The Merck p38 inhibitor produced an $IC_{50} \sim$ 7–10 nM in the MK2-EGFP translocation p38 MAPK pathway assay and IC_{50} values of 20 and 10 μM in the JNK and ERK signaling pathway assays, respectively. A number of the p38 inhibitors, including the Merck p38 inhibitor, have been withdrawn from clinical trials because of adverse toxicity profiles (English and Cobb, 2002; Fabbro *et al.*, 2002; Noble *et al.*, 2004; Regan *et al.*, 2002).

TABLE II

PROFILING OF SELECTED MK2-EGFP TRANSLOCATION INHIBITOR HIT COMPOUNDS IN DISTINCT MAPK SIGNALING PATHWAY IMAGING ASSAYS[a]

	MK2-EGFP translocation MTS hit assessment					
	MK2-GFP		cJun activation		ERK activation	
Compound	Mean % stimu at 50 μM	Mean IC_{50} μM	Mean % inhib at 50 μM	Mean IC_{50} μM	Mean % inhib at 50 μM	Mean IC_{50} μM
1	125.5	0.119	125	2	−7	>50
2	152.5	0.711	31	>50	5	>50
3	129	0.746	60	18	24	>50
4	140	0.780	30	8	−13	>50
5	109	1.028	117	4	−3	>50
6	96.5	1.331	53	40	20	>50
7	87.5	1.782	131	1	−10	>50
8	207.5	4.589	11	40	−86	>50
9	128.5	7.037	113	38	25	>50
10	110.5	14.123	9	>50	36	>50
11	50	14.916	−12	>50	−8	>50
12	10.5	>50	−21	>50	5	>50
13	12.5	>50	−2	>50	−8	>50
14	23	>50	−2	>50	−24	>50
15	16.5	>50	−9	>50	16	>50
16	15	>50	−26	>50	−11	>50

[a] Data represent the mean of duplicate independent determinations.

In contrast to staurosporin and the Merck p38 MAPK inhibitor, the PI3-kinase inhibitor wortmannin did not inhibit any of the three MAPK signaling pathway assays (Table I), and the JNK-1/2 inhibitor SP 600125 only inhibited the c-Jun activation assay, perhaps indicating some degree of MAPK pathway selectivity (Table I). The MEK-1/2 inhibitor UO126 produced IC_{50} values ranging between 1.2 and 6.9 μM in the ERK activation assay and an IC_{50} of 44 μM in the IL-1β-induced c-Jun activation assay. In contrast, the MEK-1/2 inhibitor PD 98059 did not inhibit either the c-Jun or the ERK activation assay. The MEK-1/2 inhibitors UO126 and PD 98059 produced IC_{50} values ranging between 25 and 39 μM in the MK2-EGFP translocation p38 MAPK assay, perhaps indicating that ERK1/2 may be capable of phosphorylating MK2 as well as MK1 or perhaps reflecting the cross talk between the different modules of MAPK signaling pathways (Cowan and Storey, 2003; Garrington and Johnson, 1999). All five p38 inhibitors produced IC_{50} values for inhibition of anisomycin- and TNFα-induced MK2-EGFP translocation <200 nM (Table I). As discussed earlier, the Merck p38 inhibitor also inhibited the c-Jun and ERK activation assays with IC_{50} values of 20 and 10 μM, respectively, likely due to the apparent cytotoxicity at these concentrations interfering with the image analysis algorithm. The RWJ 68354 p38 inhibitor inhibited anisomycin- and TNFα-induced MK2-EGFP translocation with IC_{50} values between 74 and 87 nM and the c-Jun activation assay with IC_{50} values between 20 and 38 μM. The SB203580 p38 inhibitor inhibited anisomycin- and TNFα-induced MK2-EGFP translocation with IC_{50} values between 53 and 220 nM and the c-Jun and ERK activation assays with IC_{50} values of 41 and 26 μM, respectively. In general, therefore, selective MAPK inhibitors exhibited pathway selectivity by producing significantly lower IC_{50} values for their respective ERK, JNK, or p38 signaling pathway assays. However, each of the MAPK pathways assays was also susceptible to nonselective kinase inhibitors such as staurosporin and inhibitors that inhibit upstream MKKKs and MKKs in the MAPK signaling pathway, especially those involved in the cross talk between different modules of the MAPK signaling pathways.

p38 Inhibitor Hit Assessment

Williams *et al.* (2006) and Trask *et al.* (2006) described the development and optimization of the MK2-EGFP translocation assay and the subsequent use of this assay to screen a 32K kinase biased library to identify p38 MAPK inhibitors. As part of the follow-up hit assessment for the screen, selected hits and related structural analogs were profiled in the three MAPK signaling pathway assays described here (Table II). Eleven of the 16 hits and related analogs produced an IC_{50} in the anisomycin-induced MK2-EGFP translocation assay: 4 in the submicromolar range, 5 in the 1 to 10 μM range, and 2 at $\sim$14 μM (Table II). Five of the 16 hits and related analogs produced an IC_{50} in the anisomycin-induced c-Jun activation assay: 3 in the 1 to 10 μM range, 1 at $\sim$18 μM, and 1 at 38 μM (Table II). None of the 16 hits and related analogs produced an IC_{50} in the OSM-induced ERK activation assay (Table II). Most of the hits produced significantly lower IC_{50} values in the MK2-EGFP translocation assay relative to the c-Jun activation assay, and 3 of the hit scaffolds identified in the MK2-EGFP translocation HCS were selected for p38a inhibitor hit-to-lead chemistry structure activity relationship (SAR).

p38a Inhibitor Profiling

To further evaluate the ability of the three MAPK signaling pathway assays to discriminate activity across these pathways, we compiled a set of

40 compounds directed at p38 inhibition. All 40 compounds produced IC_{50} values in the p38 cell-based MK2-EGFP translocation assay: 29 in the submicromolar range, 10 in the 1 to 5 μM range, and 1 at ~21.7 μM (Table III). Three of the 40 tested produced IC_{50} values in the cell-based ERK activation assay: 3 in the 30 to 40 μM range (Table III), suggesting little or no interference with the ERK MAPK pathway. In contrast, 22 of the 40 p38 inhibitors also inhibited the c-Jun activation assay, albeit with significantly less potency than in the MK2-EGFP translocation assay: 1 in the submicromolar range, 5 in the 1 to 10 μM range, and 16 in the 10 to 40 μM range (Table III). Although most of the compounds exhibited low nanomolar potency against p38 isoforms *in vitro*, a number of these compounds also exhibited submicromolar or low nanomolar potency against JNK isoforms *in vitro* (data not shown). In the majority of cases, however, despite exhibiting low to mid-nanomolar potency in the MK2-EGFP translocation assay, these mixed p38-JNK inhibitors typically only produced 5 to 30 μM potencies in the c-Jun activation assay. Because all three MAPK pathway assays were performed in the HeLa cell background, it is unlikely that differential cell permeability is contributing to the apparent differences in cellular potencies between mixed p38-JNK inhibitors with similar *in vitro* potencies. In general, the apparent MAPK selectivity profile from the *in vitro* kinase assays was reflected in the cell-based MAPK assays; however, the apparent p38 selectivity of the compounds was typically more pronounced in the cell-based assays.

JNK Inhibitor Profiling

To further characterize the selectivity of the three MAPK signaling pathway assays, a set of 40 JNK inhibitors was assembled (Table IV). Twenty-three of the 40 compounds produced IC_{50} values in the c-Jun activation cell-based assay: 4 in the sub-micromolar range, 3 in the 1 to 10 μM range, and 16 in the 10 to 40 μM range (Table IV). However, despite exhibiting potencies in the mid-nanomolar or sub-micromolar IC_{50} values in the JNK *in vitro* kinase assays (data not shown), with only four exceptions the compounds typically produced cellular activity in the 5 to 40 μM range. Fifteen of the 40 tested also produced IC_{50} values in the cell-based ERK activation assay: 3 in the 1 to 10 μM range and 12 in the 10 to 40 μM range (Table IV). The ERK MAPK module responds primarily to growth factors and mitogens and is involved in the regulation of cell division, migration, and survival (Cowan and Storey, 2003; Garrington and Johnson, 1999; Johnson and Lapadat, 2002), and inhibition of this pathway may have an adverse effect that could impact the therapeutic index of these JNK inhibitors. Conversely, there appears to be relatively little interference of the p38 MAPK signaling

TABLE III
PROFILING OF SELECTED p38α INHIBITOR COMPOUNDS IN
DISTINCT MAPK SIGNALING PATHWAY IMAGING ASSAYS[a]

Selected p38α inhibitors

Cell-based imaging assay IC_{50}

Compound	MK2-EGFP	cJun	ERK
1	0.011	17.91	>50
2	0.024	31.07	>50
3	0.024	>50	>50
4	0.036	31.20	>50
5	0.038	10.25	>50
6	0.043	0.05	>50
7	0.044	5.02	>50
8	0.076	10.10	>50
9	0.077	32.95	>50
10	0.092	14.04	>50
11	0.118	29.63	>50
12	0.127	15.55	>50
13	0.147	30.15	38.32
14	0.207	>50	>50
15	0.227	5.37	>50
16	0.255	5.05	>50
17	0.259	14.94	>50
18	0.262	15.34	>50
19	0.269	>50	>50
20	0.284	>50	>50
21	0.354	23.36	>50
22	0.386	37.24	>50
23	0.452	>50	>50
24	0.469	2.02	>50
25	0.477	18.85	>50
26	0.487	>50	>50
27	0.790	>50	>50
28	0.853	19.92	>50
29	0.987	>50	>50
30	1.259	>50	>50
31	1.354	9.25	>50
32	1.390	>50	>50
33	1.564	>50	>50
34	1.648	>50	>50
35	2.061	>50	>50
36	2.865	>50	38.67
37	3.177	>50	>50
38	3.675	>50	>50
39	4.277	>50	>50
40	21.730	>50	33.50

[a] Cell-based data represent the mean of duplicate independent determinations.

TABLE IV
PROFILING OF SELECTED JNK INHIBITOR COMPOUNDS IN DISTINCT MAPK SIGNALING
PATHWAY IMAGING ASSAYS[a]

	Selected JNK inhibitor compounds					
	Cell-based imaging assays					
	cJun activation		MK2-GFP		ERK activation	
Compound	Mean % inhib at 50 μM	Mean IC$_{50}$ μM	Mean % stimu at 50 μM	Mean IC$_{50}$ μM	Mean % inhib at 50 μM	Mean IC$_{50}$ μM
1	24	0.575	17	>50	99.5	14.080
2	109	10.335	36.5	>50	−40	>50
3	157	0.707	98	19.789	111	7.503
4	−2.5	>50	−1.5	>50	−38.5	>50
5	136	0.508	65	11.951	94	4.581
6	114	5.891	22	>50	−20	>50
7	138	0.830	116	9.742	98	1.931
8	99	10.957	24	>50	−50.5	>50
9	17	>50	4.5	>50	−12.5	>50
10	68	32.009	−3	>50	94.5	31.939
11	101	9.110	53	40.208	86.5	15.777
12	61	26.373	25	>50	34.5	>50
13	84	16.446	6	>50	78	27.825
14	73	21.465	8	>50	45.5	>50
15	46.5	9.413	21	>50	71	36.620
16	25.5	>50	−1.5	>50	37.5	34.098
17	68.5	27.925	32.5	>50	93.5	15.332
18	29	>50	21	>50	88.5	19.944
19	−18.5	>50	15	>50	14	>50
20	99	12.776	46	24.543	4	>50
21	109	10.769	24.5	>50	21.5	43.802
22	88.5	13.439	40	43.156	70.5	38.125
23	56	40.990	92.5	24.965	65.5	35.000
24	61	40.089	28	>50	−62	>50
25	84	18.839	58	38.146	1	>50
26	40	>50	21	>50	−42	>50
27	46	34.161	54	35.865	−23.5	>50
28	21	>50	54	28.192	−15	>50
29	−5.5	>50	10	>50	−25	>50
30	2	>50	11.5	>50	15.5	>50
31	61	40.842	17	>50	−48	>50
32	25	>50	5	>50	8	>50
33	17	>50	24	>50	−33.5	>50
34	0	>50	18	>50	−1	>50
35	−21	>50	18	>50	−3.5	>50
36	11	>50	24	>50	−17.5	>50
37	1	>50	16	>50	−0.5	>50
38	20	>50	24	>50	10.5	40.846
39	8	>50	19	>50	−11.5	>50
40	60	19.354	25	>50	−4	>50

[a] Cell-based data represent the mean of duplicate independent determinations.

pathway by these JNK inhibitors, with only 10 of the 40 tested producing IC_{50} values in the cell-based MK2-EGFP translocation p38 assay: 1 in the 1 to 10 μM range and 9 in the 10 to 40 μM range (Table IV).

Discussion

This chapter and ones by Williams *et al.* (2006) and Trask *et al.* (2006) described the development and implementation of the three distinct high-content cell-based MAPK signaling pathway assays to profile kinase inhibitors for MAPK module selectivity. The apparent redundancy of the kinase isoforms and functions in the MAPK signaling modules presents a challenge for kinase inhibitor selectivity and potency. For ATP-competitive inhibitors, these selectivity concerns may be exacerbated further by the degree of conservation of the ATP-binding site within the kinase family, the millimolar intracellular concentrations of ATP that may significantly affect cellular potency, and the potential for off-target effects due to binding to many other cellular proteins that bind and/or utilize ATP (English and Cobb, 2002; Noble *et al.*, 2004). The development of imaging assays for the major MAPK signaling modules, p38, JNK, and ERK, provided a strategy to profile kinase inhibitors for signaling pathway selectivity while simultaneously measuring effects on cell morphology or toxicity end points. These assays provided a mechanism to screen kinase inhibitors against targets that were expressed in the context of their endogenous substrates and scaffolding proteins in the appropriate cellular localization and that had to perform their functions in a complex environment involving cross talk between signaling pathways, phosphatase regulatory control, and intracellular ATP concentrations. The MAPK pathways assays were susceptible to nonselective kinase inhibitors such as staurosporin and inhibitors that inhibit upstream MKKKs and MKKs in the MAPK signaling pathway, especially those involved in cross talk between the pathways. In general, however, selective MAPK inhibitors exhibited pathway selectivity as evidenced by significantly lower IC_{50} values for their respective p38, JNK, or ERK signaling pathway assays.

References

Almholt, D. L., Loechel, F., Nielsen, S. J., Krog-Jensen, C., Terry, R., Bjorn, S. P., Pedersen, H. C., Praestegaard, M., Moller, S., Heide, M., Pagliaro, L., Mason, A. J., Butcher, S., and Dahl, S. W. (2004). Nuclear export inhibitors and kinase inhibitors identified using a MAPK-activated protein kinase 2 redistribution screen. *Assay Drug Dev. Technol.* **2**, 7–20.

Cowan, K. J., and Storey, K. B. (2003). Mitogen-activated protein kinases: New signaling pathways functioning in cellular responses to environmental stress. *J. Exp. Biol.* **206**, 1107–1115.

English, J. M., and Cobb, M. H. (2002). Pharmacological inhibitors of MAPK pathways. *Trends Pharmacol. Sci.* **23**, 40–45.

Fabbro, D., Ruetz, S., Buchdunger, E., Cowan-Jacob, S. W., Fendrich, G., Liebetanz, J., Mestan, J., O'Reilly, T., Traxler, P., Chaudhuri, B., Fretz, H., Zimmermann, J., Meyer, T., Caravatti, G., Furet, P., and Manley, P. W. (2002). Protein kinases as targets for anticancer agents: From inhibitors to useful drugs. *Pharmacol. Ther.* **93**, 79–98.

Garrington, T. P., and Johnson, G. L. (1999). Organization and regulation of mitogen activated protein kinase signaling pathways. *Curr. Opin. Cell Biol.* **11**, 211–218.

Giuliano, K. A., DeBiasio, R. L., Dunlay, R. T., Gough, A., Volosky, J. M., Zock, J., Pavlakis, G. N., and Taylor, D. L. (1997). High-content screening: A new approach to easing key bottlenecks in the drug discovery process. *J. Biomol. Screen.* **2**, 249–259.

Johnson, G. L., and Lapadat, R. (2002). Mitogen-activated protein kinase pathways mediated by ERK, JNK and p38 protein kinases. *Science* **298**, 1911–1912.

Mitchison, T. J. (2005). Small-molecule screening and profiling by using automated microscopy. *Chembiochem.* **6**, 33–39.

Noble, M. E. M., Endicott, J. A., and Johnson, L. N. (2004). Protein kinase inhibitors: Insights into drug design and structure. *Science* **303**, 1800–1805.

Ono, K., and Han, J. (2000). The p38 signal transduction pathway, activation and function. *Cell. Signal.* **12**, 1–13.

Regan, J., Breitfelder, S., Cirillo, P., Gilmore, T., Graham, A. G., Hickey, E., Klaus, B., Madwed, J., Moriak, M., Moss, N., Pargellis, C., Pav, S., Proto, A., Swinamer, A., Tong, L., and Torcellini, C. (2002). Pyrazole urea-based inhibitors of p38 MAP kinase: From lead compound to clinical candidate. *J. Med. Chem.* **45**, 2994–3008.

Trask, O. J., Jr., Baker, A., Williams, R. G., Kickischer, D., Kandasamy, R., Laethem, C., Johnston, P. A., and Johnston, P. A. (2006). Assay development and case history of a 32K-biased library high-content MK2-EGFP translocation screen to identify p38 MAPK inhibitors on the ArrayScan 3.1 imaging platform. *Methods Enzymol.* **414** (this volume).

Williams, R. G., Kandasamy, R., Nickischer, D., Trask, O. J., Jr., Laethem, C., Johnston, P. A., and Johnston, P. A. (2006). Generation and characterization of a stable MK2-EGFP cell line and subsequent development of a high-content imaging assay on the Cellomics ArrayScan platform to screen for p38 mitogen-activated protein kinase inhibitors. *Methods Enzymol.* **414** (this volume).

Zhang, J. H., Chung, T. D., and Oldenburg, K. R. (1999). A simple statistical parameter for use in evaluation and validation of high throughput screening assays. *J. Biomol. Screen.* **4**, 67–73.

[23]   Assay Development and Case History of a
32K-Biased Library High-Content MK2-EGFP
Translocation Screen to Identify p38 Mitogen-Activated
Protein Kinase Inhibitors on the ArrayScan 3.1
Imaging Platform

By Oscar J. Trask, Jr., Audrey Baker, Rhonda Gates Williams,
Debra Nickischer, Ramani Kandasamy, Carmen Laethem,
Patricia A. Johnston, and Paul A. Johnston

Abstract

This chapter describes the conversion and assay development of a
96-well MK2-EGFP translocation assay into a higher density 384-well
format high-content assay to be screened on the ArrayScan 3.1 imaging
platform. The assay takes advantage of the well-substantiated hypothesis
that mitogen-activated protein kinase-activating protein kinase-2 (MK2)
is a substrate of p38 MAPK kinase and that p38-induced phosphorylation
of MK-2 induces a nucleus-to-cytoplasm translocation. This chapter also
presents a case history of the performance of the MK2-EGFP translocation assay, run as a "high-content" screen of a 32K kinase-biased library to
identify p38 inhibitors. The assay performed very well and a number of
putative p38 inhibitor hits were identified. Through the use of multiparameter data provided by the nuclear translocation algorithm and by checking
images, a number of compounds were identified that were potential artifacts
due to interference with the imaging format. These included fluorescent compounds, or compounds that dramatically reduced cell numbers due to cytotoxicity or by disrupting cell adherence. A total of 145 compounds produced IC_{50}
values <50.0 μM in the MK2-EGFP translocation assay, and a cross target
query of the Lilly-RTP HTS database confirmed their inhibitory activity against
in vitro kinase targets, including p38a. Compounds were confirmed structurally
by LCMS analysis and profiled in cell-based imaging assays for MAPK signaling pathway selectivity. Three of the hit scaffolds identified in the MK2-
EGFP translocation HCS run on the ArrayScan were selected for a p38a
inhibitor hit-to-lead structure activity relationship (SAR) chemistry effort.

Introduction

The ability to automate the capture and analysis of fluorescent images
of thousands of cells in tens of thousands of wells of microtiter plates has

METHODS IN ENZYMOLOGY, VOL. 414   0076-6879/06 $35.00
         DOI: 10.1016/S0076-6879(06)14023-9

made fluorescence microscopy, one of the premier tools of cell biology, compatible with drug discovery (Almholt *et al.*, 2004; Giuliano *et al.*, 1997; Lundholt *et al.*, 2005; Mitchison, 2005; Oakley *et al.*, 2002; Ramm *et al.*, 2003; Thomas and Goodyer, 2003). All high-content screening (HCS) platforms require a process for the input and output of multiple microtiter plates, mechanisms to position plates on a stage, the ability to position wells over the optics with precision and reproducibility, a method to capture quality images, image analysis applications (algorithms), data review tools, and a protocol for data storage and management (Berlage, 2005; Giuliano *et al.*, 1997). Reagent selection for sample preparation and staining is also critical (Giuliano *et al.*, 1997). However, HCS platforms vary in the degree to which these components have been integrated. Many of the large high-throughput screening (HTS) instrument vendors have entered the HCS market by acquiring smaller imaging platform manufacturers to provide their offerings in the HCS field. For example, Cellomics, Inc. (Pittsburgh, PA), the company that introduced the ArrayScan, one of the first HCS platforms to penetrate the drug discovery market, was recently acquired by Fisher Scientific (Hampton, NH).

Williams *et al.* (2006) described the generation and characterization of a stable MK2-EGFP HeLa cell line and the subsequent development of a 96-well high-content imaging assay on the Cellomics ArrayScan platform to screen for p38 MAPK inhibitors. The assay took advantage of the well-substantiated hypothesis that mitogen-activated protein kinase-activating protein kinase-2 (MK2) is a substrate of p38 MAPK kinase and that p38-induced phosphorylation of MK2 induces a nucleus-to-cytoplasm translocation (Engel *et al.*, 1998; Neininger *et al.*, 2001; Zu *et al.*, 1995). Through a process of heterologous expression of a MK2-EGFP fusion protein in HeLa cells using retroviral infection, antibiotic selection, and flow sorting, we were able to isolate an MK2-EGFP-HeLa cell line in which the MK2-EGFP translocation response could be robustly quantified on the Cellomics ArrayScan HCS platform to provide a cell-based model for identifying p38 inhibitors (Williams *et al.*, 2006). This chapter describes the assay development process to convert this assay from a 96- to a 384-well format for screening a 32K kinase-biased library for novel p38 inhibitors.

Cellomics ArrayScan Automated Imaging Platform

The ArrayScan platform marketed by Cellomics (Fisher Scientific, Hampton, NH) is one of the most widely deployed HCS systems. The studies described in this chapter are performed on an ArrayScan II that has the software upgraded to a 3.1 version. The ArrayScan 3.1 houses a Zeiss Axiovert S100 inverted microscope outfitted with 5×/0.25 NA,

10×/0.3NA, and 20×/0.4 NA Zeiss objectives. Illumination is provided by a Xe/Hg arc lamp source (EXFO, Quebec, Canada), and fluorescence is detected by a 12-bit high sensitivity −20°-cooled CCD camera (Photometrics Quantix). The ArrayScan 3.1 provides the capability of imaging multiwavelength fluorescence by acquiring wavelength channels sequentially in which each fluorophore is excited separately and detected on the chip of a monochromatic CCD camera. Channel selection is accomplished using a fast excitation filter wheel combined with a multiband emission filter, although single band emission filters can be used to improve selectivity. The system comes with filter sets designed for the common fluorescent probes and can distinguish up to four labels in a single preparation with minimal cross talk between channels. The ArrayScan 3.1 is a wide-field imaging system that illuminates a "large" area of the specimen and directly images that area all at once. It uses an image-based autofocus system that images a fluorescent label in cells, typically fluorescently stained nuclei, but any feature of interest could be used, and an algorithm measures the relative sharpness of the image. The ArrayScan 3.1 was integrated with a Zymark Twister Robot and 80-plate stacker for fixed end point assays.

The cytoplasm-to-nuclear translocation algorithm developed by Cellomics may be used to quantify the relative distribution of a fluorescently tagged target between two cellular compartments, namely the cytoplasm and the nucleus (Giuliano et al., 1997). Labeling with a nucleic acid dye such as Hoechst 33342, DAPI, or DRAQ5 identifies the nuclear region, and this signal is used to focus the instrument and to define a nuclear mask. The mask is eroded to reduce cytoplasmic contamination within the nuclear area, and the final reduced mask is used to quantify the amount of target channel fluorescence within the nucleus. The nuclear mask is then dilated to cover as much of the cytoplasmic region as possible without going outside the cell boundary. Removal of the original nuclear region from this dilated mask creates a ring mask that covers the cytoplasmic region outside the nuclear envelope. The "Cytonuc" difference measurement is calculated as the difference of the average nuclear intensity minus the average cytoplasmic ring intensity on a per cell basis or may be reported as an overall well average value (Giuliano et al., 1997).

Conversion of the 96-Well MK2-EGFP Translocation Assay to a 384-Well Format Assay on the Arrayscan® Imager

Williams et al. (2006) described the development and optimization of a 96-well MK2-EGFP translocation assay on the ArrayScan 3.1 HCS platform that could quantify the activation and inhibition of the p38 MAPK pathway using the cytoplasm-to-nuclear translocation algorithm.

To increase the throughput for screening and reduce the demands on cell culture, we decided to convert the assay to the higher-density 384-well format. To establish the optimal cell seeding density, MK2-EGFP-A4 cells are seeded in 384-well Matrical glass plates at 2.5, 5.0, and 10.0×10^3 cells per well and on the following day are treated with 100 ng/ml anisomycin or are left untreated. Other wells are pretreated with the p38 inhibitor SB203580 at 1 μM for 12 min, followed by the addition of 100 ng/ml anisomycin (final) and a further incubation for 25 min. The ArrayScan imager was used to capture images (Fig. 1A), and the cytoplasm-to-nuclear translocation algorithm was used to quantify the MK2-EGFP translocation response at all three seeding densities (Fig. 1B). The ArrayScan imager and cytoplasm-to-nuclear translocation algorithm are able to adequately quantify the MK2-EGFP translocation at all three seeding densities (Fig. 1B). A seeding density of 2.5×10^3 cells per well was selected for further assay development to reduce the cell culture demands for the assay and to assist with the image analysis segmentation to identify discrete cells.

We next wanted to examine the time course for anisomycin-induced MK2-EGFP translocation and whether the length of preincubation of cells with p38 inhibitors prior to the addition of a maximum dose of anisomycin might affect their ability to inhibit the MK2-EGFP translocation response (Fig. 2). The amount of MK2-EGFP that translocates from the nucleus to the cytoplasm, indicated by the Cytonuc difference, appears to decrease from >300 to below 100 in a roughly linear fashion over time for up to 20 to 25 min and then remains stable for up to 60 min (Fig. 2A). Between 60 and 90 min there appears to be a small gradual increase in the Cytonuc difference, implying that the MK2-EGFP may be returning to the nucleus. However, the Cytonuc difference had not returned to the pretreatment or untreated values by 125 min, the longest time point measured (Fig. 2A). Forty minutes was selected as the standard incubation time for all further assay development. The ability of the p38 inhibitor SB203580 to inhibit anisomycin-induced MK2-EGFP translocation is not influenced by preincubation with the inhibitor prior to the addition of the activating stimulus (Fig. 2B). Indeed, co-addition of SB303580 with the anisomycin is just as effective as prior incubation with the inhibitor.

Compound screening libraries are typically solubilized in dimethyl sulfoxide (DMSO), and we wanted to evaluate the DMSO tolerance of the MK2-EGFP translocation response in HeLa-MK2-EGFP-A4 cells treated for 40 min with media or 100 ng/ml anisomycin at the indicated DMSO concentrations (Fig. 3B). The MK2-EGFP translocation response appears unaffected at DMSO concentrations $\leq 0.65\%$. However, at DMSO concentrations $\geq 1.25\%$, the Cytonuc difference increases with the DMSO concentration, independently of the treatment conditions. At DMSO

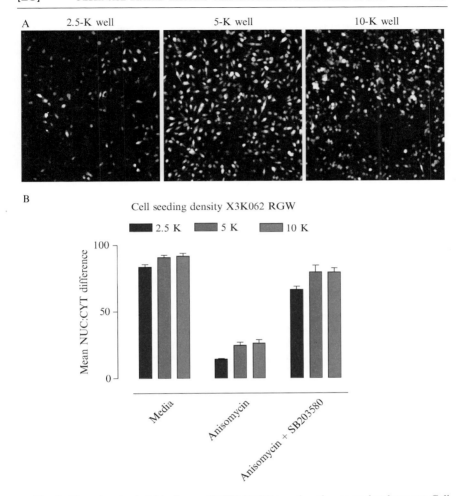

FIG. 1. Three hundred eighty-four-well MK2-EGFP translocation assay development: Cell seeding density. MK2-EGFP-A4 cells were seeded in 384-well Matrical glass plates at 2.5, 5.0, and 10.0×10^3 cells per well and on the following day were treated with 100 ng/ml anisomycin or were left untreated. Other wells were pretreated with the p38 inhibitor SB203580 at 1 μM for 12 min, followed by the addition of 100 ng/ml anisomycin (final) and a further incubation for 25 min. The ArrayScan imager was used to capture images (A), and the cytoplasm-to-nuclear translocation algorithm was used to quantify the MK2-EGFP translocation response at all three seeding densities (B).

concentrations $\geq 1.25\%$, the majority of the HeLa-MK2-EGFP-A4 population assumes a rounded cell morphology rather than the more typical well-attached, flat cell morphology (Fig. 3A). The rounded A4 cells have a much

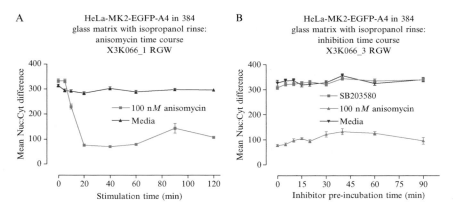

Fig. 2. Three hundred eighty-four-well MK2-EGFP translocation assay development: Activation and inhibition time courses. (A) Activation time course. HeLa-MK2-EGFP cells (2.5×10^3) from the HeLa-MK2-EGFP-A4 clone were seeded into each of the 384 wells of Matrical glass plates in EMEM + 10% FBS and incubated overnight at 37° and 5% CO_2. Cells were treated ± 200 ng/ml anisomycin for the indicated times and fixed in 3.7% formaldehyde + 2 μg/ml Hoechst dye, fluorescent images were collected on the ArrayScan imager, and the cytoplasm-to-nuclear translocation algorithm was used to quantify the MK2-EGFP translocation response. (B) Inhibition time course. HeLa-MK2-EGFP cells (2.5×10^3) from the HeLa-MK2-EGFP-A4 clone were seeded into each of the 384 wells of Matrical glass plates in EMEM + 10% FBS and were incubated overnight at 37° and 5% CO_2. The p38 inhibitor SB203580 was preincubated with the cells for the indicated times prior to the addition of 200 ng/ml anisomycin. Plates were incubated for 40 min and fixed in 3.7% formaldehyde + 2 μg/ml Hoechst dye, fluorescent images were collected on the ArrayScan imager, and the cytoplasm-to-nuclear translocation algorithm was used to quantify the MK2-EGFP translocation response.

smaller cytoplasm area than the normal well spread and attached cells, and the cytoplasm-to-nuclear translocation algorithm therefore has difficulty segmenting the cytoplasm from the nuclear areas to make the difference calculation. For analysis of rounded cells, the parameters of the algorithm would have to be modified.

Having transferred the MK2-EGFP translocation assay from a 96- to a 384-well format and completed the optimization of the assay conditions for the ArrayScan imager and nuclear translocation algorithm, we wanted to confirm the quality of the images (Fig. 4). MK2-EGFP-A4 cells were seeded in 384-well Matrical glass plates at 2.5×10^3 cells per well and on the following day were treated with the indicated doses of anisomycin and incubated for 40 min (Fig. 4). The quality of the images (Fig. 4) indicates that a 384-well MK2-EGFP translocation assay has been developed and optimized for image capture on the ArrayScan imager.

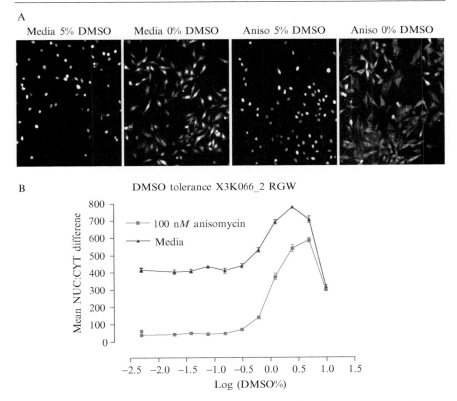

FIG. 3. Three hundred eighty-four-well MK2-EGFP translocation assay DMSO tolerance. HeLa-MK2-EGFP-A4 cells (2.5×10^3) were seeded in 384-well Matrical glass plates in EMEM + 10% FBS and were incubated overnight at 37° and 5% CO_2. Cells were treated ± 200 ng/ml anisomycin containing the indicated concentrations of DMSO for 40 min and fixed in 3.7% formaldehyde + 2 μg/ml Hoechst dye. The fluorescent images (A) were captured on the ArrayScan, and the cytoplasm-to-nuclear translocation algorithm was used to quantify the MK2-EGFP translocation response (B).

MK2-EGFP Translocation Assay Reproducibility and Signal Widow Evaluation

To evaluate the reproducibility of the MK2-EGF translocation response in HeLa-MK2-EGFP-A4 cells, cells were seeded at 2.5×10^3 cells per well and on the following day were treated with the indicated doses of anisomycin in three independent experiments run on separate days (Fig. 5). In three independent experiments, the EC_{50} for anisomycin-induced MK2-EGFP translocation ranged from 25 to 30 nM and produced on average an EC_{50} of 27.6 ± 2.3 nm, indicating a very reproducible assay.

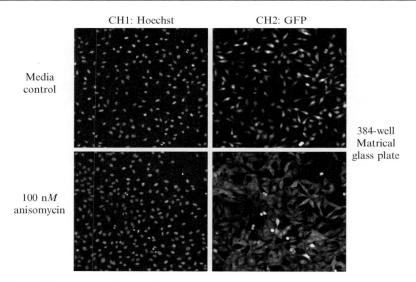

CH1: Hoechst   CH2: GFP

Media
control

384-well
Matrical
glass plate

100 nM
anisomycin

FIG. 4. Three hundred eighty-four-well MK2-EGFP translocation assay. HeLa-MK2-EGFP-A4 images captured on the ArrayScan 3.1 platform. HeLa-MK2-EGFP cells (2.5 × 10^3) from the HeLa-MK2-EGFP-A4 clone were seeded into each of the 384 wells of Matrical glass plates in EMEM + 10% FBS and were incubated overnight at 37° and 5% CO_2. Cells were treated ± 200 ng/ml anisomycin or were left untreated for 40 min and were fixed in 3.7% formaldehyde + 2 μg/ml Hoechst dye, and fluorescent images were collected on the ArrayScan imager. Images of Hoechst-stained nuclei collected in channel 1 or MK2-EGFP collected in channel 2 are both presented.

To evaluate the ability of the MK2-EGF translocation response in HeLa-MK2-EGFP-A4 cells to identify p38 inhibitors and examine the reproducibility of the assay, three independent inhibition experiments were run on separate days. Cells were seeded at 2.5 × 10^3 cells per well and on the next day the indicated doses of the p38 inhibitors SB203580 and RWJ 68543 or the JNK1/2 inhibitor SP 600125 were added simultaneously with 200 ng/ml anisomycin (final) and cells were incubated for 40 min (Fig. 6). In three independent experiments, the IC$_{50}$ for the p38 inhibitor SB203580 ranged from 98 to 106 nM and produced on average an IC$_{50}$ of 103 ± 4.2 nm. In three independent experiments, the IC$_{50}$ for the p38 inhibitor RWJ 68543 ranged from 82.6 to 139 nM and produced on average an IC$_{50}$ of 107.8 ± 28.7 nm. In three independent experiments, the JNK1/2 inhibitor SP 600125 failed to inhibit the anisomycin-induced MK2-EGFP translocation and produced an average IC$_{50}$ of >10 μM, indicating the selectivity of the assay for p38 inhibitors.

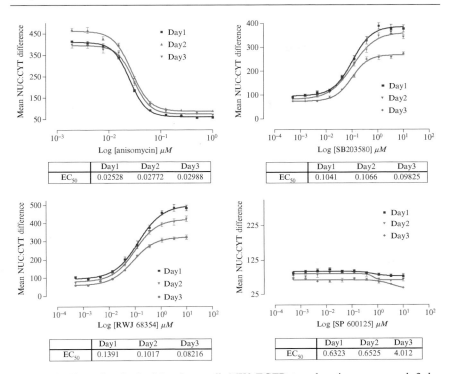

	Day1	Day2	Day3
EC_{50}	0.02528	0.02772	0.02988

	Day1	Day2	Day3
EC_{50}	0.1041	0.1066	0.09825

	Day1	Day2	Day3
EC_{50}	0.1391	0.1017	0.08216

	Day1	Day2	Day3
EC_{50}	0.6323	0.6525	4.012

FIG. 5. Three hundred eighty-four-well MK2-EGFP translocation assay and 3-day activation and inhibition curves. Three-day EC_{50} curves: 2.5×10^3 HeLa-MK2-EGFP A4 cells were seeded in 384-well Matrical glass plates in EMEM + 10% FBS and were incubated overnight at 37° and 5% CO_2. For activation of the response, cells were treated with the indicated doses of anisomycin for 40 min and fixed in 3.7% formaldehyde + 2 μg/ml Hoechst dye, and fluorescent images were collected on the ArrayScan. Data are presented from three independent experiments, each performed in triplicate wells and run on separate days. Three-day IC_{50} curves: 2.5×10^3 HeLa-MK2-EGFP A4 cells were seeded in 384-well Matrical glass plates in EMEM + 10% FBS and were incubated overnight at 37° and 5% CO_2. For inhibition of the response, the indicated doses of SB203580, RWJ 68354, or SP 600125 were added simultaneously with the addition of 200 ng/ml anisomycin (final) and plates were incubated for 40 min. Plates were fixed in 3.7% formaldehyde + 2 μg/ml Hoechst dye, and fluorescent images were collected on the ArrayScan. Data are presented from three independent experiments, each performed in triplicate wells and run on separate days. The cytoplasm-to-nuclear translocation algorithm was used to analyze the images captured on the ArrayScan and quantify the anisomycin induction and/or inhibition of the MK2-EGFP translocation response.

To further evaluate the robustness and reproducibility of the MK2-EGF translocation response in HeLa-MK2-EGFP-A4 cells, six full 384-well plates are treated under the following conditions, two plates per

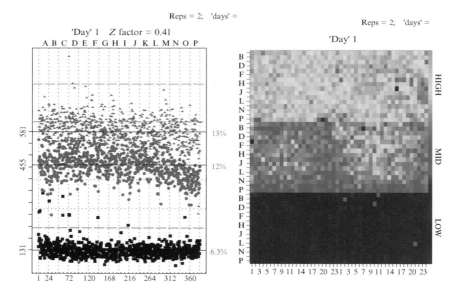

Fig. 6. Three hundred eighty-four-well MK2-EGFP translocation assay: Signal window and variability assessment. HeLa-MK2-EGFP A4 cells (2.5×10^3) were seeded in 384-well Matrical glass plates in EMEM + 10% FBS and were incubated overnight at 37° and 5% CO_2. Two full 384-well plates were treated for 40 min under the following conditions: media alone (red triangles), 200 ng/ml anisomycin (blue squares), and the p38 inhibitor SB203580 + 200 ng/ml anisomycin (green circles). Plates were fixed in 3.7% formaldehyde + 2 µg/ml Hoechst dye, and fluorescent images were collected on the ArrayScan . The cytoplasm-to-nuclear translocation algorithm was used to analyze the images and quantify the anisomycin induction and/or SB203580 inhibition of the translocation response. The Z factor was calculated according to the method of Zhang *et al.* (1999). In addition to the population scatter plots of the calculated Cytonuc differences, plate heat map views of the six plates are also presented.

condition: media alone (red triangles), 200 ng/ml anisomycin (blue squares), and p38 inhibitor + 200 ng/ml anisomycin (green circles) (Fig. 7). In addition to the population scatter plots, plate heat map views of the six plates are also presented (Fig. 7). The average Cytonuc difference for the two media control plates was 581, and the average Cytonuc difference for the two anisomycin-treated control plates was 131, producing an average assay signal to background window of 4.3-fold. A comparison of media control plate variability data to anisomycin-treated plate variability data produced a Z factor of 0.41, indicating that the assay was compatible with HTS (Zhang *et al.*, 1999).

Standardized Operation Procedure for the MK2-EGFP
  Translocation Assay

1. Wash 384-well Matrical glass plates once with isopropanol and rinse with water on an MRD8 Titertek plus plate washer to remove a cytotoxic residue due to the plate adhesive.

2. Detach 70 to 80% confluent HeLa-MK2-EGFP-A4 monolayers from tissue culture flasks with trypsin-versene and pellet by centrifugation for 5 min 800g. Cells are resuspended in complete media (EMEM + 10% FBS + 2 mM L-glutamine + penicillin/streptomycin +800 μg/ml G418) to a cell density to 6.25 $\times$ 10^4 cells/ml.

3. Seed cells into 384-well Matrical glass plates using the Multidrop (Titertek) at 40 μl/well (2500 cells/well). Incubate plates at 37°, 5% CO$_2$, and 95% humidity for 18 to 24 h.

4. Transfer 20 μl of prewarmed compounds and plate controls to the wells of assay plates using a 384-well transfer head on the Multimek (Beckman) and incubate plates at 37°, 5% CO$_2$, and 95% humidity for 40 min. For inhibitor screens, add anisomycin (200 nM final concentration) simultaneously with the compounds.

5. After the 40-min incubation, fix plates by adding 20 μl of prewarmed (37°) formaldehyde (1.0% final) + 2 μg/μl Hoechst 33342 stain using the Multidrop (Titertek) and incubate at room temperature in a fume hood for 12 min.

6. Aspirate the fixation solution, add 50 μl of phosphate-buffered saline (PBS) to wells to wash the monolayers, aspirate the PBS wash solution, add another 50 μl of PBS to the wells, and seal the plates. Perform this aspiration and wash process on a MAP C2 plate handler (Titertek).

7. Acquire images on the ArrayScan platform and analyze for the translocation of MK2-EGFP from nucleus to cytoplasm using the nuclear translocation algorithm.

8. ArrayScan 3.1 settings: Select a dual BGIp filter set for excitation and emission of the Hoechst and EGFP fluorescence. Use the XF100-filter-Hoechst filter set for the Hoechst 33342 nuclear stain collected in channel 1, and select the XF100-FITC filter set for the EGFP fluorescence collected in channel 2. Perform scans sequentially, and set up the ArrayScan to collect a minimum of 100 valid objects per well, or a maximum of 4 fields of view per well, whichever comes first.

MK2-EGFP Translocation HTS Assay for p38 Inhibitors

Two compound cassettes are selected for screening in the MK2-EGFP translocation assay to identify p38 inhibitors: a library of 854 pharmacologically active compounds obtained from RBI and a 32,000 compound

TABLE I
MK2-EGFP TRANSLOCATION SCREEN SUMMARY DATA

MK2GFP ArrayScan			% of total
Rapid MTS	# Screened	32891	100
	# >50%	483	1.47
5-point IC 50's	# tested	448	100
	# Confirmed	234	52.53
10-point IC 50's	# tested	241	100
	# Confirmed	217	90.04

TABLE II
ARRAYSCAN 3.1 INSTRUMENT PERFORMANCE DATA[a]

HTS phase	Criteria	Performance
Rapid	# Plates	110
	Total scan time	98.6 h (4 days)
	Mean scan time/384-well plate	45.4 min
	Mean scan time/well	7.1 s
	Average fields/well	1.18
	Average valid objects/well	135
5-point IC 50's	# Plates	8
	Total scan time	9.5 h
	Mean scan time/384-well plate	71.3 min
	Mean scan time/well	11.1 s
	Average fields/well	1.44
10-point IC 50's	# Plates	9
	Total scan time	8.7 h
	Mean scan time/384-well plate	57.9 min
	Mean scan time/well	9.05 s
	Average fields/well	1.41

[a] Instrument settings: 100 objects or 4 fields/well; autoexposure isodata threshold 10%.

kinase-biased cassette. Based on the assay signal window and Z-factor data (Fig. 6), the active criterion for the primary screen was set at 50% inhibition of the anisomycin-induced MK2-EGFP translocation.

In the course of the MK2-EGFP screen, a total of 32,891 compounds were tested at a final concentration of 50 μM (0.5% DMSO) for their ability to inhibit anisomycin-induced MK2-EGFP translocation. Of the 32,891 compounds tested, 483 (1.47%) met the active criterion and produced 50% inhibition of anisomycin-induced MK2-EGFP translocation (Table I).

The primary MK2-translocation HTS assay was performed over a period of 5 days of operations, screening at a rate of 20, 25, 25, 25, and 15 384-well plates per day. The total scan time for the primary screen on the ArrayScan was 98.6 h (4 days), the mean scan time/384-well plate was 45.4 min, the mean scan time/well was 7.1 s, and the average number of frames/well was 1.18 (Table II).

The primary screen performed very well and of the 474 actives identified in the primary screen (Table I), only 448 were available from the compound library for testing in the five-point IC_{50} confirmation assays. Compounds were tested in five-point IC_{50} curves, threefold dilution series, starting at a maximum concentration of 50 μM. Of the 448 compounds tested, 234 (52.2%) were confirmed as dose-dependent inhibitors of anisomycin-induced MK2-EGFP translocation with IC_{50} values <50 μM (Table I). The five-point IC_{50} confirmation assays were performed in 1 day of screening operations, and the total scan time on the ArrayScan was 9.5 h, the mean scan time/384-well plate was 71.3 min, the mean time/well was 11.1 s, and the average number of fields/well was 1.44 (Table II).

Two hundred forty-one compounds were ordered for 10-point IC_{50} curves and were tested in a threefold dilution series, starting at a maximum concentration of 50 μM. Of the 241 compounds tested, 217 (90.04%) were confirmed as inhibitors of anisomycin-induced MK2-EGFP translocation with IC_{50} values <50 μM (Table I). The 10-point IC_{50} confirmation assays were performed in 1 day of screening operations, and the total scan time on the ArrayScan was 8.7 h, the mean scan time/384-well plate was 57.9 min, the mean time/well was 9.05 s, and the average number of fields/well was 1.41 (Table II). Of the 217 compounds confirmed with IC_{50} values <50 μM, 18 produced IC_{50} values <1.0 μM, 55 produced IC_{50} values in the 1 to 10 μM range, and the remaining 144 had IC_{50} values in the 10 to 50 μM range (Table III). A total of 73 compounds were identified that reproducibly inhibited anisomycin-induced MK2-EGFP translocation with IC_{50} values <10 μM (Table III).

TABLE III
MK2-EGFP TRANSLOCATION 10-POINT IC_{50} RANGES

	ArrayScan	%
# tested	241	100
# Confirmed	217	90.04
# <1.0	18	7.47
# 1–10 μM	55	22.82
# 10–50 μM	144	59.75
>50 μM	24	9.96

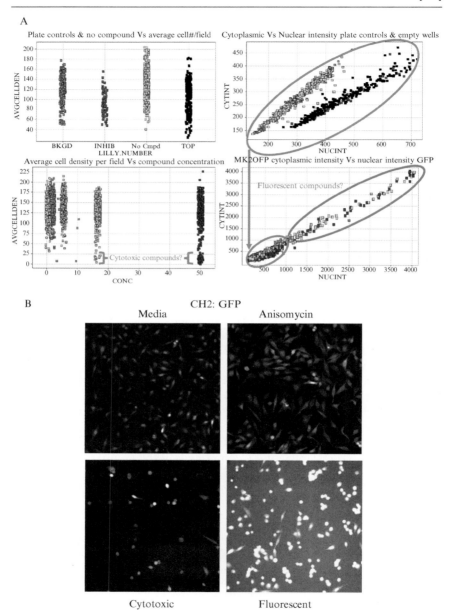

Fig. 7. Nuclear translocation algorithm secondary analysis for cytotoxicity and fluorescence. The cytoplasm-to-nuclear translocation algorithm provides data on a number of parameters that can be used to identify potential interference due to compound fluorescence

The cytoplasm-to-nuclear translocation algorithm provides data on a number of parameters that can be used to identify potential interference due to compound fluorescence or compound effects such as cytotoxicity or disruption of cell adherence. These data can be visualized with the Array-Scan view software in a variety of data plots or can be exported to data visualization software such as Spotfire (Fig. 7). A comparison of the images and secondary analysis parameters from the plate controls and empty wells versus the compound wells indicated that some of these parameters could be useful to discriminate compound effects or fluorescence interference (Fig. 7). The average cell number per field parameter indicates the number of nuclei found to pass the intensity and size filters selected in the algorithm. The ArrayScan was set up to capture a minimum of 100 cells per well or a maximum of 4 image fields per well, whichever came first. Wells with significantly <40 cells/well were likely treated with compounds that are either cytotoxic or significantly disrupt cell adherence (Fig. 7). When the average cell number per field parameter was plotted for the different plate controls, very few if any of the wells had fewer than 40 cells per well (Fig. 7A). In contrast, a plot of the average cell number per field parameter versus compound concentration indicated that a number of wells had significantly less than 25 cells per well (Fig. 7A). This was especially apparent at the two highest concentrations of the dose response, 50 and 16.67 μM. Upon checking the images from these wells (Fig. 7B), it is apparent that a number of compounds had reduced the number of cells per well, due to either a dose-dependent cytotoxicity or a dose-dependent reduction in cell adherence.

By comparing data from plate controls and empty wells in a scatter plot of the average cytoplasmic intensity/well versus average nuclear intensity/well, it was apparent there are two distinct populations apparent in data corresponding to the anisomycin-treated controls in one population and the untreated or maximally inhibited controls in the other (Fig. 7A). These two control populations do not contain any wells

---

or off-target compound effects such as cytotoxicity or disruption of cell adherence. These data were exported from the ArrayScan analysis software to Spotfire and visualized in a variety of scatter plots (A). The average number of cells per field parameter was plotted for the different plate controls or versus compound concentration to assess cytotoxicity or a reduction in cell adherence. A scatter plot of the average cytoplasmic intensity/well versus average nuclear intensity/well in the target channel (EGFP) was used to identify fluorescent compounds. Representative images from control wells and compound wells that exhibited either a dose-dependent cytotoxicity/reduction in cell adherence or a dose-dependent fluorescence are presented (B).

that exceed thresholds of cytoplasmic intensity >500 and/or nuclear intensity >750. However, when the compound wells are included in the analysis, even though the majority of wells fall below these thresholds, there were a significant number of wells that exceed the thresholds of cytoplasmic intensity >1000 and/or nuclear intensity >1000. This was especially apparent at two of the highest concentrations of the dose response, 50 and 16.67 μM. Upon checking the images from these wells (Fig. 7B), it is apparent that a number of compounds were likely fluorescent.

Of the 217 actives confirmed in the 10-point IC_{50} assays, 59 (27%) exhibited average cell numbers/field data with significantly <40 cells/well, indicating that they were likely cytotoxic or significantly disrupted cell adherence (Table IV). Thirteen (6.0%) of the compounds exceeded thresholds of cytoplamic intensity >1000 and/or nuclear intensity >1000, indicating that they were likely fluorescent (Table IV).

Seventy-three of the confirmed actives produced IC_{50} values <10.0 μM in the MK2-EGFP translocation assay (Table III). A cross target query of the Lilly-RTP HTS database for the confirmed active compounds in the MK2-EGFP translocation assay revealed that the most prevalent targets with confirmed IC_{50} values in the database were in vitro kinase targets, including p38a. Around 40 compounds were profiled in cell-based imaging assays for MAPK signaling pathway selectivity; MK2-EGFP translocation, cJun activation, and ERK activation (see Nickischer et al., 2006). Compounds were also structurally confirmed by LCMS analysis (data not shown). In sum, upon completion of the hit assessment for the p38 inhibitors identified in the MK2-EGFP translocation assay, a new structural class of p38a inhibitor had been identified (data not shown) and three of the hits were selected as potential p38a inhibitor hit-to-lead scaffolds.

TABLE IV

MK2-EGFP TRANSLOCATION 10-POINT IC_{50} SECONDARY ANALYSIS HIT ASSESSMENT

Secondary analysis	ArrayScan	
Total # tested	241	%
>50 μM	24	9.96
<50 μM	217	90.04
Cytox	59	27.19
Fluorescent	13	5.99
<50 μM	145	66.82

Discussion

This chapter described the conversion and assay development of a 96-well MK2-EGFP translocation assay developed on the ArrayScan imaging platform into a higher-density 384-well format HCS assay (Figs. 1–6). This chapter also presented a case history of the performance of the MK2-EGFP translocation assay run on the ArrayScan imager as a "high-content" screen of a 32K kinase-biased library to identify p38 inhibitors (Fig. 7, Tables I–IV). As described in this chapter, the MK2-EGFP translocation assay performed very well on the ArrayScan platform and a number of p38 inhibitor hits were selected for hit-to-lead follow-up SAR.

The ArrayScan 3.1 is a wide-field imaging system that illuminates a "large" area of the specimen and directly images that area all at once. It uses an image-based autofocus system that images a fluorescent label in cells, typically fluorescently stained nuclei, but any feature of interest could be used, and an algorithm measures the relative sharpness of the image. Although many imaging applications perform well on wide-field HCS imagers, there is a perception that confocal capability is desirable, perhaps because many HCS assays have their genesis on stand-alone confocal microscope platforms. Confocal scanning systems work by illuminating the specimen in one or more small regions (spots or lines) and building up an image by scanning the illumination through the specimen while measuring the emission in synchrony with the scanning. Confocal HCS systems can be further divided based on illumination scan design, with systems available that use point scanning, line scanning, and multipoint scanning (e.g., spinning disk). Confocal imaging systems have a definite advantage in rejecting background fluorescence from material outside the plane of focus, either due to the specimen being significantly thicker than the depth of field or due to some fluorescent component in the well plate or surrounding media, such as excess label. For assays with a high solution background or with thick, multilayer cell preparations, confocal imaging will certainly be an advantage. However, confocal HCS imagers are more complex to build and therefore are typically more expensive than wide-field HCS systems such as the ArrayScan, INCell 1000, or Image Express 5000. The same assay plates that were described in this chapter were also read on the Incell 3000 confocal line-scanning platform (GE-Healthcare) with equivalent results, except that the scan times were on average ~2-fold shorter (data not shown). When comparing throughputs on different imaging platforms it is good practice to run as many distinct biologies and combinations of fluorophores as possible. The nature and quality of the sample preparations will have a significant impact on the performance of any imaging platform.

p38 has been a major target for drug discovery by the pharmaceutical industry, as indicated by the numerous patent applications and small molecule inhibitors that have been developed (English and Cobb, 2002; Fabbro *et al.*, 2002; Noble *et al.*, 2004; Regan *et al.*, 2002). Many of these p38 inhibitors have exhibited significant efficacy in cellular assays and animal disease models, and several have progressed into human clinical trails for the treatment of inflammation and cancer (English and Cobb, 2002; Fabbro *et al.*, 2002; Noble *et al.*, 2004; Regan *et al.*, 2002). For example, SB203580 and BIRB 796 and their analogs have been shown to be potent and selective inhibitors of p38 MAPK (English and Cobb, 2002; Fabbro *et al.*, 2002; Noble *et al.*, 2004; Regan *et al.*, 2002). SB203580 and BIRB 796 produced IC_{50} values of 60 and 18 nM, respectively in a THP-1 tumor necrosis factor-α production cell model (Regan *et al.*, 2002). In the MK2-EGFP translocation assay described here, SB203580 produced an average IC_{50} of 103 ± 4 nM and RWJ 68543 produced on average an IC_{50} of 108 ± 28 nm. As an indicator of selectivity the JNK1/2 inhibitor SP 600125 failed to inhibit the anisomycin-induced MK2-EGFP translocation response. In a similar MK2-GFP redistribution assay developed in a BHK-1 cell background and quantified on the Incell 3000 imaging platform, SB203580 produced an IC_{50} of 3.4 μM for anisomycin-induced redistribution (Almholt *et al.*, 2004).

A total of 32,891 compounds, including a 32K kinase-biased library, were tested at a final concentration of 50 μM (0.5% DMSO) for their ability to inhibit anisomycin-induced MK2-EGFP translocation on the ArrayScan platform. Of the 32,891 compounds tested, 483 (1.47%) met the active criterion and produced $\geq$50% inhibition of anisomycin-induced MK2-EGFP translocation (Table I). Of the compounds available from the compound library, 448 were tested in five-point IC_{50} curves, and 234 (52.23%) were confirmed as dose-dependent inhibitors of anisomycin-induced MK2-EGFP translocation with IC_{50} values $<$50 μM (Table I). Two hundred forty-one compounds were tested in 10-point IC_{50} curves, and 217 (90.04%) were confirmed as inhibitors of anisomycin-induced MK2-EGFP translocation with IC_{50} values $<$50 μM (Table I). Through use of multiparameter data provided by the cytoplasm-to-nuclear translocation algorithm and by checking images, a number of compounds were identified that were potential artifacts due to interference with the imaging format. These included fluorescent compounds, or compounds that dramatically reduced cell numbers due to cytotoxicity or by disrupting cell adherence. A total of 145 compounds that were not fluorescent or cytotoxic produced IC_{50} values $<$50.0 μM in the MK2-EGFP translocation assay (Table V). Of the confirmed actives, 73 (30%) produced IC_{50} values

<10.0 μM in the MK2-EGFP translocation assay (Table III), and a cross target query of the Lilly-RTP HTS database confirmed their inhibitory activity against *in vitro* kinase targets, including p38a (data not shown). Compounds were confirmed structurally by LCMS analysis and profiled in cell-based imaging assays for MAPK signaling pathway selectivity: MK2-EGFP translocation, cJun activation, and ERK activation (see Nickischer *et al.*, 2006). The hit assessment of the p38 inhibitors identified in the MK2-EGFP HCS run on the ArrayScan identified a new structural class of p38a inhibitor (data not shown), and three of the hit scaffolds were selected for p38a inhibitor hit-to-lead chemistry SAR.

The previously described BHK-1 MK2-GFP redistribution assay (Almholt *et al.*, 2004) was utilized for an HTS of 183,375 compounds that were screened at 10 μM for inhibitors of MK2-GFP redistribution. A throughput of 19,000 compounds per day was achieved, and 1960 (1.1%) inhibited redistribution >35%. The secondary analysis parameters of the Incell 3000 nuclear trafficking module were used to eliminate cytotoxic and fluorescent compounds, dropping the hit rate to 1109 compounds (0.6%) (Almholt *et al.*, 2004). Six hundred thirty-four hits were selected for retesting, and 206 (0.11%) were confirmed at exhibiting >25% inhibition of MK2-GFP redistribution without causing cell rounding or producing fluorescence (Almholt *et al.*, 2004). Hits from the primary screen were categorized via a number of secondary assays, and two main classes of compounds were identified: direct inhibitors of the MK2 nuclear export process and inhibitors of the upstream p38 MAPK pathway (Almholt *et al.*, 2004). One of the confirmed actives was shown to resemble a p38 kinase inhibitor structurally and functionally (Almholt *et al.*, 2004).

The MK2-EGFP screen described here differed from the previously described redistribution screen in three main ways: the MK2-EGFP assay was ~30-fold more sensitive to inhibition by the p38 inhibitor SB 203580 than the redistribution assay, a 32K kinase-biased library was screened rather than a 185K random library, and compounds were screened at 50 μM rather than at 10 μM. Nevertheless, both assays successfully identified p38 inhibitors and demonstrated the utility of HCS assays for screening compounds that interfere with therapeutically important signaling pathways.

There are some 518 protein kinases encoded in the human genome that share a catalytic domain, the ATP-binding site, conserved in sequence and structure (English and Cobb, 2002; Noble *et al.*, 2004). The vast majority of known kinase inhibitors were found to be competitive with ATP, and thus are believed to interact within the ATP-binding site (English and Cobb, 2002; Noble *et al.*, 2004). The conservation of the ATP-binding site within the kinase family, the fact that a significant number of the other enzymes in cells use ATP, and the millimolar intracellular concentration of ATP raise significant concerns about inhibitor potency in cellular assays, kinase selectivity, and adverse off-target effects (English and Cobb, 2002; Noble *et al.*, 2004). There have been some examples of kinase inhibitors that appear to bind preferentially to the inactive (unphosphorylated) form of the enzyme, thereby blocking activation of the kinase, or bind to an allosteric site spatially distinct from the ATP pocket that causes a shift in the conformation of the kinase that is incompatible with ATP binding (English and Cobb, 2002; Noble *et al.*, 2004; Regan *et al.*, 2002). While it may be technically challenging to design an *in vitro* kinase assay to measure inactive enzyme or allosteric kinase inhibitors, cellular assays may be configured to test both these and ATP competitive kinase inhibitors. Image-based cellular assays provide multi-parameter quantitative and qualitative information beyond the single parameter target data typical of most other assay formats, and thus are termed "high content." HCS may be configured for simultaneous multiple target readouts (multiplexing) and can provide information on cellular morphology, population distributions, and subcellular localizations and relationships. By selection of the appropriate probes (antibodies, fluorescent protein fusion partners, biosensors, and stains), HCS can be designed to measure and discriminate between on-target and off-target effects of lead compounds and can provide cell-based models that have the potential to improve the conversion rate of drug candidates to successful drugs.

References

Almholt, D. L., Loechel, F., Nielsen, S. J., Krog-Jensen, C., Terry, R., Bjorn, S. P., Pedersen, H. C., Praestegaard, M., Moller, S., Heide, M., Pagliaro, L., Mason, A. J., Butcher, S., and Dahl, S. W. (2004). Nuclear export inhibitors and kinase inhibitors identified using a MAPK-activated protein kinase 2 redistribution screen. *Assay Drug Dev. Technol.* **2**, 7–20.

Berlage, T. (2005). Analyzing and mining image databases. *Drug Discov. Today* **10**, 795–802.

Engel, K., Kotlyarov, A., and Gaestel, M. (1998). Leptomycin B-sensitive nuclear export of MAPKAP kinase 2 is regulated by phosphorylation. *EMBO J.* **17**, 3363–3371.

English, J. M., and Cobb, M. H. (2002). Pharmacological inhibitors of MAPK pathways. *Trends Pharmacol. Sci.* **23**, 40–45.

Fabbro, D., Ruetz, S., Buchdunger, E., Cowan-Jacob, S. W., Fendrich, G., Liebetanz, J., Mestan, J., O'Reilly, T., Traxler, P., Chaudhuri, B., Fretz, H., Zimmermann, J., Meyer, T., Caravatti, G., Furet, P., and Manley, P. W. (2002). Protein kinases as targets for anticancer agents: From inhibitors to useful drugs. *Pharmacol. Ther.* **93**, 79–98.

Giuliano, K. A., DeBiasio, R. L., Dunlay, R. T., Gough, A., Volosky, J. M., Zock, J., Pavlakis, G. N., and Taylor, D. L. (1997). High-content screening: A new approach to easing key bottlenecks in the drug discovery process. *J. Biomol. Screen.* **2**, 249–259.

Lundholt, B. K., Linde, V., Loechel, F., Pedersen, H. C., Moller, S., Praestegaard, M., Mikkelsen, I., Scudder, K., Bjorn, S. P., Heide, M., Arkhammar, P. O., Terry, R., and Nielsen, S. J. (2005). Identification of Akt pathway inhibitors using redistribution screening on the FLIPR and the Incell 3000 analyzer. *J. Biomol. Screen.* **10**, 20–29.

Mitchison, T. J. (2005). Small-molecule screening and profiling by using automated microscopy. *Chembiochem.* **6**, 33–39.

Neininger, A., Thielemann, H., and Gaestel, M. (2001). FRET-based detection of different conformations of MK2. *EMBO Rep.* **2**, 703–708.

Nickischer, D., Laethem, C., Trask, O. J., Jr., Williams, R. G., Kandasamy, R., Johnston, P. A., and Johnston, P. A. (2006). Development and implementation of three mitogen-activated protein kinase (MAPK) signaling pathway imaging assays to provide MAPK module selectivity profiling for kinase inhibitors: MK2-EGFP translocation, c-Jun, and ERK activation. *Methods Enzymol.* **414** (this volume).

Noble, M. E. M., Endicott, J. A., and Johnson, L. N. (2004). Protein kinase inhibitors: Insights into drug design and structure. *Science* **303**, 1800–1805.

Oakley, R. H., Hudson, C. C., Cruickshank, R. D., Meyers, D. M., Payne, R. E., Jr., Rhem, S. M., and Loomis, C. R. (2002). The cellular distribution of fluorescently labeled arrestins provides a robust, sensitive, and universal assay for screening G protein-coupled receptors. *Assay Drug Dev. Technol.* **1**, 21–30.

Ramm, P., Alexandrov, Y., Cholewinski, A., Cybuch, Y., Nadon, R., and Soltys, B. J. (2003). Automated screening of neurite outgrowth. *J. Biomol. Screen.* **8**, 7–18.

Regan, J., Breitfelder, S., Cirillo, P., Gilmore, T., Graham, A. G., Hickey, E., Klaus, B., Madwed, J., Moriak, M., Moss, N., Pargellis, C., Pav, S., Proto, A., Swinamer, A., Tong, L., and Torcellini, C. (2002). Pyrazole urea-based inhibitors of p38 MAP kinase: From lead compound to clinical candidate. *J. Med. Chem.* **45**, 2994–3008.

Thomas, N., and Goodyer, D. (2003). Stealth sensors: Real-time monitoring of the cell cycle. *Targets* **2**, 26–33.

Williams, R. G., Kandasamy, R., Nickischer, D., Trask, O. J., Jr., Laethem, C., Johnston, P. A., and Johnston, P. A. (2006). Generation and characterization of a stable MK2-EGFP cell line and subsequent development of a high-content imaging assay on the Cellomics ArrayScan platform to screen for p38 mitogen-activated protein kinase inhibitors. *Methods Enzymol.* **414** (this volume).

Zhang, J. H., Chung, T. D., and Oldenburg, K. R. (1999). A simple statistical parameter for use in evaluation and validation of high throughput screening assays. *J. Biomol. Screen.* **4**, 67–73.

Zu, Y. L., Ai, Y., and Huang, C. K. (1995). Characterization of an autoinhibitory domain in human mitogen-activated protein kinase-activated protein kinase 2. *J. Biol. Chem.* **270**, 202–206.

[24] Compound Classification Using Image-Based Cellular Phenotypes

By CYNTHIA L. ADAMS, VADIM KUTSYY, DANIEL A. COLEMAN, GE CONG,
ANNE MOON CROMPTON, KATHLEEN A. ELIAS, DONALD R. OESTREICHER,
JAY K. TRAUTMAN, and EUGENI VAISBERG

Abstract

Compounds with similar target specificities and modes of inhibition cause similar cellular phenotypes. Based on this observation, we hypothesized that we could quantitatively classify compounds with diverse mechanisms of action using cellular phenotypes and identify compounds with unintended cellular activities within a chemical series. We have developed Cytometrix™ technologies, a highly automated image-based system capable of quantifying, clustering, and classifying changes in cellular phenotypes for this purpose. Using this system, 45 out of 51 known compounds were accurately classified into 12 distinct mechanisms of action. We also demonstrate microtubule-binding activity in one of seven related cytochalasin actin poisons. This technology can be used for a variety of drug discovery applications, including high-throughput primary screening of chemical and siRNA libraries and as a secondary assay to detect unintended activities and toxicities.

Introduction

Recent advances in high-throughput screening, chemistry, and genomics have resulted in a proliferation of new drug candidates. However, candidate drugs often fail in the clinic due to unintended activity (Prentis *et al.*, 1988). Crucial to successful drug discovery is a robust approach that reduces the number of compounds tested in animals by identifying compounds that have unintended activities *in vitro*. Several approaches to assess the broader biological effects of candidate drugs prior to clinical testing have been developed, such as panels of biochemical and cell assays, gene expression analyses, and proteomic profiling (Pritchard *et al.*, 2003). These assays, however, often are limited to a few compound conditions tested late in the discovery process.

Cell biology largely remains an unexploited tool in the attrition management of early stage compounds. At the gene and protein levels, many variables can be difficult to track, including posttranslational modifications,

METHODS IN ENZYMOLOGY, VOL. 414   0076-6879/06 $35.00
DOI: 10.1016/S0076-6879(06)14024-0

protein turnover, and functional redundancy, which can often be assessed by monitoring cytoskeletal, cell cycle, and organelle responses to perturbations in the pathways that affect these systems. Cytoskeletal biology provides a rich source of markers for monitoring effects of compounds of interest on cellular structures and processes long known to be indicative of cellular integrity and general health of the cell, including the cytoskeleton, signal transduction pathways, and protein trafficking machinery.

Mechanisms of action (MOA) of a compound in a cell are dependent on many parameters, including concentration, time of incubation, the genotype, and environment of the cell. A phenotypic signature of a compound can be collected using cell morphology changes measured systematically across many conditions. To do this we developed Cytometrix technologies to (1) automatically and broadly measure effects of compounds on individual cells and (2) use multivariate statistics to analyze those effects across multiple conditions and compounds. This system is applicable at the earliest stages of drug discovery, enabling identification and prioritization of compounds that are highly selective for their intended targets and that do not display other cellular effects, thus increasing the chances of success during clinical testing. We have reported on an application of the Cytometrix technologies wherein an unbiased cell morphology-based screen of 107 small molecules comprising four different kinase inhibitor scaffolds was used to identify an unexpected activity in one of the compounds deriving from its role as an inhibitor of carbonyl reductase 1 (Tanaka et al., 2005).

Cytometrix technologies is a system designed to broadly characterize the effects of compounds on cells. It includes fluidic automation for handling cells and fluorescent probes, automated microscopes for imaging relevant cellular structures, and software for image and data analysis. This chapter shows that cellular morphology changes are well correlated with compound MOA and that multivariate quantification of morphological changes across many experimental variables allows for clustering and classification of different mechanistic classes as well as for the detection of secondary mechanisms within a single class. This chapter provides general guidelines for collecting and analyzing such data, as well as detailed protocols for one particular type of experiment.

Quantifying Cellular Morphology Changes

Changes in cell cycle, cell shape, cytoskeleton organization, and protein trafficking machinery in response to compounds often result in concomitant cellular morphology changes. Accordingly, we label cells after they have been treated with a compound with a panel of fluorescent markers that broadly provide surrogate measures of these events. Hoechst, a DNA marker, reveals the morphology of the cell nucleus and cell cycle status. Antitubulin antibody DM1α is used to assess cell shape and cytoskeletal organization (Blose *et al.*, 1984). The lectin from *Lens culinaris* labels the Golgi apparatus, whose morphology is sensitive to changes in directed intracellular traffic between the nucleus and the plasma membrane (Polishchuk and Mironov, 2004). Measurements of changes in cellular and organelle morphology are taken with segmentation and data reduction algorithms that identify cell and nuclear boundaries based on the DNA and microtubule markers.

Cell types of differing origins respond differently to compounds acting by different MOAs depending on the expression levels and posttranslational modifications of their proteomes. It follows that monitoring changes across many cell types with a few markers may provide a more complex compound signature than using a single cell type with a large number of markers, as has been described (Perlman *et al.*, 2004). The example that follows used six cell lines with three markers, but the methodology is not limited to these conditions.

Cell Culture, Compound Addition, and Image Acquisition

Cell Culture

SKOV3 (ovarian epithelial cancer), A498 (kidney epithelial cancer), A549 (lung epithelial cancer), and SF268 (central nervous system epithelial cancer) are grown and maintained in RPMI media (Mediatech, Inc., Herndon, VA) with 5% fetal calf serum (FBS, HyClone, Logan, UT). DU145 (prostate epithelial cancer) cells (ATTC, Manassas, VA) are maintained in MEM with 5% FBS. HUVEC cells (VEC Technologies, Inc., Rensselaer, NY) are maintained in MCDB131 media (VEC Technologies) with 10% FBS.

Cell Plating

1. Plate 1000–1800 trypsinized (Mediatech) cells per well into 384-well plates (Costar) using a Multidrop (Thermo Labsystems, Beverly, MA). Make 3 cell plates for each drug plate to be tested, plus three control plates with no compound.

2. Incubate for 24 h at 37°, 5% CO_2.

Preparation of Compounds Plates

Compound stocks are maintained in dimethyl sulfoxide (DMSO). Compounds are serially diluted in DMSO in separate 384-well drug plates to achieve eight concentrations in 3× dilution steps using a Multimek

(Beckman Coulter Inc., Fullerton, CA). Wells are reserved on every drug plate for negative and positive controls. Negative control wells contain DMSO only and positive control wells receive eight, threefold dilutions of paclitaxel (Sigma-Aldrich-Fluka, St. Louis, MO) starting at a high concentration of 1 μM on cells.

Compound Addition

1. Add media to entire compound plate and mix using a multichannel device.
2. Add compound to three cell plates per cell line (18 cell plates total) from the diluted compound plate using a PlateTrak (CCS Packard, Torrance, CA). Compound mixed with media is added to the cells to achieve a final 0.4% DMSO concentration.
3. Incubate for another 24 h at $37°$, 5% CO_2.

Immunocytochemistry

All of the following steps are done using an automated pipetting and washing system.

1. Wash cells once with TBS buffer (Teknova, Half Moon Bay, CA).
2. Fix with 4% formaldehyde (Polysciences, Inc., Warrington, PA) made up in TBS for 1 h.
3. Wash three times with TBS buffer.
4. Block in 0.01% Triton X-100 (ICN Biomedicals, Inc., Irvine, CA) and 1% bovine serum albumin (BSA) in TBS (Teknova) for 1 h.
5. Stain with 0.01% Triton-X 100, 0.1% BSA, 5 $\mu g/ml$ Hoechst 33342 (Invitrogen, Eugene, OR), 5 $\mu g/ml$ FITC-lectin *L. culinaris* (Sigma-Aldrich-Fluka), and 3 $\mu g/ml$ rhodamine red-labeled monoclonal antibody DM1α (courtesy of Tim Stearns) for 1 h.
6. Wash three times in TBS.

Microscopy

Cells are imaged on an inverted Axiovert 100 M epifluorescence microscope (Carl Zeiss Inc., Oberkochen, Germany) with a 5× objective and a xenon lamp (Sutter Instruments, Novato, CA). Metamorph (Universal Imaging, Downingtown, PA) is used to control the motorized x, y, z stage (Prior Scientific, Rockland, MA) that moves the plate to each well, autofocuses, and takes three successive fluorescent images with an Orca 100 camera (Hamamatsu, Shizuoka Pref., Japan). Exposure times are set to minimize the number of saturated pixels in the image.

Pseudocolor overlays of the images are made in Adobe Photoshop using channel layers. The three-color overlay of the entire field of view is shown in Fig. 1A. Adobe can be used to zoom in on a portion of the field for illustrative purposes (Fig. 1B).

In the presence of compounds with different MOAs, the attributes of a cell line change is characteristic of each MOA. Several examples are shown in Fig. 2. HUVEC cells are stained with markers for the nucleus, the Golgi apparatus, and tubulin as described earlier. In the presence of DMSO control, HUVEC cells show smooth yet differently sized nuclei, slightly compact Golgi, and slightly elongated and overlapping cell shape as revealed by microtubule organization. Upon addition of compound, cellular organization changes significantly. Folimycin, a vacuolar-type ATPase inhibitor, causes an increase in the overall staining and distribution of Golgi. SB 203580, a p38 MAP kinase inhibitor, causes elongation of the cells. Vincristine, a microtubule-binding compound, causes formation of microtubule paracrystals and loss of Golgi organization. Paclitaxel, at subtoxic concentrations, causes the Golgi to compact and microtubules to reorganize. Cytochalasin D (CD), an actin poison, causes cells to collapse with two nuclei and results in the appearance of many retraction fibers. All of these morphology changes are unique and mechanism dependent.

Image and Data Reduction

Image Segmentation

Custom software is used to segment objects and extract attributes, but other commercially available software packages (such as MetaMorph, Axon ImageExpress, MatLab, and others) can achieve similar results. Segmentation of nuclei is based on a gradient method for edge detection (Cong and Vaisberg, 2005), the cells are segmented using a modified watershed algorithm on microtubule marker, and perinuclear area containing the Golgi complex is defined by expanding the mask from nuclei (Vaisberg et al., 2002) (Fig.1C).

Image Analysis

Image masks are used to locate cell and nuclear areas for multiple shape-, texture-, and intensity-related measurements. For each object identified by segmentation algorithms, collected attributes include object location, area, perimeter, form factor, and axis ratio, as well as sum, mean, variance, moment, and kurtosis of pixel intensities.

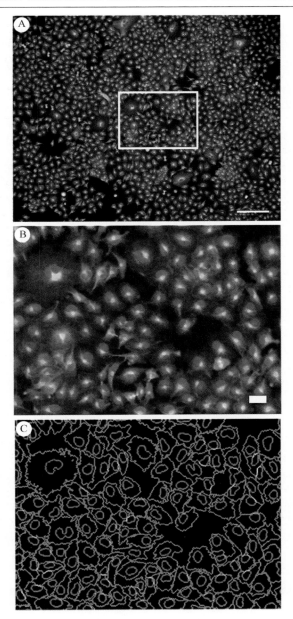

FIG. 1. Fluorescent markers and cell segmentation. (A) Image of A549 cells treated with 0.4% DMSO. The DNA marker fluorescence is colored blue, the Golgi is green, and the tubulin is red. The boxed region in A is enlarged in B. Segmentation of the same region is shown in C with DNA and tubulin segmentation outlines in red and green, respectively. Scale bars: (A) 200 μm and (B) 30 μm.

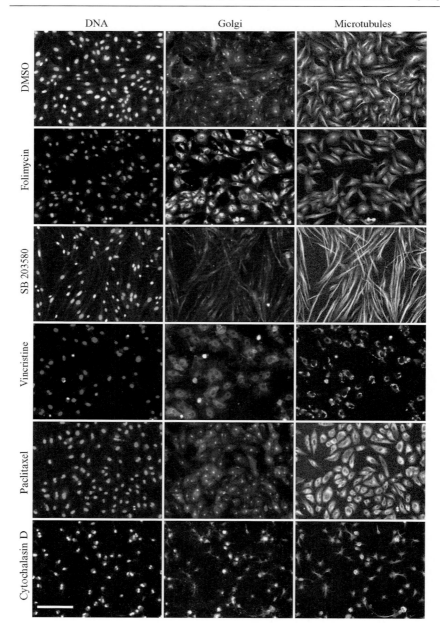

FIG. 2. Examples of compound-induced cell morphology changes. HUVEC cells in the presence of 0.4% DMSO control, folimycin (4.8 nM), SB 203580 (0.1 mM), vincristine

Object Attributes and Classification

After multiple features of individual objects are collected, each feature is summarized by a set of population statistics (called attributes) that describe distributions of a given feature across given populations of cells. In most cases we use attributes derived from subpopulations of cells in different phases of the cell cycle. Characterization of cell cycle status of each cell is performed by two algorithms that use features of cell nuclei for classification (Vaisberg and Coleman, 2002). The cell cycle algorithm classifies each cell by total intensity of its nuclei (DNA content) as gap 1, synthesis, or gap 2. The condensation algorithm classifies each cell as having condensed or not-condensed DNA using the mean intensity and area of the nuclei. We use DNA condensation as a surrogate marker for mitosis.

Attributes used for further analysis in this paper are listed in Table I. These attributes were chosen for their biological information content and relatively low cross-correlation. There are many other attributes that can be extracted from the object or image, but we have found that the set described in Table I captures most of the significant morphological reorganizations.

Data Storage

Image, feature, and attribute data, as well as experimental information, such as cell type, compound concentrations, and plate bar codes, are stored in a custom Oracle database. Actual images are stored on a file server with pointers stored in the database. Many such configurations can be designed using custom or commercial software packages, and some have been described (Swedlow et al., 2003).

Analysis of Quantitative Cellular Phenotypes across Cell Lines

A broader inspection of compound-induced cellular effects can be achieved by examining attribute changes across multiple cell lines. The compounds used in Fig. 2 can produce different effects in different cell types. Figure 3 shows quantification of image attributes for the same five compounds and concentrations on six cell lines: A498, A549, DU145, HUVEC, SF268, and SKOV3. These cell lines were chosen for their broad diversity of tissue source and genetic makeup (Scherf et al., 2000), but

---

(14 μM), paclitaxel (0.37 nM), and cytochalasin D (4.0 μM). For each condition, subregions of the entire images are shown for DNA, Golgi, and microtubule markers. Compound concentrations were chosen to highlight diverse morphology changes seen in a single cell line. Scale bar (200 μm) applies to all images.

TABLE I
Names and Descriptions of Attributes Measured

Attribute	Description
Adhesion coefficient	Relative change in cell counts upon extensive washing of wells
Average area	Average area of interphase cells in a population
Average area, mitotic	Average area of mitotic cells in a population
Average cell axes ratio	Average ratio of short axis of an ellipse that has the same second moment as a cell mask to its long axis in interphase cells
Average cell axes ratio, mitotic	Average ratio of short axis of an ellipse that has the same second moment as a cell mask to its long axis in all mitotic cells
Average cell axes ratio, preanaphase	Average ratio of short axis of an ellipse that has the same second moment as a cell mask to its long axis in preanaphase mitotic cells
Average cell form factor	Average form factor (ratio of area to square of perimeter) for all interphase cells in a population
Average cell form factor, mitotic	Average form factor (ratio of area to square of perimeter) for all mitotic cells in a population
Average cell form factor, preanaphase mitotic	Average form factor (ratio of area to square of perimeter) for preanaphase mitotic cells in a population
Average cell moment	Average moment of masks of all interphase cells in a population
Average cell moment, mitotic	Average moment of masks of all mitotic cells in a population
Average cell moment, preanaphase mitotic	Average moment of masks of all preanaphase mitotic cells in a population
Average Golgi intensity	Average of mean lectin pixel intensities for all interphase cells in a population
Average Golgi intensity, mitotic	Average of mean lectin pixel intensities for all mitotic cells in a population
Average Golgi kurtosis	Average kurtosis, or shape of the probability distribution, of lectin in interphase cells.
Average Golgi kurtosis, mitotic	Average kurtosis, or shape of the probability distribution, of lectin in mitotic cells.
Average nuclei area	Average area of nuclei of interphase cells
Average nuclei axes ratio	Average ratio of short axis of an ellipse that has the same second moment as a nuclear mask to its long axis in interphase cells
Average nuclei intensity	Average of mean Hoechst pixel intensities within nuclear masks for all interphase cells in a population
Average total tubulin intensity over DNA mask, mitotic	Average of the sum of $Dm1\alpha$ pixel intensities for all mitotic cells in a population. Only pixels contained within nuclear masks are included in computation
Average total tubulin intensity, interphase	Average of the sum of $Dm1\alpha$ pixel intensities for all interphase cells in a population

(continued)

TABLE I (*continued*)

Attribute	Description
Average total tubulin intensity, mitotic	Average of the sum of Dm1α pixel intensities for all mitotic cells in a population
Average total tubulin intensity, preanaphase mitotic	Average of the sum of Dm1α pixel intensities for preanaphase mitotic cells in a population
Average tubulin intensity	Average of mean Dm1α pixel intensities for all interphase cells in a population
Average tubulin kurtosis	Average kurtosis, or shape of the probability distribution, of Dm1α intensity in interphase cells
Average tubulin kurtosis, mitotic	Average kurtosis, or shape of the probability distribution, of Dm1α intensity in mitotic cells
Average tubulin kurtosis, preanaphase mitotic	Average kurtosis, or shape of the probability distribution, of Dm1α intensity in preanaphase mitotic cells
G1 proportion	Proportion of cells in G1 phase of cell cycle in a population
G2 proportion	Proportion of cells in G2 phase of cell cycle in a population
Mitotic index	Proportion of mitotic cells in a population
Postanaphase proportion	Proportion of postanaphase mitotic cells in a population
Preanaphase proportion	Proportion of preanaphase mitotic cells in a population
Proportion of branched cells, interphase	Proportion of interphase cells with three or more well-formed projections in a population of all cells
Proportion of branched cells, mitotic	Proportion of mitotic cells with three or more well-formed projections in a population of all cells
Relative number of cells	Relative number of cells with respect to DMSO control cells
S proportion	Proportion of cells in S phase of cell cycle in a population
Standard deviation nuclei intensity	Average standard deviation of Hoechst pixel intensities within nuclear masks for all interphase cells in a population
Standard deviation tubulin intensity	Average standard deviation of Dm1αa pixel intensities for all interphase cells in a population

many other adherent cell types may be used. Average Golgi intensity changes in HUVEC cells seen in the presence of folimycin (yellow line and arrow, Fig. 3) are seen in all other cell lines except DU145 cells. Golgi intensity changes are significant in all CD-treated cells (red line, Fig. 3) except HUVEC cells (red arrow, Fig. 3). In SB 203580-treated cells, the average cell area of A498, HUVEC, and SKOV3 cells increases, whereas A549, DU145, and SF268 remain unchanged compared to DMSO (black arrow, Fig. 3). Subtle changes in morphology are also resolved. For example, differences in Golgi morphology between DMSO and paclitaxel seen in Fig. 2 are evident in the quantification shown in Fig. 3 (green arrow). Overall, each compound has a unique signature, or combination of attribute changes, across multiple cell lines.

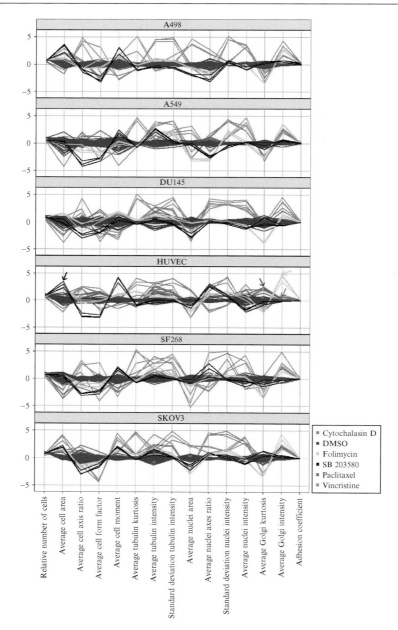

Fig. 3. Attribute profiles. A profile plot showing the relative changes in many attributes in six cell lines for six different compounds at a single concentration of each, where HUVEC

Comparing Attribute Changes Across Cell Types

1. Normalize attributes to the negative control by expressing their value as standard deviations from the mean of the DMSO control on each plate. Standard deviation of DMSO is usually stable from plate to plate; as a result pooled standard deviation from the experiment is used for the normalization.

2. Choose representative conditions for comparison, for example:

   a. One compound at the same level of growth inhibition in different cell types

   b. Multiple compounds at the same level of growth inhibition in a single cell type

   c. One compound at the same concentration in multiple cell types

3. Use SpotFire^{TM} or other graphical program to visualize changes (Fig. 3).

Additional information comes from analyzing dose–response curves (Fig. 4). Dose–response curves are best viewed using normalized features. Attributes change in both drug and cell line-dependent ways with respect to increasing doses of compound. For example, the cell axis ratio becomes larger (or the cells become rounder) in CD-treated cells in a dose-dependent manner when compared to DMSO-treated cells. A498 and HUVEC cells exhibit changes in the cell axis ratio at lowest concentrations of compound, whereas SF268 exhibit changes only after 1 μM, even though the relative effect on cell number is largely the same for all cell lines. If the cell axis ratio had been compared on different cell lines at a concentration lower than 1 μM, the morphology change of SF268 would not have been apparent.

Cell morphology changes are often detected at concentrations much lower than those needed to induce reduction in cell number. For example, SB 203580 shows subtle effects on cell number, but does change the cell axis ratio in a cell line-specific way (Fig. 4). SB 203580 causes the cell axis ratio to decrease as cells become elongated in HUVEC and A498 cells well before there are any significant changes in cell number. DU145 cell counts were most sensitive to paclitaxel treatment, but HUVEC cells were more

---

data are from the images seen in Fig. 2. The x axis lists representative attributes calculated using automated algorithms. Lines connect the values between attributes, but are themselves irrelevant. The relative difference of each attribute compared to DMSO is plotted on the y axis as the number of standard deviations from the mean of DMSO for each cell line. Individual values of DMSO deviate from the mean values for each cell line and are indicative of the noise in that feature. Colored arrows highlight feature changes in HUVEC cells for different compounds compared to other cell lines. Red, cytochalasin D (4.0 μM); dark blue, DMSO control; yellow, folimycin (4.8 nM); black, SB 203580 (0.1 mM); green, paclitaxel (0.37 nM); and light blue, vincristine (14 μM).

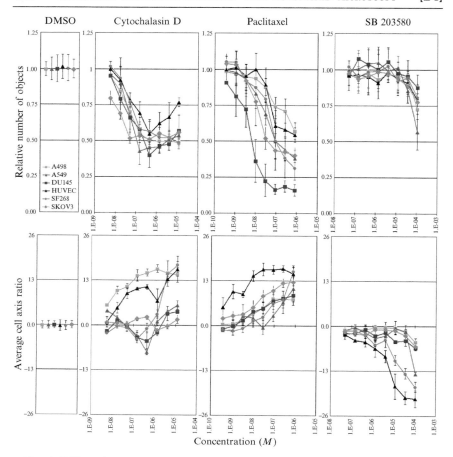

FIG. 4. Differential cell line response. The top row shows plots of the relative number of objects, compared to DMSO, against multiple concentrations of CD, paclitaxel, and SB 203580. The bottom row shows dose–response curves of the average cell axis ratio for the same concentrations of drug, where the results from each well are normalized by dividing by the median axis ratio obtained from eight DMSO control wells on the same plate. The DMSO controls are shown in the far left column. Each graph shows dose–response curves for six cell lines. Each data point is the average value of three replicate wells. Error bars show the standard deviation. Yellow square, A498; green triangle, A549; blue square, DU145; black triangle, HUVEC; red circle, SF268; purple diamond, SKOV3.

sensitive to changes in the cell axis ratio compared to the other cell lines. These data show that, depending on the mechanism of action of the compound, different cells types respond differentially. They also show that analysis of morphological attributes allows detection of effects of compounds at concentrations lower than those that cause noticeable changes in growth. The underlying causes of the cell line-specific changes may not be clear, but we hypothesize that differential responses across multiple attributes and cell types are related to mechanisms of action of compounds that generate these responses.

To test whether a combination of attributes can be used to describe MOA-related changes in cells, we applied a set of 51 commercially available compounds (Table II) that broadly represent 12 MOA classes (Table III) to six cell lines using the protocols outlined earlier. We combined normalized values of attributes collected at a single concentration of compound into a vector, or signature. In order to visually investigate patterns in the multivariate signature data set, we used principal component analysis (PCA) on signature data (Fig. 5). This linear transformation converts a number of correlated variables into an equal number of uncorrelated variables, or principal components, in such a way that first components account for as much of the variability in the data set as possible (Johnson and Wichern, 1998).

Signature Construction and Visualization

1. Collect well level attributes for all conditions (all attributes in Table I from all cell lines were chosen for the signature used to construct the PCA plots in this example).

2. Normalize attributes to negative controls and build a table of attribute data for all cell lines of interest (columns) for each compound at each concentration (rows). One can report replicate data (multiple rows at each compound concentration) or use median results (single row for a compound concentration) at this step.

3. Compute principal components.

4. Plot first two or three principal components using a graphical rendering program.

In this data set the first three principal components account for 37% of the variation. Each point represents a signature, or a compound at a single concentration. A solid line connects the signatures of each compound according to concentration and forms a multivariate dose–response curve. The multivariate dose response of a compound is the piecewise linear interpolation of its signatures ordered by concentration. In Fig. 5 the compounds are color coded by the biochemical mechanism of action listed in Table II. The dose–response curves emanate from the origin of the untreated cells' signatures. Closely adjacent points in PCA space have similar values for most of the attributes across all of the cell lines. Although adjacent points for different compounds may represent substantially different absolute

TABLE II
TEST COMPOUNDS[a]

Compound (abbreviation)	Class	Highest concentration used (μM)
2,5-Di-*t*-butylhydroquinone (DBHG)	ER-ATPase	20.6
Amsacrine	Topo 2	10.7
Butyrolactone I	Cdc2/cyclin B	10.8
Calmidazolium chloride (calmidazolium)	Calmodulin	6.6
CCCP	Mitochondria	18.0
Colcemid	Tubulin	0.6
Cyclopiazonic acid	ER-ATPase	13.6
Cytochalasin A (CA)	Actin	9.6
Cytochalasin B	Actin	9.5
Cytochalasin C	Actin	9.0
Cytochalasin D (CD)	Actin	1.4
Cytochalasin E	Actin	0.1
Cytochalasin H	Actin	2.1
Cytochalasin J	Actin	10.1
DMSO	Control	0
Dolostatin	Tubulin	5.4
Doxorubicin	Topo 2	4.7
Ellipticine	Topo 2	18.6
Epirubicin	Topo 2	2.4
Etoposide	Topo 2	7.8
Fluphenazine-*N*-2-chloroethane (FCE)	Calmodulin	8.7
Latrunculin B	Actin	11.6
Mastoparan, *Polistes jadwagae* (Mas PJ)	Gi and Go	2.8
Mastoparan, *Vespa xanthoptera* (Mas VX)	Gi and Go	2.9
Mastoparan, *Vespula lewisii* (Mas VL)	Gi and Go	3.1
Mitoxantrone	Topo 2	0.003
Nocodazole	Tubulin	0.03
Olomoucine	Cdc2/cyclin B	15.3
Phomopsin A	Tubulin	5.8
Phorbol 12,13-dibutyrate	PKC	0.7
Phorbol 12,13-didecanoate	PKC	0.3
Phorbol 12-myristate 13-acetate (PMA)	PKC	7.4
Purvalanol A	Cdc2/cyclin B	11.8
Roscovitine	Cdc2/cyclin B	12.9
Paclitaxel	Paclitaxel	2.9
Thapsigargin	ER-ATPase	7.0
Tubulozole-C	Tubulin	9.0
Vinblastine	Tubulin	0.4
Vincristine	Tubulin	1.6
GGTI-2147	GGTase1	3.4
GGTI-298	GGTase1	3.3
GGTI-286	GGTase1	3.7

(*continued*)

TABLE II (*continued*)

Compound (abbreviation)	Class	Highest concentration used (μM)
W-13	Calmodulin	13.1
W-12	Calmodulin	14.5
W-7	Calmodulin	12.2
W-5	Calmodulin	13.4
Mas 7	Gi and Go	3.1
Tyrphostin 9	Mitochondria	16.2
SKF 86002	p38 MAPK	15.4
PD 169316	p38 MAPK	12.7
SB 203580	p38 MAPK	12.1

[a] Compounds used, the abbreviations used in the text and figures, assigned MOA class, and highest concentrations used in the assay.

TABLE III
MECHANISM OF ACTION CLASSES OF TEST COMPOUNDS[a]

MOA classes	Abbreviation	Number of representatives
Actin poisons	Actin	8
Calmodulin antagonists	Calmodulin	6
Endoplasmic reticulum Ca^{2+}-ATPase inhibitors	ER-ATPase	3
G protein-coupled receptor: $G\alpha i$ and $G\alpha o$ activators	Gi and Go	4
Geranylgeranyltransferase 1 inhibitors	GGTase1	3
Oxidative phosphorylation uncoupler	Mitochondria	2
p34 cdc2/cyclin B inhibitors	Cdc2/cyclin B	4
p38 MAP kinase inhibitors	p38 MAPK	3
PKC activator	PKC	3
Topoisomerase II inhibitors	Topo II	6
Tubulin destabilizers	Tubulin	7
Tubulin stabilizer	Paclitaxel	1 (two replications)

[a] MOA classes, the abbreviations used in the text and figures, and the number of representative compounds in each class.

concentrations, the effects of compounds at those concentrations on all attributes are similar and thus likely have similar overall effects on all cell types and measurements used. Conversely, distant points represent distinct signatures, and thus attribute values, which reflect differences in effects of compounds. On the whole, signature data in PCA space show patterns of similarity and differences between mechanisms.

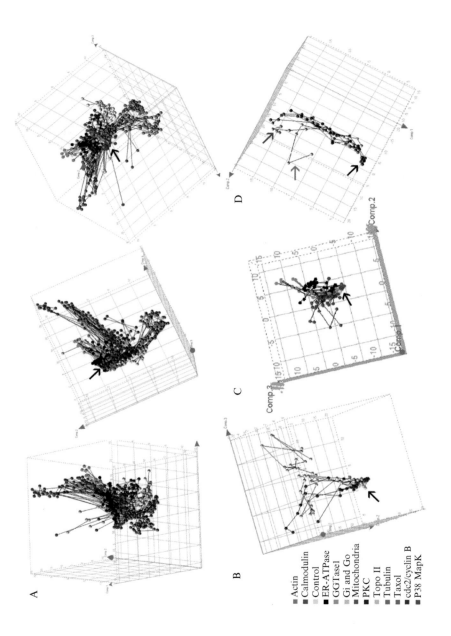

Actin
Calmodulin
Control
ER-ATPase
GGTaseI
Gi and Go
Mitochondria
PKC
Topo II
Tubulin
Taxol
cdc2/cyclin B
P38 MapK

On the PCA plot, differences among compound classes increase with concentration. Clusters of compounds corresponding to MOA are apparent. For example, compounds known to inhibit tubulin, actin, mitochondria, GGTase, and Topo II are shown to form discrete clusters in principal component space as the concentrations of the compounds increase (Fig. 5A). The classes Topo II, cdc2/cyclinB, ER-ATPase, GGTase, and mitochondria inhibitors are difficult to visualize in Fig. 5A despite the three different angles, as their signatures are masked by signatures of other compounds. When shown separately, they have unique trajectories as well (Fig. 5B and C). Overall, the distinct clusters show that cellular phenotyping can discriminate different MOAs.

Compounds can affect a common, primary target by different molecular mechanisms. This is exemplified by application of the microtubule stabilizer paclitaxel and microtubule destabilizers (Jordan and Wilson, 2004) (Fig. 5A). Changes in overall cellular morphologies at increasing concentrations of these compounds cause the classes to diverge in the PCA view. The related attributes responsible for the divergence include differences in tubulin attributes, as expected (Peterson and Mitchison, 2002) (data not shown). This example shows that cellular phenotyping can precisely distinguish compounds that affect the same target, but in different ways.

Not all compounds assigned to a particular class may actually be specific for the target of interest. For example, cytochalasin A (CA) is similar to the other actin inhibitors at low concentrations, but deviates at higher concentrations (red arrow, Fig. 5D). CA is known also to inhibit microtubules (Himes and Houston, 1976), unlike cytochalasins B, D, E, H, or J. Consistent with this broader specificity, attributes relating to tubulin, cell shape, and adhesion are differentially affected by CA at high concentrations. Latrunculin inhibits actin by binding only to monomeric actin, whereas cytochalasins (Spector *et al.*, 1989) bind to both monomeric actin and

---

FIG. 5. Principal component plots of multivariate signatures. Color legend applies to A, B, and C. Black arrows point to location of DMSO control data points in all panels. (A) Three different views of the three-dimensional plots of the first three principal components of a multivariate signature made up of 32 attributes from each cell line (198 attributes total) for replicate data sets for the compounds listed in Table III. Data points for which the relative number of cells for any cell line was less than 50% of control are not shown. Lines connect increasing concentrations of a single compound in a single well. Dose–response curves are colored by MOA as listed in the color legend. (B) All compounds from the Topo II (white), cdc2/cyclin B (blue), and control (yellow) classes. (C) All compounds from the calmodulin (dark blue), ER-ATPase (black), GGTase1 (light blue), mitochondria (purple), and control (yellow) classes. (D) Actin MOA cluster. Each color is a different actin inhibitor, with DMSO at the bottom (black arrow). Red arrow points to cytochalasin A. Blue arrow points to latrunculin.

barbed ends of actin filaments. Latrunculin clusters with actin inhibitors, but it follows a slightly deviant trajectory (blue arrow, Fig. 5D). If the mechanisms of CA or latrunculin had been unknown, cellular phenotyping would have alerted the researcher to the probability that these compounds have unintended activities or deviant mechanisms compared to the rest of the compounds in that class.

While the markers used in this experiment label only tubulin, DNA, and Golgi, the test set of compounds broadly represents MOAs of interest in drug discovery, many of which affect pathways not related to cellular components visualized by the markers. For example, the actin inhibitor class is a well-defined cluster in the PCA graph even though the organization of the actin cytoskeleton was not measured directly. In fact, most of the molecular targets of the compounds tested were not measured directly, yet inhibition of those targets causes distinct morphological changes in the cell that were detected using surrogate markers.

Clustering and Classification of Compounds

Concentration-dependent differences can also be used to form a single phenotypic descriptor. Differences between the effects of a single compound at two different concentrations can be greater than differences between the effects caused by compounds from different classes. Moreover, different compounds might cause similar changes in cells at substantially different concentrations. We thus developed a quantitative measure that describes the biological effects of a compound regardless of the efficacy (i.e., an absolute concentration that induces a particular phenotype) and that is concentration independent using an average angle dissimilarity measure. This measure aligns multivariate dose responses by their potencies, or distances from control. For every pair of aligned points on two curves, the angle between points with the origin as the vertex is computed. The dissimilarity, d, is the average of these angles, mathematically:

$$d(D_1, D_2) = \frac{1}{b-a} \int_a^b angle(D_1(p), D_2(p)) dp$$

where $D_i(p)$ is the ith dose response at potency p and a and b are the minimum and maximum common potency obtained by the two drugs. The dissimilarity is scaled between 0 and p. A dissimilarity of 0 is obtained for dose responses that are identical over their common potency range. While there are multiple signatures, there is only one dissimilarity value per compound pair per dose response. Compared to PCA, this measure is more sensitive to the subtle changes that occur at lower concentrations of compounds, which are hard to visualize by PCA. Further, the angle dissimilarity measure allows reduction of the effects of a compound on multiple cell types, attributes, and compound concentrations to a single value.

Calculation and Presentation of Angle Dissimilarity Measure Data

1. Calculate dissimilarity using normalized signature data.
2. Visualize using heat map ordered by hierarchical clustering derived from the angle dissimilarities for all pairs of compounds with the Ward clustering technique (Ward, 1963) (Fig. 6).

High correlation is seen within compound classes, as well as some similarities between compound classes. The microtubule, actin, G_i and G_o, PKC, and p38 mitogen-activated protein kinase classes all fall within their own branches of the tree (Fig. 6a–c, f, and g). There are two other branches that show mixed compound classes. In one branch (Fig. 6d), p34cdc2/cyclin B and Topo 2 classes intermix. In another branch (Fig. 6e), mitochondria, GGTase1, calmodulin, and ER Ca^{2+}-ATPase inhibitors intermix. Cytochalasin A does not appear very different from the actin inhibitors because the CA class deviates from the actin class only at high concentrations. However, the two classes are very similar at low concentrations, a finding that correlates with the PCA view. This type of plot would have been nearly impossible to interpret using the signatures at each concentration, as there would have been eight different branches for each compound and they would not have clustered together. Clustering compounds as shown in Fig. 6 is only possible by reducing multivariate dose–response data to a single variable.

While the heat map can be used to visualize correlations among compounds, the angle dissimilarity measure can be used to assign a single compound to a MOA class using a classifier (Tables IV and V). If one is interested in which class out of many predetermined classes a compound is most similar to (such as toxic vs nontoxic), this technique would be used. Overall, for 12 different MOAs we achieved 88% classification accuracy (45 out of 51 compounds were classified correctly). The confusion matrix for classification shows that all compounds in 9 out of 12 MOAs were classified correctly (Table V).

Classification

1. Find the optimal subset of attributes for a model to achieve <15% error using leave-one-out cross-validation (Hastie *et al.*, 2001). For the classification example presented earlier, we used mitotic index and G2 interphase proportion in SKOV3, average nuclei area in A549 and SF268,

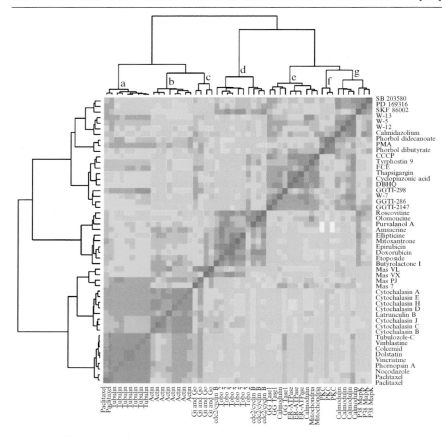

FIG. 6. Clustering using angle dissimilarity measure. Heat plot showing the correlation between the angle dissimilarity measure of each compound. The Ward clustering tree is shown on both axes. Compound names are seen along the y axis and their corresponding MOA classes are listed along the x axis. The color scale from red to yellow indicates compounds that are more or less correlated. Colors are scaled by row.

G1 mitotic proportion in DU145, average Golgi kurtosis in SF268, and average cell axis ratio in interphase cells in A498 cells.

2. Run leave-one-out cross-validation using the model to classify each compound using all other compounds as a training set (Hastie *et al.*, 2001).

  a. Calculate the square distance of each compound to the mean of each MOA class (Table IV).

  b. Assign classification of compound to class with minimum square distance.

3. Visualize using confusion matrix (Table V).

Many compounds are known to affect more than one target, and our analysis shows that at least some of the compounds tested have multiple activities. The compounds that were misclassified in Table V are calmidazolium, FCE, W-7, butyrolactone 1, amsacrine, and doxorubicin. FCE and W-7 are not tightly clustered with the other calmodulin members in the dendogram, and their distances are larger as well. Review of the PCA plot (Fig. 5D) shows that the calmodulin class of compounds is not tightly clustered and has only one or two data points clearly distinguishable from control. There may not have been a sufficiently pronounced cellular response using this assay to allow robust classification of this MOA because we observed only weak responses for all attributes and cell types tested. However, if these compounds represented a chemical series that was predicted to be specific for calmodulin, these data would warn the researcher that perhaps the compounds have additional targets. Calmidazolium, amsacrine, butyrolactone, and doxorubicin are very narrowly misclassified, as judged by the fact that the difference in the distances to the misclassified class to their assigned class is less than 0.2 (compared to a median separation distance of 2.5 ± 1.8 and a max of p^2 for all compounds to all classes) and they are tightly clustered with the other members in their class in the dendogram. Misclassification of the cdc2/cyclin B and Topo II inhibitors, butyrolactone 1, amsacrine, and doxorubicin, is expected, as compounds from both mechanisms intermix in both the clustering (Fig. 6) and the PCA plot (Fig. 5C) and may indicate the these compounds inhibit the same pathway, albeit via different targets or mechanisms. Indeed, cdc2 and topoisomerase II have been shown to interact (Escargueil et al., 2001).

Conclusions

We have shown that it is possible to successfully classify inhibitors of a diverse array of targets using Cytometrix technologies using relatively generic morphological changes in a number of different cell types by successfully classifying 45 out of 51 known compounds into 12 distinct MOA classes. We show that compounds that inhibit the same target via different MOAs can be distinguished, for example, the microtubule-binding taxanes and vinca alkaloids. Moreover, Cytometrix technologies can identify unexpected activities in members of a closely related compound series, for example, microtubule binding in the actin poison cytochalasin A. In a previous publication, Cytometrix technologies were used to discover a novel inhibitor of carbonyl reductase 1 (Tanaka et al., 2005). While the techniques outlined are specific for this one experiment, they are applicable to many different biological assay designs where cell lines, markers, time points, and reagents are varied.

TABLE IV
CLASSIFICATION RESULTS[a]

Treatment name	Assigned MOA	Classification	Correctness	Actin	Calmodulin	Cdc2/cyclin B	ER-ATPase	GGTase1	Gi and Go	Mitochondria	p38 MAPK	PKC	Paclitaxel	Topo 2	Tubulin
Cytochalasin A	Actin	Actin	Correct	0.2	3.1	3.3	5.3	2.8	1.6	2.3	3.6	1.7	0.7	4.5	1.1
Cytochalasin B	Actin	Actin	Correct	0.2	3.7	2.5	6.2	3.9	1.2	2.5	3.2	2.5	1.1	3.2	1.9
Cytochalasin C	Actin	Actin	Correct	0.3	4.6	2.1	6.8	4.2	1.4	3.4	3.0	3.3	1.5	2.5	3.1
Cytochalasin D	Actin	Actin	Correct	0.1	3.0	2.0	6.4	3.1	1.4	2.3	2.7	2.9	0.5	2.9	0.8
Cytochalasin E	Actin	Actin	Correct	0.2	3.2	3.1	4.8	2.9	1.3	2.5	3.3	2.4	0.6	3.7	0.9
Cytochalasin H	Actin	Actin	Correct	0.1	3.3	1.8	6.5	3.1	1.3	2.4	3.0	3.2	0.5	2.7	0.8
Cytochalasin J	Actin	Actin	Correct	0.2	4.4	2.9	6.7	4.0	1.4	3.1	3.0	2.2	1.2	3.1	2.3
Latrunculin B	Actin	Actin	Correct	0.7	3.5	3.1	3.0	2.9	1.0	1.6	3.4	3.2	0.9	3.5	1.4
W-12	Calmodulin	Calmodulin	Correct	4.1	0.8	3.4	2.5	1.4	2.0	1.5	1.6	2.1	1.5	5.3	4.6
W-13	Calmodulin	Calmodulin	Correct	3.1	0.1	2.8	1.5	0.5	1.7	0.4	1.1	1.8	0.6	4.5	1.3
W-5	Calmodulin	Calmodulin	Correct	6.2	0.3	3.5	1.8	0.6	2.6	1.0	1.4	2.6	1.3	5.2	3.2
Olomoucine	Cdc2/cyclin B	Cdc2/cyclin B	Correct	7.0	2.8	0.7	2.5	2.5	3.8	1.1	1.4	5.6	2.5	1.6	7.8
Purvalanol A	Cdc2/cyclin B	Cdc2/cyclin B	Correct	3.8	3.9	0.6	3.7	2.6	2.8	1.5	2.1	6.9	1.8	1.3	4.5
Roscovitine	Cdc2/cyclin B	Cdc2/cyclin B	Correct	5.7	3.9	0.3	3.8	3.3	4.1	2.1	1.2	6.7	2.6	1.0	7.8
Cyclopiazonic acid	ER-ATPase	ER-ATPase	Correct	10.0	1.5	3.6	0.2	1.2	2.6	0.6	2.8	2.9	3.8	6.3	7.3
DBHQ	ER-ATPase	ER-ATPase	Correct	8.8	1.4	4.3	0.2	0.9	2.3	0.7	2.9	2.6	3.2	6.3	5.9
Thapsigargin	ER-ATPase	ER-ATPase	Correct	8.3	1.5	3.2	0.2	0.9	2.6	0.4	3.0	3.2	3.4	5.4	6.5
GGTI-2147	GGTase1	GGTase1	Correct	6.3	0.8	3.4	1.2	0.4	2.8	0.7	1.2	2.6	2.6	5.7	5.3
GGTI-286	GGTase1	GGTase1	Correct	8.9	0.9	4.1	0.9	0.4	3.3	0.8	1.8	2.3	2.5	6.4	5.2
GGTI-298	GGTase1	GGTase1	Correct	4.1	1.3	3.5	2.0	0.9	1.9	1.3	2.7	3.0	1.0	4.4	1.2
Mas 7	Gi and Go	Gi and Go	Correct	3.5	1.1	3.3	1.6	1.3	0.7	1.2	3.2	2.6	1.1	4.1	1.8
Mas PJ	Gi and Go	Gi and Go	Correct	1.1	1.8	4.2	2.8	1.7	0.6	1.4	3.8	1.7	0.6	4.5	0.7
Mas VL	Gi and Go	Gi and Go	Correct	2.5	3.0	3.2	2.8	3.0	0.3	2.1	4.0	3.3	2.8	3.2	4.6

Compound			Assessment												
Mas VX	Gi and Go	Gi and Go	Correct	3.1	3.8	2.0	4.1	3.9	1.4	2.6	3.7	3.6	2.1	1.8	5.2
CCCP	Mitochondria	Mitochondria	Correct	6.6	1.2	2.8	0.8	0.9	2.2	0.2	2.5	3.2	2.8	4.8	6.1
Tyrphostin 9	Mitochondria	Mitochondria	Correct	4.2	1.0	2.1	0.7	0.9	2.1	0.2	2.6	2.9	1.2	4.2	2.5
PD 169316	P38 MapK	P38 MapK	Correct	5.7	1.6	1.6	3.5	1.6	4.2	2.2	0.4	3.6	1.2	2.6	7.5
SB 203580	P38 MapK	P38 MapK	Correct	6.5	1.4	2.4	2.5	1.7	3.4	2.0	0.5	2.4	1.6	3.7	8.6
SKF 86002	P38 MapK	P38 MapK	Correct	5.2	3.1	0.9	4.2	2.1	4.6	2.5	0.6	5.1	1.5	1.3	7.4
Phorbol dibutyrate	PKC	PKC	Correct	7.8	2.9	6.2	2.3	2.5	3.0	2.5	2.9	0.6	3.8	8.7	8.5
Phorbol didecanoate	PKC	PKC	Correct	4.2	2.7	6.4	3.4	2.5	2.6	2.6	3.3	0.2	2.4	9.6	4.6
PMA	PKC	PKC	Correct	2.0	2.4	6.3	3.5	1.8	2.3	1.9	3.5	0.1	1.3	7.7	1.8
Paclitaxel	Paclitaxel	Paclitaxel	Correct	1.5	1.7	1.9	4.4	1.8	1.6	1.7	2.1	3.3	0.0	2.3	0.3
Paclitaxel	Paclitaxel	Paclitaxel	Correct	1.2	1.3	2.9	4.2	1.8	1.3	1.6	1.4	2.1	0.0	3.2	0.2
Ellipticine	Topo 2	Topo 2	Correct	3.3	5.7	1.1	5.6	5.0	2.5	3.5	1.5	5.7	2.5	0.9	7.5
Epirubicin	Topo 2	Topo 2	Correct	4.4	6.0	0.8	6.0	5.6	2.2	3.5	2.1	6.7	2.0	0.3	5.6
Etoposide	Topo 2	Topo 2	Correct	2.8	5.4	0.6	5.3	4.2	2.4	2.8	2.0	8.1	1.7	0.3	5.4
Mitoxantrone	Topo 2	Topo 2	Correct	3.3	3.9	1.2	5.0	3.7	1.8	3.2	1.4	5.2	1.3	0.9	4.6
Colcemid	Tubulin	Tubulin	Correct	1.1	1.7	3.5	3.8	1.3	1.3	1.7	3.5	2.0	0.1	4.3	0.0
Dolostatin	Tubulin	Tubulin	Correct	1.1	1.9	3.0	3.6	1.6	1.1	1.7	3.7	2.9	0.1	3.6	0.0
Nocodazole	Tubulin	Tubulin	Correct	0.8	2.1	3.8	3.5	1.6	0.8	1.5	5.0	1.7	0.2	4.3	0.1
Phomopsin A	Tubulin	Tubulin	Correct	0.7	2.1	3.6	4.5	1.9	1.0	1.9	3.6	2.0	0.2	4.3	0.1
Tubulozole-C	Tubulin	Tubulin	Correct	1.2	1.7	3.2	3.4	1.3	1.2	1.5	3.8	2.5	0.1	4.1	0.0
Vinblastine	Tubulin	Tubulin	Correct	1.2	1.8	3.3	3.3	1.4	1.2	1.6	3.9	2.6	0.1	4.1	0.0
Vincristine	Tubulin	Tubulin	Correct	1.1	1.9	3.1	3.5	1.6	1.2	1.6	3.7	2.8	0.1	3.7	0.0
Calmidazolium	Calmodulin	p38 MAPK	Incorrect	6.0	1.1	3.4	1.9	1.5	1.8	1.5	1.0	1.6	1.7	5.5	6.8
FCE	Calmodulin	Mitochondria	Incorrect	3.8	1.0	2.8	0.8	0.5	1.5	0.4	2.2	2.8	1.0	4.2	1.8
W-7	Calmodulin	GGTase1	Incorrect	8.0	1.0	2.8	0.8	0.7	1.9	1.1	2.2	3.7	2.5	4.3	4.8
Butyrolactone I	Cdc2/cyclin B	Topo 2	Incorrect	2.5	4.9	1.5	6.1	5.3	2.2	4.0	1.5	5.5	2.2	1.2	6.2
Amsacrine	Topo 2	Cdc2/cyclin B	Incorrect	2.3	3.3	1.0	3.6	3.1	1.8	2.0	1.6	6.0	1.7	1.1	4.6
Doxorubicin	Topo 2	Cdc2/cyclin B	Incorrect	5.4	4.5	0.4	4.9	4.0	2.6	2.7	1.6	8.0	2.0	0.6	5.8

[a] The classification assignment of each compound and assessment of correctness and the square distance of each compound to the mean of each MOA class are shown. Lower numbers denote more similarity.

TABLE V
CONFUSION MATRIX OF CLASSIFICATION[a]

	Actin	Calmodulin	Cdc2/cyclin B	ER-ATPase	GGTase1	Gi and Go	Mitochondria	p38 MAPK	PKC	Paclitaxel	Topo 2	Tubulin	Grand total
Actin	8												8
Calmodulin		3											3
Cdc2/cyclin B			3								2		5
ER-ATPase				3									3
GGTase1		1			3								4
Gi and Go						4							4
Mitochondria		1					2						3
p38 MAPK		1						3					4
PKC									3				3
Paclitaxel										2			2
Topo 2			1								4		5
Tubulin												7	7
Grand total	8	6	4	3	3	4	2	3	3	2	6	7	51

[a] Summary of Table IV such that compounds in the MOA classes listed in the first row are scored as having been classified as any of the other MOAs listed in the first column. A correct classification is one where the compound being classified is classified as its assigned MOA, or along the diagonal. Misclassifications are seen off the diagonal.

The multivariate approach to analyzing image data has many additional applications beyond the data set discussed in this chapter, including but not limited to siRNA and compound screening, toxicity profiling, and clinical pharmacodynamics. There have been many studies in the literature to which this approach may be applied, for example, to identify new compounds similar to desired control compounds or with novel phenotypes (Haggarty *et al.*, 2000; Mayer *et al*, 1999; Yarrow *et al.*, 2003) or the approach could have been applied to characterize morphological changes observed in siRNA screens (Goshima and Vale, 2003; Kiger *et al.*, 2003) or switch-of-function mutation screens (Heo and Meyer, 2003). The approach can be applied using siRNA phenotypes as control morphologies to suggest primary screening targets, as has been done manually (Eggert *et al.*, 2004). When coupled with toxicity data and profiles, appropriate assay combinations may enable safety profiling of compounds.

Other high-content imaging assays have been described and most share similar approaches to data collection and image quantification as those described here (Abraham *et al.*, 2004). However, most of these data are collected using one cell line with many markers, and it is analyzed primarily using dose responses of single features, which tend to be simple averages of image data that do not distinguish subpopulations of cells. Perlman *et al.* (2004) tested 100 compounds on HeLa cells using 11 markers and presented clustering data using a different method than that described here. While they used a different set of compounds, it is noteworthy to state that in this publication only 61 of the compounds showed any morphological changes, and we suspect that additional cell lines would have increased the number of responding compounds. Also, while they used different analytical techniques, some of the mechanistic classes did not cluster together, including the actin inhibitors, which clustered robustly in our study. It remains to be seen how the results from both studies would change using the different analyses.

Most cell-based assays that measure cell death or gene or protein expression quantify averages for cell populations or the concentration at which a singular effect is achieved. A cell-based screening approach that uses multiple GI50 measurements was developed at the National Cancer Institute (Weinstein *et al.*, 1997). This approach is based on constructing profiles of compounds based on their growth inhibition effects in a panel of 60 cell lines and then using that multivariate profile to compare mechanisms. The inherent limitations of this method are the lack of activity seen for compounds that do not cause significant growth inhibition and the logistics of handling many cell lines. The strength of our approach lies in the discovery that morphological changes in cells can precede measurable growth inhibition by over two orders of magnitude and that these changes are cell line dependent. Thus, quantification of the morphology of cellular structures may be more predictive of an *in vivo* response.

Microarray technology has been used for discovering molecular targets and for classifying compounds (Gunther *et al.*, 2003; Hughes *et al.*, 2000; Marton *et al.*, 1998). One of the differences between microarrays and Cytometrix technologies is that the former approach generally relies on a measure of the population average. In contrast, Cytometrix algorithms measure the response of individual cells and thus are sensitive to changes in as low as 10% of the population (data not shown). A 10% change may be in the noise level for microarray analysis. Ultimately, we expect Cytometrix technologies and microarray analysis to be complementary. Integration of data sets from these two forms of analysis therefore likely will have additional advantages.

In conclusion, we have introduced a new approach for generically quantifying the effects of compounds on cells. Cytometrix technologies are highly flexible and can be configured to characterize compounds on multiple cell lines, time points, markers, and concentrations, depending on the application. Cellular compound activity can be measured quantitatively using a multivariate experimental and analytical approach. We showed that the differential effects of compounds on cells of different genetic backgrounds facilitate discrimination among MOAs and that measuring a full range of biological activity provides a powerful approach to novel compound characterization and identification.

Acknowledgments

We gratefully acknowledge G. Alexander, J. Armstrong, R. de la Rosa, C. Elliot, J. Finer, A. Fritsch, E. Kiros, S. Kriz, P. Julius, C. Lei, P. Ling, P. McMahon, R. Moody, B. Muller, K. Penhall-Wilson, D. Platz, L. Reddy, A. Rao, C. Shumate, G. Singh, D. Snyder, K. Suekawa, B. Tang, R. Venkat, W. Zink, S. Wu for their technical contributions to the project; C. Beraud, L. Goldstein, B. Jack, F. Malik, J. Sabry, S. Smith, J. Spudich, R. Vale, and K. Wood for their support and D. Drubin, C. Hazuka, D. Lenzi, M. Maxon, S. Ramchandani, and R. Sakowicz for their support and critical reading of the manuscript.

References

Abraham, V. C., Taylor, D. L., and Haskins, J. R. (2004). High content screening applied to large-scale cell biology. *Trends Biotechnol.* **22,** 15–22.

Blose, S. H., Meltzer, D. I., and Feramisco, J. R. (1984). 10-nm filaments are induced to collapse in living cells microinjected with monoclonal and polyclonal antibodies against tubulin. *J. Cell Biol.* **98,** 847–858.

Cong, G., and Vaisberg, E. (2005). Extracting shape information contained in cell images. U.S. Patent 6,956,961.

Eggert, U. S., Kiger, A. A., Richter, C., Perlman, Z. E., Perrimon, N., Mitchison, T. J., and Field, C. M. (2004). Parallel chemical genetic and genome-wide RNAi screens identify cytokinesis inhibitors and targets. *PLoS Biol.* **2,** 0001–0009.

Escargueil, A. E., Plisov, S. Y., Skladanowski, A., Borgne, A., Jeijer, L., Gorbsky, G. J., and Larsen, A. K. (2001). Recruitment of cdc2 kinase by DNA topoisomerase II is coupled to chromatin remodeling. *FASEB J.* **15,** 2288–2290.

Goshima, G., and Vale, R. D. (2003). The roles of microtubule-based motor proteins in mitosis: Comprehensive RNAi analysis in the *Drosophila* S2 cell line. *J. Cell Biol.* **162,** 1003–1016.

Gunther, E. C., Stone, D. J., Gerwien, R. W., Bento, P., and Heyes, M. P. (2003). Prediction of clinical drug efficacy by classification of drug-induced genomic expression profiles *in vitro*. *Proc. Natl. Acad. Sci. USA* **100,** 9608–9613.

Haggarty, S. J., Mayer, T. U., Miyamoto, D. T., Fathi, R., King, R. W., Mitchison, T. J., and Schreiber, S. L. (2000). Dissecting cellular processes using small molecules: Identification of colchicine-like, paclitaxel-like and other small molecules that perturb mitosis. *Chem. Biol.* **7,** 275–286.

Hastie, T., Tibshirani, R., and Friedman, J. H. (2001). Model assessment and selection. *In* "The Elements of Statistical Learning: Data Mining, Inference, and Prediction," pp. 214–217. Springer-Verlag, New York.

Heo, W. D., and Meyer, T. (2003). Switch-of-function mutants based on morphology classification of Ras superfamily small GTPases. *Cell* **113,** 315–328.

Himes, R. H., and Houston, L. L. (1976). The action of cytochalasin A on the *in vitro* polymerization of brain tubulin and muscle G-actin. *J. Supramol. Struct.* **5,** 81–90.

Hughes, T. R., Marton, M. J., Jones, A. R., Roberts, C. J., Stoughton, R., Armour, C. D., Bennett, H. A., Coffey, E., Dai, H., He, Y. D., Kidd, M. J., King, A. M., Meyer, M. R., Slade, D., Lum, P. Y., Stepaniants, S. B., Shoemaker, D. D., Gachotte, D., Chakraburtty, K., Simon, J., Bard, M., and Fried, S. H. (2000). Functional discovery via a compendium of expression profiles. *Cell* **102**, 109–126.

Johnson, R. A., and Wichern, D. W. (1998). Principal components. *In* "Applied Multivariate Statistical Analysis," 3rd Ed. pp. 356–380. Prentice-Hall, Upper Saddle River, NJ.

Jordan, M. A., and Wilson, L. (2004). Microtubules as a target for anticancer drugs. *Nature Rev. Cancer* **4**, 253–265.

Kiger, A. A., Baum, B., Jones, S., Jones, M. R., Coulson, A., Echeverri, C., and Perrimon, N. (2003). A functional genomic analysis of cell morphology using RNA interference. *J. Biol.* **2**, 27.1–27.15.

Marton, M. J., DeRisi, J. L., Bennett, H. A., Iyer, V. R., Meyer, M. R., Roberts, C. J., Stoughton, R., Burchard, J., Slade, D., Dai, H., Bassett, D. E., Jr., Hartwell, L. H., Brown, P. O., and Friend, S. H. (1998). Drug target validation and identification of secondary drug target effects using DNA microarrays. *Nature Med.* **4**, 1293–1301.

Mayer, T. U., Kapoor, T. M., Haggarty, S. J., King, R. W., Schreiber, S. L., and Mitchison, T. J. (1999). Small molecule inhibitor of mitotic spindle bipolarity identified in a phenotype-based screen. *Science* **286**, 971–974.

Perlman, Z. E., Slack, M. D., Feng, Y., Mitchison, T. J., Wu, L. F., and Altschuler, S. J. (2004). Multidimensional drug profiling by automated microscopy. *Science* **306**, 1194–1198.

Peterson, J. R., and Mitchison, T. J. (2002). Small molecules, big impact: A history of chemical inhibitors and the cytoskeleton. *Chem. Biol.* **12**, 1275–1285.

Polishchuk, R. S., and Mironov, A. A. (2004). Structural aspects of Golgi function. *Cell. Mol. Life Sci.* **61**, 146–158.

Prentis, R. A., Lis, Y., and Walker, S. R. (1988). Pharmaceutical innovation by the seven UK-owned pharmaceutical companies. *Br. J. Clin. Pharmacol.* **3**, 387–396.

Pritchard, J. F., Jurima-Romet, M., Reimer, M. L., Mortimer, E., Rolfe, B., and Cayen, M. N. (2003). Making better drugs: Decision gates in non-clinical drug development. *Nature Rev. Drug Discov.* **7**, 542–553.

Scherf, U., Ross, D. T., Waltham, M., Smith, L. H., Lee, J. K., Tanabe, L., Kohn, K. W., Reinhold, W. C., Myers, T. G., Andrews, D. T., Scudiero, D. A., Eisen, M. B., Sausville, E. A., Pommier, Y., Bostein, D., Brown, P. O., and Weinstein, J. N. (2000). A gene expression database for the molecular pharmacology of cancer. *Nature Genet.* **24**, 236–244.

Spector, I., Shochet, N. R., Blasberger, D., and Kashman, Y. (1989). Latrunculins: Novel marine macrolides that disrupt microfilament organization and affect cell growth. I. Comparison with cytochalasin D. *Cell Motil. Cytoskel.* **13**, 127–144.

Swedlow, J. R., Goldberg, I., Brauner, E., and Sorger, P. K. (2003). Informatics and quantitative analysis in biological imaging. *Science* **300**, 100–102.

Tanaka, M., Bateman, R., Rauh, D., Vaisberg, E., Ramchandani, S., Zhang, C., Hansen, K. C., Burlingame, A. L., Trautman, J. K., Shokat, K. M., and Adams, C. L. (2005). An unbiased cell morphology-based screen for new, biologically active small molecules. *PloS Biol.* **3**, 0764–0776.

Vaisberg, E., and Coleman, D. A. (2002). Classifying cells based on information contained in cell images U.S. Patent 6,876,760..

Vaisberg, E. A., Cong, G., and Hsien-Hsun, W. (2002). Image analysis of the Golgi complex. World International Property Organization 02/067195.

Ward, J. H. (1963). Hierarchical grouping to optimize an objective function. *J. Am. Stat. Assoc.* **58**, 236–244.

Weinstein, J. N., Myers, T. G., O'Connor, P. M., Friend, S. H., Fornace, A. J., Jr., Kohn, K. W., Fojo, T., Bates, S. E., Rubinstein, L. V., Anderson, N. L., Buolamwini, J. K., van Osdol, W. W., Monks, A. P., Scudiero, D. A., Sausville, E. A., Zaherevitz, D. W., Bunow, B., Viswanadhan, V. N., Johnson, G. S., Wittes, R. E., and Paull, K. D. (1997). An information-intensive approach to the molecular pharmacology of cancer. *Science* **275,** 343–349.

Yarrow, J. C., Feng, Y., Perlman, Z. E., Kirchhausen, T., and Mitchison, T. J. (2003). Phenotypic screening of small molecule libraries by high throughput cell imaging. *Comb. Chem. High Throughput Screen.* **6,** 279–286.

[25]   High-Content Screening: Emerging Hardware and Software Technologies

By Seungtaek Lee and Bonnie J. Howell

Abstract

The field of high-content screening has flourished since 2000 with advancements in automated fluorescence microscopy technologies, fluorescent labeling techniques, and sophisticated image analysis software. Through the use of these technologies, researchers can now monitor cellular and molecular events in individual cells *in vitro* following drug treatment or RNAi and rapidly screen compound and siRNA libraries. This chapter discusses current and next-generation hardware and software features and capabilities.

Introduction

High-content screening (HCS) has gained tremendous popularity in the past few years in the drug discovery industry from advances in fluorescence microscopy and automated screening technologies. HCS provides an opportunity to rapidly screen chemical or siRNA libraries by imaging subcellular and molecular events of individual cells with an automated fluorescent microscope. HCS gets its name from the rich and diverse set of information provided by analyzing the phenotype of whole cells. The ability to multiplex with multiple markers or fluorescent probes and even multiple cell lines simultaneously allows for more efficient screens for target identification, target validation, lead identification, and lead optimization. Although fluorescence microscopy and digital image analysis have been around for decades, the complete integration and automation

METHODS IN ENZYMOLOGY, VOL. 414
0076-6879/06 $35.00
DOI: 10.1016/S0076-6879(06)14025-2

of cell-based assays, image capture, and image analysis exploded into the pharmaceutical scene in the late 1990s. The use of this technology has also been fueled by advances in molecular cloning, fluorescent proteins, for example, green fluorescent protein (GFP), and vast arrays of immunolabeling kits. Currently, more than 10 vendors worldwide provide sophisticated HCS systems, ranging from low-throughput assay development platforms to ultrahigh-throughput screening systems. These systems have become a rising trend in drug discovery as exploratory platforms as well as primary and secondary screening tools.

There are four main components to successful, reproducible HCS campaigns: cell preparation and labeling, image acquisition, image analysis, and data management. Each component has an important role and scientists must carefully design assays to ensure adequate image capture, quantification of biological events of interest, and discrimination between on-target and off-target effects. Important factors include use of validated reference/control molecules, choice of cell line(s), kinetic vs end-point readout, fluorescent labeling techniques, imaging approaches (wide-field fluorescence, confocal, bright field, etc.), image analysis modules, and data presentation. This chapter outlines the importance of these four HCS components, using the GPCR Transfluor assay as an example/case study.

Cellular Assay and Imaging Preparation

An important factor for HCS success is cell line choice and cell preparation. As with most cellular assays, adherent cells are ideal but HCS is also amendable to suspension or semisuspension formats. For adherent-based assays, cells are first dispensed into the wells of microtiter plates (96- or 384-well plates) and allowed to adhere to the bottom surface. Optimal cell attachment is not only important for proper biological response, but also for ease of capturing a biological process within a single imaging plane. Occasionally, plates are coated with poly-D-lysine, fibronectin, collagen, or other extracellular matrices to promote cell adherence and cell spreading. For semisuspension or suspension cells or to minimize cell loss due to reagent addition/wash steps, centrifugation steps can be added to the routine.

For a chemical library screen, compounds are transferred to the assay or cell plates and incubated at 37° for an appropriate duration. Depending on the assay, some reagents may be added before or after compound addition. For an end-point assay, cells are fixed after drug treatment to retain physiological state and antigen (protein) distribution. In some cases, permeabilization is performed to allow for antibodies to reach the antigens within the cell. Two of the most common fixation methods include paraformaldehyde or ice-cold methanol. A typical immunofluorescence protocol involves numerous blocking and washing before, in between, and after primary and secondary binding steps. This conventional staining protocol is very time-consuming and difficult to miniaturize for robotic screening systems. Some of these steps have been combined and truncated to conserve time and to make the assay robotic friendly. Advantages of fixed-cell kinetic studies include multiplexing capabilities, stable signal for convenient imaging, flexibility in cell line choice, and large availability of antibodies. Some disadvantages include fixation and permeabilization artifacts, static information, and laborious labeling protocols.

Fluorescence microscopy, when applied to living cells, provides critical insight into how proteins behave in time and space and offers a clearer understanding of protein function and biological processes. Recent advances in molecular cloning, optical imaging, and the discovery of fluorescent proteins have fueled live-cell imaging and offer a nondestructive method for studying dynamic processes *in vivo*. An example is GFP-β-arrestin Transfluor technology from Norak Biosciences (now Molecular Devices) (Oakley *et al.*, 2002). This assay can be used to detect and screen for compound activity against G-protein-coupled receptor (GPCR) targets. The assay is based on the desensitization and association of GPCRs with β-arrestin upon ligand binding. When a GPCR is activated, it binds to a cytoplasmic protein called β-arrestin. This interaction inactivates or desensitizes the receptor signaling, which is followed by receptor internalization for recycling. Tagging β-arrestin with GFP allows for live-cell monitoring of GPCR activation with fluorescence microscopy. Before GPCR activation, β-arrestin is distributed diffusely in the cytoplasm. Upon receptor activation, β-arrestin translocates to the plasma membrane and binds activated receptor where it can be visualized as punctate pits. Depending on the binding affinity between the receptor and β-arrestin, internalization of the receptor–β-arrestin complex via endocytic vesicles can be visualized as well. This assay can also be combined with conventional immunofluorescence microscopy as the fluorescence of GFP remains after fixation of cells with paraformaldehyde.

Image Acquisition

A second component to successful HCS campaigns is hardware choice and image acquisition. Most HCS instruments are based on an inverted fluorescent microscope; however, various components and peripherals make each machine unique. Users should consider the following features when purchasing an imaging system and defining image quality: wide field versus confocality, excitation source (lamp or laser), objectives for magnification and collection of emitted fluorescence, excitation/emission filters and detectors for proper light separation, autofocus mechanism, stage precision, environmental control, robotic load, throughput, and software for controlling the hardware. Detailed discussions of these principal hardware features, as well as caveats to keep in mind when selecting a system to maximize image quality, are discussed next. Please refer to Table I for a summary of these features.

Wide-Field Versus Confocal Systems

There are several factors to consider when selecting an imaging platform for generating HCS data: sensitivity of detection, flexibility in wavelengths, speed of acquisition, and specimen viability. In wide-field microscopy, the entire specimen in the field of view is illuminated with light from a mercury or xenon lamp source, and the emitted fluorescence is captured by the objective (Fig. 1). All fluorescence emitted from the sample (from both the focal plane and the out-of-focus area below and above) is projected onto a charge-coupled device (CCD) chip for a set amount of exposure time. Hence, the throughput of the imager is somewhat dependent on the exposure time and the platform may not be suitable for thick specimens. These platforms, however, offer flexibility in wavelength choice, fast image acquisition, and low cost. Also, due to their wide depth of field, wide field has an advantage when imaging heterogeneous cell population with varying cell thickness. A number of wide-field systems are available commercially, including but not limited to the IN Cell Analyzer 1000 from GE (www.amersham.com) and the Arrayscan from Cellomics (www.cellomics.com).

In confocal microscope systems, blurry or out-of-focus light is removed by two changes to the conventional wide-field microscope: (1) laser illumination and (2) a confocal pinhole or a barrier to block out-of-focus light. The former provides better illumination of the specimen, whereas the latter prevents out-of-focus light from entering the hole/slit and reaching the detector. Because of these features, confocal microscopy provides sharper images and improved subcellular resolution through decreased background fluorescence and shallow depth of field but decreased throughput, as the field of view must be raster scanned with a single point of a laser and collected point by point. Fortunately, this issue has been improved through adjustments and/or additions to the microscope. For example, one variation is the use of a line-scanning technique as with IN Cell 3000 from GE (www.amersham.com). With this system, a horizontal laser line is used to excite the field in the Y direction and an adjustable slit is used to block out-of-focus light. Although some vertical or Z resolution is sacrificed (Cox, 2002), the increase in scanning rate is improved significantly

TABLE I

ADVANTAGES AND DISADVANTAGES OF THE VARIOUS HCS SYSTEM FEATURES/PERIPHERALS

Features	Options	Advantages	Disadvantages
Fluorescent technology	GFP or other fluorescent proteins	Live-cell imaging, easy assay protocol, temporal and space information	Time to develop stable cell line, heterogeneous population and activity interference
	Immunofluorescence	Availability of antibodies, ability to multiplex, stable signal, flexibility in cell line choice	Difficult assays steps, more washes, non-specific staining or artifacts, no temporal information
Excitation source	Lasers	Stronger illumination for confocal imaging	Limited wavelengths
	Lamp (mercury/xenon)	Wide range of wavelengths available	Longer exposure time
Objectives	Air/dry	Can image plastic plates with various thickness	Lower NA (numerical aperture)
	Water/oil	Higher NA; images are brighter and sharper	Only thin, cover glass thickness plates can be used
Autofocus	Laser based	Very fast and independent of well content	Cannot change from well to well
	Image based	Optimizes focus on well content	Significantly lowers throughput when wells are empty
Microscopy	Wide field	Better imaging of heterogeneous cell population	Poor Z resolution, high background, long exposure times
	Line scanning	High throughput with much larger field of view	Semiconfocal
	Nipkow spinning disk	High throughput with true confocality	Longer exposure times
CCD camera	No binning (full resolution)	Best resolution	Longer acquisition time
	Binning (2 × 2, 4 × 4, etc.)	Faster acquisition with shorter exposure time	Loss in resolution
Plate type	Plastic	Great cell attachment and viability, less costly	Loss is image quality
	Glass/film	Optimal for bright and sharp images	Poor cell viability, expensive

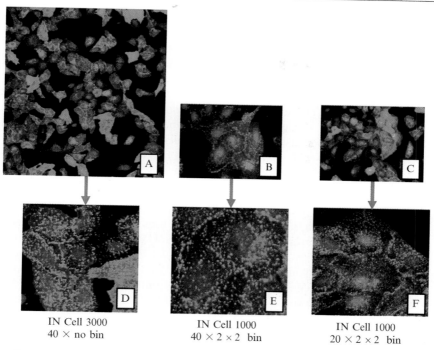

IN Cell 3000   IN Cell 1000   IN Cell 1000
40 × no bin   40 × 2 × 2 bin   20 × 2 × 2 bin

FIG. 1. Sample images of Transfluor technology taken from plastic bottom plates using wide-field and confocal (line-scanning) microscopes. Images A and D were captured with the IN Cell 3000 line-scanning confocal imaging system with its 40× objective with no binning. Image A is a full-field view (zoomed out), and image D is a cropped full-resolution image (100% zoomed). IN Cell 3000 can acquire around 200 cells per image with very fine resolution of 1-μm-sized clathrin-coated pits (shown in green). Images B and E were captured with the IN Cell 1000 wide-field fluorescent imaging system with a 40× object with binning of 2 × 2. Image B is a full-field view (zoomed out), and image E is a cropped full-resolution image (100% zoomed). With a 40× objective, only about 12 cells can be acquired per image but the pits can be resolved easily. Images C and F were also captured with the IN Cell 1000 but with a 20× objective with binning of 2 × 2. Image C is a full-field view, and image F is a cropped full-resolution image. With the 20× objective, around 50 cells can be acquired per image with sufficient resolution of the pits.

(Fig. 1). A second approach has been to include a Nipkow spinning disk as seen on the Opera from Evotec (www.evotec-technologies.com) and Pathway HT from BD Biosciences (www.atto.com). This technology utilizes a spinning disk with a large number of pinholes in which the light source passes through to excite the specimen (Cox, 2002). Multiple points are excited at once and multiple foci are generated and captured simultaneously with a CCD camera. The emitted light from the sample returns through the same set of pinholes in which the out-of-focus light is blocked from reentering. Therefore, a large number of points are collected at a time, and the spinning property of the system allows fast scanning and acquisition of the field of view, resulting in near real-time image capture and decreased photobleaching. Unfortunately, illumination strength is sacrificed greatly when the excitation light is mostly hindered by the holes on the spinning disk (Cox, 2002). Therefore, longer exposure time may be necessary, thereby impacting throughput. In both the line-scanning system and the Nipkow spinning disk system, the acquisition rates are sufficient for live-cell imaging, as well as high-throughput fixed-cell applications.

Excitation Source and Filters

With a lamp, one could have a wide range of selection for their excitation wavelengths with the appropriate exciter filters. Xenon lamps release a spectrally uniform intensity profile from the UV range (300) to far red (>700). Mercury light sources releases energy of similar profile but is much stronger at the following wavelengths: 365, 400, 440, 546, and 580 nm (Herman, 1998). Exciter filters are used in front of a lamp to select the wavelength of choice for transmission while blocking the rest of the spectrum from passing the filter. HCS systems that utilize lamps for excitation will have multiple positions (in a filter wheel) for a range of exciter filters that can be moved into position automatically for sequential multicolor capture.

Laser-based excitation systems are bound to the specified wavelengths and intensities of the lasers. Lasers can be replaced and/or added but they are serious projects involving realignment of optics and other components as well as high cost. Also, due to the space restrictions, systems may only be installed with three or four laser lines. Each laser line outputs only a single wavelength, such as 488 argon ion and 633 Red-HeNe laser systems. However, because there are multiple excitation sources, multiple color fluorochromes may be excited at once, permitting simultaneous multicolor image acquisition (with appropriate dichroic, emission filters, and multiple detectors). Therefore, laser-based imagers have an advantage in throughput when it comes to multicolor imaging.

Objectives

Most HCS systems have objectives ranging from 4 to 60× magnification from various microscope manufacturers (i.e., Nikon and Olympus). The objective(s) one requires will depend on their particular application, but in general, 10× objective (0.25 NA) is a minimum in order to observe and monitor subcellular components and molecular interactions. Higher objectives with a higher numerical aperture (NA) can be used to capture brighter and sharper images; however, throughput (number of cells captured per second) will be sacrificed greatly. Sufficient data points (cells) are required in order to correctly represent the response of the cells in the well, partly due to the heterogeneous cell population and diverse cell phenotype. Accordingly, number of cells captured per image is a significant factor in throughput. Some systems offer only a single objective, either 20 or 40×. Generally, assuming semiconfluent population of cells, around 50 cells are captured per image using a 20× objective (0.45 NA) while only around 12 cells are captured per image using a 40× (0.6 NA) objective. Differences in the full-field image and a fully zoomed image between 20 and 40× objectives are shown in Fig. 1. With higher objectives and/or underconfluent wells, more images or frames are captured per well to collect sufficient cell data.

Detector

The most commonly found detectors in HCS systems are highly light-sensitive, cooled CCD cameras. CCDs are basically an array of photodiodes (referred to as pixels or picture elements) that collect and convert the accumulated photons from emitted fluorescence from the sample to electronic charge. The converter translates the analog signal to digital values that are analyzed by the computer (optimal microscopy primer; http://www.micro.magnet.fsu.edu/primer/index.html). Twelve-bit CCD cameras will provide 4096 grayscale values at each pixel whereas 8 bit provides only 256 values. Therefore, a higher bit-depth implies greater range of image data so that the digital values can represent the actual image more accurately. Most HCS systems will offer a 12-bit camera(s). Also, most cameras have the option to bin the pixels, or combine signals from multiple pixels. Binning results in a higher signal-to-noise ratio, greater sensitivity to weak fluorescence, and faster acquisition rate, but with the sacrifice of reducing image resolution.

Autofocus Mechanisms

Autofocus of specimens is achieved either through contrast- or laser-based approaches. In the image-based approach, images of cells at different Z planes are acquired and analyzed quickly, and the optimal plane with the sharpest focus of the cells or features of interest is selected. This technique allows for very accurate focus at each well independent of the plate irregularity and unevenness. It is also able to sharply image heterogeneous population (cells with different thickness or focus plane) from well to well, as the focus is based on the content of each well. However, disadvantages include slower throughput and inability to focus on wells with a low cell count. Cell Lab IC 100 from Beckman Coulter (www.beckman.com) utilizes this technology.

A laser-based autofocus uses an IR beam and a detector to identify the interface between the plate surface and the solution in the well where the reflective index changes. With this method, the autofocus quality is not determined by the content in the wells. In cases where there is a lack of cells in the wells due to toxicity or low cell count, image-based systems will fail while attempting to find cells to focus on. Hence, the laser-based autofocus has a significant advantage in this regard as cells are prone to detaching from the plate when they are undergoing mitosis or apoptosis. IN Cell Analyzer 1000, 3000, and Evotec Opera are some of the instruments that utilize this technology.

Environmental Control/Kinetic Imaging

Many systems offer environmental control on the instrument for live cell and kinetic imaging. The instruments are usually enclosed by a housing to regulate temperature and CO_2 levels. These systems emulate the environment of an incubator to allow for cells to grow and respond to treatments under optimal conditions while imaging simultaneously. Nontoxic, cell-permeable dyes or cell lines with fluorescently tagged protein (i.e., GFP) will be required for this application. Liquid dispensers are also available within the instrument to perform complete kinetic assays all within the system. Live cellular images can be acquired immediately before and after reagent dispense. These tools allow for live monitoring of molecular interactions within complex cellular systems under optimal conditions.

Image Analysis

A third component to successful HCS campaigns is image analysis. Each imager is supplied with software that includes image analysis capability in one way or another. In most cases, vendors supply a list of "canned" modules that perform analysis for a specific assay (receptor trafficking, object intensity, translocation, cell cycle analysis, etc.). An image analysis algorithm in an HCS context is a step-by-step procedure of analyzing the contents of the image for a particular event(s) in each cell. Canned modules are designed to be turnkey solutions for quantifying a particular biological event or changes in intensity.

As HCS assays become more complex, tools must be available for modifying existing modules and creating new algorithms that allow flexibility in measuring phenotypic results. Presently, some software packages have an open environment where programmers can access the code and commands of the algorithms in a specified programming environment (C++, MATLAB, etc.). Also, many packages provide Application Programming Interface (API), which allows third-party software (i.e., database, image capture software) to access tools or parts of the software for further modification, addition, and automation. A few provide scripting (macro) abilities where nonprogrammers can create a new protocol/algorithm by using the available image analysis tools, e.g., Acapella by Evotec (www. evotec-technologies.com), Developer by GE Healthcare (www.amersham. com.), and Cellenger by Definiens (www.definiens.com). The latter option is preferred, as scientists without programming experience can modify or even create algorithms for their specific applications.

When the images are analyzed, numerous outputs, hence the high content, are generated for each image or well. Simple and common outputs include nuclei/cell count, size, and shape, average intensity at the region of interest, and other attributes from the segmented object are generated for each well. In addition to these data, more complex phenomenon can be represented by numerical outputs such as number of internalized vesicles per cell, cell toxicity, apoptosis, cell cycle phase, and many others. Therefore, each well can generate many numerical outputs that may be useful in a given assay.

Image analysis can be performed on the individual cell level or on the whole field level. Cell-by-cell or per cell analysis provides measurement and statistics on the feature of interest for each cell. Whole-field analysis simply calculates the average value of the cellular event in the entire field by dividing the total sum value of the event by the total number of cells. For example, when obtaining the average number of vesicles per cell, whole-field analysis will simply count the total number of vesicles in the image and divide by the total number of cells. In individual cellular analysis, the number of vesicles in each cell will be obtained before calculating the field average (Fig. 2). The cell-by-cell analysis provides numerous advantages, such as subpopulation evaluation, cell filtering or selection, individual cell statistics for detailed analysis, and many others. In the case of Fig. 2, cells expressing little or no GFP and cells with very small nucleus (dead or apoptotic cells) were excluded in the cell by cell analysis; therefore, only viable cells that were filtered through were analyzed for receptor activation. Analyzing only the relevant subpopulation in the well or image provides better numerical representation of the biological activity under investigation.

The proper image analysis algorithm must be applied to measure the biological activity of interest correctly. The appropriately applied algorithm can provide improved statistics (i.e., lower coefficient of variance

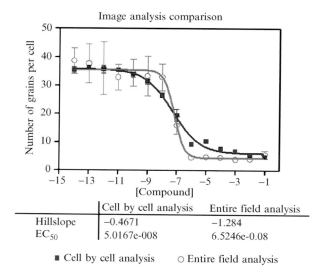

	Cell by cell analysis	Entire field analysis
Hillslope	−0.4671	−1.284
EC_{50}	5.0167e-008	6.5246e-0.08

■ Cell by cell analysis   ○ Entire field analysis

FIG. 2. Individual cell vs whole-field image analysis comparison. IN Cell 3000 software was used to perform the individual cell analysis, and Opera software was used to perform the entire field spot-counting analysis. (Opera software has the flexibility to perform both individual and whole-field analysis.) Curve represents a dose response of an inhibitor in the presence of ligand. The Y axis represents the average number of grains per cell, and the X axis represents the concentration of a compound. Whole-field analysis resulted in a much steeper hillslope and wider error bars. Both instruments acquired two images per well with Opera capturing 90 cells/well and the IN Cell 3000 capturing 290 cells/well.

and higher Z factor and S/N) and better assessment of compound potency using EC_{50} or IC_{50} values (Fig. 3). The algorithm may also be applied to a kinetic data set or stacks of images. A movie of receptor internalization (Transfluor assay) can be analyzed to quantify the temporal activity (Fig. 4).

Image Database and Data Visualization Tools

After completing the first three stages of HCS—cellular assay, image acquisition, and image analysis—one is left with a plethora of data to mine and explore. As with any other screens, there is much assay information that needs to be kept, such as cell type(s), protocols, and reagents (antibodies, fluorescent markers, nuclear stain, etc.). During a screen, numerous metadata are produced as with any typical HTS campaign, including, but not limited to, compound or siRNA ID, well ID, plate bar code ID, and the location of controls within each plate. The number of compounds screened,

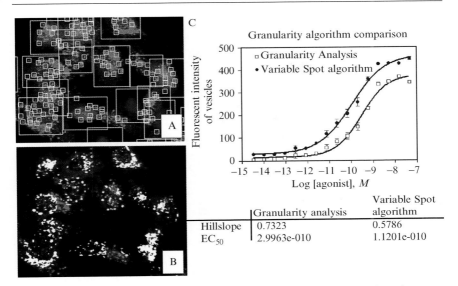

The following data appears in the figure:

	Granularity analysis	Variable Spot algorithm
Hillslope	0.7323	0.5786
EC_{50}	2.9963e-010	1.1201e-010

FIG. 3. Comparison of two spot/vesicle counting algorithms. Images of activated receptor with Transfluor technology were captured and analyzed by the IN Cell 3000 software, Raven. (A) A canned algorithm called Granularity Analysis was used to identify the internalized vesicles and measure their average fluorescent intensity per cell. The Granularity Analysis captures vesicles of specified size and intensity; therefore, aggregated vesicles or smaller vesicles were not identified. This issue can be resolved using the modified Varible Spot algorithm (B) where vesicles of any size can be identified as long as their intensity level is above the specified threshold value. (C) EC_{50} values were determined by analyzing the same set of images from a 16-point agonist titration using both algorithms. The Variable Spot algorithm provided a larger window and a slightly left-shifted EC_{50} value.

number of images (frames) taken per well, and number of color channels per frame determine the amount of raw image data taken for the screen. Each raw image must be associated with the proper compound ID and control ID. These IDs must also be associated with the numerical outputs from image analysis. Linking all of the results and metadata not only requires a comprehensive data and image management solution, but also a large and upgradeable storage capacity.

File size of a single high-resolution image can range from 1 to 5 megabytes, depending on resolution and size (height and width) of the image. Imaging an entire 384-well plate would produce anywhere from 384 to nearly 2000 megabytes or 2 gigabytes, and accordingly a 10,000 compound screen would produce at least 10 to 50 gigabytes of data. Thus, a typical computer hard drive would not have the capacity to hold images for a screening campaign.

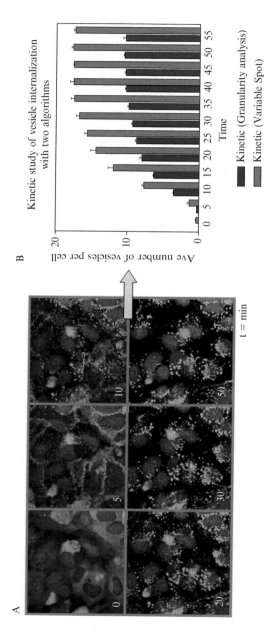

FIG. 4. Time-lapse image capture and analysis of receptor internalization (Transfluor technology) after agonist addition. The movie was captured with IN Cell 3000 upon activation of receptor with a known agonist for 55 min. The movie was converted into individual frames (A) using the IN Cell Movie Conversion tool and analyzed with the Granularity Analysis and the Variable Spot algorithm. The average number of vesicles per cell was measured for selected time points and plotted (B). The Variable Spot algorithm provided a larger signal window.

There are numerous options for storing terabytes of data. One option is to keep the images locally, on the computer(s). Media storage (i.e., DVD jukebox and tape) is also a possibility, but slow speed and management of hundreds of media are some of the issues associated with this option. A more popular solution is the storage area network (SAN). These storage devices have multiterabyte capacity and can be scalable to tens of terabytes. Commonly, data on the SAN are backed up on tape as well. Next, a database, such as Oracle, with an image management application would be required to organize and retrieve data efficiently. Cellomics, Inc. has developed a software platform called High Content Informatics (HCi) that provides a complete solution for image and data storage, retrieval, and interpretation. Bioimagene (www.bioimagene.com) has also released a product called CellMine HCS, which is a Scientific Image Management System tailored for HCS.

Conclusion

Four components to an HCS campaign include the following: assay design (cell line, probes, kinetic, or end point), image acquisition (confocal vs wide-field, hardware specifications), image analysis (canned or custom algorithms), and data interpretation and image management (storage space, database, and visualization). Technology associated with each component must be selected and assessed properly in order to carry out a successful HCS campaign. With the proper tools, HCS can be an invaluable tool to exploit the complexity of cellular signaling pathways at a single cell level in a high-throughput manner.

Protocols

The Transfluor assay is used as a case study to examine the protocols involved in an HCS campaign. Protocols in cell preparation, hardware selection, and image analysis are outlined in this section. Note that the GFP-β-arrestin Transfluor technology for GPCRs described here is offered by Molecular Devices.

GFP-β-Arrestin Fixed-Cell Assay

Protocol for the Transfluor Assay to Screen for Small Molecule Inhibitors

1. Plate Transfluor cell line (U2OS cells) at 3000 cells/well (30 μl) in 384-well plates.
2. Incubate overnight to allow cell adherence.

3. Transfer 150 nl of 2 m*M* compound stock from compound source plates.
4. Incubate for 10 min at 37°.
5. Dispense 5 μl of agonist from 7× stock concentration.
6. Incubate at 37° for 60 min.
7. Dispense 35 μl of the staining solution. Staining solution contains 4% paraformaldehyde and 1 μM DRAQ5 [DNA labeling dye from www.biostatus.com, catalog number DRAQ5 (5 m*M*) HTS] in phosphate-buffered saline (PBS).
8. Incubate for 10 min at room temperature.
9. Remove staining solution and wash three times with PBS.
10. Perform image acquisition.

Protocol for Image Acquisition (Transfluor Assay) Using Confocal Microscopy

1. Select the appropriate plate format and type from the software and set the correction collar on the objective to the plate thickness.
2. The pinhole diameter or the slit width is usually fixed on these imaging systems.
3. Select the appropriate lasers and emission filters:
   a. For DRAQ5, use the 633-nm laser with 695/55-nm emission filter.
   b. For GFP, use the 488-nm laser with 535/45-nm emission filter.
4. Set the exposure time (in milliseconds), laser power, neutral density filters, gain, and other signal intensity settings for each channel. Ensure that the image is not overexposed (resulting in saturation) or underexposed (resulting in insufficient range of intensity values).
5. Set the correct autofocus Z offset (offset distance in um in the Z direction from the plate/liquid interface).
6. Assign the appropriate cameras to each channel (multiple cameras for simultaneous multicolor acquisition).
7. Enter the number of frames or images to be captured per well.
8. Begin image acquisition.

Protocol for Image Analysis Routine (Granularity Analysis Algorithm from IN Cell Analyzer 3000) on the Transfluor Assay (Fig. 5)

1. Select the Granularity Analysis algorithm from the list of analysis modules.
2. Assign the red channel (nuclear stain or DRAQ5) as the marker channel and assign the green channel (GFP-β-arrestin) as the signal channel.

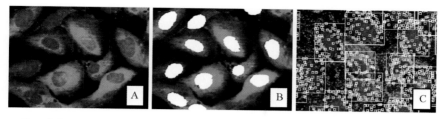

FIG. 5. Image analysis of Transfluor technology (GFP-labeled β-arrestin). (A) Arrestin (green) is distributed evenly in the cytoplasm before treatment with an agonist. The nuclear marker DRAQ5 is shown in red. (B) Nuclei are identified or segmented (labeled by a white bit map) on the red channel by selecting pixel intensity levels above the background. The nuclei bit map can be used to identify individual cells as well as to measure nuclear size and fluorescent intensity. (C) A bounding box of specified size can be drawn around the cell body to measure grains or spots of specified intensity and size within a single cell, shown by the smaller spot boxes.

3. Open a sample image and adjust the intensity levels (brightness enhancement) to clearly see the DRAQ5 and GFP signals.
4. Activate the marker bit map (white overlay on top of the image) showing the thresholding result of the DRAQ5 channel. Adjust the threshold accordingly to accurately identify the full nuclei.
5. Activate the granules bit map (large boxes surrounding the cells and small boxes surrounding the granules). Adjust the size of the bounding box around the cell to outline the cell appropriately. Adjust the parameters defining the granules within a cell (spot intensity and size).
6. Begin batch image analysis with the aforementioned settings.

References

Cox, G. C. (2002). Biological confocal microscopy. *Materials Today* 34–41.
Herman, B. (1998). "Fluorescence Microscopy," 2nd Ed. BIOS Scientific Publishers Limited.
Oakley, R., Hudson, C., Cruickshank, R., Meyers, D., Payne, R., Jr., Rhem, S., and Loomis, C. (2002). The cellular distribution of fluorescently labeled arrestins provides a robust, sensitive, and universal assay for screening G protein-coupled receptors. *Assay Drug Dev. Technol.* **1,** 21–30.

[26] An Infrastructure for High-Throughput Microscopy: Instrumentation, Informatics, and Integration

By Eugeni A. Vaisberg, David Lenzi, Richard L. Hansen, Brigitte H. Keon, and Jeffrey T. Finer

Abstract

High-throughput, image-based cell assays are rapidly emerging as valuable tools for the pharmaceutical industry and academic laboratories for use in both drug discovery and basic cell biology research. Access to commercially available assay reagents and automated microscope systems has made it relatively straightforward for a laboratory to begin running assays and collecting image-based cell assay data, but doing so on a large scale can be more challenging. Challenges include process bottlenecks with sample preparation, image acquisition, and data analysis as well as day-to-day assay consistency, managing unprecedented quantities of image data, and fully extracting useful information from the primary assay data. This chapter considers many of the decisions needed to build a robust infrastructure that addresses these challenges. Infrastructure components described include integrated laboratory automation systems for sample preparation and imaging, as well as an informatics infrastructure for multi-level image and data analysis. Throughout the chapter we describe a variety of strategies that emphasize building processes that are scaleable, highly efficient, and rigorously quality controlled.

Introduction

The application of high-throughput image-based cell assays to early stage drug discovery began in the late 1990s, and their use in both industry and academia continues to steadily increase (Abraham, 2004; Mitchison, 2005; Taylor, 2001; Yarrow, 2003). This trend has been driven by the realization that image-based cell assays can play a valuable role throughout the drug discovery process from target validation to primary screening to late-stage lead optimization. Access to commercially available assay reagents and automated microscope systems has further fueled adoption of these technologies.

Although it can be relatively straightforward to begin running assays and collecting image-based cell data, obtaining controlled, reproducible data on a large scale can have numerous challenges and pitfalls. First-order

METHODS IN ENZYMOLOGY, VOL. 414
0076-6879/06 $35.00
DOI: 10.1016/S0076-6879(06)14026-4

challenges include process bottlenecks, such as those associated with sample preparation, image acquisition, and data analysis. Potential pitfalls include problems with assay consistency and day-to-day reproducibility that can result from either suboptimal data acquisition or insufficient quality control. As these challenges are overcome, second-order challenges arise in managing and digesting unprecedented quantities of data, including terabytes of image files and databases filled with extracted image features and biological readouts.

The aim of this chapter is to provide a framework for addressing these challenges and avoiding these pitfalls through the development of a robust scalable infrastructure and implementing processes aimed at maximizing assay quality. The components of this framework highlighted in this chapter include instrumentation and processes for assay plate creation and image acquisition, strategies for image analysis, and a database and software infrastructure built to allow data integration at multiple levels. The final component described is an approach to quality control (QC), which is important during each step of both assay processes and postassay data analysis. Cell culture (Freshney, 2005), compound handling (Chan and Hueso-Rodriguez, 2002; Gosnell *et al.*, 1997), and specific assay readouts are also critical to the success of image-based cell assays; however, these topics are not covered in depth in this chapter as they have been addressed by several others in this volume and the existing literature. The assay processes described throughout this chapter apply directly to end point (i.e., fixed cell) assays. Although live cell assays are not discussed, many of the same strategies, infrastructure components, and practices can be applied. This system has broad utility and has been used in a variety of contexts from primary phenotype-based screening, to compound classification and clustering (Adams *et al.*, 2006), to applications in compound profiling as part of drug discovery lead-optimization campaigns (see, e.g., Tanaka *et al.*, 2005). In order to put the assay and analysis processes and infrastructure components into context, an automated cell cycle assay is used as a common example throughout the sections of this chapter.

One of the keys to building a high-performance and flexible infrastructure for high-throughput image-based assays is attention to selection of components and details of process design. The systems that we describe have been built with careful attention to selection of the most appropriate hardware and software components and integration of them into a common framework. In most cases the selected components were available commercially; however, custom software and databases were built to optimize performance and process efficiency. A detailed description of the custom software is beyond the scope of this chapter, although several important design considerations are provided. For example, opportunities for parallel processing have been utilized whenever possible. This design feature contributes to the scalability and flexibility of the system for running a variety of different assays either independently or concurrently, and it also allows for the development of efficient processes where the throughput of various process steps can be matched for optimal performance.

An overview of a generalized assay process is shown schematically in Fig. 1 and is referred to and elaborated on throughout the chapter. This process flow diagram illustrates both processes involving manipulations of assay plates, as well as data analysis and the interplay of multiple databases. At the point of the initiation of an experiment, plate streams are

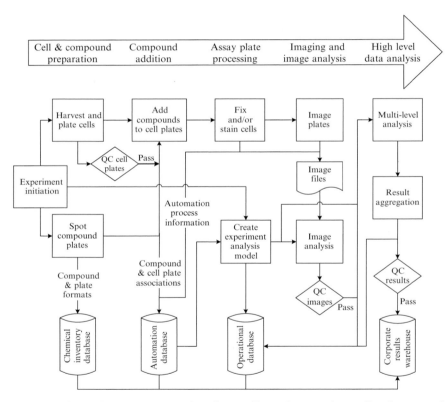

FIG. 1. Generalized assay process flow diagram illustrating steps from cell and compound plate preparation through assay processing, imaging, and data analysis. This is a simplified view of a complex process involving the coordination of assay plate manipulations, data analysis processes, quality control steps, and data flow between multiple databases. Several additional process steps, analysis processes, and quality control steps have been omitted for simplification. See text for additional description.

created for both cell plates and compound plates, and an analysis model is prepared to execute multilevel analysis steps specific to the experiment. The two plate streams are then brought together in a compound addition step. After cells are treated with compounds for a specific period of time, the assay plates are prepared for imaging through a series of steps that typically involve cell fixation and staining with fluorescent dyes for cellular components. Following plate imaging, which is used to collect raw image files, data analysis steps occur first at the level of individual cellular features followed by multiple levels of higher-order analysis. The results of an experiment are subsequently aggregated and stored in a data warehouse. Quality control processes are inserted at multiple points throughout the process flow of each experiment to ensure high-quality results and to provide opportunities to abort process streams when necessary without further investment of valuable reagents or resources. Process information is also collected and stored in a separate automation database to allow for additional quality checks and to simplify troubleshooting when problems arise.

Assay Processing

As detailed in Fig. 1, this section discusses infrastructure supporting creation of cell plates, treatment of the cells with compounds, and preparation of the cell plates for imaging. Each of these steps is generally amenable to and can benefit greatly from automation. Aside from the obvious time-saving benefit of automating routine and repetitive tasks, data quality and consistency can be improved greatly. Data consistency and process stability are important during prolonged primary screening campaigns where hundreds of thousands to millions of compounds are tested over weeks or months, but it is also important for assays that are routinely run over an extended period of time, such as those used in drug discovery lead-optimization campaigns, which may span years.

It is clear that automating an assay requires instrumentation for plate and liquid handling and software to control and coordinate these devices. An automated imaging system is also of clear benefit as is a robust image analysis platform, both of which are discussed in detail later in the chapter. However, an informatics infrastructure to track and characterize processes, samples, and reagents is often overlooked. In order to create the most robust and informative results, care must be taken to capture the specific details about how each sample was generated, handled, and treated. The goal of process tracking is to enable results to be linked back to instrumentation, reagents, and protocols used during assay preparation. Additionally, process information should allow results to be linked to specific cell samples and include lineage, culture, and passaging information. This level of information is valuable for understanding variation in large data sets and becomes invaluable during troubleshooting. Process-tracking databases are most effective when tied in with chemical inventory and image result databases. Typically the unique identifiers enabling all assay information to be tied together across multiple databases are assay plate barcodes. A robust informatics infrastructure facilitates assay processing and tracking greatly and becomes crucial when running multiple assays in parallel simultaneously. Throughout this section we highlight examples of relevant process information.

Cell and Compound Plate Preparation

At the initiation of an experiment, two plate streams are created. One contains compounds for testing and the other contains cultured cells. Consistent delivery large numbers of cell samples for large-scale assays can be challenging. Large-scale cell culture processes have been automated successfully, including cell passaging, cell harvesting, and cell plating. For example, the SelecT (The Automation Partnership) can maintain multiple cell lines using unique passaging protocols for each cell line. The device can also prepare assay plates at specified seeding densities in a variety of plate formats. Automation of routine cell culture activities ensures consistency in cell handling by establishing routine cell maintenance schedules, which minimize the effect of personnel and work-week schedule conflicts. Each passaging and plating event is recorded in a process database. In this way, robust tracking of sample lineage, passage number, growth rates, and viability can be achieved.

When cells are harvested from culture flasks, they can be plated efficiently from a stirred suspension with a simple manifold device such as the Multi-drop (Thermo). We use such an instrument to dispense cells into microplates with minimal cell-shearing forces. During plating, a sufficient number of cell plates for each cell line should be generated to satisfy the experiment design and supply several additional sentinel plates for assessing plating density and uniformity and other process QC tests. Although immediate visual inspection of the plated cells provides a rough idea of plating quality, a short-term imaging experiment to count stained nuclei is more quantitative. Typically cells are fixed after the cells have adhered to the plate and stained with a simple nuclear marker such as Hoechst 33342. This short-term count can also serve as a baseline for growth rate measurements during the rest of the assay. A second cell count is performed just prior to adding compounds to the cell plates. This QC step serves as a pass/fail checkpoint; if the samples do not pass QC, valuable materials (i.e., compounds and staining reagents) are not committed to the assay. In our cell cycle assay we have optimized the number of cells delivered per well for each human tumor cell line based on the cell size and growth rates so that cells are in exponential growth during the compound treatment period. As one example, for the ovarian tumor cell line SKOV3, we deliver approximately 1200 cells per well (in a 384-well plate), with the expectation that without compound, they will double approximately twice during the 48-h assay.

Compounds are typically dissolved and stored in dimethyl sulfoxide (DMSO), and compound plates are prepared in an assay-ready format with low-volume spots of high compound concentration. Compound plates are typically registered in a chemical inventory database that provides compound-to-well mappings for use during analysis. At assay time, cell growth medium is added to the compound plates to reduce the compound and DMSO concentration to acceptable levels (usually <0.4% DMSO in the final cell assay). A portion of this diluted compound solution is then transferred to one or more cell assay plates.

Assay formats can range from single concentration tests, as are often used in primary library screening, to complex multiconcentration dose–response formats. Compound-handling instrumentation can readily accommodate advantageous well groupings and dilution schemes. For example, we and others have determined that there are sufficient edge well effects to warrant skipping use of the outer perimeter wells of microtiter plates. Most importantly, any standardized plate format should accommodate control compounds.

Control or reference compounds can be used to qualify assay performance in several ways. During development and optimization, assays are characterized to determine if they can report on a specific cellular activity with statistical significance. Known reference compounds are especially valuable for assay validation. Once an assay is qualified and implemented for a high-throughput application, control compounds are used to monitor assay performance through comparison of control data with historically accumulated values. Controls are most informative if they are chosen to span the dynamic range of the readouts of the assay and should include solvent-only or null treatment wells. It is important to process controls alongside other test compounds to best gauge the effectiveness of plate-processing steps.

Another role for control compounds is to assist with data analysis. If an assay is inherently noisy or run over an extended period of time, it may be necessary to normalize or bound data on a per-plate or per-batch basis. For example, the dynamic range of the biological response for a given assay instance can be calibrated by reference compounds. Controls are also important, as discussed in a later section for building classification models in which individual cells are grouped into subpopulations. For example, in our cell cycle assay, we use DMSO-only treatment as a negative control and dilution series of paclitaxel (peaking at 800 nM) as a positive control to arrest cells in mitosis. Test compounds are ranked relative to the assay bounds determined by these controls. This strategy can help minimize excess variance introduced by deterministic noise sources such as low cell growth rates, marginal assay staining, or other hard-to-identify environmental influences.

Cell plate and compound plate streams meet after the cells have had sufficient time to adhere to the plate surface, stabilize their morphology, and resume growth (often 24 h following cell plate preparation). One compound plate can usually supply compounds sufficient to treat multiple cell plates. This has the advantage of reducing assay variance because all replicates spanning multiple cell lines or even several different assays receive the same compound treatment. Similar to including reference compounds in an assay, reference cell lines may also be included to differentiate or categorize compound response. For example, control cell lines may be refractory to certain compound classes or display characteristic phenotypes or response ranges. As compounds are added to one or more cell plates, barcode associations are created in the process database, which will enable assay results to be linked to compounds.

Automated Assay Processing

Efficient processing of compound and cell plates requires at minimum automated pipetting and is typically best accomplished with integrated systems that operate on a static schedule. A number of fully integrated single vendor solutions for cell processing are currently on the market. For example, a BioCell (Velocity11) can be configured to prepare live cell assays. Vendor solutions provide complete systems with little need for automation expertise, although the choices of adorning instruments and configurations are limited to what the vendor can make or is willing to support. The advantage of custom integration is that systems with the best options can be created. The best-suited liquid-handling devices, storage devices, and readers or imagers can all be incorporated regardless of manufacturer. A second advantage is that the system can be tailored to quickly meet the needs of shifting assay requirements.

Assay processing steps can be abstracted to several classes. Plates must be introduced into a system, possibly from a device that can support live cells, they must be moved from device to device, and liquids must be added and removed from the plates. Ideally the processed plates leave the system sealed and stable for imaging. Figure 2 illustrates the layout of a custom-integrated cell processing system that meets these

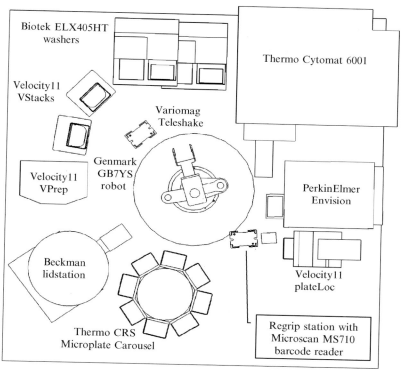

FIG. 2. Schematic of integrated cell processing system. This system is designed specifically for processing cell assay plates. In a typical assay, plates start in a random access CO_2 incubator Cytomat 6001 (Thermo) and move around the system by a robot adapted from the semiconductor industry, GB7 (Genmark Automation). Plates are fixed and receive staining reagents via a parallel pipetting device, VPrep (Velocity11). Plates can be washed on one of two plate washers, ELX405HT (BioTek), and incubated with reagent on an open random access device, CRS Microplate Carousel (Thermo). When finished, plates are sealed with a thermal plate sealer, PlateLoc (Velocity11), and moved to stackers, VStacks (Velocity11). Curtains (not shown) shield the system from room light to minimize bleaching of photo-labile reagents.

general requirements. The components have all been chosen for their speed, quality, and performance.

The central component of the system is the plate-handling robot. This system uses a robot adapted from the semiconductor industry, GB7 (Genmark Automation). This six-axis robot is fast and was designed to carry small silicon wafer payloads, similar in weight to microtiter plates. This type of robot is very easy to work with and will stall easily (and appropriately) during a collision. A second option for robotics in this type of system is an industrial robot adapted from manufacturing applications. These robots are very reliable and typically have much higher working payloads as compared to the semiconductor robots. Examples of such robots include the CRS F3 (Thermo) and CRS CataLyst5 (Thermo). Additional care must be taken when working with the larger manufacturing robots to avoid collisions and situations that occur when the arm is commanded to move through a physically impossible trajectory in order to achieve a position. The result can be an unexpected and often rapid move, resulting in a potentially damaging collision with other system components.

The system allows plates to enter, exit, or incubate in one of several devices. Each has been purchased or configured to operate preferentially in landscape plate orientation. Two plate stackers, VStacks (Velocity11), are included, which are typically used to introduce empty labware or collect processed plates from the system. A random access CO_2 incubator, Cytomat 6001 (Thermo), holds live cell plates on the system, and a random access carousel, CRS Microplate Carousel (Thermo), provides 120 locations to park in-process plates. One advantage of this carousel is that the robot can place and retrieve plates directly from its nests, speeding plate access greatly. A lid-handling station, Sagian Lidding Station (Beckman Coulter), allows lid removal and replacement of up to five plates at a time. Any plates that enter the system have their barcodes read and captured in the process database.

An important piece of equipment from an assay quality and speed perspective is the parallel pipetting device. We chose the VPrep (Velocity11) for several reasons. First, it was designed for use in a radial system such as ours, with a small footprint and vertically integrated design. The system allows users to easily change between 96 and 384 pipetting heads as needed and can deliver good pipetting precision and accuracy with either head. The device has 8 pipetting shelves, which allow simultaneous delivery of plates to the device while the unit is performing pipetting tasks at another shelf position. Another important device for creating quality cell assay plates is the plate washer. We included two duplicate washers, ELX405HT (BioTek), in the system for added throughput. Care must be taken when establishing washer settings to aspirate and deliver wash buffer at the correct speed and height above the plate bottom in order to not affect the sample adversely. For example, in our cell cycle assay, washing too vigorously may remove loosely adhered mitotic cells, whereas insufficient washing or poor removal of wash buffer can lead to suboptimal staining.

The system is equipped with a plate sealer, PlateLoc (Velocity11), and homemade low-volume autofilling reservoirs to work on the VPrep. These devices allow processes to run into the night or over the weekend without compromising assay quality. The system also contains a multilabel plate reader, Envision (Perkin Elmer), to accommodate nonimage-based assays such as end-point cell viability assays.

Process Control Software

The present system is controlled by custom-built software that allows users to model assay processes and generate static schedules for execution. A static schedule means that all steps have been choreographed based on sequence and timing constraints and scheduled prior to the assay. This is in contrast to a dynamic scheduling system, which processes plates as soon as the next instrument resource becomes available. Consider the following fix-and-stain protocol for compound-treated cell plates.

1. Remove live cell plates from incubator, read barcode, and add formaldehyde fixative.
2. Incubate 1.5 h.
3. Remove fixative and culture medium, wash, and add blocking solution.
4. Incubate 1.5 h.
5. Remove blocking solution, wash, and add primary antibody.
6. Incubate 4 h.
7. Remove primary antibody, wash, and add secondary antibody solution.
8. Incubate 1.5 h.
9. Remove secondary antibody and wash.
10. Seal plates.

Computing a static schedule for this process requires designing a process model (assay script) complete with timings for each step, including all robotic plate movements. A scheduling engine creates an ordered set of execution tasks to most efficiently process the plates and avoid resource conflicts. The static schedule order will be adhered to even if the process does not follow exact timings. Figure 3 shows the static schedule used for fixing and staining up to 112 plates for the aforementioned protocol, a process that takes over 10 h. When a process involves lengthy steps such as incubation with an antibody, care should be taken to match the timings of early and late incubations; this will allow the largest batch sizes to be achieved during which any plate in the system, regardless of number, will experience consistent timings. When a suitable schedule is developed allowing maximal batch sizes, the schedule is saved to a process database. The master schedule is adhered to for processing any number of plates up to the batch size limit.

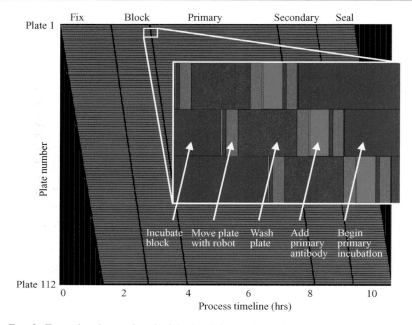

Fig. 3. Example of a static schedule for fixing and staining cell plates. Depicted is a schedule for processing plates on the system in Fig. 2 for the fixation and stain protocol given in the text. The schedule is compiled for 112 assay plates and runs over approximately 10.5 h. Each line on the *y* axis is a graphic representation of the instrument tasks required to process a single plate; the tasks are color coded by device. For example, all incubations utilizing the random access carousel are blue, robot moves are pink, plate washes are red, and pipetting steps are green. The entire schedule contains 7392 individual tasks or 66 for each plate. (Inset) A blowup of the processing steps for removal of blocking solution, washing, adding primary antibody, and starting incubation with primary antibody. In brief, the plate is removed from the carousel by the robot, washed on a plate washer, moved by a robot to the pipetting device, has primary antibody added, and is returned to the carousel by the robot for an extended incubation.

One advantage of a statically scheduled process is that other processes upstream or downstream can be timed to run at a similar rate. For example, if one plate can be fixed every minute then the rate at which compounds are added to plates the previous day should also be at one per minute. In this way if hundreds of plates are processed in the same order, a common compound exposure (e.g., 24-h exposure in our cell cycle assay) is achieved even across a large assay run. As plates are processed on the system, each processing task is recorded to the database, including information about which device instance was used. This process record creates an audit trail for the plates, which is useful during QC and troubleshooting activities.

Image Acquisition

Once cells have been treated with compounds, fixed, and stained, the next step is to image them to acquire data for subsequent analysis (Fig. 1). Imaging should have an appropriate throughput to match both upstream and downstream processing, reliable plate handling, and provide consistent, robust data. How can these requirements be met? Whether making a single biological measurement (such as translocation of a protein) or capturing many metrics per cell (morphology, intensity), imaging a single wavelength, or several channels per field of view, the same general considerations apply: choice of hardware, single vs multiple microscopes, the trade-off between image resolution and field of view, and developing strategies to make data comparable within and between experiments. This section discusses these instrument and imaging choices.

Imaging System Instrumentation

As with the cell processing system described earlier, we chose to custom integrate commercially available instrument components with custom-built control software to build our imaging system. Although the system includes robotics for plate handling and storage, the heart of the system is the automated microscope. Primary considerations for choosing an automated microscope system are speed, flexibility, and field of view size. Speed is important because imaging is typically rate limiting to the entire process flow, and flexibility is important because multiple assays are likely to be implemented on the system. Magnification at the image, or pixel resolution, is determined by the microscope objective, downstream optics, and the pixel dimensions of the detector (often a CCD camera). There is always a trade-off between field of view and resolution of features. Use of the lowest magnification objective that can resolve the subcellular features of interest for a given assay can provide a throughput benefit by enabling the largest number of cells to be imaged simultaneously, thus minimizing the need to acquire multiple fields per well.

There are many microscope choices, from motorized research instruments to specialized screening imagers, but the greater speed of the dedicated machines makes them the best choice, especially if data turnaround times need to be minimized, as often is the case in a drug discovery effort. Screening microscopes (Smith and Eisenstein, 2005) fall into two broad classes, wide-field imagers (including Cellomics ArrayScan, Molecular Devices ImagExpress and Discovery-1, and GE Healthcare IN Cell Analyzer 1000) and confocal imagers, either line scan (GE Healthcare IN Cell Analyzer 3000) or Nipkow disk (BD Pathway; Evotec Opera). Although improved z-axis resolution and contrast favor confocal microscopes in some cases, their cost and complexity are not required for many immuno-cytochemistry assays. Wide-field imaging is often capable of capturing images with sufficient contrast for reliable automated image analysis.

We built our current system around three identical Molecular Devices Discovery-1s. Although distributing imaging across three instruments poses a challenge for data normalization, advantages include a tripling of throughput, the ability to swap parts for fault diagnosis, and the capability of running different experiments on different machines. This is particularly useful for imaging high-priority quality control plates such as those used to determine plating density or growth curves, interleaved with assay plates. Furthermore, because the system is modular, we can add additional micro-scopes easily with minimal effect to the control software.

Regardless of microscope selection, imaging large numbers of plates is generally time-consuming so automated plate handling must be reliable enough for unattended operation around the clock. Figure 4 shows the layout of our current imaging system. Again, as described for the cell processing system, we have favored the submillimeter precision and near-zero failure rates of industrial robots. Here, a CRS CataLyst-5 (Thermo) five-axis robot, which also carries a barcode reader, is mounted on a track and shuttles plates between a plate hotel and the microscopes. Upgrading the stages to a closed-loop design with rotary encoders also greatly im-proved the reliability of plate positioning during imaging. While each Discovery-1 has its own computer, a fourth master computer controls plate traffic and instructs each microscope on which imaging protocol to run for each plate. Unlike the static scheduling implemented on the cell processing system, the imaging system runs instead on a dynamic schedule, sending plates to microscopes as they become available. Application of dynamic scheduling is possible because of the end-point nature of fixed cell assays, and it enables on-the-fly addition of plates to the imaging queue.

Imaging Parameter Optimization

Images can be acquired in many ways, varying camera gain, the number of pixels, exposure and magnification to name a few variables. For many assays we find an advantage in acquiring images at low magnification in order to increase sample size. In our cell cycle assays, for example, where we describe the distribution of cells across the phases of the cell cycle, we require a sufficient sample of cells in each phase, including mitosis, which is short lived and therefore rare. Imaging with a $4 \times /0.2$ NA objective captures 1000 to 2000 cells per image, sufficient to compute accurate DNA classification models and to accurately measure the mitotic proportion in the population without having to image multiple sites per well. Nevertheless, with a 1380×1024 pixel

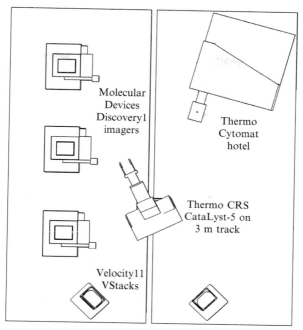

Fɪɢ. 4. Schematic of robotic imaging system. A CRS CataLyst-5 track robot system (Thermo) carrying a barcode reader shuttles plates among three Discovery-1 (Molecular Devices) microscopes, input and output stackers VStacks (Velocity11), and a plate hotel Cytomat (Thermo). Microscope controllers, powers supplies, and computers sit in the bays below each instrument, with the xenon lamps beneath the table connected to the imagers by liquid light guides. Control software maintains an inventory of plates in the hotel, interacts with the database to set imaging parameters, and feeds plates to the next available imager.

camera, CoolSnap HQ (Roper), pixel resolution is 1.6 μm/pixel (in x and y), which is sufficient to resolve subcellular morphology.

A rarely discussed issue in image-based screening is exposure setting. Although it is obviously important when comparing data from multiple microscopes, developing an exposure setting strategy is helpful even with a single imager in order to compare data across experiments. While perhaps less important when the goal of the assay is to measure the presence or absence of a signal, it becomes more significant when signal intensity contains information such as DNA content for cell-cycle analysis, or in a phenotype assay where many metrics, including intensity distributions of markers, are measured in each cell (Adams *et al.*, 2006). The goal of the exposure setting is to compensate for sources of variation not related to the biology of interest, including variability of staining reagents lots, fluidics processing, and time-dependent changes in microscope lamps and filters. Several approaches are possible, but all require imaging a control sample, which often consists of stained, untreated cells. The simplest approach is to use the pixel intensity histogram and assume that some statistic is a reliable index of upstream processing variation. For example, the mode of the distribution could be assumed to represent the background signal, or the 99th percentile of the distribution to represent the cell pixels. While this approach is the simplest, it is indirect. A more robust method is to measure the features of interest directly. We have therefore developed an exposure setting method that identifies the cells in the image and measures their brightness using the same image analysis methods used to compute assay results. At the beginning of an imaging run, we acquire 72 test images, three exposures at each of four wavelengths in six replicate wells. After analysis, including background flattening and object segmentation, object intensity is plotted against exposure and a linear regression is computed for each wavelength (or marker). Finally, the exposure required to achieve a given target intensity is calculated for each wavelength (Fig. 5A). Target intensities are determined during assay development by examining cell intensities and contrast (Fig. 5B) in positive and negative control conditions to choose the lowest target intensity where sufficient contrast is achieved in a negative control, and few pixels are saturated in the positive control. In this way exposures are as short as possible, which speeds throughput, but images remain in the linear region of the dynamic range of the camera. Exposures can be computed this way for each plate, although for many assays, plate-to-plate cell feature intensity variations are sufficiently small if exposures are determined from the first plate only, even with multiple cell lines included in the batch.

Quality control metrics monitored during imaging are principally exposure lengths and differences between microscopes. Extended exposures typically indicate the need for replacement of the main wear items, such as the lamps and, more rarely, excitation filters and light guides. Xenon arc lamps gradually decay in brightness until exposures become impractically long, typically at around 1000 h of use.

Imaging Speed Optimization

Typically, the microscope spends a considerable amount of time focusing, so minimizing the focus range is important for throughput. However, the focus range must be broad enough to compensate for plate curvature, stage leveling, and variation in plate positioning by the robot. Additional time can also be saved if refocusing is not required between wavelengths. At 4× magnification on our system, focusing is only performed on the first

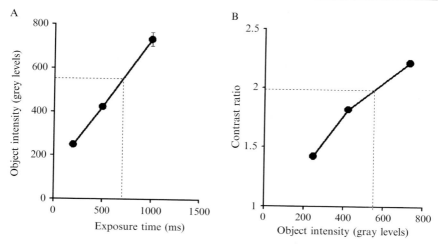

Fig. 5. Exposure calculation for Hoechst-stained nuclei. Six wells were each imaged at three exposures and nuclei were segmented using a custom algorithm. After correction for uneven illumination, object (nuclei) and background (mode of the pixel intensity histogram) intensities were computed for each image. (A) Average object intensity for the three test exposures. An exposure of 708 ms, computed by linear regression, was required to achieve a target object intensity of 550 (dotted lines). (B) Average contrast (object intensity divided by the background) is approximately linear with object intensity. During assay development, a target object gray level of 550 (of 4095) was chosen as providing sufficient contrast without saturation under most assay conditions. A similar computation is used for each fluorescence channel. Error bars show standard deviation, but are only visible if larger than the plotted symbol.

of four wavelengths. If focus is not identical for all channels, sometimes a subset of the wavelengths can still be imaged at the same focus or, alternatively, a fixed z offset defined between wavelengths. Furthermore, only a small focus range usually needs to be explored within a field of view so microscope software that can fine-focus within a well but have a coarser focus range between wells will realize a benefit in speed. With these parameters optimized, our imaging system typically averages one 384-well plate imaged at four wavelengths every 20 min (60 min per plate per imager), with a batch size of up to 120 plates. The system routinely runs overnight and almost never mishandles a plate.

Image and Data Analysis

With a robust infrastructure for assay processing and image acquisition, it is important to have an equally robust and flexible infrastructure for image and data analysis. A productive high-throughput image analysis system for a drug discovery environment ideally has several important features. It should be flexible enough to adapt to a variety of assays and assay formats, which are rapidly emerging and changing, it should produce consistent and reliable results, despite inevitable variations in image inputs, and it should be built to have a scalable throughput. The large scale of experiments makes user intervention to adjust processing parameters difficult, if not impossible, thus analyses should be robust and capable of automatically adjusting to these variations. Finally, and most importantly, the analysis system should produce measurements that adequately represent the complex biological phenomena of interest.

We designed our analysis system based on several general principles:

- Modularity of analyses
- Multiple levels of analyses
- Importance of detailed analysis of individual cells and characterization of cell subpopulations
- Rigorous quality control of analysis results
- Robustness of analysis results vis-à-vis variations in cell populations, cell processing, and image acquisition.

This section focuses on the application of these principles to building a robust and flexible image and data analysis infrastructure, as well as describing the analysis software components that underlie the capabilities of the system.

Analysis Modularity

For our purposes, a modular analysis system implies that analyses are implemented as a loosely coupled library of image and data analysis algorithms. Multiple algorithms are implemented for image segmentation, feature extraction, and aggregation of results. Individual algorithms can be assembled into a variety of "programs" for use in a specific assay or experiment. Most of the algorithms can be tuned to better adapt to specifics of a particular assay, and an algorithm with an established set of tuning parameters can be reused in different programs. Analysis programs can also be adjusted and expanded by the addition of new algorithms and by adjusting parameters of existing algorithms. Outputs of each algorithm are stored in a database and can be used for subsequent analyses at a later time without the need to rerun the entire analysis program.

Basic Feature Extraction/Cell-Level Analysis

Data processing includes analyses at multiple levels of data aggregation. Typically, the first stage of an analysis pipeline is aimed at image preparation for subsequent quantitative analyses. This stage might include correction of uneven field illumination, image smoothing, noise reduction, and similar types of image preprocessing. This image preparation step is highly dependent on specific types of imaging equipment and assay conditions. In some cases, if very high-quality images are acquired consistently, this step may be omitted.

The next stage of analysis is to identify individual cells and subcellular compartments. Our library of segmentation algorithms includes edge detection-based segmentation and a modified version of a watershed algorithm that is used for segmentation of low-contrast objects. These algorithms are used to identify cells and subcellular compartments, such as cell nuclei and peripheral areas of the cytoplasm. Results of the segmentation analysis ("segmentation masks"), as well as relationships between objects ("belongs to," "is neighbor of," etc.), are stored in a database and are available for use by subsequent feature extraction algorithms and for review by investigators.

Feature extraction algorithms are applied to segmentation masks (possibly, in conjunction with preprocessed images) to produce a number of shape, intensity, and texture features. A broad review of methods and techniques for image preparation and analysis has been presented in Russ (2002) and Soille (1999). An important characteristic of our system is that segmentation algorithms are uncoupled from feature extraction algorithms so that intensity and texture-related features for multiple markers can be extracted for each of the segmented cellular components.

Cell Classification/Well-Level Population Analysis

The analysis steps described so far are similar to those used in a number of image analysis systems used in high-throughput and high-content screening: individual objects are identified in images and multiple measurements of these objects are collected. A typical subsequent step in analysis is to generate aggregated metrics for an image or a well, for example, average and standard deviation values for each of the features. Our experience shows that there is rich biological information reflected in changes of individual subpopulations of cells and objects rather than global changes of cell populations. This information is rarely fully exploited. We utilize this information by classifying objects identified in images into groups of interest. To that end, our system contains a number of binary and multiclass classifiers that use individual object features and their combinations as inputs. Most of the classifiers are implemented as two algorithms: one for model building and one for classifying objects based on a model. Classification models are built automatically for each assay plate using information from positive and negative controls on the plate. This approach allows us to design classification algorithms that can adjust to and compensate for inevitable plate-to-plate variations in image properties due to variations in cell staining and imaging.

Examples of implemented classifiers for our automated cell cycle assay include:

- Identification of dust particles and segmentation artifacts (based on size and intensity)
- Identification of cell debris (based on DNA content)
- DNA content classifier (based on DNA stain)
- Interphase/mitotic classifiers (based on cell cycle marker staining, or shape and DNA content)
- Identification of live and dead cells
- Identification of out-of-focus cells
- Identification of cells with multiple projections

Most of these classifiers utilize just two algorithms: Expectation Maximization algorithms with mixture model (Dempster *et al.*, 1977; McLachlan and Krishnan, 1997) and Cytokinetics' proprietary heuristic thresholding algorithm.

Identification of multiple subclasses of cells allows aggregation of object features to be performed for specific subpopulations of cells. For example, DNA content classification allows us to compute the average size of nuclei at each of the stages of the cell cycle. Such "group-specific" aggregation is performed for all subclasses and all features of objects. Classification results produced by each of these algorithms can be used not only on their own, but in any combination. For example, after completion of a DNA content classifier, an interphase/mitotic classifier (see Fig. 6), and a classifier that identifies cells with long projections, a user can then automatically define derivative classes, such as "interphase cells without projections and with 4N DNA content," for further analysis.

Multiwell and Plate-Level Analysis

Our analysis system includes a number of algorithms that can integrate data across multiple sets of images or wells on a plate. Different groupings of wells are defined as a part of a plate format definition. This feature allows for a flexible use of plate controls. For example, algorithms that generate models for object classification require a large number of objects of each class as a training set. In some cases the required size of a training set can be achieved by pooling object data from multiple control wells. Plate-level analysis is also used extensively for quality control purposes. QC algorithms pool data from control wells on each plate and compute

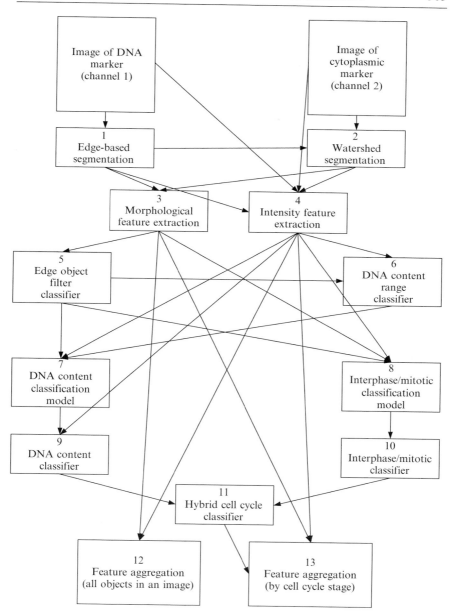

Fig. 6. Program for analysis of two-channel assays with basic feature extraction, artifact removal, cell cycle classification, and feature aggregation. Each box represents an individual algorithm. Arrows show data dependencies for algorithms. See text for details.

statistics, such as average and trimmed average attribute values for each control type and variation in values in control wells.

Experiment-Level Analysis

The next level of analyses are at the experiment level and include many different data aggregation and summarization methods—from aggregation of replicate data to advanced data mining. Typical processing in a high-throughput setting includes automatic aggregation of data for replicate wells, which are run on different plates, and, when appropriate, characterization of dose–response relationships for test compounds. An example of a more elaborate high-content compound profiling application of experiment-level analyses using our system is presented in Adams *et al.* (2006).

Analysis Program Example

Figure 6 shows an example of a simple analysis program for a two-marker assay where cell nuclei and cytoplasm are identified by distinct fluorophores. Each box in the diagram represents an individual algorithm, and arrows indicate data dependencies for algorithms. Numbers in each box indicate the possible order of the algorithms' execution within the program.

Analysis starts with edge detection-based segmentation applied to nuclei images (1). Results of this segmentation are then used by a watershed segmentation algorithm that identifies cells (2). Resulting object masks are used to produce a number of morphological object features (3) and, in conjunction with images, are used to generate intensity and texture-related features for each object (4). Objects that are too close to or intersect with image edges are identified by an edge object filter (5). Cell debris and segmentation artifacts are identified by a DNA content range classifier (6). Objects detected at steps 5 and 6 are excluded from subsequent analyses. Next, a classification model for discrimination of cells with sub-2N, 2N, 4N, and greater than 4N DNA is built based on controls on the plate (7). This model is used for DNA content classification of each cell in every image (9). Similarly, an interphase/mitotic classification model is built based on the morphology and intensity of a DNA marker in control wells (8) and applied to every cell in every image (10). The hybrid cell cycle classifier combines classifications from steps 9 and 10 to characterize each cell as being in G1, S, G2, or M phase of the cell cycle (11). Feature aggregation algorithms (12,13) compute average, median, standard deviation, and quantiles for each of the object features extracted at steps 3 and 4. These statistics are computed for all cells in an image (12), as well as for subpopulations at each stage of the cell cycle (13). Results of this aggregation can be used for higher-level analyses, such as those described in Adams *et al.* (2006).

Analysis Software Components

Our analysis infrastructure was designed to support analysis and integration of a very large number of images and extracted features and classifiers. It is a modular, distributed system, where each component is responsible for a specific task. Figure 7 shows the major software and hardware components of the infrastructure.

Experiment manager: A client application that allows users to perform a number of the following functions.

1. Create/update/query experiments. Plates can be configured and registered in the database and later associated with or removed from an experiment.
2. Create and manage assay protocols that include information about cell lines, sets of markers, staining procedures, incubation times, etc.
3. Monitor analysis progression. Users can restart and abandon analysis tasks, and manage allocation of analysis servers for specific tasks.
4. Create/update/query assay meta-data elements, such as markers, cell lines, and plate types, as well as protocols for plating, staining, imaging, and analyzing data.
5. Define and associate plate formats with assay plates.
6. Create/update/query analysis programs composed of analysis algorithms and their parameters.
7. Export/import selected meta-data items.

Image loader: A service that registers images appearing on a temporary storage file server and moves them into a permanent storage location for subsequent analysis and review.

Analysis servers: Several individual servers dedicated to run analysis algorithms. These are conventional Windows-based computers with specifications common for modern desktop machines. The system is scalable to allow for additional servers to be added for increased analysis throughput. The last analysis step typically starts the warehouse builder, which transfers operational aggregated data to a data warehouse.

S-Plus analysis server/report generator: An analysis server dedicated to experiment-level statistical analysis. This server connects to an Oracle database via Open Database Connectivity protocol. Computation results are stored either in a database or as files in a file system and are available for review and further analysis in Spotfire via a Decision Site server.

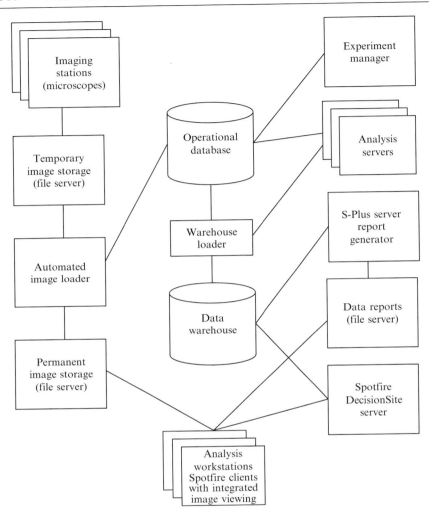

FIG. 7. Image analysis infrastructure. Boxes represent components of the analysis system responsible for image registration, analysis, and presentation of results. Lines indicate network connections between components. See text for a description of each of the components.

Operational database: Stores assay and analysis data (such as cell line and marker information, analysis programs and their parameters, execution status) and intermediate analysis results.

Data warehouse: Stores results of analyses, assay information, and compound information in a format optimized for data retrieval and report generation.

Spotfire DecisionSite server and Spotfire clients: Automated Spotfire (www.spotfire.com) guides provide access to QC metrics and analysis results (see next section).

These individual software components are loosely coupled and interact primarily through the database. The modularity of the system allows us to "mix and match" custom and commercial software components.

Analysis algorithms are implemented in Java and C++. Computations are performed in parallel in a distributed system on multiple analysis servers. This distributed and parallel analysis strategy removes a restriction that some commercial systems put on analysis performance, namely requiring image processing speeds to match image acquisition speeds. This allows us to use complex and time-consuming algorithms and ensures that the system is scalable so that any desired throughput can be achieved by simply using an appropriate number of analysis computers. Typically a single analysis server running one analysis session processes images from one assay plate at a time.

Data Review and Quality Control

Automated microscopy systems such as the one described here are capable of generating terabytes of data, and the scalable nature of the infrastructure will enable the creation of even larger data sets in the future. It is important to have in place an equally scalable data review process that allows distillation of complex data sets into meaningful information. There are two types of information to extract: information about the integrity of the data (data QC) and information about the biological activity of interest (results). As we have emphasized throughout this chapter, building high-quality processes is critical, and this section describes an approach to ensuring the integrity of data that will ultimately be used to drive future experiments and decisions.

Key concepts associated with an efficient data review process include keeping the intended use of the results in perspective, understanding factors that can affect the quality of data, defining metrics that report on problems associated with data quality, and assembling tools to aid in the filtering and visualization of the results. The development of a QC process needs to occur in concert with the development of the assay; this will ensure that controls are appropriately positioned throughout the process and that the image and data analysis approach provides metrics to detect problems. Data QC processes can vary tremendously in their complexity, their stringency, their cost, and their effectiveness. It is therefore important to identify an appropriate level of scrutiny to apply to a particular data set and acceptable failure rates when designing a QC process. For example, in a primary screening application an appropriate QC process may be aimed at minimizing false-positive and false-negative rates while putting much less emphasis on the absolute accuracy of results. In contrast, for assays used to rank compounds as part of a lead-optimization campaign, the absolute accuracy of the results is much more important. Because applications can vary significantly, our discussion here is restricted to a general approach to identifying potential problems, building appropriate QC metrics, and a description of software tools that allow the data review process to be completed efficiently.

Identifying Potential Problems That May Affect Data Integrity

Sources of potential assay problems include the biology (cell health and behavior), cell handling, liquid handling and reagents, plate handling, and imaging. Problems can be inherent to a particular cell line or assay reagent or they can be process based, caused by a particular instrument or process. In addition, no matter how well controlled a process is, the risk of human error will always exist. By generating data and gaining experience with a given assay, knowledge of its vulnerabilities can be determined and used to design strategic control steps to robustly report on problems. Effective use of controls should isolate variables and report potential problems as quickly as possible. In our automated cell cycle assay, for example, we use both negative controls (DMSO-only treatment) and positive controls (paclitaxel) on each plate to flag a variety of problems with cells, reagents, and fluidics. In-process QC steps, such as the cell counts on sentinel plates described in the assay processing section, should be used whenever possible; however, the majority of assay problems will not be detected until postprocessing at the data review and QC stage.

Problems with cell health, such as abnormal morphology and growth performance, can be detected by monitoring negative controls (DMSO-only treatment) that are included on every plate to assess effects of the process on the cells. In our automated cell cycle assay, we assess changes in cell cycle and cell morphology metrics in these controls. For example, we have detected drift in nuclear size and variance, which traced back to a cell source. Abnormal values for these and other metrics, however, may result from other problems, including instrumentation or processing issues. For example, a clogged washer tip can alter image brightness by preventing stain removal. An integrated process tracking database links the results to instrumentation and process components, making it possible to look for correlations between the problematic results and specific events in the assay process. For example, if an aberrant value is identified in association with a subset of the plates, we can query the database to determine if a specific plate washer unit or a specific imager correlates with the problem. If a problem with a new assay cannot be associated with cell health, reagent formulation, or a process or instrument malfunction, then a problem with an analysis algorithm should be considered. Manual inspection of images and object masks may provide some insight into potential problems, as algorithm failures can stem from errors in object classification.

Not all problems that can occur are easily categorized or anticipated. It is therefore important to complement strategic QC approaches with an unbiased approach to identifying problems. This can be achieved by looking for changes in patterns across a broad set of image analysis metrics. A clear understanding of the relevant metrics and features will then allow for a logical mapping to potential sources of variation.

Identifying and Validating QC Metrics

Once we have defined what could go wrong with an assay we need to determine what image analysis metrics are the most sensitive indicators of specific problems and define their acceptance ranges for data QC. During assay development, this can be determined experimentally by staging problems and assessing whether the chosen metrics flag the problem by crossing a QC threshold. Baseline values and variances should be established for each cell line within the context of the assay. Acceptable levels of variation can be determined by threshold or tolerance testing to establish performance criteria. These metrics can be reviewed and updated periodically by examining historical data.

As an example, to monitor cell health we assess nuclear and cell morphology descriptors, cell number and derived growth rates, and cell cycle distribution metrics. To monitor problems associated with liquid handling we assess background fluorescence levels and metrics that report on the spatial distribution of cell subpopulations within each image. For monitoring staining and image quality we assess contrast and intensity, as well as saturation levels associated with each imaging channel.

Software Tools

As with other aspects of our infrastructure the software applications used for our data review and QC processes are a hybrid of commercially available components and custom-built tools. We have chosen to build our data review process around Spotfire DecisionSite (Fig. 8). Customized

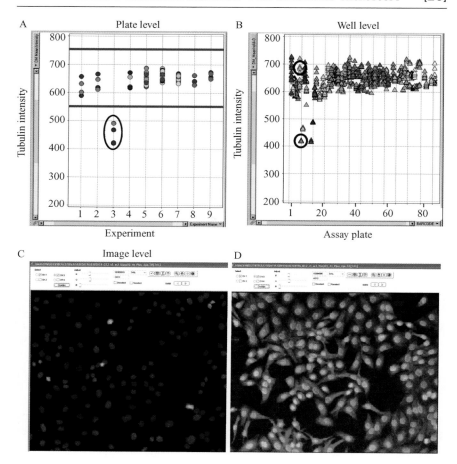

FIG. 8. Spotfire analysis of immunofluorescence data quality at multiple levels. (A) Experiment-level analysis of antitubulin intensity (gray level; range 0–4095). Points plot the mean cell intensity for each plate, computed from 14 DMSO-treated control wells per plate. Points are colored by barcode, grouped in columns by experiment date, and shown in relation to predetermined QC limits (blue lines). Circled points show plates in an outlier experiment where antitubulin staining was dim. (B) Well-level analysis of antitubulin intensity. Symbols show the mean intensity for all cells in one well. Points are colored and also grouped in columns by plate bar code. The custom-built image visualization tool enables the interactive selection of points (circled) and the display of images associated with those wells (C and D). (C and D) Two images, displayed at the same contrast, reveal noticeably different antitubulin intensity, confirming the difference seen in B. Images associated with lower (C) and upper (D) points circled in B, respectively. Only a subregion of each image is displayed. The nuclear channel (Hoechst, in red), is overlaid with the antitubulin stain channel (green). Plots show screenshots from an interactive Spotfire session; similar analyses can be performed for other channels and metrics.

Spotfire guides (automated workflows) have been created specifically for the data QC process. For example, we have generated an assay plate QC guide that loads plate-level summarization data, creates multiple visualizations for a number of metrics that report on marker intensity, cell number, image contrast, and so on, and allows querying on a preselected list of variables. Thresholds for each metric have been built into the graphical presentations to allow easy identification of outlier data. These data can be flagged automatically and marked for follow-up review. The querying tools aid troubleshooting efforts by establishing potential correlations to various experimental variables. For example, if a pattern of outlier values for multiple plates was detected, it would be possible to query on the microscope used to generate the images to determine if a particular instrument correlated with the problem.

We have also integrated custom tools with the Spotfire DecisionSite application. For example, we have integrated a custom image visualization tool that retrieves images and displays them within the Spotfire data review session. Individual wells or entire plates of images can be rejected at any point in the process before experiment-level results are finalized. In some cases, automated data filters eliminate poor-quality data, such as rejection of wells with dust particles or large cell clumps. In other cases, data are rejected as part of the semimanual data QC process using Spotfire and our custom-built image viewer.

The image analysis process reduces data at several levels, as described in detail in the previous section. Ideally, data should be reviewed at the highest level that allows direct access to the source of the problem. With increasing levels of summarization the granularity of data decreases. Figure 8 illustrates a typical set of data views encountered when progressing from plate-level review to image review. In this example we used the assay plate QC guide to assess immunofluorescence quality for a tubulin marker. As illustrated by the example described in Fig. 8, we start our review process at the plate level and work our way down to the image level leading to a triaged list of images for manual review. We typically only review data at the individual cell level if a problem cannot be resolved at any other level.

Summary

This chapter has outlined many of the issues encountered in building an infrastructure to support high-throughput microscopy. Our guiding strategy has been integration so that all steps in cell-based assays are linked by both informatics and process, including compound and cell handling, image acquisition, image analysis, and data storage and reporting. The ability to query all steps in the process at multiple levels of data analysis has made it possible to build powerful quality control tools and to generate robust assay data. Integration also has been one of the keys to building a highly efficient system, capable of running large numbers of assays weekly. Our preference for modular and scalable hardware and software will ensure that our infrastructure will adapt and scale to our future needs.

Acknowledgments

We thank Lane Conn, Daniel Pierce, and Jay Trautman for helpful discussions and comments on the manuscript. We also thank the many people at Cytokinetics who have contributed to developing and building our image-based screening infrastructure.

References

Abraham, V., Taylor, D., and Haskins, J. (2004). High content screening applied to large-scale cell biology. *Trends Biotechnol.* **22**, 15–22.

Adams, C., Kutsyy, V., Coleman, D., Cong, G., Crompton, A., Elias, K., Oestreicher, D., Trautman, J., and Vaisberg, E. (2006). Compound classification using image-based cellular phenotypes. *Methods Enzymol.* **414** (this volume).

Chan, J., and Hueso-Rodriguez, J. (2002). Compound library management. *Methods Mol. Biol.* **190**, 117–127.

Dempster, A., Laird, N., and Rubin, D. (1977). Maximum likelihood from incomplete data via the EM algorithm. *J. R. Stat. Soc. Ser. B* **39**, 1–38.

Freshney, R. (2005). "Culture of Animal Cells: A Manual of Basic Technique." Wiley, Hoboken, NJ.

Gosnell, P., Hilton, A., Anderson, L., Wilkins, L., and Janzen, W. (1997). Compound library management in high throughput screening. *J. Biomol. Screen.* **2**, 99–102.

McLachlan, G., and Krishnan, T. (1997). "The EM Algorithm and Extensions." Wiley, Hoboken, NJ.

Mitchison, T. (2005). Small-molecule screening and profiling by using automated microscopy. *ChemBioChem.* **6**, 33–39.

Russ, P. (2002). "The Image Processing Handbook," 4th Ed. CRC Press, Boca Raton, FL.

Smith, C., and Eisenstein, M. (2005). Automated imaging: Data as far as the eye can see. *Nature Methods* **2**, 547–555.

Soille, P. (1999). "Morphological Image Analysis: Principles and Applications." Springer-Verlag, Berlin.

Tanaka, M., Bateman, R., Rauh, D., Vaisberg, E., Ramachandani, S., Zhang, C., Hansen, K., Burlingame, A., Trautman, J., Shokat, K., and Adams, C. (2005). An unbiased cell morphology-based screen for new, biologically active small molecules. *PLoS Biol.* **3**, 764–776(e128).

Taylor, D., Woo, E., and Giuliano, K. (2001). Real-time molecular and cellular analysis: The new frontier of drug discovery. *Curr. Opin. Biotechnol.* **12**, 75–81.

Yarrow, J., Feng, Y., Perlman, Z., Kirchhausen, T., and Mitchison, T. (2003). Phenotypic screening of small molecule libraries by high throughput cell imaging. *Comb. Chem. High Throughput Screen.* **6**, 279–286.

[27] Protein Translocation Assays: Key Tools for Accessing New Biological Information with High-Throughput Microscopy

By Arne Heydorn, Betina K. Lundholt, Morten Praestegaard, and Len Pagliaro

Abstract

Redistribution technology is a cell-based assay technology that uses protein translocation as the primary readout for the activity of cellular signaling pathways and other intracellular events. Protein targets are labeled with the green fluorescent protein, and stably transfected cell lines are generated. The assays are read using a high-throughput, optical microscope-based instrument, several of which have become available commercially. Protein translocation assays can be formatted as agonist assays, in which compounds are tested for their ability to promote protein translocation, or as antagonist assays, in which compounds are tested for their ability to inhibit protein translocation caused by a known agonist. Protein translocation assays are high-content, high-throughput assays primarily used for profiling of lead series, primary screening of compound libraries, and as readouts for gene-silencing studies using siRNAs. This chapter describes two novel high-content Redistribution assay technologies: (1) The p53:hdm2 GRIP interaction assay, in which one high-content image feature is used for detection of primary hits, whereas a different feature is used to deselect compounds with unwanted mode of action, and (2) application of siRNAs to Redistribution assays, exemplified by knockdown of Akt isoforms in a FKHR translocation assay reporting on the PI3 kinase signaling pathway.

Pathway Screening Using BioImage Redistribution Technology

Imaging of cells by fluorescence microscopy has been an important tool in cell biology for more than a decade now, and recent years have seen rapidly growing interest in the application of high-throughput microscopy for studying cellular signaling pathways. It has been estimated that the number of high-content screens will increase by 50% over the coming years (Comley, 2005).

Cell-based protein translocation assays can be used to probe cellular signaling pathways that are otherwise difficult to study using traditional biochemical assays. Cellular proteins that translocate in response to external stimuli, such as transcription factors, receptors, kinases, and scaffolding

METHODS IN ENZYMOLOGY, VOL. 414
0076-6879/06 $35.00
DOI: 10.1016/S0076-6879(06)14027-6

proteins, can be tagged with fluorescent molecules and traced by fluorescence microscopy. BioImage Redistribution is a technology whereby a target protein is labeled with the green fluorescent protein (GFP) and expressed in a stable cell line. Redistribution assays allow for a "pathway" approach to screening in which an entire pathway is screened using a downstream target protein for fusion to GFP. When performing such a screen, the cell "selects" the actual target among the proteins in the pathway, which offers the possibility to discover entirely new classes of compounds with new modes of action, including protein translocation modulators and cell surface receptor modulators, as well as modulators of intracellular enzymes.

The ERF1 Redistribution assay is an elegant example illustrating the way in which a signaling pathway that has historically been difficult to study in high-throughput mode using biochemical methods can be probed with a protein translocation assay. ERF1 is a ubiquitously expressed membrane-distal reporter of the Ras–mitogen-activated protein kinase (MAPK) signaling pathway (Le Gallic et al., 1999). The Ras-MAPK pathway is activated by a wide variety of receptors and is involved in growth and differentiation (Robinson and Cobb, 1997). The pathway is deregulated in various forms of cancer, and compounds inhibiting the pathway are already in clinical development (Hilger et al., 2002). In short, activation of receptor tyrosine kinases, integrins, or ion channels leads to activation of Ras. Ras plays an essential role in the activation of Raf kinase, which directly phosphorylates and activates MEK1 and MEK2. MEK1/2 phosphorylates ERK1 and ERK2, which translocate to the nucleus, where they phosphorylate downstream targets, including ERF1. Unphosphorylated ERF1 is present in the nucleus, and the nuclear export of ERF requires phosphorylation by ERK1/2 (Le Gallic et al., 1999). T24 cells, a human bladder cancer cell line, signal constitutively through the Ras-MAPK pathway due to a single mutation in HRas (Capon et al., 1983). Thus, in T24 cells ERF1 is phosphorylated continuously by ERK1/2 and is exported from the nucleus (see Fig. 1A). If signaling through the Ras-MAPK pathway is prevented downstream of HRAS by the MEK inhibitor UO126 or by the Raf inhibitor BAY-43-9006, ERF1 is no longer phosphorylated by MEK1/2, and therefore accumulates in the nucleus (see Fig. 1B). In addition to inhibitors of the Ras-MAPK pathway, inhibition of the nuclear export machinery by leptomycin B also results in nuclear accumulation of ERF1. By measuring the translocation of the ERF1-GFP fusion from cytoplasm to nucleus in stably transfected T24 cells, compounds modulating the Ras/MAPK pathway downstream of HRAS and compounds modulating nuclear export can be identified. The translocation can be quantified by taking the logarithm of the ratio between fluorescence intensity in the nucleus and the cytoplasm (Log NucCyt Ratio). Figure 1C shows concentration–response curves of UO126, BAY-43-9006,

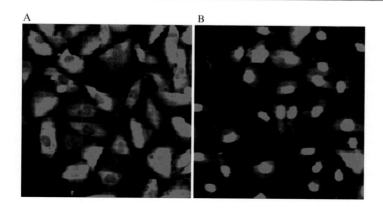

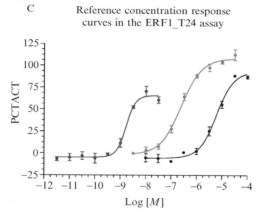

FIG. 1. (A and B) Images of T24 cells stably expressing GFP-ERF1. (A) Untreated cells. (B) Cells treated with 10 μM UO126 for 1 h at 37°. GFP-ERF1 accumulates in nuclei of cells treated with UO126, which can be measured as an increase in the image variable Log NucCyt ratio. (C) Concentration–response curve of UO126 (red), leptomycin B (blue), and BAY-43-9006 (black) in the ERF1 T24 Redistribution assay using Log NucCyt ratio as the response variable ($n = 3$). Scales are normalized to 100% activity for 10 μM UO126.

and leptomycin B when added to T24 cells expressing ERF1-GFP using Log NucCyt ratio as the primary response variable. The ERF1 Redistribution assay has been screened at BioImage using a 265,000 compound library, resulting in 1739 primary hits with an activity above 25%. Fluorescent compounds and toxic compounds causing cell rounding were removed by a second round of image analysis on the same set of images, and the chemical stability and drug-like structure of the hits were evaluated, leaving 718 compounds for retesting and secondary profiling. Two compounds were eventually identified that exhibit activity in the ERF1 Redistribution assay and selectively inhibit the proliferation of Ras-MAPK pathway-dependent malignant melanoma cells (Grånäs et al., 2006).

BioImage has developed a collection of commercially available protein translocation assays for probing diverse cellular signaling pathways and cellular functions. These include assays probing the MAKP/MEK/ERK pathway, the p39 MAPK pathway, the PI3 kinase pathway, the Wnt/β-catenin pathway, and the TGF-β pathway, as well as assays for GPCR internalization and nuclear hormone receptor translocation. In addition to the assays described earlier, we are developing novel assay types that expand the use of protein translocation using GFP. Two of these new techniques are presented in the following sections: (1) GRIP technology that allows for probing of specific protein–protein interactions outside of a signaling pathway context and (2) the use of siRNA to measure the effect of protein knockdown on cellular signaling pathways as measured in a Redistribution assay. These last two examples are detailed in this chapter.

p53-Hdm2 Protein–Protein Interaction Assay

The p53 tumor–suppressor plays a critical role in the prevention of human cancer. In the absence of cellular stress, the p53 protein is maintained at low steady-state levels and exerts very little effect on cell fate. In the presence of cellular stress, posttranslational modifications of p53 cause increased transcriptional activity and elevated protein levels, resulting in cellular changes such as cell cycle arrest, cellular senescence, and apoptosis (Gregorieff and Clevers, 2005). The antitumor activity of p53 is controlled by its negative regulator, Hdm2, through a feedback mechanism. Hdm2, which is overproduced in many tumors, binds p53 and inhibits its function by modulating its transcriptional activity and stability. Activation of p53 in tumor cells by inhibition of its physical interaction with Hdm2 is therefore a focus of cancer drug discovery.

The p53-Hdm2 GRIP assay is designed to measure the protein–protein interaction between p53 and Hdm2. GRIP is a protein–protein interaction discovery system that can be used to generate truly high-throughput-compatible cellular assays that allow screening for inhibitors of protein interactions. The technology is based on translocation of the human cAMP phosphodiesterase PDE4A4 (Terry et al., 2003) to distinct cytoplasmic foci, where a bait protein (in this case Hdm2) is fused to the cAMP phosphodiesterase isoform PDE4A4B. PDE4A4 acts as an inducible anchor protein and the prey protein (in this case p53) is labeled with GFP. Figure 2 illustrates the configuration of the assay system: Treatment with the PDE4A4 Redistribution agonist RS25344 (Ashton et al., 1994) leads to localization

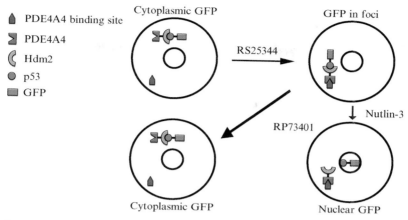

FIG. 2. Illustration of the p53–Hdm2 interaction assay. Fusion constructs of p53-GFP and PDE4A4-HDM2 are stably transfected into the same CHO parental cell line. Aggregation of PDE4A4 at punctate cytoplasmic foci in the presence of the inhibitor RS25344 provides a functional control for the assay. GFP fluorescence only enters the cell in the presence of compounds that disrupt the protein–protein interaction between p53 and HDM2.

of PDE4A4 into compact foci, and through the interaction between p53 and Hdm2, the GFP tag (on p53) is recruited to PDE4A4-binding sites, resulting in GFP-labeled foci. The PDE4A4 inhibitor RP73401 antagonizes the effect of RS25344, causing focus dispersal, thereby serving as a universal reference compound for the assay system (Alvarez *et al.*, 1995).

Nutlin-3 is a small molecule p53–Hdm2 interaction inhibitor, developed by a combination of screening approaches and rational drug design (Vassilev *et al.*, 2004), which serves as a reference compound in the assay. Nutlin-3 causes loss of GFP fluorescence from the PDE4A4 foci (Fig. 2) by dissociating the interaction between PDE4A4-Hdm2[1–124] and GFP-p53 [1–312]. This can be measured through image analysis software that first identifies spots and subsequently calculates the primary response. In this assay, the primary response (quantified by the Fgrains parameter) is the average in each image of a dimensionless number defined for each spot as the ratio between the sum of spot pixel values divided by the sum of pixel values in a region surrounding the spot. The Fgrains variable was found to produce Z factors values around 0.5 to 0.8 for the p53-Hdm2 assay and was superior to other spot features, such as number of spots per cell, total area of spots per cell, and average intensity of spots.

Whereas Nutlin-3 specifically dissociates the p53–Hdm2 interaction, other compounds, such as RP73401, dissociate the PDE4A4 spots (PDE4A4 dislocators) without breaking the p53–Hdm2 interaction. Both types of compounds, that is, p53–Hdm2 interactors and PDE4A4 dislocators, resulted in spot dispersal, and by only using the Fgrains variable, we were unable to distinguish between the two modes of action. However, it is well known that Hdm2 is a negative regulator of p53 and that treatment with Nutlin-3 results in nuclear accumulation of the p53 transcription factor (Vassilev *et al.*, 2004). This suggested that monitoring GFP-p53[1–312] translocation to the nucleus might be useful for discriminating between p53–Hdm2 interactors and PDE4A4 dislocators. This turned out to be true. Using the logarithm of the ratio between fluorescence intensity in the nucleus and the cytoplasm (Log NucCyt Ratio) to quantify GFP-p53 [1–312] translocation resulted in a significant response when cells were treated with Nutlin-3, whereas treatment with RP73401 showed no response. Figure 3A illustrates the translocation events described earlier, and Fig. 3B and C show concentration–response curves of Nutlin-3 and RP73401 for the output variables Log NucCyt Ratio and Fgrains on normalized scales. The EC_{50} value for Nutlin-3 was ~1.5 μM when Fgrains or Log NucCyt ratio was used as an output variable, whereas the EC_{50} for RP73401 was ~20 nM when using Fgrains. Thus, by using multiple readouts from one assay and one set of images, it was possible to accelerate the compound deselection process during a screen. In addition, we eliminated the need for running a PDE4A4 counterscreening assay, thereby saving time and resources.

Protocol for Generation of p53-Hdm2_CHO Assay Cell Line

CHOhIR cells are stably cotransfected with GFP-p53[1–312] and PDE4A4-Hdm2[1–124] plasmids using the FuGene6 reagent and maintained with 0.5 mg/ml geneticin/G418 and 1 mg/ml zeocin in Ham's F-12 nutrient mixture containing glutamax-1 supplemented with 10% fetal bovine serum (FBS) and 100 IU/mL penicillin, 100 μg/mL streptomycin, Stable transfectants are FACS sorted and cloned to obtain a homogeneous population of cells. p53-Hdm2_CHO cells are grown in a humidified incubator at 37° and 5% CO_2 and are passaged twice weekly by resuspension with 0.25% trypsin/0.53 mM EDTA solution for 3 min at room temperature.

Assay Protocol

1. The day before the assay is performed, seed p53-Hdm2_CHO cells in 96-well Packard ViewPlates at a density of 7000 cells/well in 100 μl Nut. mix.F-12 (HAM) medium containing 1 μM RS25344, 10% by volume FBS, 100 IU/mL penicillin, 100 μg/mL streptomycin, 0.5 mg/ml geneticin, and 1 mg/ml zeocin. Incubate for 24 h at 37°, 5% CO_2 with lid.

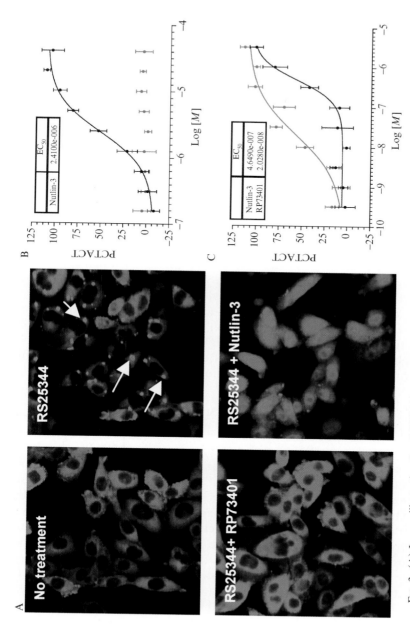

FIG. 3. (A) Images illustrating that RS25344-induced focus formation is dispersed by both Nutlin-3 and RP73401; only Nutlin-3 results in nuclear fluorescence. (B and C) Concentration–response curves of Nutlin-3 (black) and RP73401 (red) for the output variables Log NucCyt ratio (B) and Fgrains (C) on normalized scales ($n = 5$).

2. The next day inspect cells under a microscope to ensure that they look healthy and uncontaminated and that there is no "edge effect," that is, uneven distribution of cells in the edge wells.

3. Prepare assay buffer: Ham's Nutrient Mixture F12 medium containing 10% by volume FBS, 10 mM HEPES, and 100 IU/mL penicillin, 100 μg/mL streptomycin.

4. Prepare 400× controls: Negative control: 100% dimethyl sulfoxide (DMSO). Positive control: 4 mM Nutlin-3 in DMSO. Final concentrations in assay is 10 μM Nutlin-3/0.25% DMSO.

5. Pipette 1 μl of the controls in the wells of a 96-well polystyrene compound plate; the negative control is placed in wells [A1;D1] and [E12; H12] and the positive control in wells [E1;H1] and [A12;D12].

6. Add 199 μl assay buffer to the compound plate and mix on a shaker table.

7. Transfer 100 μl from compound plate to cell plate.

8. Incubate cell plate with lid for 2 h in a CO_2 incubator.

9. Gently decant assay buffer.

10. Add fix solution (4% formaldehyde in phosphate buffer [pH = 7.0], 150 μL/well).

11. Fix cells for 10 min at room temperature.

12. Wash cells four times with PBS (300 μl/well).

13. Decant PBS and stain cell nuclei by adding 100 μl 1 μM DRAQ5 (Biostatus) in PBS containing 1 mM MgSO$_4$.

14. Image and analyze assay plates on a suitable imaging system by measuring Fgrains and Log NucCyt ratio.

Use of High-Content Assays in RNAi Studies

High-content imaging assays have the potential for being the preferred technology format for RNAi studies (for a review about RNAi, see Mittal, 2004). Imaging assays have the advantage, compared to alternative technologies, such as various forms of reporter and ELISA assays, that cells are visualized directly. In addition to the primary functional readout, for example, protein translocation, high-content imaging assays enable the user to acquire and evaluate secondary data from the cells. This is a very valuable feature considering that RNAi studies are quite prone to both false-negative and false-positive results. For instance, cell morphology data can be used to estimate toxicity arising from procedural disturbance derived from siRNA transfection or shRNA expression. In general, transfection efficiency, knockdown efficiency, and transfection-related toxicity are common problems in siRNA studies, but are relatively straightforward to control when using high-content imaging assays. For instance, fluorescence-tagged siRNAs have been widely used to estimate transfection efficiency by simple fluorescence microscopy of transfected cells. Although useful, this method has a tendency to overestimate the percentage of cells having sufficient knockdown to generate a functional effect. When applying siRNAs to GFP-based translocation assays, we have successfully used a GFP-specific siRNA as a control for transfection efficiency as well as an indirect measure of knockdown level (see Fig. 5). In this case, the level of GFP fusion protein expression following transfection of GFP-specific siRNA is assumed to be representative for the general transfection efficiency and indicative for general knockdown effects. For target discovery purposes, a good control would be to employ a control siRNA that is related directly to the primary functional assay readout, such as H-Ras siRNA in the ERF Redistribution assay mentioned earlier.

When conducting siRNA studies using high-content imaging assays, it is quite evident that a cell population transfected with siRNA is not a uniform population, but rather a distribution of cells ranging from full knockdown to no effect. This is obviously also true for reporter-type assays in which, however, the cell population cannot be evaluated as individual cells, but only as a whole. This is a very common problem in siRNA studies. Transfection of siRNA may result in close to 100% knockdown in a fraction of transfected cells, but if only 50% of the cells are efficiently transfected, the final assay activity profile will be hidden in background noise. This prompts the question: Is there a way to use high-content imaging assays to only include cells with a certain level of knockdown or transfection efficiency in the image analysis? For large siRNA studies, such as target discovery screens, one suggestion is to express an unstable fluorescent protein in the assay cell line that does not interfere with the functional assay readout, for example, express a red or blue unstable fluorescent protein in a translocation assay cell line using a green fluorescent protein as the primary readout. If siRNA directed against the unstable fluorescent protein is included as a fraction in all siRNA transfections, the intensity of this fluorophore represents transfection efficiency and can be used to set a cutoff for cells that should be included in the image analysis.

Translocation assays based on expression of fluorescent fusion proteins can be performed on live cells, creating an opportunity to do multiple measurements in time. This can be beneficial when effects of siRNAs against proteins with divergent or unknown half-life are studied. The time from transfection to effective siRNA-mediated knockdown is normally 2 to 3 days, but can easily vary from overnight to several days and should, in principle, be optimized for each siRNA. Considering the often long response time to siRNA-mediated knockdown, the assay itself should preferably be reasonably short to obtain a reproducible readout. High-content imaging assays, including translocation assays, have the advantage that most assay time lines range from minutes to 1 to 2 h, whereas reporter assays usually are performed overnight.

By far the broadest application of high-content imaging assays is currently small academic research studies of cell signaling pathways where siRNAs are used to link protein activities to biological functions. However, several academic laboratories and commercial institutions have commenced doing focused target discovery using RNAi technologies. Large libraries of siRNAs or retroviral vectors expressing shRNA have been screened to identify new members of signaling pathways and new targets for drug discovery (Berns et al., 2004; Brummelkamp et al., 2004). For the reasons mentioned earlier, high-content imaging assays are good candidate assays for such broad RNAi screens.

Use of siRNA-Mediated Knockdown to Validate Akt Isoform Dependency of a FKHR Redistribution Assay

The FKHR Redistribution assay is an example of a high-content imaging assay that has been biologically validated with Western blots, reference compound testing, and siRNA studies. The mammalian transcription factors FKHR, FKHRL1, and AFX function as key regulators of insulin signaling, cell cycle progression, and apoptosis downstream of phosphoinositide 3-kinase (PI3K). In growing cells, Forkheads are kept inactive through Akt-mediated phosphorylation of three conserved threonines. Phosphorylation of these threonines causes binding to 14–3-3 proteins in the nucleus followed by nuclear export and cytoplasmic retention. Nuclear export of Forkhead proteins is dependent on the classical NES/Crm1 pathway. When PI3K/Akt signaling is inhibited, FKHR translocates to the nucleus, where it activates target genes involved in growth arrest and apoptosis (Burgering and Kops, 2002).

We tested a range of reference compounds inhibiting PI3K, Akt, and nuclear export to validate that the FKHR Redistribution assay detects inhibition of individual components of the PI3K pathway upstream of FKHR. As shown in Fig. 4A, all of these reference compounds have expected activities in the FKHR Redistribution assay. Furthermore, a PI3K inhibitor (wortmannin) and an Akt inhibitor (compound 13 in Barnett et al., 2005) inhibit phosphorylation of Akt and FKHR at relevant serine residues (Fig. 4B).

The FKHR Redistribution assay has been further characterized by transfection of Akt isoform-selective siRNAs (see later for a general protocol for siRNA transfection). Figure 5A shows that transfection of a GFP-specific siRNA prevents expression of FKHR-GFP, although some residual expression is detectable in a fraction of cells, indicating that the

A

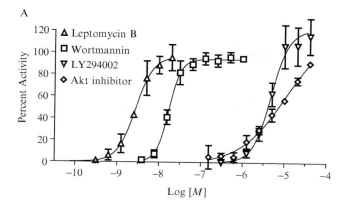

B

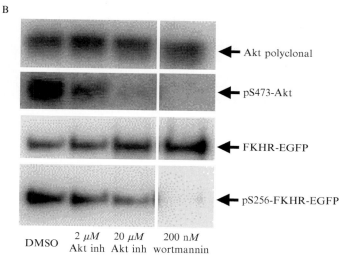

FIG. 4. Validation of FKHR Redistribution assay by reference compound testing and Western blotting. (A) Concentration–response curves of PI3K inhibitors (wortmannin and LY294002), a reference Akt inhibitor [compound 13 in Barnett *et al.* (2005)], and the nuclear export inhibitor leptomycin B. Protein translocation is detected as the logarithm of the ratio between nuclear and cytoplasmic fluorescence (Log NucCyt ratio) in FKHR-GFP expressing U2OS cells and normalized to the activity of 150 nM wortmannin (100% activity). (B) Western blots on cell lysates from the FKHR Redistribution assay cell line. Activation of Akt by phosphorylation of Ser473 and deactivation of FKHR by phosphorylation of Ser256 can be inhibited by wortmannin (PI3K inhibitor) and an Akt inhibitor [compound 13 in Barnett *et al.* (2005)].

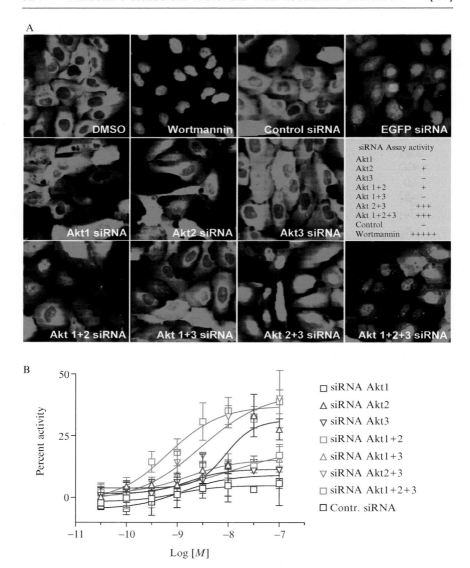

FIG. 5. Activity of Akt isoform-specific siRNAs in FKHR-GFP Redistribution assay. (A) Representative images showing effects of Akt siRNA transfection. The FKHR-GFP assay is performed 72 h posttransfection of Akt isoform-selective siRNAs (20 nM per siRNA). DRAQ5-stained red nuclei in cells transfected with EGFP siRNA show that cells are present in the image, despite knockdown of FKHR-GFP expression. Selective knockdown of Akt isoforms has been verified by TaqMan PCR (done in collaboration with ALTANA Pharma AG). (B) Concentration–response curves derived from analysis of images acquired from Akt siRNA transfections performed in triplicate. Concentrations are plotted as total siRNA concentrations. Percentage activity is calculated relative to the activity of 150 nM wortmannin, which is defined as 100% activity.

transfection is successful, but knockdown is not 100% effective. Transfection of a nontargeting control siRNA has no effect on FKHR-GFP localization (compared to DMSO-treated cells). This is also the case for transfection of Akt1- and Akt3-specific siRNAs, whereas transfection of Akt2 siRNA leads to slight redistribution of FKHR from the cytoplasm to the nucleus. When the Akt1-, Akt2-, and Akt3-specific siRNAs are mixed before transfection either as mixtures of two siRNAs or as a mix of all three siRNAs, only the combination of Akt2 and Akt3 siRNA and the combination of all the Akt siRNAs give rise to significant translocation of FKHR-GFP to the nucleus. This indicates that Akt2 and Akt3 are important for FKHR regulation, at least in the assay cell line, whereas Akt1 is not. All shown siRNA knockdowns have been validated by quantitative polymerase chain reaction (PCR) on RNA purified from parallel transfections (data not shown). Figure 5B shows concentration–response curves of siRNA transfections in the FKHR assay using the Log NucCyt ratio as a variable on a normalized scale (100% activity is treatment with 150 nM wortmannin for 1 h). These data suggest that transfection of Akt siRNAs is not as effective in terms of FKHR redistribution as treatment with reference compounds such as wortmannin. For example, transfection of a mix of all three Akt siRNAs results in 40% activity, as compared to the 100% activity achieved with 150 nM wortmannin. This is probably due to the fact that reference compounds (inhibition of enzymatic activity) act differently than siRNAs (knockdown of expression). Knockdown mediated by siRNA is rarely 100% effective, and the remaining amount of protein may be sufficient to partly rescue protein function, whereas inhibition of enzymatic activity directly blocks protein function for the total pool of the relevant enzyme. We conclude that translocation of FKHR-GFP is regulated by Akt2 and Akt3 but not Akt1 in the FKHR Redistribution assay cell line presented here.

Protocol: Test of Functional Effects of siRNAs in Redistribution or Translocation Assays

General Protocol for siRNA Transfection in 96-Well Plates

Reagents: siRNA in 20 μM stock solutions, Lipofectamine 2000 (Invitrogen), Opti-MEM medium (Invitrogen)

1. The day before transfection plate cells in 96-well Packard View-Plates at a density of 1500 to 2000 cells/well in 100 μl/well growth medium without penicillin–streptomycin. Incubate for 24 h at 37°, 5% CO_2 with lid.

2. At the day of transfection dilute at 5× concentration of final siRNA concentration in 15 μl/well Opti-MEM. The optimal final siRNA

concentration for transfection of standard cell lines such as HeLa and U2OS cells is around 10 nM and a good range for concentration response studies is 0.1 to 100 nM. A few siRNAs can be mixed, but the total concentration should not exceed 50 to 100 nM depending on the cell type. Stocks of siRNA should be handled with RNase-free filter tips as other types of RNA.

3. Dilute Lipofectamine 2000 reagent in Opti-MEM medium. Use 0.15 μl/well lipofectamine 2000 diluted in 10 μl/well Opti-MEM. Mix by pipetting and incubate for 5 to 10 min at room temperature.

4. Mix siRNA and Lipofectamine 2000 dilutions (15 + 10 = 25 μl/well).

5. Incubate at room temperature for 20 min.

6. Add 25 μl/well siRNA and Lipofectamine 2000 transfection mix to cell plates.

7. Incubate plates 24 h.

8. Change medium on plates to fresh growth medium.

9. Incubate for 48 h.

10. Run Redistribution or translocation assay according to relevant assay protocol.

11. Image and analyze assay plates on a suitable imaging system.

Assay-Specific Cell-to-Cell Heterogeneity Plays a Role in Assay Quality

Development of Redistribution assays comprises transfection and selection of high-performing clonal cell lines, followed by thorough optimization of assay conditions, image acquisition, and image analysis. Details of clone generation and assay condition optimization have been reviewed elsewhere (Loechel et al., 2006). Here, we describe an often overlooked factor in the development of high-content assays: the number of cells to analyze in order to achieve the best possible Z factor.

Most traditional biochemical assays performed on cells in microtiter plates, such as reporter assays and ELISA assays, result in a single assay response per well, with no opportunity to assess the heterogeneity of the cell population in individual wells or to remove the contribution of nonresponders in order to lower assay noise. In contrast, high-content imaging assays allow assessment of the variation between cells in the same well, thus enabling assay developers to sort cells based on primary response or secondary image features, as well as to adjust the number of analyzed cells to achieve the best possible Z factor (Zhang et al., 1999). The standard deviation (SD) of So and Smax used for computing the Z factor is calculated as the SD between the average response in the So wells and Smax wells, respectively, the so-called interwell variation. However, even though the

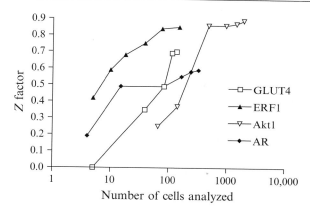

FIG. 6. Assay quality (Z factor) as a function of the number of cells analyzed. Images were acquired on an IN Cell Analyzer 3000 (GE Healthcare). The number of cells in an image was reduced by changing the scan width and scan length, thereby reducing the field of view. The number of cells analyzed was increased by acquiring multiple images from different positions of the same well. The GLUT4 assay and the Akt1 assay are membrane translocation assays, i.e., their primary response is the translocation of GLUT4 or Akt1 from the cytoplasm to the membrane as measured by the area of fluorescent spots in the cell membrane. The ERF1 assay and the androgen receptor (AR) assay are nuclear translocation assays, whose primary output is the translocation of ERF1 or AR from cytoplasm to nucleus as measured by the variable Log NucCyt.

intrawell variation does not appear directly in the Z factor, it has a direct influence on the precision of the average response from each well. In other words, the more variation between individual cells in the same well (intrawell variation), the more cells need to be quantified per well to achieve the same assay Z factor. We find that the intrawell variation in Redistribution assays depends strongly on the identity of the assay. This is illustrated in Fig. 6, which shows the relationship between number of analyzed cells and Z factor of various Redistribution assays. The ERF1 Redistribution assay (described earlier) and the androgen receptor (AR) Redistribution assay are assays whose primary output is the translocation of ERF1 or AR from cytoplasm to nucleus as measured by the variable Log NucCyt. Both assays yield acceptable Z factors, even when relatively few cells per well are analyzed. The ERF1 assay gave a Z factor of 0.42 when only an average of 5 cells per well was analyzed, whereas the AR assay had a Z factor of 0.49 when an average of 15 cells per well was analyzed. Both assays showed rapidly increasing Z factors as the number of analyzed cells increases up to around 100 to 150 cells. The GLUT4 assay and the Akt1 assay are both membrane translocation assays, that is, their primary output is the translocation of GLUT4 or Akt1 from the cytoplasm to the membrane as measured by the area of fluorescent spots in the cell membrane. It is our experience that membrane translocation assays generally have higher intrawell variation compared to nuclear translocation assays, as exemplified by the GLUT4 and Akt1 assays. The GLUT4 assay needed around 40 cells, and the Akt1 assay needed around 140 cells to achieve Z factors above 0.3.

In practice, the number of cells analyzed in high-content assays is often determined by the field of view of the microscope and lenses used for image acquisition. Thus, optimization of the number of cells needed for analysis is sometimes deemphasized as an important factor for overall assay quality because of technical limitations. Certainly, it is always best to optimize assay conditions, such as serum concentration, incubation time, and temperature, in order to develop robust assays that require a fairly low number of analyzed cells to achieve good assay quality. However, if a large variation between cells is caused by the endogenous biology of the cells, such as the cells being in different growth phases, having different shapes, and so on, the only way to achieve a satisfactory Z factor can be to increase the number of analyzed cells. Most high-content platforms include the possibility of acquiring multiple images from different positions of the same well, thereby increasing the number of analyzed cells. Because of the increase in image acquisition time per well, this is often undesirable when developing high-content assays for screening purposes, but it can be the only way to achieve a good Z factor for translocation assays characterized by high cell-to-cell variation.

Future Developments

The ability to assess biological responses at the cellular level using optical microscopy as a primary tool has a long and productive legacy, dating as far back as early work by pioneers such as Leevenhoek and Koch (Wilson, 1995). Recent decades have seen dramatic improvements in our abilities to monitor and manipulate biological systems with tools, including fluorescent analog cytochemistry (Wang, 1989), novel fluorescent dyes (Haugland, 2005), and fluorescent proteins (van Roessel and Brand, 2002). As valuable as these advances have been from a research perspective, their integration into industrialized settings, such as drug discovery, has been limited by the lack of tools for rapidly acquiring and analyzing the enormous quantity of image information that must be processed to evaluate biological responses at the cellular level. These aspects of the field have changed, and continue to change, rapidly, with the introduction of instruments designed for high-speed, optical microscope-based acquisition of cellular images from multiwell plates. A number of such instruments are currently available commercially and are reviewed elsewhere in this volume. These instruments have quickly eased the bottleneck for data acquisition that presented challenges to the field for a number of years and have instead moved the bottleneck to data analysis. Work in this field is progressing rapidly, and the coming generation of instrument platforms promises to provide both the hardware and the software support necessary to realize the full potential of protein translocation assays.

References

Alvarez, R., Sette, C., Yang, D., Eglen, R. M., Wilhelm, R., Shelton, E. R., and Conti, M. (1995). Activation and selective inhibition of a cyclic AMP-specific phosphodiesterase, PDE-4D3. *Mol. Pharmacol.* **48**, 616–622.

Ashton, M. J., Cook, D. C., Fenton, G., Karlsson, J. A., Palfreyman, M. N., Raeburn, D., Ratcliffe, A. J., Souness, J. E., Thurairatnam, S., and Vicker, N. (1994). Selective type IV phosphodiesterase inhibitors as antiasthmatic agents: The syntheses and biological activities of 3-(cyclopentyloxy)-4-methoxybenzamides and analogues. *J. Med. Chem.* **37**, 1696–1703.

Barnett, S. F., Bilodeau, M. T., and Lindsley, C. W. (2005). The Akt/PKB family of protein kinases: A review of small molecule inhibitors and progress towards target validation. *Curr. Top. Med. Chem.* **5**, 109–125.

Berns, K., Hijmans, E. M., Mullenders, J., Brummelkamp, T. R., Velds, A., Heimerikx, M., Kerkhoven, R. M., Madiredjo, M., Nijkamp, W., Weigelt, B., Agami, R., Ge, W., Cavet, G., Linsley, P. S., Beijersbergen, R. L., and Bernards, R. (2004). A large-scale RNAi screen in human cells identifies new components of the p53 pathway. *Nature* **428**, 431–437.

Brummelkamp, T. R., Berns, K., Hijmans, E. M., Mullenders, J., Fabius, A., Heimerikx, M., Velds, A., Kerkhoven, R. M., Madiredjo, M., Bernards, R., and Beijersbergen, R. L. (2004). Functional identification of cancer-relevant genes through large-scale RNA interference screens in mammalian cells. *Cold Spring. Harb. Symp. Quant. Biol.* **69**, 439–445.

Burgering, B. M., and Kops, G. J. (2002). Cell cycle and death control: Long live Forkheads. *Trends Biochem. Sci.* **27**, 352–360.

Capon, D. J., Chen, E. Y., Levinson, A. D., Seeburg, P. H., and Goeddel, D. V. (1983). Complete nucleotide sequences of the T24 human bladder carcinoma oncogene and its normal homologue. *Nature* **302**, 33–37.

Comley, J. (2005). High content screening: Emerging importance of novel reagents/probes and pathway analysis. *Drug Disc. World* **3**, 1–53.

Grånäs, C., Lundholt, B. K., Loechel, F., Pedersen, H.-C., Bjørn, S. P., Linde, V., Krogh-Jensen, C., Nielsen, E.-M. D., Præstegaard, M., and Nielsen, S. J. (2006). Identification of RAS-mitogen activated protein kinase signaling pathway modulators in an ERF1 Redistribution screen. *J. Biomol. Screen.* **11**, 423–434.

Gregorieff, A., and Clevers, H. (2005). Wnt signaling in the intestinal epithelium: From endoderm to cancer. *Genes. Dev.* **19**, 877–890.

Haugland, R. P. (2005). "The Handbook: A Guide to Fluorescent Probes and Labeling Technologies," 10 Ed. Invitrogen Corp., Carlsbad, CA.

Hilger, R. A., Scheulen, M. E., and Strumberg, D. (2002). The Ras-Raf-MEK-ERK pathway in the treatment of cancer. *Onkologie.* **25**, 511–518.

Le Gallic, L., Sgouras, D., Beal, G., Jr., and Mavrothalassitis, G. (1999). Transcriptional repressor ERF is a Ras/mitogen-activated protein kinase target that regulates cellular proliferation. *Mol. Cell. Biol.* **19**, 4121–4133.

Loechel, F., Bjorn, S., Linde, V., Prastegaard, M., and Pagliaro, L. (2006). High content translocation assays for pathway profiling. *Methods Mol. Biol.* **356**, 401–414.

Mittal, V. (2004). Improving the efficiency of RNA interference in mammals. *Nature Rev. Genet.* **5**, 355–365.

Robinson, M. K., and Cobb, M. H. (1997). Mitogen-activated protein kinase pathways. *Curr. Opin. Cell. Biol.* **9**, 180–186.

Terry, R., Cheung, Y. F., Praestegaard, M., Baillie, G. S., Huston, E., Gall, I., Adams, D. R., and Houslay, M. D. (2003). Occupancy of the catalytic site of the PDE4A4 cyclic AMP phosphodiesterase by rolipram triggers the dynamic redistribution of this specific isoform in living cells through a cyclic AMP independent process. *Cell. Signal.* **15**, 955–971.

van Roessel, P., and Brand, A. H. (2002). Imaging into the future: Visualizing gene expression and protein interactions with fluorescent proteins. *Nature Cell Biol.* **4**, E15–E20.

Vassilev, L. T., Vu, B. T., Graves, B., Carvajal, D., Podlaski, F., Filipovic, Z., Kong, N., Kammlott, U., Lukacs, C., Klein, C., Fotouhi, N., and Liu, E. A. (2004). *In vivo* activation of the p53 pathway by small-molecule antagonists of MDM2. *Science* **603**, 844–848.

Wang, Y. L. (1989). Fluorescent analog cytochemistry: Tracing functional protein components in living cells. *Methods Cell Biol.* **29**, 1–12.

Wilson, C. (1995). "The Invisible World: Early Modern Philosophy and the Invention of the Microscope." Princeton Univ. Press, Princeton, NJ.

Zhang, J. H., Chung, T. D., and Oldenburg, K. R. (1999). A simple statistical parameter for use in evaluation and validation of high throughput screening assays. *J. Biomol. Screen.* **4**, 67–73.

[28]   High-Content Screening of Functional Genomic Libraries

By Daniel R. Rines, Buu Tu, Loren Miraglia, Genevieve L. Welch, Jia Zhang, Mitchell V. Hull, Anthony P. Orth, and Sumit K. Chanda

Abstract

Recent advances in functional genomics have enabled genome-wide genetic studies in mammalian cells. These include the establishment of high-throughput transfection and viral propagation methodologies, the production of large-scale cDNA and siRNA libraries, and the development of sensitive assay detection processes and instrumentation. The latter has been significantly facilitated by the implementation of automated microscopy and quantitative image analysis, collectively referred to as high-content screening (HCS), toward cell-based functional genomics application. This technology can be applied to whole genome analysis of discrete molecular and phenotypic events at the level of individual cells and promises to

METHODS IN ENZYMOLOGY, VOL. 414
Copyright 2006, Elsevier Inc. All rights reserved.

0076-6879/06 $35.00
DOI: 10.1016/S0076-6879(06)14028-8

significantly expand the scope of functional genomic analyses in mammalian cells. This chapter provides a comprehensive guide for curating and preparing function genomics libraries and performing HCS at the level of the genome.

Introduction

A recent census of the human genome sequence estimated that our cells contain between 20,000 and 25,000 protein-encoding genes (Pennisi, 2005). If this tally is accurate, then the genetic complexity of *Homo sapiens* lies somewhere in between that of a roundworm (*Caenorhabditis elegans* ~18,000) and a flowering plant (*Arabidopis thalisala* ~25,000). If we consider ourselves at the pinnacle of the evolutionary process, then it stands to reason that gene function, not gene number, at least partially dictates biological complexity. However, our understanding of genome function remains sparse. The last five decades of molecular biology research have led to the functional characterization of less than half of the genes in our cells (Su *et al.*, 2004). A major challenge for the biological community in the coming years will be the complete functional annotation of the human genome.

A number of technologies have emerged that facilitate the systematic assessment of various aspects of gene function at the level of the genome. These include expression (microarray) profiling and immunoprecipitation mass spectrometry (IPMS) (Chanda and Caldwell, 2003). Microarray studies can detect spatial or temporal changes in gene regulation on a global scale, whereas IPMS can be utilized to understand protein–protein associations at the level of the proteome. Until recently, genome-scale genetic experimentation was limited to studies in tractable model organisms, such as yeast, Drosophila, or *C. elegans*. However, a number of advances in the manipulation of mammalian cells have made genome-wide gain-of-function and loss-of-function analyses possible. The development of high-throughput transfection technologies allowed for the introduction of large-scale arrayed libraries of nucleic acids into mammalian cells in a rapid and economical fashion (Chanda *et al.*, 2003; Ziauddin and Sabatini, 2001). Thus, one could monitor the effects of gene dosage on particular cellular phenotypes for hundreds to thousands of genes in a single experiment. Furthermore, discovery of the RNA interference (RNAi) pathway, which allows systematic suppression of gene expression in mammalian cells, has greatly facilitated loss of function studies (Elbashir *et al.*, 2001). Large collections of cDNAs, transcribed siRNAs, and chemically synthesized siRNAs are currently being compiled and distributed. These arrayed libraries can be used to interrogate various cellular processes, including differentiation, apoptosis, oncogenesis, and inflammation. The repertoire of these assays is only limited by the ability to detect a particular molecular or cellular phenotype.

High-throughput microscopy (HCS) is a developing technology that enables the rapid capture and analysis of cellular images and is a promising tool that can be used to significantly expand the scope of high-throughput cellular genetic analysis. It offers many of the capabilities and advantages of fluorescent-activated cell sorting (FACS) analysis with a few important exceptions. First, adherent cells do not require from their growth surface. This reduces intricate sample preparation procedures, which may be rate limiting in large-scale assays, and also enables kinetic reads. Furthermore, HCS allows for the analysis of subcellular events such as cytoplasmic to nuclear translocation, which is currently untenable through FACS measurements. Thus, the application of high-throughput microscopy toward large-scale cellular genetic experimentation will facilitate the interrogation of a large number of biological processes that have previously not been addressable in mammalian systems. This chapter addresses considerations and provides methodologies found to be critical in this area of functional genomics investigation.

Development of Large-Scale Genomic Libraries

A number of academic and commercial institutions are currently building and distributing mammalian genome-scale libraries, or sublibraries, that target families of proteins (i.e., kinases) (Table I). Until more enabling technologies are established, the availability of these reagents largely obviates the need for the construction of these libraries for screening. Libraries can be categorized by the delivery methodologies utilized to introduce them to mammalian cells. These libraries can be transfected utilizing various high-throughput and/or automated processes or, alternatively, cells that are recalcitrant to the transfection process may be infected with various viruses harboring short-hairpin RNAs or cDNAs. These libraries have distinct advantages and limitations, which are discussed next.

cDNA Libraries

Performing HCS gain-of-function screens using cDNAs enables the identification of proteins, which, when expressed ectopically, overcomes a rate-limiting step in the biological process being investigated (Chanda *et al.*, 2003; Harada *et al.*, 2005; Ziauddin and Sabatini, 2001). While multiple libraries of full-length cDNAs are available commercially, only those cloned into vectors bearing a mammalian expression promoter such

TABLE I

COMMERCIALLY AVAILABLE LIBRARIES FOR GENOME-SCALE FUNCTIONAL GENOMIC STUDIES

Vendor	Web site	Collection	Species	Reagent Type	Number of Genes Targeted	Coverage	Comments
Ambion	http://www.ambion.com/	Druggable genome	Human/mouse	siRNA	5495	3 siRNAs/gene	Genome wide Libraries In process
Sigma	http://www.sigmaaldrich.com/	MISSION TRC shRNA	Human/mouse	Lentiviral shRNA	9651	5 shRNAs/gene	5x Coverage of Human & Mouse Genome in process. Also available from OpenBiosystems
Open Biosystems	http://www.openbiosystems.com	shMir libraries	Human/mouse	Lentiviral and retroviral	~35,000	~6 shRNAs/gene	
Qiagen	http://www.qiagen.com	Genome wide	Human/mouse	siRNA	28,500	4 siRNAs/gene	
Invitrogen	http://www.invitrogen.com	Protein families	Human/mouse	siRNA	49–636	3 siRNAs/gene	
Dharmacon	http://www.dharmacon.com	Genome wide	Human/mouse/rat	siRNA	~22,000	4 siRNAs/gene	siRNAs in pooled format
IDT	http://www.idtdna.com	Protein families	Human/mouse	siRNA	TBD	TBD	27-mer oligos, protein family libraries in process
Open Biosystems	http://www.openbiosystems.com	Assay-ready MGC	Human/mouse	cDNA	16,000	N/A	

as the CMV promoter can be directly introduced by transfection into cells with the expectation that the encoded protein will be produced. Collections of CMV-driven, fully sequenced mammalian clones in arrayed-well format have been produced by the Mammalian Genome Consortium (Strausberg *et al.*, 1999, 2002) and are available commercially from several vendors, including ATCC, Invitrogen, and Openbiosystems. Notably, ready-to-screen cDNA collections are available from the latter company.

From a technical standpoint, gain-of-function HCS assays based on forced expression of cDNAs have some advantages and disadvantages relative to loss-of-function assay conducted using RNAi. cDNAs are more difficult to produce in assay-ready format than synthetic siRNAs, as DNA requires preparation from living bacteria. Unlike siRNAs, however, cDNAs can be produced inexpensively once acquired in virtually limitless quantities, can be moved into viral vectors for delivery into difficult-to-transfect cell types, and suffer less experimental ambiguity associated with the persistent problem of off-target effects observed when using RNAi. However, results from cDNA screens may not necessarily be physiologically relevant, as these results are often prone to artifacts that result from simple overexpression of a protein.

Chemically Synthesized siRNAs

Synthetic oligonucleotide libraries that target the entire human genome, or subsets thereof, are currently available from a number of vendors (Table I). These sets usually contain two to four siRNAs per gene. siRNAs are typically 21 to 27 bp in length, although it is currently unclear which length is most advantageous for screening. Once introduced within cells, double-stranded siRNAs are unwound by the activated RISC complex in an ATP-dependent process, and the antisense strand directs recognition of the target mRNAs sequences (Hannon and Rossi, 2004). The mRNA homologous to the introduced siRNA is then subsequently degraded, resulting in a loss of gene function (Hannon and Rossi, 2004).

Advantages of chemically synthesized siRNA libraries are that their synthesis is relatively standardized, which results in little variation in the purity of individual siRNAs in a collection. This, and the (comparatively) small size of these molecules, enables robust and consistent transfection throughout a large-scale screen. From a workflow standpoint, these libraries are much easier to deploy for screening than are either cDNAs or hairpin-encoding RNAi libraries, as they do not require significant investments in automation and technical expertise required for high throughput DNA preparation and normalization. With these collections, the reagents are not renewable. Thus, ordering additional RNAi for validation and confirmation studies rapidly becomes an expensive prospect. Alternatively, siRNAs can be produced *in vitro* by a number of enzymatic methodologies (Betz, 2003; Kittler *et al.*, 2004; Luo *et al.*, 2004; Sen *et al.*, 2004), which can reduce the costs of large collections substantially.

The two major drawbacks of RNAi oligonucleotide libraries are the requirement that they be transfected, limiting their use to highly transfectable cell types, and the fact that gene suppression will occur transiently, limiting assay phenotypes to those that occur over the course of a few days. While long-term depletion of some genes may occur in some cell types with chemically synthesized siRNAs, robust cell division will gradually dilute siRNAs, and intracellular RNase activity will gradually contribute to siRNA metabolism as well.

Plasmid Short-Hairpin siRNAs

It is also possible to clone DNA sequences that, when transcribed, will produce a single-stranded, hairpin-forming RNA giving rise after processing by the RNAi machinery to a functional siRNA. These short-hairpin RNAs (shRNAs) can be introduced into cells to mimic the effects of chemically synthesized siRNAs (Brummelkamp *et al.*, 2002). Briefly, the cloned double-stranded DNA oligonucleotide contains sense and anti-sense sequences targeting the gene of interest separated by a hairpin loop. Once inside the cell, a promoter—typically an RNA polymerase III promoter—drives the expression of an RNA, which forms a hairpin-like structure, which is modified by a protein called Dicer to produce nucleotide fragments analogous to chemically synthesized siRNAs (Hannon and Rossi, 2004). It has been shown that polymerase II promoters can also be utilized to drive the transcription of short hairpin siRNAs (Stegmeier *et al.*, 2005).

Due to the larger size of the vectors that harbor these short hairpins (typically >3 kb), they are more difficult to transfect than their chemically synthesized counterparts. However, this method of delivering RNAi constructs has several advantages over chemically synthesized siRNAs. First, by engineering these hairpins into viral vectors, such as lenti-, retro-, or adenoviral vectors, it is possible to introduce shRNAs into cell types refractive to transfection and also achieve long-term siRNA-mediated reduction in mRNA levels (Tiscornia *et al.*, 2003; Xia *et al.*, 2002). Second, because these vector-based libraries can be propagated in bacteria, they represent a renewable resource, which facilitates a reduction in long-term screening and validation costs. Short hairpin technology can additionally be utilized in *in vivo* model systems—either by direct viral injection or the creation of transgenic organisms (Tiscornia *et al.*, 2003). Although the methodology is not currently robust enough to replace conventional genetic approaches, it offers significant promise for characterization of organismal gene function in a rapid and robust manner.

Formats for RNAi Arrays

The number of individual siRNAs to screen per well has been a matter of considerable debate. siRNAs may be arrayed such that there is one siRNA or multiple siRNAs targeting a single gene in a well (Aza-Blanc *et al.*, 2003; Li *et al.*, 2004). Alternatively, strategies employing multiple siRNAs targeting many different genes in a single well have also been used for successful loss-of-function screens (Berns *et al.*, 2004; Paddison *et al.*, 2004). The latter two approaches are variations of what is referred to as a pooling strategy. Overwhelming opinion currently mandates that, in order to be considered a validated activity, a phenotype must be observed using two nonoverlapping siRNAs targeting the same gene. This is largely due to the preponderance of nonspecific (or "off-target") RNAi effects in which short stretches of sequence homology can trigger the degradation or translational repression of an unintended message (Jackson and Linsley, 2004). Thus, to enable rapid discrimination between validated and potential off-target phenotype-to-target relationships it would be ideal to screen individual siRNAs in a well, with the library containing multiple siRNA per gene (called a "high coverage" library). However, some have hypothesized that pooling siRNAs against the same gene increases the efficiency of mRNA knockdown, arguing that since each RNAi is at approximately 25% of the concentration of a single siRNA transfection, there may be a corresponding reduction in potential off-target effects. If a pooled strategy is employed for screening, however, pools should subsequently be split and retested as individual siRNAs to identify those responsible for the activity seen in the original screen. If only a single active siRNA is obtained, it would be necessary to confirm the observed phenotype with a second, independent (non-overlapping) siRNA. Ultimately, decisions on assay format are dependent on both the goals of the project and the budget and infrastructure available for screening.

Maintaining Large Arrayed-Well Plasmid cDNA or shRNA Clone Libraries

Clone library maintenance is extremely important and often involves a substantial investment of time (Fig. 1). Arrayed-well libraries, those with one clone species per well, are frequently costly to generate or obtain commercially. Therefore it is imperative to ensure that collections are free of phage, contaminant bacterial species, or fungi. Additionally, it is

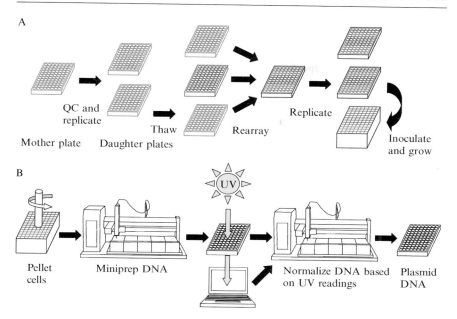

FIG. 1. Replication and preparation of clone collections used in functional genomic assays.

necessary to replicate archival copies both for general use and to provide a contaminant-free source should contamination with such organisms arise. Phage contamination, typically with T4 or T7 phage, can be detected by plaque assay (Sambrook, 1989) and, if detected early, can be contained by discarding affected plates and thorough bleaching (10%) of laboratory areas and instruments. Collections generated internally should be propagated in phage-resistant *Escherichia coli* strains such as DH10B-T1 (Invitrogen). Additionally, when propagating plasmid collections known to recombine, such as those containing inverted repeat sequences (i.e., viral LTRs), it is strongly recommended to use a recombination-deficient bacterial strains, such as STBL3 (Invitrogen) or Sure (Promega).

Contamination of libraries by bacterial species other than *E. coli* can be more difficult to detect and contain than contamination by phage. Contaminant bacteria such as *Pseudomonas aeruginosa* and *Bacillus cereus* may possess characteristics of *E. coli*, are frequently capable of outcompeting *E. coli* due to inherited, rather than episomal, antibiotic resistance, and may sporulate and contaminate neighboring cultures via aerosolization. Unusual colony morphology, color, smell, unusually large bacterial pellets, or failure to produce plasmid DNA may all indicate the presence of contaminant bacterial species. However, nonturbid cultures failing to yield plasmid DNA are typically more indicative of phage. Methods for detecting contaminant bacteria include the spotting of inoculate using disposable pin replicators from plates onto solid, antibiotic-containing media, streaking out of sentinel wells on solid media, isolation and gram staining, and a number of other, more rigorous tests aimed at characterizing the contaminants (Krieg, 1984). Plates suspected of bacterial contamination should be discarded immediately and the laboratory area and instruments cleaned with a 10% bleach solution. Laboratory benches and equipment should be wiped and the wipes dotted onto antibiotic-containing solid media to test for the continued presence of live bacteria or bacterial spores.

Collection Replication

Once satisfied that a collection is safe to use, at least two glycerol stock replicate (daughter) copies should be made as soon as practical (Fig. 1A). Generating ready-to-assay cDNA/shRNA and sublibraries for use in assay reconfirmation requires multiple freeze–thaw cycles tolerated well by some *E. coli* strains but less well by others. As a general rule, library copies should be tested for freeze/thaw tolerance and daughter copies discarded as soon as some wells begin to grow poorly. At any given time, one or more unthawed copies of a library should be reserved as a master copy. This copy should not be thawed until the next time the library needs to be replicated, and only a single copy should serve as a working copy. It is helpful to record which wells are empty in the original library plates in order to determine when a particular copy begins to become unviable. Additionally, it may be necessary to select (or "hitpick") specific clones from the original collections (i.e., full-length cDNAs, protein families, active genes, etc.). This rearray step enables the construction of customized collections (Fig. 1A).

Bacterial glycerol stocks in a 96- or 384-well format can be replicated using a variety of devices, all of which transfer liquid inoculate from one plate to another. The most robust and also most laborious method employs liquid inoculation using a multichannel pipettor fitted with disposable filter tips. Automated pipetting instruments can also be used if fitted for use of such tips. Fixed probe devices should be not be employed to pipette solutions containing bacteria, as probes can be corroded by frequent bleaching and heavy cultures may contain sticky bacterial colloids resistant to sterilization, thus facilitating the spread of contaminating organisms through the library if present anywhere in the clonal population. Fixed-pin inoculation employing either 96- or 384-pin disposable plastic pin replicators (Genetix), 96- or 384-pin reusable metallic pin replicators, or automated devices fitted with pin heads such as the QBot (Genetix) represents the least expensive and labor-intensive method for glycerol stock replication. Furthermore, unlike fixed probe pipettors, fixed-pin devices are relatively easy to sterilize between inoculate steps, and most pin-tool heads, whether manual or automated, can be autoclaved periodically. The following protocol details how to make glycerol stock copies of 96- or 384-well bacterial libraries.

1. Before replicating, prepare sterile LB broth and, when cooled, supplement with appropriate antibiotic.

2. Add media to prelabeled, flat-bottomed 96- or 384-well plates fitted with lids to no more than half the total well volume.

3. Mother plates, if frozen, should be removed from 80° and unfoiled immediately (but relidded) to prevent liquid on the foil seal from thawing and dripping into adjacent wells during unfoiling, thereby contaminating adjacent wells. Once thawed, culture media from mother plates should be carefully transferred to daughter plates containing fresh media either by dipping pins from a pin replicator from one to the other or by aspirating and dispensing 5 to 10 μl using a pipetting device fitted with disposable filter tips.

4. Inoculated daughter plates should be lidded but left unfoiled, stacked two plates high, wrapped loosely with plastic film to prevent desiccation, and incubated overnight at 37° or until turbid without shaking or agitation.

5. Mother plates should be refoiled and replaced at −80°.

6. Once turbid, daughter plates should be cataloged for empty wells, supplemented with 50% the original media volume of a 1:1 glycerol/sterile water solution (16.7% final glycerol concentration), foiled, shaken slightly to mix, placed at −80°, and retained there until thawed for subsequent use.

If a collection proves particularly hardy, it is also possible to inoculate into daughter plates filled with media already containing glycerol, thereby avoiding the additional pipetting step. A test inoculation into glycerol-containing media will usually indicate whether this is possible.

Growth of Bacterial Cultures in High-Throughput Format for DNA Preparation

Plasmid DNA is most commonly prepared in a 96-well plate format. Most assay applications require submicrogram quantities of DNA and thus 384-well miniprep formats may require unrealistically frequent preparation of DNA. The following protocol, illustrated schematically in Fig. 1, can be followed to prepare 96-well bacterial cultures for purposes of preparing transfection-grade plasmid DNA:

1. Fill 96-well deep-well blocks with 1.5 ml of Terrific broth (Sambrook, 1989) containing antibiotic. If automation is to be employed in a miniprep DNA preparation, ensure that the chosen deep-well blocks are compatible with the automated preparation instrument.

2. Source glycerol stocks are removed from −80° and unfoiled immediately but relidded thereafter to prevent frozen liquid adhering to the foil from thawing and sliding into neighboring wells during unfoiling. Once thawed, inoculate the deep-well blocks containing TB media using either a disposable pin replicator or by transferring liquid using a multichannel pipettor or liquid handler fitted with disposable, filtered tips.

3. Cover the block with a film that is permeable to air but not liquid or bacteria and place in a shaking incubator fastened tightly into an angle tube rack holder. Poor bacterial growth may indicate insufficient oxygen, in which case forced oxygen may be required.

4. Incubate with shaking at 270 rpm at the appropriate temperature until the culture reaches an OD_{600} reading that is within the recommendations of the miniprep kit that will be used for plasmid isolation.

While some libraries require only an overnight growth period, it is not uncommon for libraries containing low-copy expression vectors or those propagated in slow-growing strains of bacteria to take 32 to 48 h to reach proper turbidity. To minimize the number of low-yielding or dead wells, blocks should be cultivated at least 24 h to allow slowest-growing cultures to reach the plateau growth phase. This is particularly important for cDNA libraries, as larger inserts may grow slower than smaller ones and leaky expression from eukaryotic promoters may cause bacterial toxicity.

Preparing DNA from 96-Well Deep-Well Block Cultures

In a 96-well format, plasmid DNA to be employed in transfection-based assays or used for producing virus is most commonly separated from lysed bacteria using silica-based 96-well filterware or, less commonly, magnetic beads. Silica-based technologies are currently favored due to their versatility—the same filterware can usually be used manually or adapted to automated preparation instruments—and to the high yields of protein-, RNA-, lipid-, and genomic DNA-free plasmid DNA produced. Magnetic bead-based DNA purification, although usually less expensive on a per-well basis, frequently suffers from poorer yields or higher levels of endotoxin and is not discussed in detail here. High-throughput, filtration-based plasmid DNA miniprep kits are available from a number of commercial vendors, as are automated instruments to run them. The following protocol, illustrated schematically in Fig. 1B, has been adapted from Macherey-Nagel 11/2003/Rev. 02, and applies specifically to the Macherey-Nagel Nucleospin Robot 96 plasmid kit run either using a vacuum manifold or an MWG 2500 RoboPrep instrument. However, it could apply with minor modification to similar products from Qiagen, Eppendorf, Fisher, and other major reagent providers deployed on such instruments as the Qiagen-BioRobot9600, the Beckman-Coulter-Biomek2000, the Beckman-Counter FX, the Tecan Freedom EVO, and others.

1. Centrifuge the deep-well blocks containing the bacterial cultures at 1000g for 10 min. After centrifugation, remove the membrane and immediately decant culture supernatant to a container for proper disposal. Invert and gently tap the blocks on several layers of paper towels to remove residual culture supernatant.

2. If performing minipreps manually using a vacuum manifold, we recommend performing liquid-handling steps using a high-quality, large-volume, eight-channel digital electronic pipettor such as the Thermo Electron Corporation Finnpipette Novus. While nonmotorized multichannel pipettors can be employed, for large collections their use may increase the risk to laboratory staff of developing repetitive motion injuries. If performing minipreps using automated instruments, program the instruments to perform the following steps, with the exception of the resuspension step, which is typically performed before blocks are subjected to automated handling.

3. Resuspend bacterial pellets by adding the recommended volume of resuspension buffer containing RNase (buffer A1) to each well, usually 250 μl, and then mixing vigorously using a plate shaker until the bacterial cells are totally resuspended. Clumped bacteria do not lyse well and incomplete lysis can degrade DNA yields; therefore light vortexing can be employed to resuspend clumps. Vigorous vortexing is to be avoided as it will shear the bacterial genomic DNA, causing it to remain in solution with the plasmid DNA instead of being filtered out with the bacterial cell wall components to which it would otherwise adhere.

4. Lyse the resuspended bacteria by adding the appropriate volume of alkaline lysis buffer (buffer A2) to each well, usually 250 μl, and incubating 2 min at room temperature.

5. Neutralize the lysis solution by adding the appropriate volume of neutralization buffer (buffer A3), usually 350 μl.

6. Transfer the neutralized crude lysates onto the lysate filter module. This module, which stacks above the filter plate but rests on the rubber gasket of the manifold, permitting a seal to form when vacuum is applied. It is designed to filter bacterial cell-wall components, including the attached bacterial chromosomal DNA, permitting the plasmid DNA and soluble bacterial components to flow through to the binding plate below. Incubate for 1 min and then apply vacuum at 750 mbar for 2.5 min to draw the cleared crude lysate into the binding plate. If using a vacuum manifold attached to a house vacuum, take care to predetermine the amount of vacuum to apply, as the application of an excessively strong vacuum can rupture the lysis filter

plate or permit lysate to be aspirated through an intact filter, either of which will contaminate the cleared lysate with undesired bacterial components.

7. Disassemble the manifold, discard the filter plate, and place the plasmid DNA-binding module containing the cleared lysate atop the manifold, taking great care to avoid spilling the cleared crude lysate or sloshing it into neighboring wells.

8. Apply vacuum at 750 mbar for 2 min to aspirate the cleared lysate through the DNA-binding silica resin.

9. Wash the plasmid DNA adhering to the silica membrane by adding and pulling through first 600 μl of AW wash buffer, then 900 μl of A4 wash buffer, and finally another 900 μl of A4 wash buffer by applying vacuum at 750 mbar for 45 s to allow buffer to pass the columns.

10. Dry the plamid-binding module by applying 10 to 15 min of maximum vacuum. Even minor amounts of residual ethanol can have a negative impact on some downstream applications, including transfection into certain cell types, and in these cases it is wise to centrifuge the binding modules placed atop blotting paper at 1000g for 2 min.

11. Elute plasmid DNA into an ultraviolet (UV)-transparent 96-well plate with 100 to 200 μl Tris buffer or TE buffer. Apply vacuum at 500 mbar for 5 min.

Normalization of Plasmid DNA

DNA yields from 1.5-ml TB cultures grown to saturation and prepared using 96-well miniprep kits can exceed 15 μg/well, but more typically only reach 7 to 10 μg/well or, for 100 μl of eluate, approximately 100 ng/μl. However, these concentrations frequently vary considerably from well to well. Assay concentrations for transfected plasmid DNA range between 20 and 60 ng/well in a 384-well format depending on application but generally should be fixed for all wells for a given assay to ensure consistent transfection efficiencies. Superior assay results require that DNA be normalized to a target DNA concentration, but unlike the preceding steps, which can be performed manually, the need to dilute each well differently to obtain a single concentration makes manual normalization virtually impossible to perform at any scale exceeding several 96-well plates. Normalization therefore requires at a minimum a 96-well spectrophotometer and a liquid handler capable of handling microliter levels of liquid (Fig. 1B). The following protocol illustrates a general method for normalizing plasmid DNA.

1. Blank the 96-well spectrophotometer using a dummy UV-transparent elution plate filled with the chosen elution buffer at levels representative of the contents of the actual elution plates. This represents the background UV correction figure.

2. For each elution plate, measure the OD 280 nm and 260 nm twice, saving all values. If the spectrophotometer has an automated path length correction feature relying on readings in the 900-nm range to normalize for slight variations in well volume, it should be enabled, although minor variation (~10%) around 100 μl will not seriously affect quantitation.

3. Take mean values for OD 260 and 280 and calculate based on the OD 260 the amount of DNA per well, noting that for a 1-cm path length, 50 μg of double-stranded DNA has a OD 260 nm of 1. Many 96-well spectrophotometers can be precalibrated to a dummy elution plate, providing one or more log files for each elution plate containing DNA concentrations.

4. Calculate the volume of eluate and 5 mM Tris-HCl (no EDTA if proceeding to sequencing) required to achieve for each well a final concentration at the assay level, nominally 40 ng/μl.

5. Format a file for import to the liquid handler capable of interpreting these volumes and import to the liquid handler. For transfection-based assays, fixed probe devices incorporating water tip flushing can be employed readily without concern regarding well-to-well carryover of DNA due to the low relative lability of double-stranded DNA in a solution of Tris-HCl. Downstream applications involving the more sensitive polymerase chain reaction or other amplification strategies should always avoid the use of fixed probes, however.

6. To achieve a final concentration per well of 40 ng/μl, program the liquid handler to transfer DNA eluate containing 3.2 μg of DNA per well to a U- or V-bottom 96-well plate, adding the amount of 5 mM Tris-HCl required to bring the total final volume to 80 μl. U- or V-bottom destination plates should always be employed, as this permits most of the DNA to be removed during the creation of assay-ready 384-well plates. Flat-bottom plates will form a meniscus such that the last 20 μl of DNA eluation cannot be removed reliably.

7. Foil the normalized destination plates and store indefinitely at -20 or $-80°$. Alternatively, samples may be stored in sealed plasmid plates for several weeks in the refrigerator.

8. The precision and accuracy of the entire normalization process can be assessed using a variety of intercalating DNA dyes, such as Quant-iT, according to the manufacturer's specifications (see following section).

Arraying Collections into High-Throughput Assay Plates

Large-scale genetic screens in mammalian cells are usually conducted in 384-well assay plates, as this format provides a balance between the economies of high-throughput assay formats and the ability to interrogate a sufficient number of cellular events to extract biologically meaningful data. Depending on assays and cell types intended to be utilized, between 15 to 60 ng of DNA and 5 to 20 ng of siRNA are arrayed in each well. However, virtually all siRNA libraries and prepared cDNA libraries exist in the 96-well format, necessitating reformatting from a 96- to a 384-well format for those HCS assays that can be executed in a 384-well format (Fig. 2). These 96-well (source plates) to 384-well plate (destination plate) transfers are accomplished most efficiently using a reformatting device, such as the Perkin Elmer EP3 or Beckman FX. These instruments can be used to accurately and consistently deliver between 500 and 3000 nl of liquid to a dry 384-well plate ("dry touch off"). Before embarking on this step, it is essential that the reformatting instrument be calibrated properly. This can be accomplished using limiting dilutions of fluorescein isothiocyanate dye, followed by measurement on a standard fluorescent plate reader. Once the coefficient of variance (CV) and standard deviations for each tip are determined to be in an acceptable range (10% or less CV), multiple screening sets may be arrayed in a single run and kept in appropriate storage conditions ($-80°$) for future use.

For each spotting run, a final quality control step should be undertaken to ensure that the process yielded plated sets of acceptable screening quality. To accomplish this, it is necessary to use two to three of the plated sets to measure the concentration of siRNA or cDNA, utilizing reagents such as Ribogreen (Promega) or Quanit-it (Invitrogen), respectively, across the collection. An example of how to use the latter to assess DNA preparation is shown next.

To Determine the DNA Concentration

1. Spot the DNA plates to quality control/validate final spotted DNA amounts/well.
2. Set up standard curves in 384-well white-bottom plate as follows.

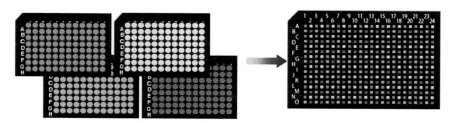

FIG. 2. Rearraying from 96-well to 384-well format.

    a. Set up the standard curve using a known concentration DNA at 0, 5, 10, 20, 40, 60, 80, and 100 ng/well with a final volume of 20 μl/well in distilled water.

    b. Set up the Quant-iT standard curve using the kit provided λ DNA dilutions with the same setup as step 2a.

3. Remove the two spotted plates to be assayed from 20°/80° and equilibrate to room temperature.

4. Make a working solution by diluting Quant-iT DNA BR reagent 1:200 in Quant-iT DNA BR buffer.

5. Add 80 μl of diluted Quant-iT solution to each well and assay for fluorescence.

Generally, a screening set should have <20% CVs in concentration over 90% of the wells in a 384-well plate. If plated concentration varies significantly from these specifications, it is prudent to reprepare those plates.

High-Throughput Transfections

For assay optimization, a cell line must be chosen that can be transfected at an adequate efficiency. This can be determined using a surrogate reporter such as green fluorescent protein (GFP) or a fluorescently labeled siRNA. In general, transfection efficiencies above 25% will yield acceptable results in HCS assays. Transfectability of cell lines varies, as do the efficiencies of different transfection reagents when combined with these cell lines. Performing transfections with a matrix of both different cell densities and transfection reagent/plasmid ratios to determine optimal conditions is recommended. Table II lists a number of reagents that may be initially tested for efficacy. Once a cell line/transfection reagent pair has been selected, it is necessary to determine if these conditions are compatible with high-throughput (or "reverse") transfection (Fig. 3).

Reagents

    FuGENE 6 transfection reagent or equivalent (Roche)
    DMEM +10% fetal bovine serum (FBS) with antibiotics
    DMEM
    HEK 293T cells
    384-well black clear-bottom plate (Greiner)
    Sterile microfuge tubes
    pCMV-GFP (or equivalent)

TABLE II

A SELECT LIST OF TRANSFECTION REAGENTS THAT CAN BE UTILIZED TO OPTIMIZE
HIGH-THROUGHPUT TRANSFECTION CONDITIONS[a]

Reagent	Vendor	Library
Fugene	**Roche**	**DNA/viral**
TransFectin	Bio-Rad	DNA/viral
CLONfectin	Clontech	DNA/viral
DreamFect	OZ Biosciences	DNA/viral
TransFast	Promega	DNA/viral
Escort	Sigma-Aldrich	DNA/viral
LipoGen	InvivoGen	DNA/viral
Transit-Express	**Mirus**	**DNA/viral**
GeneJuice	Novagen	DNA/viral
SuperFect	Qiagen	DNA/viral
GeneJammer	Strategene	DNA/viral
Lipofectamine2000	**Invitrogen**	**RNAi**
X-tremeGENE	Roche	RNAi
siIMPORTER	Upstate	RNAi
Block-it	Invitrogen	RNAi
RNAifect	Qiagen	RNAi
GeneEraser	Strategene	RNAi
RiboJuice	Novagen	RNAi
HiPerFect	Qiagen	RNAi
GeneSilencer	Genlantis	RNAi
siPORT	Ambion	RNAi
siLentFEC	Bio-Rad	RNAi
siFECTOR	B-Bridge	RNAi
TransIT-siQUEST	Mirus	RNAi
TransIT-TKO	Mirus	RNAi
jetSI	Polyplus	RNAi
Codebreaker	Promega	RNAi

[a] Recommended starting reagents are in bold.

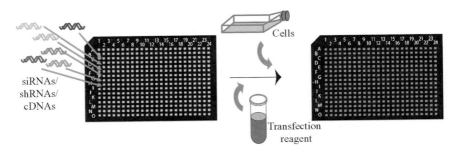

FIG. 3. Reverse transfection of siRNAs and cDNAs in multiwell plate format.

Protocol

1. Spot GFP plasmid at a desired concentration in a 384-well plate.
2. Prepare transfection cocktails in sterile microcentrifuge tubes by first adding FuGENE 6 to DMEM (20 μl/well) at a concentration of 3 μl per μg of spotted DNA and then mix gently by inverting the tubes, that is, 40 ng/well = 150 ng $\times$ 3 μl/μg DNA = 0.36 μl FuGENE 6.
3. Dispense 20 μl of reporter cocktail to each well.
4. Incubate at room temperature for 30 min.
5. While complexes are forming, prepare HEK 293T cells by harvesting and resuspending at a concentration of 5 $\times$ 10^5 cells/ml. Dispense 20 μl of cells/well.
6. Incubate for 24 to 48 h at 37°, 5% CO_2 depending on assay.
7. Analyze on a fluorescent plate reader or microscope (see p. 558).

Note: for a successful experiment, intrasample CVs should not exceed 20%.

High-Throughput Retroviral/Lentivial Packaging

Large-scale production of lentiviruses or retroviruses harboring cDNAs or siRNAs can be accomplished using techniques based on the aforementioned high-throughput transfection protocol and transfection (assay)-grade DNA. Experimental methodologies for producing either lentivirus or retrovirus are effectively the same and only differ in the packaging constructs used (Fig. 4). Additionally, the choice of envelopes for viral pseudotyping is dependent on the types of assays that are intended to be run. However, we find that utilization of the vesicular stomatitis virus G protein (VSV-g) envelope enables the infection of a broad range of cell types and provides the most stability for viral manipulation and storage.

Finally, adequate safety precautions, which are, at minimum, in compliance with CDC recommended BSL-2 safety standards, should be in place before embarking on high-throughput production protocols. These viral particles are infectious to human cells, and the scale of the production runs presents additional risk factors not normally encountered in small-scale laboratory viral production experiments. Hazards can be reduced substantially through the use of automation, which limits contact with virus.

Transfection Protocol (Lentiviral Packaging)

1. Using a reformatter (i.e., Packard EP3), spot library viral DNA containing GFP into Greiner 96-well plates at a final concentration of 100 ng.

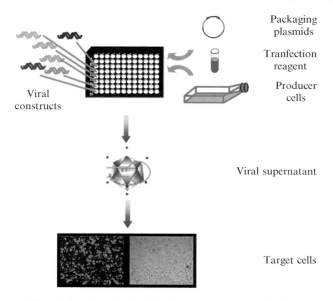

FIG. 4. A schematic for high throughput viral production.

2. Prepare the following master transfection mix based on per well:
   a. 50 μl of serum free-DMEM.
   b. 0.87 μl of FuGENE 6.
   c. 83.3 ng lentiviral packaging plasmid containing gag and pol.
   d. 32.1 ng lentiviral packaging plasmid containing rev.
   e. 44.9 ng envelope plasmid (i.e., VSV-g).
3. Using a Multidrop (or similar instrument), add 50 μl of the trans-fection mix.
4. Allow the plate to incubate at room temperature for at least 30 min.
5. Resuspend the 293T cells in 20% FBS DMEM, 2% p/s, 20 mM of nonessential amino acid, 20 mM HEPES at a concentration of 6.6 × 10^5 cells/ml.
6. After the transfection plate has incubated at room temperature for at least 30 min, dispense 50 μl of cells into the plate using the Multidrop or similar instrument.
7. Incubate the plate at 37° with 5% CO_2 for 24 h.
8. At 24 h posttransfection, replace media.
9. 96-well packaging plates should be spun for 10 min at high speed to ensure that non- or weakly-adherent 293T producer cells are pelleted (optional), after which the viral supernatant can be

transferred via an automated pipettor, such as a MiniTrack (Packard Instruments) fitted with filter tips into a 384-well plate.

Infection Protocol for Titering Purposes

1. Spot dilutions of virus (1 μL to 30 μL) from step 9 or step 10 into a 384-well plate.
2. Prepare HeLa cells at a concentration of 1.5×10^6 cells/ml.
3. Using the multiwell dispenser, dispense 10 μl of cells into the plate.
4. Using the multiwell dispenser, add 0.5 μl of 82.4 μg/ml of protamine sulfate in H_2O.
5. Incubate the plate at 37° with 5% CO_2 for 24 h.
6. After a 24-h incubation, add 80 μl of fresh media.
7. Incubate the plate at 37° with 5% CO_2 for another 24 h.
8. Analyze by FACS or HCS for marker expression. If the marker is a selectable marker, such as puromycin or neomycin, a selection step should be introduced after transduction of the HeLa cells.

To determine final titer, use the following formula:

$$([\%GFP + cells] * [total\ cell\ number]) / (volume\ of\ virus\ added\ [mL]) = infectious\ particles/ml$$

Instrumentation Required for Functional Genomics Screening

Automation required for functional genomic screen is largely based on the number of assays and depth of libraries intended to be screened. Small libraries (i.e., siRNAs against all kinases) can be interrogated with very limited investment in instrumentation. However, larger-scale efforts, including those undertaken by "core" facilities, require significantly greater investment in automation. Solutions range from integration of "off-the-shelf" instruments to fully customized robotic suites specifically designed for high-throughput functional genomics analysis and high-content screening (Table III). Advantages of partially or fully automated approaches include substantial increases in throughput, as well as data quality and reproducibility.

Automated Microscopy

Morphological analysis of cell populations treated with cDNA or RNAi libraries requires automated microscopes, sophisticated image processing tools, and significant data storage. The second half of this chapter focuses on

TABLE III

A PARTIAL LIST OF INSTRUMENTS THAT CAN BE UTILIZED TO ENABLE HIGH-THROUGHPUT FUNCTION GENOMIC ANALYSES

Procedure	Process	Instruments
DNA prep	Deep-well media dispense	μFill microplate dispenser (BioTek)
		FlexDrop IV Exi (Perkin Elmer)
		Multidrop DW (Titertek)
		CyBi-Drop (Cybio)
	Minipreps	Biomek FX (Beckman Coulter)
		MultiPROBE II (Perkin Elmer)
		Freedom EVO (Tecan)
		Biomek 2000 (Beckman Coulter)
		BioRobot 9600 (Qiagen)
Reformatter	Rearray DNA from 96- to 384-well plates	Evolution EP3 (Perkin Elmer)
		Bravo (Velocity11)
		Janus (Perkin Elmer)
		Beckman FX (Beckman Coulter)
Transfection	Microplate dispenser	BottleValve (GNF)
		μFill microplate dispenser (BioTek)
		NanoQuot microplate dispenser (BioTek)
		BioRAPTR FRD (Aurora Discovery)
	Media changes	BottleValve (GNF)
		ELx405 microplate washer (BioTek)
		AquaMax DW4 (Molecular Devices)
Viral preps	Reformat virus from 96 to 384	Same as DNA reformatter
	Seal plates for storage	PlateLoc (Velocity11)
		ALPS 300 (Abgene)
		Stacking heat sealer (Tecan)
Viral production and siRNA screening	Automated suite for high-throughput viral production, siRNA/cDNA transfection, cell fixation, and high-content screening	ALVS robotic suite (GNFSystems)

the capabilities of these tools, provides protocols for sample preparation, and supplies image acquisition techniques to leverage library screening. Depending on the scientific application, HCS can be done with either live or fixed cell samples. Some microscope systems can also be equipped with environmental control chambers that regulate temperature and atmospheric CO_2 concentrations. Here we focus on the preparation and imaging of fixed time-point samples, but the techniques presented are equally applicable to live cell studies. In addition, all the commercial HCS systems now include some form of robust software that quantitatively analyze image properties based on image segmentation or intelligent detection methodologies. When combined, these tools create a powerful bioinformatics platform where information can be extracted in a highly automated fashion.

High-Throughput HCS Equipment

Unlike standard research microscopes, high-content or automated screening microscopes include specialized hardware and software for dealing with multiple samples arrayed into distinct regions of either a microtiter plate or a standard microscope slide. A broad range of imaging systems have been either modified or designed specifically to work with these arrayed sample formats (Table IV). At a minimum, an automated microscope must have the ability to automatically position each and every sample in a microtiter plate above the objective lens, contain a robust autofocusing system, and have the capacity to store image data without manual intervention. Many of the commercial systems available today also offer features such as automated filter switching, sophisticated image analysis tools, simultaneous multichannel image acquisition, and robotic plate loaders, to name a few. The selection of additional features must usually be balanced between the functional requirements of a system and the funds available for the acquisition system. The complete list of features and associated capabilities is beyond the scope of this chapter. Instead, we focus briefly on those features that are most advantageous for working with genomic libraries. In addition, we also briefly discuss additional resources typically required to successfully complete functional genomic screens such as clear-bottom microtiter plates and large-scale data storage systems.

Screening microscopes are subject to fundamental limitations in optical resolution, acquisition speed, and image contrast, all of which are of practical interest in HCS applications and must be balanced to achieve the highest quality outcome. For instance, short exposures increase the sampling rate but decrease the signal-to-noise ratio (SNR). Long exposures improve SNR but increase the overall time required to complete the screen. Compared to small molecule libraries that usually contain one million or more compounds, genomic libraries are relatively small and typically contain from 5 to 250,000 elements. Thus, speed is not usually the most important factor when evaluating a HCS system. Instead, having the ability to collect more fluorescent channels and to use more colors in a single sample is often an overriding concern. Determining the transfection efficiency on a well-by-well basis in a genomic screen requires one fluorescent channel and effectively reduces the information capacity associated

TABLE IV

A Partial List of High-Content Screening Microscopes

Company	Name	Web site	Confocal
Evotec	Opera	http://www.evotec-technologies.com	Yes
Molecular Devices	ImageXpress Ultra	http://www.moleculardevices.com/	Yes
Molecular Devices	ImageXpress MICRO TM	http://www.moleculardevices.com/	No
Cellomics	ArrayScan VTI HCS Reader	http://www.cellomics.com	No
Cellomics	KineticScan HCS Reader	http://www.cellomics.com	No
Cellomics	cellWoRx	http://www.cellomics.com	No
Applied Precision	cellWoRx	http://www.api.com	No
BD BioScience	BD Pathway 415	http://www.bdbiosciences.com	Yes
BD BioScience	BD Pathway 435	http://www.bdbiosciences.com	No
Compucyte	iCyte	http://www.compucyte.com	Yes, confocal Focusing
GE Healthcare	IN Cell Analyzer 1000	http://www.amershambiosciences.com	Yes
GE Healthcare	IN Cell Analyzer 3000	http://www.amershambiosciences.com	No
MAIA Scientific	MIAS-2	http://www.maia-scientific.com	

with a biological response. Choosing a microscope capable of acquiring four or more colors per experiment, therefore, is typically justified when conducting genomic image-based screens.

The ability to select among different objective lenses when conducting genome library screens is also an important feature of a HCS system. A benefit of being able to choose between multiple magnifications is that the number of cells per image can be moderated. Some primary cell types, such as neuronal or epithelial cells, can spread out over large areas; a trade off in resolution may be essential to acquire the entire cellular region. While there is a loss in magnification when selecting lower power objectives, analyzing more cells can yield more statistically significant screening results. Conversely, higher magnification objectives with larger numerical apertures (NA) are more capable of collecting more photons per exposure, thus providing a brighter image with higher contrast. This allows for shorter exposures, faster overall acquisition times, and higher resolving capacity. However, the freedom to choose between different objectives ultimately rests with a need to measure a particular cellular phenomenon. More often than not, a $20\times$ objective is the standard choice and provides an excellent compromise among magnification, resolution, and information capacity for most screens. We have also used our $10\times$ objective extensively for many assays that do not require higher resolving capacities to obtain more statistically significant results while using fewer image frames per well.

Clear-Bottom Microtiter Plates

The introduction of 96-, 384-, and 1536-well microtiter plate formats into biological research creates an opportunity to examine many biological samples in parallel while using a limited amount of reagent. However, this format can be particularly challenging to use in an HCS environment, as the cell fixation and staining steps frequently incorporated into successful assays dictate midassay aspiration or dispensing steps that can be physically difficult to accomplish using standard bench-top equipment. In contrast to whole-well assays where aggregate activity is assayed, the distribution of cells within the well is more important in high-content screens, as it is often difficult to focus on those cells that are along the edge of the well. Thus, specialized equipment is often needed to dispense reagents into wells that without disrupting cells.

Automated Microtiter Plate Washers

Automated liquid-handling robots can consistently dispense and aspirate both media and fluorescent reagents for the best HCS assay results.

Plate washers that dispense toward the side of the well (side shooters) are usually more desirable for HCS applications than those that dispense directly into the bottom of the well (straight shooters). The ability to adjust the dispensing pressure is another important consideration when deciding on a plate washer. The ability to run the equipment at low pressure is often necessary when working with loosely adherent cell cultures. Finally, an aspiration head is also a very useful feature. During incubation, wells along the edges of the plate will often experience some significant evaporation. Thus the volume of liquid across all the wells in the plate can change dramatically depending on the time of incubation. An aspiration head provides the ability to first level the volume of media across the entire plate prior to the addition of exogenous reagents, ensuring an even concentration for all wells in the plate. We have found that this step is especially important for obtaining better z scores and ultimately more consistent screening results with fewer false negatives.

Data Storage

Conducting HCS screens also involves trade-offs among the number of optical slices, fields per well, channels acquired per exposure, and the number of cells required to reasonably determine a significant effect on a population. In addition, collecting images for every single cell per well can be time-consuming and must be balanced with the data storage available. Image data are much more information rich and also require more overall storage capacity than typical HTS data collections.

The cost of disk storage necessary for screening large-scale genome-wide libraries can be expensive. Unlike reporter gene assays, which only require megabytes (MB) to gigabytes (GB) of disk storage, HCS screens typically require GB to terabytes (TB) of storage space. These noteworthy data requirements also necessitate high-speed disk arrays and gigabit networking connections to accommodate the high rate of HCS data acquisition. Therefore, before venturing into a HCS screen, it is important to determine the data storage needed to facilitate image-based screening.

Fluorescent Biomarkers for HCS Applications

For most HCS assays, the addition of an exogenous fluorescent organelle marker is often necessary prior to imaging. The most common reagent for HCS assays is a fluorescent DNA stain. Even cell lines that have been engineered with GFP-tagged proteins usually require DAPI, Hoechst33352, or other DNA stain so that all the cells in a plate can be identified. Some systems demand the addition of the DNA signal, as they use this fluorescent channel for autofocusing. Moreover, unless the automated microscope contains an environmental chamber, it is also best to formaldehyde fix the cells, since unwanted biological effects may be introduced due to a change in temperature or reduced CO_2 concentrations.

The selection of fluorescent markers to determine organelle localization or to quantitate a cellular response is an important part of executing a successful high-content screen. Choice of markers is typically dictated by the assay readout and by the ability of the imaging equipment to discriminate between different wavelengths. Unlike standard research microscopes, automated screening microscopes tend to have a limited number of fluorescent settings. Microscopes that switch filters between each exposure tend to provide more flexibility when it comes to selecting fluorophores. However, some automated microscopes use a series of beam splitters (dichroics) in combination with multiple cameras to acquire different fluorescent channels simultaneously. While these systems can increase the rate of data acquisition rapidly, as there is no wait time for mechanical changes in filter wheel positions, they tend to limit the combination of different fluorescent reagents that can be used together.

When developing assays where multiple fluorescent markers will be employed in a single well, it is always best to select combinations where maximal emission spectra are as far apart as possible. It is also important to consider how broad excitation or emission spectra are for all the fluorescent elements used in a screen. Selecting two fluorophores that exhibit excessive overlap in wavelengths or wide emission spectra can seriously reduce the contrast of the resulting images. This loss in contrast may also extend the overall time needed to complete a screen if excessively long exposures are required to compensate for the high background signals. More importantly, overlapping signals between different channels can produce misleading results and could yield many false positives in a screen.

Biomarkers that exhibit a difference in their emission spectra based on molecular interactions can be extremely useful in high-content screening applications. Calcein AM is used commonly in HCS applications, as it is converted into a fluorescent conjugate via the activity of cellular esterases. This particular dye is permeable to cell membranes and is an excellent cytoplasmic marker. Once inside a cell, the molecule is cleaved by cellular esterases and results in a fluorescent derivative that is also nonmembrane permeable. This is ideal for live-cell applications, as there is little or no background signal from outside the cell, meaning that little or no washing is required after the dye is applied. Other fluorescent dyes depend on changes in pH for maximal fluorescence emission. These classes of dyes are quite useful for activity or viability assays. However, they may only work with live cells and, in some cases, lose their ability to selectively highlight a cellular organelle if fixatives are added after treatment. One particularly helpful source of information concerning organelle identification, fluorescent spectra, and fixation compatibility can be found on the Web site of Invitrogen (Molecular Probes, http://probes.invitrogen.com). The advantages and disadvantages must be weighed when choosing among the various options. Careful selection of organelle markers can save time and money during the screening process, not to mention a great deal of frustration.

The use of antibodies for the demarcation of protein localization in cellular studies is an important part of cellular biology. However, in order to detect intracellular proteins, cells must first be treated with a fixative and their membranes permeabilized so that the antibody can gain direct access to these proteins. This process is usually extensive and requires multiple incubation and washing steps. With the advent of automated microtiter plate washers, the use of antibodies for protein localization studies becomes more tractable. Some washers can even be programmed to complete multiple rounds of dispensing and aspiration such that little or no manual intervention is required, especially when connected directly to plate stackers that load and unload the plates. However, the use of fixatives and the number of dispensing and aspirating steps demand that the cells are treated gently, even for tightly adherent cell lines. Thus it is best to use directly conjugated fluorescent antibodies if possible.

Preparing Samples for Automated Microscopy

We have optimized a set of protocols for HCS studies using antibody-based protein localization in aldehyde-fixed cells. Cell-permeable reagents can also be used with these protocols by simply ignoring the detergent permeabilization steps. This set of protocols has been used extensively with cultures dispensed into 384-well plates (commonly used in the HTS cDNA and RNAi screens). However, given an optimized assay, these protocols could be adapted easily to other plate formats (e.g., 96- or 1536-well), with the only difference being that the amounts of reagents must be scaled up or down as necessary.

Selecting Appropriate Cell Lines for HCS Assays

Pipetting fixation or fluorescent reagents directly onto the cultured cells can often disrupt their attachment to the surface material depending on the cell line and pipetting or dispensing equipment being used. This treatment can dislodge the cells and introduce large empty spaces that prevent consistent cell counts from well to well. Thus it is important to consider the cell line being used and to optimize the reagent addition step before proceeding with a full-scale library screen.

Strongly adherent cells usually work best for HCS assays. HeLa, U2OS, MCF7, and other strongly adherent cell lines typically provide the best opportunity to conduct a successful image-based screen (Harada *et al.*, 2005). These cells grow flat, do not generally pile up on one another, and adhere to coated tissue culture plates very well. Conversely, we have found that HEK-293 or suspension cells are less desirable for HCS assays, as any disruption to the culture media may rearrange the cell distribution. Since we do not always have the luxury to work with strongly adherent cell lines, it is possible to work with loosely attached cells, although this demands that reagent addition or aspiration be done at low pressure and flow conditions, therefore extending assay duration, which, in itself, can be an important consideration for either very large screens or those with brief assay windows. Using appropriate coating of plates (i.e., poly-D-lysine) and enough care, even suspension cells can be used successfully with HCS assays.

Protocols for Applying Cell Fixatives

1. Remove the 384-well plates containing the transfected/infected cells from the CO_2 incubator after the appropriate incubation period.

2. Aspirate all but 20 μl of media from the wells using an automated plate washer, taking care not to disturb the cells, which should have adhered or settled to the bottom of the plate during incubation. (This step helps create equal volumes in each well and reduces edge effects often created from evaporation.)

3. Make an 8% (v/v) stock fixation solution of paraformaldehyde (PFA) suspended in phosphate-buffered saline (PBS).

4. Dispense 20 μl the stock PFA solution into each well using an automated liquid dispenser or plate washer. (The final concentration will be 4% paraformaldehyde in the cell culture media.)

5. Incubate the plate at room temperature for approximately 10 min.

6. Aspirate 30 μl of the PFA/cell culture media using an automated plate washer. (*Do not* turn the plate upside down and shake out media, as this will often dislodge many cells.)

7. Add 40 to 60 μl of PBS to each well using an automated liquid dispenser and incubate for approximately 5 min.

8. Repeat steps 6 and 7 two more times to wash out any remaining PFA.

Protocols for Antibody Staining

1. Add 40 to 60 μl of PBS supplemented with 0.5% Triton-X100 to each well.

2. Incubate the plate at room temperature for approximately 10 min.

3. Aspirate most of the buffer without disturbing the cells.

4. Add 40 to 60 μl of PBS supplemented with 0.1% Triton-X100 (called PBST) and 1% BSA to each well.

5. Incubate the plate at room temperature for at least 30 min.

6. Aspirate off all but 20 μl of the buffer and add the 1° antibody diluted a small volume. We typically add 10 μl of antibody solution at 3× concentration to each well. (If a 1:1500 final concentration is desired, add 10 μl of a 1:1500 stock.) The antibodies are always resuspended in PBST + 1% BSA and 0.5% sodium azide (NaN$_3$).

7. Incubate the plate for at least 60 min at room temperature or overnight at 4°.

8. Aspirate 30 μl of the buffer using an automated plate washer. (*Do not* turn the plate upside down and shake out media, as this will often dislodge many cells.)

9. Add 40 to 60 μl of PBS to each well using an automated liquid dispenser and incubate for approximately 5 min.

10. After the incubation period, repeat steps 8 and 9 three times using PBST with 10-min incubation periods between each wash.

11. After the incubation period is complete, add Hoechst 3342 (nuclear dye) to the existing buffer volume at 1:10,000 final and incubate for an additional 5 min.

12. Again, repeat steps 8 and 9 three times using PBST with 10-min incubation periods between each wash.

13. After the last wash, leave about 30 to 40 μl in the wells and seal the plate using an adhesive aluminum seal.

14. Remove any dust or large particles from the bottom of the plate before starting the imaging process.

Determining Transfection or Transduction Efficiency

Unlike chemical screening studies, functional genomic applications do not necessarily affect all the cells in a well equally. Depending on the cell type, some cells are more capable of internalizing exogenous oligonucleo-tide-based reagents or of accepting viral particles. This limitation can create a disparity in the number of cells responding to a particular treatment. When combined with fluorescent tags, however, functional genomic libraries can be analyzed more efficiently using a HCS approach as opposed to other HTS assays. Individual cells that have been transfected efficiently with a particular cDNA or chemically synthesized siRNA can first be identified from other nontransfected cells. Plasmid-based transfections such as cDNAs or lentiviral delivery systems can usually include a fluorescent protein expression cassette such as GFP or, in the case of straight plasmid transfections, involve cotransfection with a marked plasmid. Alternatively, synthetic nucleotide treatments with double-stranded RNAs can also be measured by tagging the oligonucleotides with fluorescent molecules, such as an Alexa™ fluorescent dye (Fig. 5). This may be of questionable value for several reasons. First, it may be very inconvenient or impractical to conjugate each siRNA is an arrayed library—all current commercially available libraries are unconjugated. Additionally, the effect that different conjugates may have on individual siRNA potency has not been systematically tested. Co-expressed markers in viral shRNA constructs may be more useful in marking transduced cells, since as the independently expressed marker is not likely to impact shRNA potency and shRNA function can be measured over a longer time frame than can be done with siRNA. Fluorescently labeled siRNAs are best employed during the optimization phase of an assay or for sentinel wells in live assay plates to mark plates that may have suffered from poor handling during siRNA transfection. At any rate, transfected cells can then be identified positively by thresholding on only those cells that exhibit a certain fluorescent intensity before determining the differential response of a cell to a condition or stimulus. With our siRNA-based screens, we commonly add a fluorescently tagged negative control (ds)RNA to one well per plate for establishing a baseline transfection efficiency. In several of our plasmid based libraries, we use GFP expression to measure transfection efficiency (Fig. 5). Minimum transfection efficiencies need to be empirically determined and will vary among assays. Typically, we favor cell types that have median high-throughput transfection efficiencies of 25% or more, and have found it difficult to extract statistically significant data from wells where less than 10% of the cells (~50 events/well) have been co-transduced with an shRNA/cDNA and a marker.

Quantitative Image Analysis

After establishing the assay conditions and the transfection efficiencies for a particular genomic screen, the next step is to image and analyze the biological effect on cell populations in a quantitative manner. Quantitative image analysis is typically approached in one of two different ways using either image segmentation or supervised learning algorithms. All the commercial HCS readers come equipped with some form of image segmentation software. More recently, supervised and pattern recognition techniques have

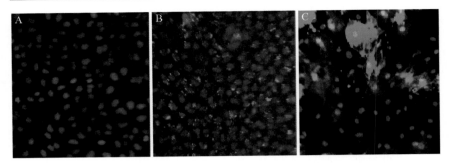

FIG. 5. Use of fluorescent transfection markers in functional genomic screens. (A) Negative control transfection well; DNA (blue, Hoechst33342). (B) siRNA tagged with Alexa-647 fluorescent molecule and transfected into HeLa cells; siRNA (red), DNA (blue). (C) Lentiviral infections with GFP expressing plasmid; GFP (green), DNA (blue).

been added to the HCS analysis packages. The latter are generally quite sophisticated and are beyond the scope of this chapter. Instead, we focus on how image segmentation is used to measure a cellular response. In particular, we present several examples that show how cell populations can be measured for cell cycle differences and cytoplasmic protein redistribution.

Image Segmentation

The most common approach to HCS analysis is done by segmenting multicolor images first on one channel, typically the nuclei signal, and then measuring intensity properties in the other color channels to determine a cellular response. Once the multicolor image is acquired, quantitative image analysis is achieved using a class of image-segmentation techniques for morphological and cellular identification (Fig. 6). The original fluorescent nuclear channel (e.g., Hoechst 33342 signal) is converted into a binary image by thresholding the grayscale image; this is then used as a mask for discriminating between nuclear and background signals. Once binarization is complete, regions of interest (ROI) are established using contour-based detection to identify the perimeter of each nucleus. Each perimeter map is converted into a geometrical feature and is designated as a nuclear ROI. These regions are filtered based on size and fluorescent intensity to remove ROIs from nonnuclear signals sometimes created from particulate or fluorescent cellular debris. The average fluorescent intensity in the other fluorescent channels is then determined (e.g., Alexa 647 signal) and gating on only those nuclear ROIs that have a strong positive signal, called the

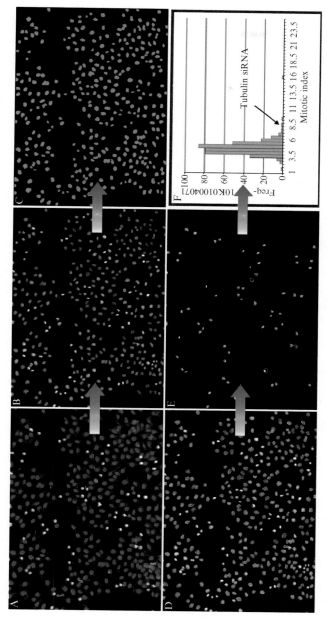

Fig. 6. Automated identification of mitotic cells.

response index. In addition to the response index, several other parameters are calculated for characteristics, including roundness of the nuclei, cellular proliferation, fluorescent DNA intensities, and the average size of the nuclear ROIs.

Protein Relocalization

The ROIs can also be manipulated to measure the fluorescent signals in other portions of the cell. For example, if each of the nuclear ROI is dilated so that it includes a portion of the cytoplasmic region, a quantitative nuclear translocation assay becomes possible. If a protein, such as a transcription factor, is marked with a fluorescent molecule, the fluorescent intensity in the expanded ROI can be compared with the original nuclear ROI based on the fluorescent channel of the protein (Fig. 7A and B). This measurement can then serve as a cellular response to external stimulus that might normally drive the transcription factor from the cytoplasm into the nucleus. A genomic library can then be screened for inhibitors or activators of this translocation event (Cho *et al.*, 2006).

Even more complex pattern changes in protein localization can be determined using HCS assays. Upon activation, some cytoplasmic proteins, such as β2-arrestin, can be driven to relocalize into cellular vesicles (Fig. 7C–E). These changes in protein localization can be assayed readily by first determining a ROI for the entire cytoplasmic region. This is usually best done with the addition of a cytoplasmic dye that does not overlap with the nuclei or protein channels, such as calein AM. Once a cytoplasmic ROI is established, then it is possible to look for discrete areas of maximal intensity, also known as punctuate spots. These can be measured and counted per cell to determine a cellular response to the stimulus. Those cells with more or brighter spots would be considered to respond stronger.

The number of HCS assays possible is virtually limitless. Assuming that a fluorescent measurement can be determined and a change in a morphological parameter established, a change in the cells response to stimuli can be output. This creates an opportunity to expand biological discovery and will also require the development of future pattern identification methods as we probe for more complex biological phenomenon.

Summary

This chapter focused on the preparation and application of genomic libraries with respect to high-content screening. These two technologies are particularly complementary and offer a powerful approach to conducting genomic analyses. The technology is robust and demonstrates the

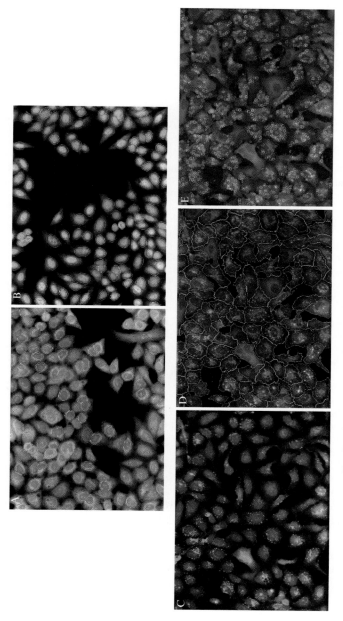

FIG. 7. High-content image analysis to determine protein redistribution.

appropriateness of image-based technology for enhancing analysis based on those cells in a population that have the highest degree of potential silencing or expression, depending on the screen. While each interrogated cellular phenotype will require additional modifications, these protocols and summaries provide a framework from which large-scale genomic screens using HCS can be achieved. Because each component of this approach involves technologies that are in their infancy, we anticipate continual advances that will transform this methodology for understanding gene function on a global scale.

References

Aza-Blanc, P., Cooper, C. L., Wagner, K., Batalov, S., Deveraux, Q. L., and Cooke, M. P. (2003). Identification of modulators of TRAIL-induced apoptosis via RNAi-based phenotypic screening. *Mol. Cell* **12,** 627–637.

Berns, K., Hijmans, E. M., Mullenders, J., Brummelkamp, T. R., Velds, A., Heimerikx, M., Kerkhoven, R. M., Madiredjo, M., Nijkamp, W., Weigelt, B., Agami, R., Ge, W., Cavet, G., Linsley, P. S., Beijersbergen, R. L., and Bernards, R. (2004). A large-scale RNAi screen in human cells identifies new components of the p53 pathway. *Nature* **428,** 431–437.

Betz, N. (2003). Produce functional siRNAs and hairpin siRNAs using the T7 RiboMAX express RNAi system. *Promega Notes* **85,** 15–18.

Brummelkamp, T. R., Bernards, R., and Agami, R. (2002). A system for stable expression of short interfering RNAs in mammalian cells. *Science* **296,** 550–553.

Chanda, S. K., and Caldwell, J. S. (2003). Fulfilling the promise: Drug discovery in the post-genomic era. *Drug Disc. Today* **8,** 168–174.

Chanda, S. K., White, S., Orth, A. P., Reisdorph, R., Miraglia, L., Thomas, R. S., DeJesus, P., Mason, D. E., Huang, Q., Vega, R., Yu, D. H., Nelson, C. G., Smith, B. M., Terry, R., Linford, A. S., Yu, Y., Chirn, G. W., Song, C., Labow, M. A., Cohen, D., King, F. J., Peters, E. C., Schultz, P. G., Vogt, P. K., Hogenesch, J. B., and Caldwell, J. S. (2003). Genome-scale functional profiling of the mammalian AP-1 signaling pathway. *Proc. Natl. Acad. Sci. USA* **100,** 12153–12158.

Cho, C. Y., Koo, S. H., Wang, Y., Callaway, S., Hedrick, S., Mak, P. A., Orth, A. P., Peters, E. C., Saez, E., Montminy, M., Schultz, P. G., and Chanda, S. K. (2006). Identification of the tyrosine phosphatase PTP-MEG2 as an antagnist of hepatic insulin signaling. *cell. Metab.* **3,** 367–378.

Elbashir, S. M., Harborth, J., Lendeckel, W., Yalcin, A., Weber, K., and Tuschl, T. (2001). Duplexes of 21-nucleotide RNAs mediate RNA interference in cultured mammalian cells. *Nature* **411,** 494–498.

Hannon, G. J., and Rossi, J. J. (2004). Unlocking the potential of the human genome with RNA interference. *Nature* **431,** 371–378.

Harada, J. N., Bower, K. E., Orth, A. P., Callaway, S., Nelson, C. G., Laris, C., Hogenesch, J. B., Vogt, P. K., and Chanda, S. K. (2005). Identification of novel mammalian growth regulatory factors by genome-scale quantitative image analysis. *Genome Res.* **15,** 1136–1144.

Jackson, A. L., and Linsley, P. S. (2004). Noise amidst the silence: Off-target effects of siRNAs? *Trends Genet.* **20,** 521–524.

Kittler, R., Putz, G., Pelletier, L., Poser, I., Heninger, A. K., Drechsel, D., Fischer, S., Konstantinova, I., Habermann, B., Grabner, H., Yaspo, M. L., Himmelbauer, H., Korn, B., Neugebauer, K., Pisabarro, M. T., and Buchholz, F. (2004). An endoribonuclease-prepared siRNA screen in human cells identifies genes essential for cell division. *Nature* **432,** 1036–1040.

Krieg, N., and Holt, J. (1984). "Bergey's Manual of Systematic Bacteriology." Williams & Wilkens, Baltimore, MD.

Li, T., Chang, C. Y., Jin, D. Y., Lin, P. J., Khvorova, A., and Stafford, D. W. (2004). Identification of the gene for vitamin K epoxide reductase. *Nature* **427,** 541–544.

Luo, B., Heard, A. D., and Lodish, H. F. (2004). Small interfering RNA production by enzymatic engineering of DNA (SPEED). *Proc. Natl. Acad. Sci. USA* **101,** 5494–5499.

Paddison, P. J., Silva, J. M., Conklin, D. S., Schlabach, M., Li, M., Aruleba, S., Balija, V., O'Shaughnessy, A., Gnoj, L., Scobie, K., Chang, K., Westbrook, T., Cleary, M., Sachidanandam, R., McCombie, W. R., Elledge, S. J., and Hannon, G. J. (2004). A resource for large-scale RNA-interference-based screens in mammals. *Nature* **428,** 427–431.

Pennisi, E. (2005). Why do humans have so few genes? *Science* **309,** 80.

Sambrook, J., Fritsch, E. F., and Maniatis, T. (1989). "Molecular Cloning, a Laboratory Manual." Cold Spring Harbor Laboratory Press, Cold Spring Harbor, NY.

Sen, G., Wehrman, T. S., Myers, J. W., and Blau, H. M. (2004). Restriction enzyme-generated siRNA (REGS) vectors and libraries. *Nature Genet.* **36,** 183–189.

Stegmeier, F., Hu, G., Rickles, R. J., Hannon, G. J., and Elledge, S. J. (2005). A lentiviral microRNA-based system for single-copy polymerase II-regulated RNA interference in mammalian cells. *Proc. Natl. Acad. Sci. USA* **102,** 13212–13217.

Strausberg, R. L., Feingold, E. A., Grouse, L. H., Derge, J. G., Klausner, R. D., Collins, F. S., Wagner, L., Shenmen, C. M., Schuler, G. D., Altschul, S. F., Zeeberg, B., Buetow, K. H., Schaefer, C. F., Bhat, N. K., Hopkins, R. F., Jordan, H., Moore, T., Max, S. I., Wang, J., Hsieh, F., Diatchenko, L., Marusina, K., Farmer, A. A., Rubin, G. M., Hong, L., Stapleton, M., Soares, M. B., Bonaldo, M. F., Casavant, T. L., Scheetz, T. E., Brownstein, M. J., Usdin, T. B., Toshiyuki, S., Carninci, P., Prange, C., Raha, S. S., Loquellano, N. A., Peters, G. J., Abramson, R. D., Mullahy, S. J., Bosak, S. A., McEwan, P. J., McKernan, K. J., Malek, J. A., Gunaratne, P. H., Richards, S., Worley, K. C., Hale, S., Garcia, A. M., Gay, L. J., Hulyk, S. W., Villalon, D. K., Muzny, D. M., Sodergren, E. J., Lu, X., Gibbs, R. A., Fahey, J., Helton, E., Ketteman, M., Madan, A., Rodrigues, S., Sanchez, A., Whiting, M., Young, A. C., Shevchenko, Y., Bouffard, G. G., Blakesley, R. W., Touchman, J. W., Green, E. D., Dickson, M. C., Rodriguez, A. C., Grimwood, J., Schmutz, J., Myers, R. M., Butterfield, Y. S., Krzywinski, M. I., Skalska, U., Smailus, D. E., Schnerch, A., Schein, J. E., Jones, S. J., and Marra, M. A. Mammalian Gene Collection Program (2002). Generation and initial analysis of more than 15,000 full-length human and mouse cDNA sequences. *Proc. Natl. Acad. Sci. USA* **99,** 16899–16903.

Strausberg, R. L., Feingold, E. A., Klausner, R. D., and Collins, F. S. (1999). The mammalian gene collection. *Science* **286,** 455–457.

Su, A. I., Wiltshire, T., Batalov, S., Lapp, H., Ching, K. A., Block, D., Zhang, J., Soden, R., Hayakawa, M., Kreiman, G., Cooke, M. P., Walker, J. R., and Hogenesch, J. B. (2004). A gene atlas of the mouse and human protein-encoding transcriptomes. *Proc. Natl. Acad. Sci. USA* **101,** 6062–6067.

Tiscornia, G., Singer, O., Ikawa, M., and Verma, I. M. (2003). A general method for gene knockdown in mice by using lentiviral vectors expressing small interfering RNA. *Proc. Natl. Acad. Sci. USA* **100,** 1844–1848.

Xia, H., Mao, Q., Paulson, H. L., and Davidson, B. L. (2002). siRNA-mediated gene silencing *in vitro* and *in vivo*. *Nature Biotechnol.* **20,** 1006–1010.

Ziauddin, J., and Sabatini, D. M. (2001). Microarrays of cells expressing defined cDNAs. *Nature* **411,** 107–110.

[29]   Fluorescent Protein-Based Cellular Assays Analyzed by Laser-Scanning Microplate Cytometry in 1536-Well Plate Format

By Douglas S. Auld, Ronald L. Johnson, Ya-qin Zhang,
Henrike Veith, Ajit Jadhav, Adam Yasgar, Anton Simeonov,
Wei Zheng, Elisabeth D. Martinez, John K. Westwick,
Christopher P. Austin, and James Inglese

Abstract

Microtiter plate readers have evolved from photomultiplier and charged-coupled device-based readers, where a population-averaged signal is detected from each well, to microscope-based imaging systems, where cellular characteristics from individual cells are measured. For these systems, speed and ease of data analysis are inversely proportional to the amount of data collected from each well. Microplate laser cytometry is a technology compatible with a 1536-well plate format and capable of population distribution analysis. Microplate cytometers such as the Acumen Explorer can monitor up to four fluorescent signals from single objects in microtiter plates with densities as high as 1536 wells. These instruments can measure changes in fluorescent protein expression, cell shape, or simple cellular redistribution events such as cytoplasmic to nuclear translocation. To develop high-throughput screening applications using laser-scanning microplate cytometry, we used green fluorescent protein- and yellow fluorescent protein-expressing cell lines designed to measure diverse biological functions such as nuclear translocation, epigenetic signaling, and G protein-coupled receptor activation. This chapter illustrates the application of microplate laser cytometry to these assays in a manner that is suitable for screening large compound collections in high throughput.

Introduction

Cell-based fluorescent assays that use green fluorescent protein (GFP) as a reporter are a well-validated format for measuring a variety of biological phenomenon (Shaner *et al.*, 2005; van Roessel and Brand, 2002). The adaptation of these systems to microtiter plates has been a central goal within the biopharmaceutical industry because the microtiter plate is the standard container format for compound screening (Carroll *et al.*, 2004). A challenge for some instrument developers has been to accommodate the use of high-density microtiter plates, such as the 1536-well plate. Application of

METHODS IN ENZYMOLOGY, VOL. 414   0076-6879/06 $35.00
DOI: 10.1016/S0076-6879(06)14029-X

GFP-based cellular assays to large-scale compound screening depends on a robust and stable signal from the GFP reporter and high-throughput handling and reading of microtiter plates. The use of GFP detection in 1536-well plates enables the large-scale screening of compound collections against a diverse number of optimized GFP cell lines now available.

Using laser-scanning cytometry, we developed methodologies to screen three cellular assays based on inherently fluorescent GFP or fragments based on GFP variants (Fig. 1A). The first is a Redistribution assay (BioImage A/S) in which a glucocorticoid receptor–GFP (GR-GFP) fusion

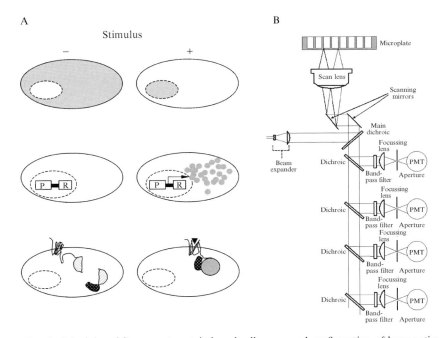

FIG. 1. Principles of fluorescent protein-based cell assays and configuration of laser optics within the Acumen Explorer. (A) GFP fusion proteins can be constitutively produced and their subcellular distribution modulated as in the GR-GFP assay where cytosolic GR-GFP is translocated into the nucleus upon ligand binding (top). GFP production can report transcriptional activation as in the LDR assay where derepression of a transcriptionally silent locus induces GFP expression (middle). Nonfluorescent fragments of YFP can interact noncovalently to reconstitute fluorescence as in the βARR:β2AR assay where fragments fused to interacting partners, β-arrestin and β_2-adrenergic receptor, associate upon receptor stimulation (bottom). (B) Laser light is directed to the F-theta scan lens by a dichroic beam splitter and galvanometric mirrors. The scan lens focuses the light onto the microplate bottom, and the resulting epifluorescence is directed to four PMT detectors by a series of mirrors and dichroics. Band-pass filters select specific ranges of wavelengths for each PMT. Galvanometric mirrors guide the laser line across a 20-mm^2 area at 5 m/s (diagram adopted from Bowen *et al.*, 2006).

is used to measure ligand-dependent translocation from the cytoplasm to the nucleus in U2OS cells (Jung-Testas and Baulieu, 1983; Levinson et al., 1972). The second, a locus derepression (LDR) assay, is a mouse mammary cell line, derived from C127 cells, where GFP is stably integrated into a transcriptionally silent region of the genome. Compounds that activate this chromosomal locus (loci) can be identified by enumerating GFP-positive cells after treatment. The third is a protein fragment complementation assay (PCA) in HEK293T cells whereby fragments of an intensely fluorescent YFP mutant are fused to the β-adrenergic G protein-coupled receptor and β-arrestin. Following activation of the β-adrenergic receptor, the association of β-arrestin enables yellow fluorescent protein (YFP) fragment association, folding, and generation of a fluorescent signal (MacDonald, 2006; Remy and Michnick, 2003; Yu et al., 2003).

The Principle of Laser-Scanning Microplate Cytometers

Current microplate readers vary with respect to their speed of plate reading and the amount of information that is extracted. On one end of the spectrum are very fast plate readers such as the ViewLux (Perkin Elmer), a cooled charge-coupled device (CCD)-based imager that captures plate data in under a minute for typical luminescence or fluorescence assays. Imagers such as these offer the additional advantage of data collection speed that does not depend on the plate well density. However, such readers only measure the total signal arising from the cell population in each well. Additionally, the excitation source and detection systems in such readers are not sufficiently sensitive to reliably measure weakly fluorescent molecules such as GFP (unpublished observations).

On the other end of the spectrum are high-content screening (HCS) imagers, such as the ArrayScan (Cellomics, Inc.; see, e.g., Williams et al., 2006), the Opera (Evotec; see, e.g, Garippa et al., 2006), the Pathway BioImager (BD Biosciences; see, e.g,, Richards et al., 2006), and the IN Cell imagers (GE Healthcare; see, e.g., Haasen et al., 2006) that capture high-resolution images in subcellular detail from numerous cells per well. While HCS readers provide a wealth of data on every well, they have relatively slow scan times compared to population-averaged readers, generate large data files (MB per well are common), and data collection speeds are dependent on plate well density.

Between these two approaches resides the laser-scanning microplate cytometer that images fluorescent beads or cells within microtiter wells independent of microscope optics. The latest generation of these instruments is the Acumen Explorer (TTP Labtech) that provides read times sufficient for high-throughput screening (HTS) (up to 300,000 samples per day in 1536-well plates) while making multiparameter measurements on individual cells or other micrometer size particles (Bowen and Wylie, 2006).

The Acumen Explorer collects data by focusing a laser excitation beam on the bottom of the microtiter plate and collecting the resulting epifluorescence using photomultiplier tubes (PMT) for specific wavelength ranges (Fig. 1B). Fluorescent characteristics of individual objects can then be characterized and enumerated into various populations. The 488-nm laser is optimal for excitation of GFP-based assays, and the optics and plate positioning are sufficiently fast to read a 1536-well plate in approximately 10 min (200 whole-well scans/minute at 1×8-μm resolution). The resolution along the x axis is set at predetermined intervals (0.1, 0.5, 1, 2.5, or 5 μm) while the y axis is set by the user, typically between 1 and 10 μm.

The amount of image data saved is determined by the need of the process; for identifying the critical object characteristics of an assay phenotype, the assay development mode is used to capture high-density information. However, for screening, the HTS mode is used to collect only the critical assay parameters, thereby reducing the file size by several hundredfold (<200 KB for 1×8 μm whole well scan of a 1536-well plate). As plate file size is kept small, microplate cytometers facilitate HTS of large chemical libraries against various cell-based assays. The methods for performing 1536-well assays using the Acumen Explorer, as well as proof-of-principle screening data, are shown next.

General Methods

Preparation of the Library of Pharmacologically Active Compounds (LOPAC) Titration Series

The LOPAC 1280 (Sigma Aldrich) collection is screened as a series of interplate titrations to generate concentration–response curves for each compound. The generation of titration–response curves enables accurate assessment of the AC_{50} of each compound (concentration for half-maximal activity) and eliminates false positives and negatives (Inglese, 2006).

The collection is received in 96-well tube plates as dimethyl sulfoxide (DMSO) solutions at 10 mM and compressed into 384-well Kalypsys polypropylene plates via interleaved quadrant transfer using a CyBi-Disk Vario system equipped with a 96-tip head (CyBio AG). Titration and subsequent compression of 384-well plates to 1536-well plates are performed by an Evolution P^3 (Perkin Elmer). In the final 1536-well plates, the four left-most columns are unfilled. Fifteen 2.24-fold and seven 5-fold dilutions of the LOPAC library are prepared with each plate containing a copy of the library at successive dilutions. For long-term storage, plates are heat sealed on a PlateLoc thermal plate sealer equipped with a BenchCell 2× stacker system (Velocity 11). For a more detailed description of library preparation, see Inglese (2006).

Configuration of the Acumen Explorer

The Acumen Explorer uses a 25-mW argon laser as an excitation source and four PMTs that collect the following emission wavelengths: channel 1, 500–530 nm; channel 2, 530–585 nm; channel 3, 575–640 nm; and channel 4, 655–706 nm. The instrument is controlled by a Hewlett Packard Intel XEON CPU (1.8 GHz) workstation containing 3 GB of RAM and a MDAQ data acquisition card. In this configuration, the maximum scan speed is 10 m/s with a maximum sampling rate per channel of 10 MHz.

1536-Well Plate Liquid Handling

Cell suspensions are dispensed into 1536-well plates using a Flying Reagent Dispenser [FRD, Aurora Discovery (Niles and Coassin, 2005)] or a Kalypsys 1536-well plate washer/dispenser. These dispensers use pressurized bottles (6 to 9 psi) to dispense liquid through a solenoid valve and can accurately dispense volumes as low as 1 μl. Bulk reagent addition and aspiration are performed using a Kalypsys washer/dispenser.

The compound is transferred by a Kalypsys pin tool workstation equipped with 1536 10-nl slotted pins [0.457 mm diameter, 50.8 mm long; VPN Scientific (Cleveland and Koutz, 2005)]. The pin-tool transfer is validated using fluorescein-containing DMSO solution that confirms transfer of 23 nl from the compound plate to an assay plate containing 5 μl of buffer. The pin transfer protocol for cell-based assays uses one pin immersion in the compound plate and three immersions in the assay plate, with the pins not touching the assay plate bottom to avoid disrupting the cell monolayer. For screening, two compound plates are transferred to each assay plate, one containing library compounds followed by one containing assay-specific controls. For all incubations, assay plates are lidded with stainless-steel Kalypsys plate lids that contain a rubber gasket to seal the edges of the plate and pin holes across the middle of the plate allow gas exchange (Mainquist, 2003).

The Kalypsys washer/dispenser has a 32-channel stainless-steel head to aspirate an entire column from 1536-well plates. Aspiration depends on several factors: the vacuum flow level, the aspiration-tip well depth, and the dwell time. The vacuum flow level is controlled by manually adjusting a needle valve that on our instrument is opened two full turns. The aspiration well depth is set to 2.75 mm from the top of the well and the dwell time to 200 ms. These settings result in 1.5 μl/well of residual liquid as determined by gravimetric measurement of test plates. At the start and finish of all procedures, the aspirator head is cleaned by a 5-min aspiration of 10% bleach followed by distilled water. If the head is used frequently, then at least once per month, remove the head, soak overnight in Coulter Clenz (Coulter Corporation), and then sonicate for several hours.

The accuracy and precision of the aspirator head are tested as follows: 5 μl/well of 5 μM fluorescein in buffer is dispensed into a tared 1536-well plate, the plate is weighed, and the fluorescent signal is measured by a plate reader. The coefficient of variation (CV) of the plate should be below 5%. Following one wash cycle, the residual volume is checked gravimetrically and the CV is measured by plate reader. The CV should be below 10% following a single aspiration cycle.

Data Processing

Screening data are corrected and normalized using Assay Analyzer (GeneData). Control assay plates containing vehicle (DMSO) only are used to monitor background systematic variation. Correction factors are generated from these variations and applied to each assay plate to correct systematic errors. Percentage activity is calculated by normalization to the median values of the activator and neutral controls present on each plate. Plate and compound mapping is performed to correlate percentage activity with the corresponding concentration for each sample to generate a titration-response series for each compound. Concentration–response graphs are displayed using Prism (Graph Pad) and Origin (OriginLab).

GR-GFP Nuclear Translocation Assay

The glucocorticoid receptor (GR) Redistribution assay enables the visualization of GR cytoplasmic to nuclear translocation by the use of a GR-GFP fusion (BioImage). GR is normally cytosolic; however, ligands such as dexamethasone cause nuclear translocation where the protein binds to response elements and interacts with various cofactors to modulate transcription (Bai et al., 2003; Zhou and Cidlowski, 2005). Because both functional agonists and antagonists can induce nuclear translocation (Htun et al., 1996), this assay can detect ligands regardless of their effects on gene expression. For this reason, the measurement of nuclear translocation is a desirable assay for broadly screening compound libraries for nuclear receptor ligands.

Solutions and Materials

GR-GFP expressing U2OS cells are maintained in DMEM with Glutamax and high glucose (Invitrogen), 1% (v/v) penicillin–streptomycin (Invitrogen), and 0.5 mg/ml Geneticin (Invitrogen). For cell culture, 10%

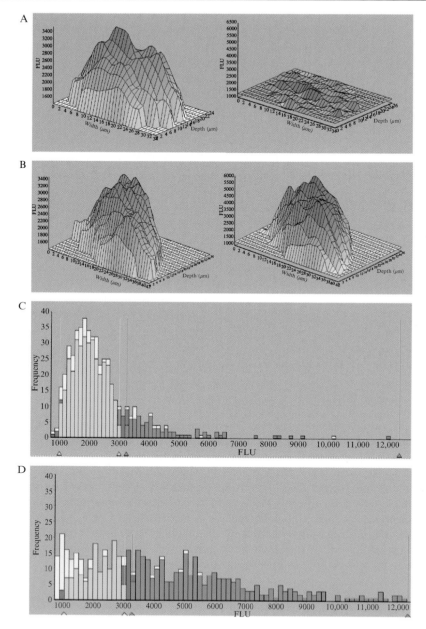

FIG. 2. Three-dimensional representation and population histogram of translocated and untranslocated objects from the GR-GFP assay. The PI (red) and GFP (green) signals of a

heat-denatured fetal bovine serum (FBS) is used (Invitrogen). For cell plating, 10% dextran, charcoal-stripped FBS (Hyclone SH30068.03) is used. Dimethyl sulfoxide, certified A.C.S. grade, is from Fisher. Black polystyrene, clear-bottom, 1536-well plates (#K1536SBCN, Hi Base plates) are manufactured by Greiner and purchased from Kalypsys. Propidium iodide (PI) is from Molecular Probes. Dexamethasone is from Calbiochem.

High-Throughput Screening Protocol

1. U2OS cells are trypsin treated, suspended in medium, and passed through a 40-μm basket filter (BD Falcon). The cells are seeded at 600 cells/5 μl/well into 1536-well plates and incubated for approximately 20 h at 37° and in 5% CO_2.

2. The compound is transferred to assay plates by pin tool, resulting in a 217-fold dilution of compound and 0.4% DMSO final concentration. The control plate contains a titration of dexamethasone in the first two columns (starting at 0.2 mM and diluting by 2-fold for 16 rows), DMSO in the third column, 0.2 mM dexamethasone in the fourth column, and the remaining columns are empty.

3. The assay plates are incubated at 37° in 5% CO_2 for 2 h.

4. Cells are fixed using the Kalypsys washer/dispenser. The protocol involves aspiration of the cell media, addition of 6 μl/well of phosphate-buffered saline (PBS), aspiration of PBS, addition of 5 μl/well of 100% methanol, aspiration of methanol, and addition of 3 μl/well PBS with 1.5 μM PI.

5. The plates are read by the Acumen Explorer using the following settings: 6 mW laser power, 650 and 690 V for the GFP (channel 1) and PI (channel 3) PMTs, respectively, 2 standard deviations (SD) above background fluorescence threshold for data collection, and 1 × 4 μm[1] scan of the whole well area. PI fluorescence triggers the collection of PI and GFP signals. A "nuclei" population is defined by a 10- to 100-μm width and depth filter that eliminates small and large fluorescent particles (Fig. 2). Two subpopulations

---

[1] We have obtained equivalent data with 1 × 8-μm scans.

---

nucleus within an untreated cell (A) and a cell treated with 50 nM dexamethasone (B) are shown. Histograms depicting object number versus fluorescent intensity for untreated (C) and dexamethasone-treated cells (D). Nuclei were classified as objects between 10 and 100 μm width and depth. Objects outside this size range are shaded pale yellow. Nuclei subpopulations were categorized by the level of GR-GFP fluorescence (translocated, blue; untranslocated, yellow; and unclassified, green).

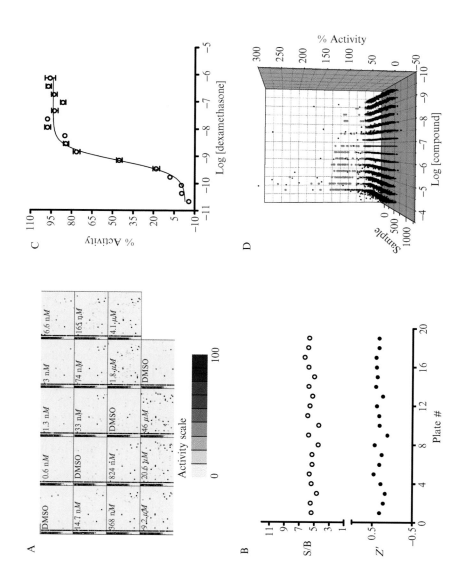

of "nuclei" are defined by peak intensity of GFP fluorescence, where translocated objects are between 4000 and 15,000 fluorescence units (FLU) and untranslocated objects are between 1000 and 3999 FLU.

Validation of the 1536-Well GR Nuclear Translocation Assay

To validate the 1536-well protocol, we screened the LOPAC library that contains known glucocorticoid and steroidal agonist and antagonists. Nineteen 1536-well plates were run in one batch (Fig. 3A). With the optimized protocol, we obtained a 10% CV for the number of nuclei/well, indicating there was little spurious loss of cells from plate washing. The Z' calculated from the enumerated translocated objects averaged approximately 0.3 with a signal to background (S/B) of 6 (Fig. 3B). Intraplate dexamethasone titrations also showed reproducible performance with an average AC_{50} of 6 nM (Fig. 3C), in agreement with published observations (Martinez et al., 2005). Displaying the entire titration-based screen as a scatter plot revealed the presence of wells with spurious high values (Fig. 3D). However, as the LOPAC library was screened as titration, artifacts showing a response at a single concentration were easily distinguished from genuine actives that showed a concentration–response relationship.

Example concentration–response curves obtained from the screen were of high quality (Fig. 4). Twenty-seven actives were identified with an AC_{50} under 12 μM. All 10 compounds annotated as glucocorticoids in the LOPAC collection[2] were found as actives, including both agonists and antagonists. At the higher concentrations, additional classes of active compounds were found that included compounds annotated as phosphatase inhibitors and G protein-coupled receptor (GPCR) ligands. These compounds could be targeting pathway components involved in GR function or translocation.

---

[2] LOPAC annotates 34 compounds as nuclear receptor hormone ligands with 10 of these annotated as glucocorticoids, including cortisol and its analogs.

FIG. 3. Validation screen of the LOPAC library against the GR-GFP assay. (A) A heat map is shown of the LOPAC library screened at 15 concentrations plus four DMSO blanks used for background correction. Carryover of potent library compounds into the DMSO plates can be seen. Controls are arrayed on the left side of each plate. (B) Average S/B and Z' calculated from control wells for all plates. (C) Titration–response curves for the dexamethasone control present on all plates indicated an average AC_{50} of 6 nM. (D) A three-dimensional scatter plot displaying inactive (black) and active (red) samples that showed a concentration–response relationship.

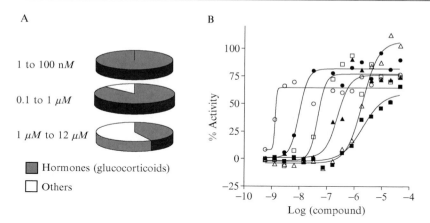

FIG. 4. Results of the LOPAC screen against the GR-GFP assay. (A) LOPAC annotation of glucocorticoid (shaded) and nonglucocorticoid (unshaded) actives identified in the GR-GFP screen binned by AC50: 1 to 100 nM ($n = 5$), 0.1 to 1 μM ($n = 8$), 1 to 12 μM ($n = 14$). (B) Example concentration–response curves obtained from the screen: LOPAC-B-7777 (budesonide, O), LOPAC-H-4001 (hydrocortisone, ●), LOPAC-S-3378 (spironolactone, □), LOPAC-T-6376 (fluoxyprednisolone, ▲), LOPAC-B-7880 (8-bromo-cAMP, △), and LOPAC-H-2270 (hydrocortisone 21, ■).

Locus Derepression Assay

The LDR assay detects the derepression of a GFP reporter that is stably integrated in a transcriptionally silent region of the mouse genome (see Chapter 2, this volume). In the vector, GFP transcription is controlled by a CMV promoter, which normally is strong and constitutively active. However, this line was selected for lack of constitutive expression of the GFP reporter presumably due to epigenetic silencing of the integration locus and/or methylation of the CMV promoter, which contains a large CpG island. GFP transcription can be induced by incubating the cells with histone deacetylase (HDAC) inhibitors such as butyrate or trichostatin A (TSA), which remodel the chromatin from a closed to open conformation to allow transcription (Thiagalingam et al., 2003) or with inhibitors of DNA methyltransferases such as 5-aza-deoxycytidine. A more thorough discussion of this cell line and assay can be found in Martinez et al. (2006).

In the LDR assay, activity is measured by the number of GFP-positive cells in each well, where cells show little or no fluorescence without treatment, but many cells become fluorescent after treatment with HDAC inhibitors (Fig. 5). We used this assay to screen for compounds that induce GFP in the absence of HDAC inhibitors. This approach has the advantage of detecting an increase in signal (the number of positive GFP cells) from a

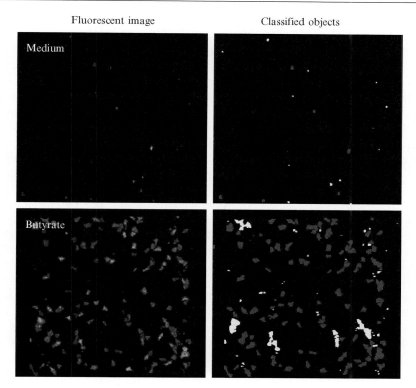

FIG. 5. Classification of GFP-positive cells in the LDR assay. Cells in a 1536-well plate were treated with culture medium minus (top) or plus (bottom) 25 mM sodium butyrate for 29 h at 37°, washed twice with PBS, and imaged by the Acumen Explorer. Live (left) and false-colored (right) images of a representative well treated with medium or butyrate are shown. Objects classified between 20 and 120 μm in width and depth are shaded gray, whereas objects outside this range are shaded white.

very low background and therefore the number of false positives is low. Furthermore, cytotoxic compounds are not identified, as dead or dying cells will not induce GFP production. Hence, the screen is very selective for compounds that are specific activators of GFP expression.

Assay Optimization for 1536-Well Plate Format

For the LDR assay to be sensitive, the GFP signal must be strong, the background fluorescence low, and the number of GFP-positive cells high. To achieve this sensitivity in a 1536-well plate format, several parameters required optimizing: cell density, incubation time, and cell preparation for imaging.

Cell density was optimized in order for the Acumen Explorer to accurately score individual cells within a well. GFP-positive cells were classified as fluorescent objects between the size of 20 to 120 μm width and depth. This filter selects objects in the size range of single cells to clusters of 2 to 3 cells. When too many cells are plated, cell foci form that are larger than the selected size range and are thus excluded from enumeration, resulting in fewer positive cells counted (Fig. 5). When too few cells are plated, single cells are mostly present but the total cell number is too low to differ significantly from background. For this assay, 250 to 300 cells/well in 1536-well plates yielded single cells or small cell clusters within the desired size range.

Several conditions were examined to determine the optimal imaging procedure. The simplest approach was to image live cells in medium following compound incubation, thus minimizing the number of steps in the protocol and reducing assay variability. Under these conditions, only the brightest GFP signals were detected and the resulting cell images appeared fragmented (Fig. 6). Because many of these image fragments were below the specified size range (20–120 μm width and depth), they were excluded from enumeration. This effect likely arises from fluorescent component(s) in the medium creating a high background that is then subtracted from the scan. Consequently, any low-level GFP signal arising from cells is also eliminated. While we have not identified the medium or cell components contributing to background fluorescence, phenol red, a medium constituent, is not the source (data not shown).

Image quality improved markedly following removal of the medium and two PBS washes (Fig. 6). Given that a brighter GFP signal was detected and the cell images were not fragmented, the number of cells enumerated increased by two thirds. The cell signal remained unchanged in PBS after 1 h at 37°, and even after an overnight incubation at 37°, the cells remained adherent and GFP positive, although the signal quality was diminished considerably (data not shown). Following PBS washes, cells could be fixed with methanol, although GFP fluorescence was lower and some cell images appeared fragmented (Fig. 6). While methanol fixation results in lower cell object counts, plates can be stored for later reading or reference.

The length of treatment time was optimized to determine the minimum incubation time to induce a strong GFP signal. Overnight incubation with sodium butyrate or TSA was sufficient to induce GFP but only in a minority of cells. Longer compound incubation times of 24 and 29 h with butyrate induced GFP at higher levels and in the majority of cells.

Solutions and Materials

C127-derived stably transfected mouse mammary cells are grown in DMEM with L-glutamine and high glucose (Invitrogen), 1 mM sodium

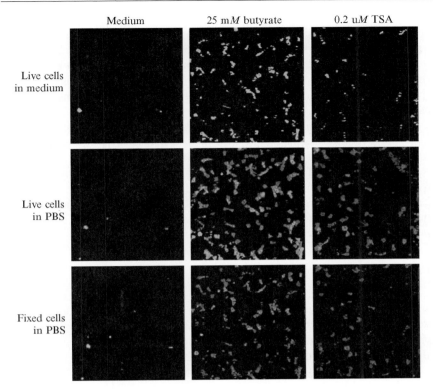

FIG. 6. Effect of wash conditions on LDR assay quality in a 1536-well plate format. Cells were treated with medium alone (left column), 25 mM sodium butyrate (middle column), or 0.2 μM trichostatin A (right column) for 24 h at 37°, and wells were imaged in culture medium (top row), after two PBS washes (center row), and after methanol fixation (bottom row). In each column, the same well is imaged following sequential wash and fix steps.

pyruvate (Invitrogen), 0.1 mM MEM nonessential amino acids (Invitrogen), 1% (v/v) penicillin–streptomycin (Invitrogen), and 10% heat-inactivated FBS (Hyclone). Sodium butyrate (Sigma-Aldrich) and TSA (Tocris) are used as positive controls.

High-Throughput Screening Protocol

1. Cells are harvested, passed through a 40-μm basket filter, and suspended at 50,000 cells/ml in growth medium. Cells are seeded at 250 cells/6 μl/well into black clear-bottom 1536-well plates using either a FRD or a Kalypsys dispenser.

2. The compound is transferred to assay plates by a pin tool. The control compound plate contains 16 2-fold titrations in duplicate of aqueous sodium butyrate beginning at 4.25 M in columns 1 and 2. Columns 3 and 4 contain water and 4.25 M sodium butyrate, respectively, and the remaining columns are empty. The final DMSO concentration is 0.4%, and the compounds are diluted by a factor of 261-fold upon transfer to the assay plate.

3. Assay plates are incubated for 30 h at 37° and 5% CO_2.

4. The medium is aspirated, and 6 μl/well of PBS is added and then replaced by a second dispense of 6 μl/well of PBS.

5. Plates are read in the Acumen Explorer using the following settings: 6 mW laser power, 660 V channel 1 PMT, 1 × 4 μm x and y scan resolution, 2.5 SD above background trigger threshold, 10-μm minimum and 100-μm maximum feature size, and sliding window of two kernel width and three kernel height for image filtering.

Validation of the 1536-Well LDR Assay

We screened the LOPAC compound collection to validate the LDR assay and characterize the number and potency of active compounds. The screen consisted of three DMSO control and seven LOPAC plates comprising a 5-fold titration series of 1280 compounds beginning at 10 μM. The signal quality across the screen was strong and stable; the S/B ranged from 66 to 199 and the Z' factor varied from 0.63 to 0.75 (Fig. 7C). Six compounds were identified as active at 10 μM and two of these showed activity at 2 μM as well (Fig. 7A and D). One identified active, 5-azacytidine, is an inhibitor of DNA methyltransferase, a known regulator of chromatin remodeling (Ben-Kasus *et al.*, 2005; Ghoshal *et al.*, 2002), indicating that the LDR assay is capable of finding known modulators of this biology. The LOPAC library does not contain known HDAC inhibitors.

β-Arrestin:β_2-Adrenergic Receptor (βARR:β2AR) Protein Fragment Complementation Assay

Protein fragment complementation assays involve two nonfluorescent GFP (or GFP variant) fragments fused to cellular proteins of interest. Upon noncovalent association of the cellular proteins, the GFP (or variant) fragments to which they are fused become fluorescent (MacDonald, 2006; Remy and Michnick, 2003; Yu *et al.*, 2003). PCA is applied to the association of β-arrestin with the β2AR whereby the C terminus of the β2AR is fused with one fragment of mutant YFP and β-arrestin is fused to the complementary fragment (Fig. 1A). Following agonist stimulation, the β2AR is

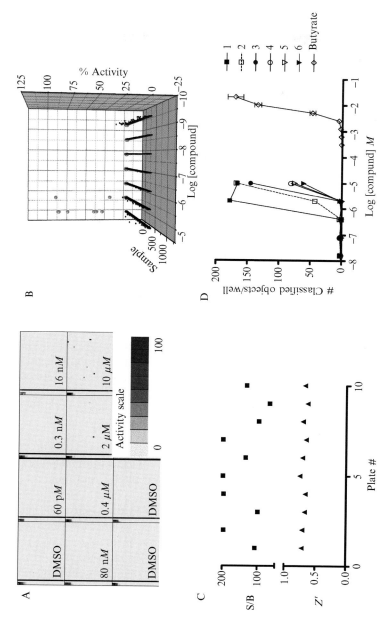

Fig. 7. Validation screen of the LOPAC library with the LDR assay. (A) A heat map depicts the LOPAC library screened at seven concentrations and includes three DMSO plates included for background correction. Controls are arrayed on the left side of each plate. (B) Three-dimensional scatter plot indicating inactive samples (black) and active (red) samples showing a concentration–response relationship. (C) Average S/B and Z' factor calculated from control wells for all plates. (D) Concentration–response plots of actives identified in the screen.

phosphorylated by GPCR kinases to cause enhanced association of the adaptor protein, β-arrestin (Ferguson *et al.*, 1998; Inglese *et al.*, 1993). In this assay, the β-arrestin and β2AR interaction enables association and folding of YFP fragments and generates a quantifiable fluorescence signal.

Solutions and Materials

βARR: β2AR cells are HEK 293T cells that stably express fusions of YFP fragments (Odyssey Thera, Inc.) maintained in DMEM with 10% FBS, 500 μg/ml of zeocin (Invitrogen) and 400 μg/ml of hygromycin B (Invitrogen). The agonist control, $(-)$-isoproterenol hydrochloride, the antagonist control, (S)-$(-)$-propranolol hydrochloride, and naphthol blue black are from Sigma-Aldrich.

Optimization of the 1536-Well Assay Protocol

Medium components can interfere with the fluorescence signal and image quality can be improved by removal of the medium (see LDR assay). βARR: β2AR cells are weakly adherent to tissue culture-treated microtiter plates and these cells cannot be washed by the Kalypsys washer/dispenser, even when grown on plates precoated with poly-L-lysine (data not shown). To circumvent the need of wash steps, a fluorescence-absorbing dye is added to decrease background fluorescence arising from the medium or unattached cells in the well. By reducing background fluorescence, the threshold value that triggers data collection (typically 2 SD above the background) is decreased and therefore the sensitivity of the fluorescence signal is increased. Cell density is kept to <700 cells/well to accurately measure background fluorescence between cells. The addition of 33 μM naphthol blue black increases the S/B by approximately threefold (Fig. 8A).

High-Throughput Screening Protocol

1. Cells are suspended, passed through a 40-μm basket filter, seeded at 700 cells/5 μl/well into black clear-bottom 1536-well plates, and incubated for approximately 20 h at 37° and 5% CO_2.

2. Compound and control samples are transferred by a pin tool. The compound control plate contains 16 2-fold dilutions in duplicate of $(-)$-isoproterenol starting at 2 mM in column 1, 20 μM $(-)$-isoproterenol (AC_{90}) in column 2, DMSO in the column 3, and 2 μM $(-)$-isoproterenol (AC_{50}) in column 4. To screen for antagonists, columns 5 through 48 contain 2 μM $(-)$-isoproterenol (AC_{50}) while when screening for agonists, these columns are empty. The final DMSO concentration is 0.5 % and the compounds are diluted 217-fold upon transfer to the assay plate.

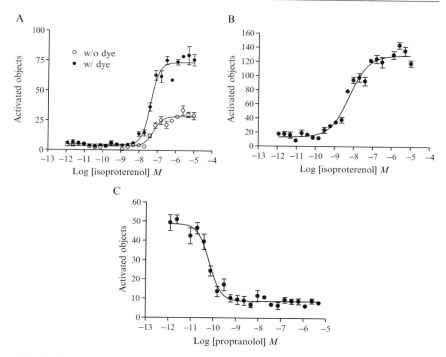

FIG. 8. Optimization of βARR:β2AR assay in agonist and antagonist formats in 1536-well plates. (A) Titration of isoproterenol in the presence or absence of 33 μM napthol blue black. Concentration–response curves for the agonist isoproterenol (B, $AC_{50} = 6$ nM) and the antagonist propranolol (C, $AC_{50} = 60$ pM) using the optimized 1536-well protocols. Cells were stimulated with 8 nM isoproterenol for the antagonist assay.

3. Assay plates are incubated at $37°$ and 5% CO_2 for 1.5 h.

4. One microliter per well of 0.2 mM naphthol blue black is added.

5. Plates are read by the Acumen Explorer using the following settings: 6 mW laser power, 630 V for the GFP (channel 1) PMT, 2 SD above background fluorescence threshold for data collection, and 1×8 μm scan of the whole-well area. Two populations, activated and unactivated, are defined by a 5- to 120-μm width and depth to eliminate small and large fluorescent particles. The activated population is classified as mean peak intensity between 1278 and 6185 FLU and total peak intensity between 21,408 and 154,242 FLU, while the unactivated population is defined as mean peak intensity between 230 and 1199 FLU and total peak intensity between 2684 and 18,464 FLU.

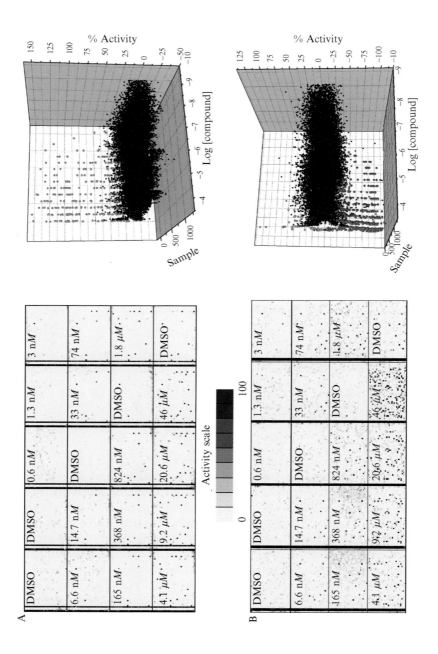

Assay Validation

The LOPAC library was screened against the βARR:β2AR assay in both agonist and antagonist formats (Fig. 9). Both assays showed suitable performance in high-throughput mode. The LOPAC library contains very potent adrenergic receptor ligands ($EC_{50} \sim 10$ nM) and for these, we observed some carryover from the pin tool on DMSO control plates (Fig. 9A and B). Because these assays were screened as a titration series, compound carryover did not complicate our analysis. However, if our assay protocol is used for single concentration screening of multiple library plates, improved pin washing will be needed.

The most potent compounds identified from the screens were known agonists or antagonists of the βAR (Fig. 10). The LOPAC library contains 47 agonists and 35 antagonists that are annotated as adrenergic receptor ligands. Of the 19 adrenergic receptor agonists identified with $AC_{50} \leq 10$ μM, 18 are annotated as β-adrenergic agonists and one as an α-adrenergic receptor agonist. Twenty-five agonists, which were not identified at the 10 μM AC_{50} threshold, are annotated as α- or $\beta 1/\beta 3$ adrenergic receptor agonists. For antagonists, 15 compounds with $AC_{50} \leq 10$ μM were identified and 14 are annotated as βAR antagonists and one annotated as an α-adrenergic receptor antagonist. Twenty antagonists were not identified at the 10 μM AC_{50} threshold, and 19 are annotated as specific for the α-adrenergic subtypes. Taken together, these results show that the pharmacology for a large number of known adrenergic ligands was defined accurately in these assays.

Theoretically, the antagonist assay can be used to identify both agonists and antagonists. However, we found that the antagonist screen did not identify βAR agonists as efficiently as the agonist screen. For example, the potent βAR agonist isoproterenol had a 16 nM AC_{50} in the agonist format but showed only a shallow increase in activation over the stimulated signal in the antagonist format (data not shown). Therefore, to ensure accurate identification of both agonists and antagonists, both formats of the assay should be performed.

FIG. 9. Validation screens of the LOPAC library against the βARR:β2AR assay. Heat maps (left) and three-dimensional scatter plots (right) of the screens against the agonist (A) and antagonist (B) formats of the βARR:β2AR assay. The LOPAC collection was screened at 15 concentrations, and five DMSO plates were included for background correction. Heat maps show carryover of some potent library compounds into the DMSO plates. Three-dimensional scatter plots indicate inactive samples (black) and active (agonist, red; antagonist, blue) samples that showed a concentration–response relationship.

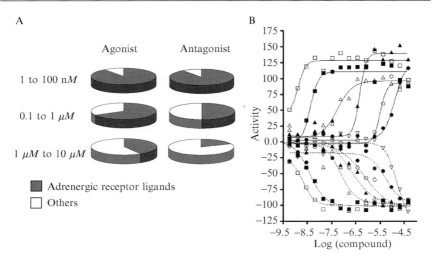

Fig. 10. Results of the LOPAC screen against the βARR:β2AR assay. (A) LOPAC annotation of compounds binned by the AC_{50} identified in the agonist format, 1 to 100 nM ($n = 10$), 0.1 to 1 μM ($n = 9$), and 1 to 10 μM ($n = 12$), and the antagonist format, 1 to 100 nM ($n = 10$), 0.1 to 1 μM ($n = 25$), and 1 to 10 μM ($n = 44$). (B) Example concentration–response curves of agonists (solid lines), LOPAC-F-9552 (formoterol, $\square$), LOPAC-I-6504(isoproterenol, $\blacksquare$), LOPAC-S-8260 (albuterol, $\triangle$), LOPAC-E-4642 [($\pm$)-epinephrine, $\blacktriangle$], LOPAC-A-0966 (4-amino-1,8-naphthalimide, $\bigcirc$), and LOPAC-S-8442 (1,3-dihydro-3-[3,5-dimethyl-1 H-pyrrol-2-yl], $\bullet$), and antagonists (dotted lines), LOPAC-P-152 [$S(-)$-pindolol, $\square$], LOPAC-P-0884 [($\pm$)-propranolol, $\blacksquare$], LOPAC-L-1011 (labetalol, $\triangle$), LOPAC-B-5683 (betaxolol, $\blacktriangle$), LOPAC-P-4670 (propafenone, $\bigcirc$), LOPAC-C-106 (7-trifluoromethyl-4 (4-methyl-1-piperazinyl)-pyrrolo[1,2-a]quinoxaline, $\bullet$), and LOPAC-T-1132 (tetraethylthiuram disulfide, $\square$), are shown.

Summary

Optimal assay performance with the Acumen Explorer requires reducing background fluorescence, maximizing signal, and choosing the proper cell density and object characteristics. We found that the greatest improvement in assay quality was achieved by lowering background fluorescence. This was accomplished by washing adherent cells in the case of the GR-GFP and LDR assays. For weakly adherent cells, as in the βARR: β2AR assay, the fluorescence-absorbing dye naphthol blue black was added to the wells.

Microtiter plate laser-scanning cytometry is a powerful technology enabling cellular analysis similar to that of conventional flow cytometry. While the plate-based method described here has fewer excitation and emission channels compared to flow cytometry [up to 17 colors can be detected (Perfetto *et al.*, 2004)], the system has important advantages for compound screening. This technique can detect fluorescent signals from adherent cells, which are the basis of numerous assays used in screening. In addition, microtiter plate cytometers can be integrated easily with standard robotic screening systems, enabling the screening of large compound libraries at greater than 200,000 samples per day.

The application of microplate cytometers can be enhanced if used in conjunction with CCD-based microscope systems by integrating both components into a screening process. In this scenario, a microplate cytometer identifies active wells in high-throughput mode; these wells are subsequently imaged at high resolution using a CCD-based microscope. In the case of GR-GFP, a microplate cytometer would identify actives based on cytosol-to-nuclear translocation, whereas a microscope-based imager would classify actives by more subtle patterns of nuclear sublocalization. It is known that different ligands cause distinct distributions of GR into nuclear subdomains (Htun *et al.*, 1996). Development of such integrated detection systems should increase the variety of phenotypic assays applied to HTS systems.

Acknowledgments

We gratefully acknowledge Dr. Pagliaro of BioImage for providing the GR-GFP cell line for this study and Eric Sprigg of TTP Labtech for technical assistance with the Acumen Explorer. This research was supported by the NIH Roadmap for Medical Research and the Intramural Research Program of the National Human Genome Research Institute, National Institutes of Health.

References

Bai, C., Schmidt, A., and Freedman, L. P. (2003). Steroid hormone receptors and drug discovery: Therapeutic opportunities and assay designs. *Assay Drug Dev. Technol.* **1,** 843–852.

Ben-Kasus, T., Ben-Zvi, Z., Marquez, V. E., Kelley, J. A., and Agbaria, R. (2005). Metabolic activation of zebularine, a novel DNA methylation inhibitor, in human bladder carcinoma cells. *Biochem. Pharmacol.* **70,** 121–133.

Bowen, W. P., and Wylie, P. G. (2006). Application of laser-scanning fluorescence microplate cytometry in high content screening. *Assay Drug Dev. Technol.* **4,** 209–221.

Carroll, S. S., Inglese, J., Mao, S.-S., and Olsen, D. B. (2004). Drug screening: Assay development issues. *In* "Molecular Cancer Therapeutics: Strategies for Drug Discovery and Development" (D. B. Prendergast, ed.), pp. 119–140. Wiley, Hoboken, NJ.

Cleveland, P. H., and Koutz, P. J. (2005). Nanoliter dispensing for uHTS using pin tools. *Assay Drug Dev. Technol.* **3,** 213–225.

Ferguson, S. S. G., Zhang, J., Barak, L. S., and Caron, M. G. (1998). Molecular mechanisms of G protein-coupled receptor desensitization and resensitization. *Life Sci.* **62,** 1561–1565.

Garippa, R. J., Hoffman, A. F., Gradl, G., and Kirsch, A. (2006). High-throughput confocal microscopy for β-arrestin–green fluorescent protein translocation G protein-coupled receptor assays using the Evotec Opera. *Methods Enzymol.* **414** (this volume).

Ghoshal, K., Datta, J., Majumder, S., Bai, S., Dong, X., Parthun, M., and Jacob, S. T. (2002). Inhibitors of histone deacetylase and DNA methyltransferase synergistically activate the methylated metallothionein I promoter by activating the transcription factor MTF-1 and forming an open chromatin structure. *Mol. Cell. Biol.* **22**, 8302–8319.

Haasen, D., Schnapp, A., Valler, M. J., and Heilker, R. (2006). G protein-coupled receptor internalization assays in the high-content screening format. *Methods Enzymol.* **414** (this volume).

Htun, H., Barsony, J., Renyi, I., Gould, D. L., and Hager, G. L. (1996). Visualization of glucocorticoid receptor translocation and intranuclear organization in living cells with a green fluorescent protein chimera. *Proc. Natl. Acad. Sci. USA* **93**, 4845–4850.

Inglese, J., Auld, D. S., Jadhav, A., Johnson, R. L., Simeonov, A., Yasgar, A., Zheng, W., and Austin, C. P. (2006). Quantitative high-throughput screening (qHTS): A titration-based approach that efficiently identifies biological activities in large chemical libraries. *Proc. Natl. Acad. Sci. USA* **103**, 11473–11478 .

Inglese, J., Freedman, N. J., Koch, W. J., and Lefkowitz, R. J. (1993). Structure and mechanism of the G protein-coupled receptor kinases. *J. Biol. Chem.* **268**, 23735–23738.

Jung-Testas, I., and Baulieu, E. E. (1983). Inhibition of glucocorticosteroid action in cultured L-929 mouse fibroblasts by RU 486, a new anti-glucocorticosteroid of high affinity for the glucocorticosteroid receptor. *Exp. Cell Res.* **147**, 177–182.

Levinson, B. B., Baxter, J. D., Rousseau, G. G., and Tomkins, G. M. (1972). Cellular site of glucocorticoid-receptor complex formation. *Science* **175**, 189–190.

MacDonald, M. L., Lamerdin, J. L., Owens, S., Keon, B. H., Bilter, G. K., Shang, Z., Huang, Z., Yu, H., Dias, J., Minami, T., Michnick, S. W., and Westwick, J. K. (2006). Identifying off-target effects and hidden phenotypes of drugs in human cells. *Nature Chem. Biol.*

Mainquist, J. K. D., Robert, C., Weselak, M. R., Meyer, A. J., Burow, K. M., Sipes, D. G., and Caldwell, J. (2003). Specimen plate lid and method of using. U.S. Patent #6534014.

Martinez, E. D., Rayasam, G. V., Dull, A. B., Walker, D. A., and Hager, G. L. (2005). An estrogen receptor chimera senses ligands by nuclear translocation. *J. Steroid Biochem. Mol. Biol.* **97**, 307–321.

Martinez, E. D., Dull, A. B., Beutter, J. A., and Hager, G. L., (2006). High content fluorescene-based screening for epigenetic modulators. *Methods Enzymol.* **414**, 21–36.

Niles, W. D., and Coassin, P. J. (2005). Piezo- and solenoid valve-based liquid dispensing for miniaturized assays. *Assay Drug. Dev. Technol.* **3**, 189–202.

Perfetto, S. P., Chattopadhyay, P. K., and Roederer, M. (2004). Seventeen-colour flow cytometry: Unravelling the immune system. *Nature Rev. Immunol.* **4**, 648–655.

Remy, I., and Michnick, S. W. (2003). Dynamic visualization of expressed gene networks. *J. Cell Physiol.* **196**, 419–429.

Richards, G. R., Jack, A. D., Platts, A., and Simpson, P. B. (2006). Measurement and analysis of calcium signaling in heterogeneous cell cultures. *Methods Enzymol.* **414**, 335–347.

Shaner, N. C., Steinbach, P. A., and Tsien, R. Y. (2005). A guide to choosing fluorescent proteins. *Nature Methods* **2**, 905–909.

Thiagalingam, S., Cheng, K. H., Lee, H. J., Mineva, N., Thiagalingam, A., and Ponte, J. F. (2003). Histone deacetylases: Unique players in shaping the epigenetic histone code. *Ann. N.Y. Acad. Sci.* **983**, 84–100.

van Roessel, P., and Brand, A. H. (2002). Imaging into the future: Visualizing gene expression and protein interactions with fluorescent proteins. *Nature Cell Biol.* **4**, E15–E20.

Williams, R. G., Kandasamy, R., Nickischer, D., Trask, O. J., Laethem, C., Johnston, P. A., and Johnston, P. A. (2006). Generation and characterization of a stable MK2-EGFP cell line and subsequent development of a high-content imaging assay on the Cellomics ArrayScan platform to screen for p38 mitogen-activated protein kinase inhibitors. *Methods Enzymol.* **414** (this volume).

Yu, H., West, M., Keon, B. H., Bilter, G. K., Owens, S., Lamerdin, J., and Westwick, J. K. (2003). Measuring drug action in the cellular context using protein-fragment complementation assays. *Assay Drug Dev. Technol.* **1**, 811–822.

Zhou, J., and Cidlowski, J. A. (2005). The human glucocorticoid receptor: One gene, multiple proteins and diverse responses. *Steroids* **70**, 407–417.

[30] High-Throughput Measurements of Biochemical Responses Using the Plate::Vision Multimode 96 Minilens Array Reader

By KUO-SEN HUANG, DAVID MARK, and FRANK ULRICH GANDENBERGER

Abstract

The plate::vision is a high-throughput multimode reader capable of reading absorbance, fluorescence, fluorescence polarization, time-resolved fluorescence, and luminescence. Its performance has been shown to be quite comparable with other readers. When the reader is integrated into the plate::explorer, an ultrahigh-throughput screening system with event-driven software and parallel plate-handling devices, it becomes possible to run complicated assays with kinetic readouts in high-density microtiter plate formats for high-throughput screening. For the past 5 years, we have used the plate::vision and the plate::explorer to run screens and have generated more than 30 million data points. Their throughput, performance, and robustness have speeded up our drug discovery process greatly.

Introduction

In the mid- to late 1990s pharmaceutical companies began to invest heavily in high-throughput screening (HTS) technologies to speed up their drug discovery processes. Early HTS platforms, typically built around track-mounted articulate robots, could handle just one task at a time and achieve throughputs of only 20,000 compounds per day. They were not fast enough to keep pace with the increasing size of compound libraries and number of screening targets. Furthermore, because many of them could not be applied to high-density microtiter plate formats (e.g., 1536 well), reagent costs became an increasing issue.

To optimize and speed up its HTS efforts, Roche approached Carl Zeiss (Jena, Germany), a company specialized in optics and micromechanics, to build an ultrahigh-throughput screening (uHTS) system. The goal was to design an easy-to-operate screening platform for parallel handling of min-iaturized samples with throughputs greater than 100,000 compounds per day. The outcome of this cooperation was the plate::explorer, a uHTS system capable of screening >200,000 compounds per day in 96-, 384-, and 1536-well formats. One of the key elements in the plate::explorer system is the plate::vision, a high-throughput multimode microtiter plate reader equipped with quasiconfocal optics capable of reading 96 wells simultaneously.[1]

Instrumentation

Plate::Vision

Prior to the introduction of the plate::vision, two kinds of readers were used in high-throughput drug discovery. One was the single-well reader, which measured one well at a time. The other was the image reader, which acquired a picture of an entire microtiter plate and then evaluated the image for light intensities resulting from individual wells. With single-well readers, the optical pathway could be kept rather simple (as just a single well needed to be illuminated and light from only one well needed to be collected and measured). As a result, these readers offered excellent sensitivity. However, because they only read one well at a time and the microtiter plate needed to move under the optical head for each read, the throughput was limited.

In contrast, image readers could realize extremely high throughputs (in excess of 1 million wells per day) with a single plate being read in just a few seconds. However, signal quality could be compromised by cross talk between neighboring wells and by background fluorescence from the plate material. As a result, the sensitivity of image readers was limited.

To combine the advantages of both single-well and image readers while circumventing the obstacles of each, Zeiss came up with a solution where 96 wells are illuminated simultaneously through a minilens array (Fig. 1). Because only 96 wells are illuminated at a time and because each of the minilens objectives generates a defined focus located in the center of the illuminated well, there is no optical cross talk between adjacent

[1] Since January 2005, both plate::explorer and plate::vision are sold, serviced, and further developed by lab automation specialist Evotec Technologies, Hamburg, Germany.

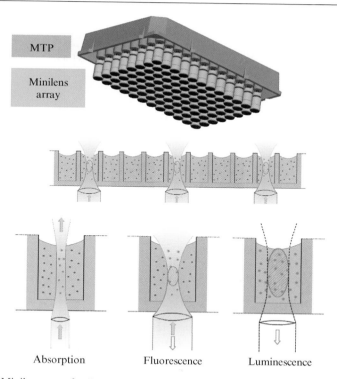

FIG. 1. Minilens array for "structured illumination" (in a 384-well plate every 4th and in a 1536h-well plate every 16th well are illuminated) and the resulting detection volumes for three main detection modes.

wells and almost no interference with background fluorescence from the plate material. This results in a reader with sensitivity as good as the single-well readers and throughputs as high as the image readers. For example, it takes only about 30 s to read a 1536-well microtiter plate. This translates into a throughput that comfortably exceeds 1 million wells per day, limiting the prospect that plate reading will become a bottleneck in the foreseeable future. Furthermore, using an array with 96 objectives allows plate reading to be synchronized with parallel liquid handling, for example, with 96 channel pipettors. That way, all wells in a microplate are treated equally, which generates high-quality results and enables multipoint readouts where individual wells are read multiple times to produce a time series.

Optical Setup. The optical pathway of the plate::vision follows that of an inverted microscope (Fig. 2). Inside the reader housing, the optical light

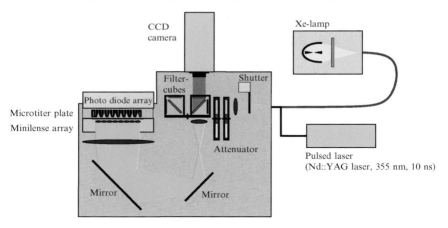

FIG. 2. Optical setup of the plate::vision.

path is folded twice, which keeps the overall instrument compact and enables easy integration into robotic environments.

Light from a 300-W high-pressure xenon arc lamp or a pulsed 355-nm Nd: YAG laser (for time-resolved fluorescence measurements) is first passed through two filter wheels fitted with attenuation filters and subsequently through a filter slider that can hold up to eight filter cubes. Each filter cube combines the necessary excitation, emission, and dichroic filters/mirrors in an easy-to-exchange metal housing. Both attenuation and fluorescence filter settings can be changed by the reader software. This enables sequential measurements at different wavelengths or with different readout methods. The excitation light is then guided via two mirrors onto a large telescope lens, which expands and homogenizes the light beam to illuminate the minilens array (which is roughly the same size as a microtiter plate). With a microtiter plate placed on top of the microlens array, the 96 miniaturized microscopy objectives, each having an aperture of approximately 0.5, will focus the light into the well, producing a quasiconfocal detection zone, which (in the case of fluorescence) has a volume of about 10 nl. With such a small fraction of the sample participating in the measurement process, the reader sensitivity becomes independent of the microtiter plate format. Thus, assays can be miniaturized to high-density microtiter plates without compromising data quality. Furthermore, as microtiter plates are measured from the bottom, artifacts originating from bubbles at the liquid surface or meniscus effects are effectively avoided. This "structured illumination" approach (in a 384-well plate every 4th, and in a 1536-well plate every 16th well are illuminated) further eliminates the strong background signal originating from the plate bottom.

To enable screening of cell- or bead-based assays, the z height of the focal zone can be adjusted inside the well. Combined with the x/y-scanning function of the microplate stage, the height adjustment allows the reader to actively scan the wells and maximize the signal generated from beads or adherent cells while facilitating the diagnosis of artifacts caused by dye molecules sticking to the well bottom or by an uneven distribution of cells or beads. All this can be done without sacrificing readout speed.

Because different measurement modes have different requirements with respect to shape and size of the detection volume, different minilens arrays are used for optimizing the optical path for the individual application (Fig. 1). For time-resolved fluorescence (TRF) and luminescence, where signal quality can be influenced by meniscus effects, an array that produces a narrow beam of light is used. However, for absorbance measurements, an array creating a large focal volume for efficient light collection is used.

Light collected from the sample is guided onto either a $-30°$ CCD camera (512×512 pixel, 12 bit) for fluorescence and luminescence measurements or a time-gated $-20°$-intensified CCD camera (16 bit) for TRF measurements. The resulting images are automatically evaluated for intensity values originating from individual wells. For absorbance measurements, a photodiode array consisting of 96 individual photodiodes located over the microtiter plate is used.

Readout Methods. The plate::vision has been designed to support almost all major assay formats (fluorescence, luminescence, absorbance, and TRF) used in HTS laboratories. TRF measures fluorophores with lifetimes in the microsecond range. These fluorophores [typically Europium(III) chelates] are measured with some time delay after a defined excitation pulse, when most of the compound-associated autofluorescence and background fluorescence originating from the assay solution or the plate material have largely decayed. The long-lived acceptor emission minimizes the short-lived background fluorescence and improves the signal-to-noise ratio significantly.

TRF is applicable both in heterogeneous (e.g., DELFIA from Perkin-Elmer) and in homogeneous (e.g., LANCE from PerkinElmer) formats. In a typical homogeneous LANCE format, one biomolecule is labeled with a europium chelate donor fluorophore and the binding partner with an appropriate acceptor dye (e.g., allophycocyanin). When the two fluorophores are in close proximity, a time-resolved fluorescence resonance energy transfer (TR to FRET) from the europium chelate (excitation at approximately 340 to 360 nm, emission typically at 615 nm) to the allophycocyanin (emission at 665 to 670 nm) is detected.

Because plate::vision utilizes a pulsed 355-nm laser light source, it can read TRF with high sensitivity and speed. Typically, it only takes less than

1 min to finish reading a 384-well microtiter plate. In contrast, single-well readers take 10 to 20 min for the same measurement.

With its powerful parallel readout concept, the plate::vision is an ideal instrument for high-throughput kinetic readouts. Because it takes less than 10 s to read absorbance of a 384-well plate, HTS assays can be carried out with kinetic readouts, which are more accurate than end-point assays. Time-course experiments can also provide a wealth of additional information and minimize artifacts generated from the reader or compounds.

Plate::Explorer

The traditional approach for plate transport in high-throughput screening systems relies on single-arm robots to pick up and transfer plates among incubators, liquid handlers, and readers. These systems have limited throughputs because only one plate can be handled at a time. To address this problem, Zeiss developed a new concept for parallel plate handling. Drawing on experience from semiconductor processing, Zeiss engineers also designed a compact turntable/conveyor belt system that moves plates quickly from one device/workstation to the next.

With this concept, up to five different devices (liquid handlers, plate washers, incubators, readers, etc.) are assembled to form a workstation. Within a workstation, microtiter plates are moved by turntables. Each turntable can hold up to four microtiter plates. Using turntables, plate handling can be parallelized with empty turntable positions acting as buffer positions for devices that are temporarily unable to accept plates. A workstation has its own power supply and computer to operate independently. To create a high-throughput screening system, multiple workstations can be joined via a bidirectional conveyor belt. With this modular approach, typically 10 to 100 assay plates can be processed in an hour, depending on the number of workstations and steps in the assay protocol. The system has been shown to be robust for daily HTS operations. Typically, more than 1000 plates can be processed without any mechanical errors.

All plate transfer tasks are controlled by intuitive software (plate::works), which controls various plate::explorer devices via one common graphical user interface. Using plate::works, screening runs can be "programmed" by simple "drag and drop" operations.

Measurements of Biochemical Responses

Absorbance Assays

Absorbance assays have been used widely for measuring biological activity. They can be set up easily by monitoring absorbance changes of physiological substrates or products. Although absorbance assays are not as sensitive as other methods, they do not require expensive antibodies or fluorophores for detection. In many cases, assay signals can also be monitored by kinetic readouts, which increase sensitivity and decrease interference by colored compounds. They can be carried out in high-density microtiter plates (e.g., 1536-well plates) to save reagent costs and increase throughputs.

In order to run an absorbance assay with kinetic readouts in a high-density format, a high-throughput reader is required. We have routinely used the plate::vision for absorbance assays in 1536-well plates. Typically, it takes only 0.1 s to complete reading 96 wells simultaneously and 2 s if the time for stage movement and data integration is included. The total reading time for a 1536-well plate is approximately 32 s. When the reader is integrated into the plate::explorer system, it becomes possible to run absorbance assays with kinetic readouts for HTS. Data can be exported and entered into ActivityBase[2] directly for calculating reaction rates based on linear regression curve fits of the time-dependent enzyme reaction.

We describe here an absorbance assay for screening ATPase (e.g., motor protein) inhibitors. One of the most common assays for measuring ATPase activity is an NADH-coupled enzyme assay utilizing pyruvate kinase and lactate dehydrogenase, as indicated by the scheme shown here:

$$ATP \xrightarrow{\text{ATPase}} ADP + Pi$$

$$ADP + Phosphoenopyruvate \xrightarrow{\text{Pyruvate kinase}} Pyruvate + ATP$$

$$Pyruvate + NADH \xrightarrow{\text{Lactate dehydrogenase}} Lactate + NAD$$

Because NADH has a high molar absorption coefficient at 340 nm, ATPase activity can be monitored by the decrease of absorbance of NADH. We have implemented a high-throughput screening method with kinetic readouts onto the plate::explorer system consisting of six workstations (Fig. 3). Individual components of the system have been described previously (Huang and Vassilev, 2005). A microtiter plate centrifuge module has been added to the system. When assays are carried out in 1536-well plates, air bubbles generated during liquid pipetting can interfere with accurate reading. We found that brief centrifugation of the assay plate removes air bubbles efficiently. One of the advantages of the plate::explorer system is the parallel process design of the software and hardware, which allows multiple assays to be run on the system simultaneously. A flowchart of a process in which we ran dual ATPase assays with kinetic readouts is shown in Fig. 4. The assays are run in a 1536-well format to save reagent costs. Because the Cybi-well pipettor has a 384-well pipetting head, each

---

[2] ActivityBase is a trademark of IDBS. It is a database managing biological and chemical information.

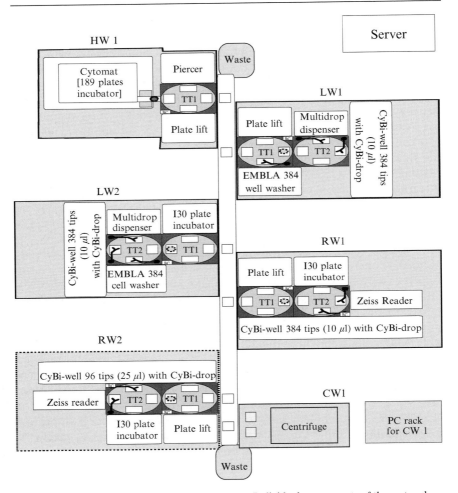

Fig. 3. A diagram of the plate::explorer system. Individual components of the system have been described previously (Huang and Vassilev, 2005) except for the addition of the microtiter plate centrifuge made by Hettich (Germany) to the system.

ATPase is run in one quadrant of the 1536-well plate (two quadrants are used for two ATPases). The 384-well plates containing test compounds in columns 3 to 24 (2 μl per well, 1 mM in dimethyl sulfoxide) are placed in the HW1 plate lift, and 1536-well assay plates are placed in the LW1 plate lift. The compound plates are then transported to the Cybi-drop in LW1, and 30 μl/well of buffer A (25 mM PIPES, pH 7.0, 2 mM MgCl$_2$, 2 mM EGTA, 0.1 mM EDTA, 25 mM KCl, and 1 mM dithiothreitol) is added by

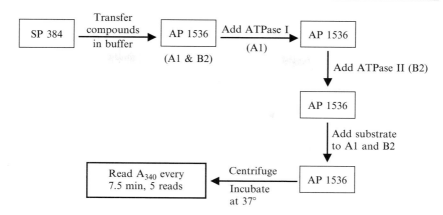

Fig. 4. ATPase assay flowchart on the plate::explorer system. SP 384 represents a compound plate in 384-well format. AP 1536 represents a 1536-well assay plate.

the Cybi-drop. Columns 1 and 2 contain assay buffer only, and columns 3 and 24 contain test compounds. Samples are mixed three times with the 384 pipettor, and 3 μl/well is transferred to quadrants A1 and B2 of the 1536-well assay plate. Enzyme solution containing ATPase I,[3] pyruvate kinase, and lactate dehydrogenase in buffer A is added to quadrant A1 (3 μl/well) of columns 1 to 48 and mixed. The assay plate is then moved to the Cybi-well 384 pipettor in LW2. Enzyme solution containing ATPase II, pyruvate kinase, and lactate dehydrogenase is added to quadrant B2 (3 μl/well) of columns 1 to 48 and mixed. The assay plate is then moved to the Cybi-well 384-well pipettor in RW1. The substrate solution containing ATP is added to quadrants A1 and B2 (3 μl/well) in columns 5 to 48. Buffer without ATP is added to the top set of wells in columns 1 to 4 as a background and KH_2PO_4 (30 μM) is added to the bottom set of the wells as an inhibitor control. After mixing, the plate is centrifuged briefly (2000 rpm, 1 min) to remove air bubbles. The plate is incubated with a lid on at 37° in an I30 incubator with 80% humidity (less than 5% of the sample volume evaporates when a lidded 1536 well-plate is incubated for 1 h). The assay plate is read kinetically on the plate::vision in RW1 every 7.5 min. By using the aforementioned process, we achieved a maximum throughput of 65,000 compounds per day for two enzymes, corresponding to 756,000 data points.

The Z' factor, a measurement for the quality of an assay (Zhang *et al.*, 1999), is shown for each assay plate in Fig. 5. The average Z' factors for the entire batch of screening plates (over 2500 compound plates) are 0.78 and

[3] ATPase I and II are general terms for two related ATPase enzymes.

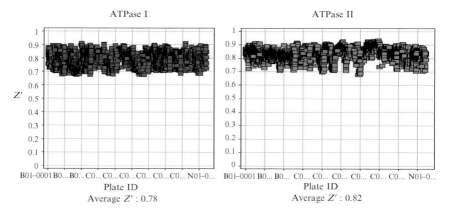

FIG. 5. Z' factors from assay plates of ATPase screens.

0.82 for ATPase I and II, respectively (an assay is considered to be good when the Z' factor is above 0.5). The frequency distribution of each ATPase screen is shown in Fig. 6. In both cases, compounds that caused greater than 35% inhibition were selected for hit confirmation.

TRF Assays

TRF assays using lanthanide chelates (e.g., Eu^{3+}, Tb^{3+}, and Sm^{3+}) as fluorophores have been used for biochemical and cell-based assays. These fluorophores exhibit an intense and long-lived fluorescence emission, making it possible to measure fluorescence after a time delay (100 to 1000 μs). Homogeneous assays based on TR-FRET have been particularly useful for high-throughput screening. They do not involve washing steps and permit the measurement of interactions between two molecules in solution under physiological conditions.

Although the TR-FRET assay is sensitive, it takes a much longer time to read a TR-FRET assay signal than other reading modes (e.g., absorbance or prompt fluorescence). Typically, it requires readings of time-delayed (50 to 400 μs) fluorescence signals at two different emission wavelengths (donor and acceptor emissions). The assay signals are then calculated by the ratio of acceptor-to-donor fluorescence intensity. Therefore, a fast reader is critical in running a TR-FRET assay for high-throughput screening. Table I compares the speed and sensitivity of different readers. Z' factors are very similar between the plate::vision and ViewLux. Z' values in both are slightly lower than in the Victor V. However, the Victor V is much slower than the other two readers.

TABLE I
COMPARISON OF SENSITIVITY AND SPEED OF DIFFERENT READERS

Sensitivity[a]	Plate::vision™	ViewLux[b]	Victor V[c]
Z' factor	0.805	0.798	0.857
% CV (signal)	4.3	3.5	4.1
% CV (blank)	5.8	7.0	11.0
Signal/blank ratio	13.9	20	181
Speed (per plate)			
384-well plate	0.9 min	1.8 min	20 min
1536-well plate	3.6 min	2.8 min	76 min

[a] Sensitivity was determined using 100 pM of Eu-labeled streptavidin and 50 pM of biotin-labeled allophycocyanin on 384-well microtiter plates.
[b] ViewLux™ is an ultrahigh-throughput microplate imager from PerkinElmer.
[c] Victor V is a multilabel, multitask plate reader from PerkinElmer.

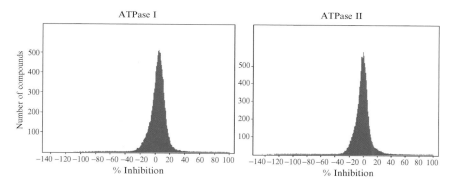

FIG. 6. Frequency distribution of ATPase screens.

We have routinely run screens for kinases, phosphatases, nuclear receptors, and protein–protein interactions in the TR-FRET assay format on the plate::explorer system and achieved throughputs of greater than 100,000 data points per day (defined as a 20-h run) by using 1536-well assay plates. Examples of screening inhibitors of protein–protein interactions have been described previously (Huang and Vassilev, 2005).

Fluorescence Intensity (FI) and Fluorescence Polarization (FP) Assays

Fluorescence intensity assays are sensitive, low cost, and easily applicable to high-throughput screening. We have developed a high-throughput assay for screening heparanase inhibitors (Huang *et al.*, 2004). Although the assay involves multiple pipetting and washing steps, the parallel process design of the plate::explorer system and the high-speed plate::vision have made it possible to screen more than 25,000 compounds per day. The sensitivity of the reader is very comparable with many other readers (e.g., Victor V, Envision, and ViewLux).

Highly fluorescent compounds and quenching compounds interfere with the FI assays. In addition to TR-FRET assays, FP assays using red-shifted fluorescent dyes (e.g., rhodamin red) can minimize compound interference. FP is a simple, homogeneous, and relatively low-cost assay. Because two measurements (parallel and perpendicular fluorescence intensities) are required for calculating polarization for each data point, the reading time is longer than in FI assays. A high-throughput reader can increase the throughput of an FP assay greatly, particularly in high-density microtiter plate formats.

A new FP assay utilizing IMAP beads has been developed (Molecular Devices). Because it does not require specific antibodies for detection, it is widely applicable for screening for inhibitors of kinases and phosphatases (Gaudet et al., 2003). It is also very compatible with high-throughput robotics and can be applied to the 1536-well format. We have evaluated the performance of the plate::vision in FP reading in IMAP assays and found that it is comparable with that of other readers, for example, View-Lux, Envision (PerkinElmer), and Acquest (Molecular Devices). Because the plate::vision reads signals from the bottom of the microtiter plates, we have been using black glass-bottom plates to avoid the light polarization that occurs in polystyrene plates. More recently, we found that μ clear (black clear bottom) polystyrene microtiter plates from Greiner Bio-One can also be used for FP reading on the plate::vision. The Z' factor (about 0.8 for an assay with a 100-mP change) is very similar to that for glass plates.

Acknowledgments

We thank John L. Myer, Xiaolei Zhang, Kathie Huang, and Janna Holmgren for their contributions.

References

Gaudet, E. A., Huang, K.-S., Zhang, Y., Huang, W., Mark, D., and Sportman, J. R. (2003). A homogeneous fluorescence polarization assay adaptable for a range of protein serine/threonine and tyrosine kinases. *J. Biomol. Screen.* **8,** 164–175.

Huang, K.-S., Holmgren, J., Reik, L., Lucas-McGady, D., Robert, J., Liu, C.-M., and Levin, W. (2004). High-throughput methods for measuring heparanase activity and screening potential antimetastatic and anti-inflammatory agents. *Anal. Biochem.* **333,** 389–398.

Huang, K.-S., and Vassilev, L. T. (2005). High throughput screening for inhibitors of the Cks1-Skp2 interaction. *Methods Enzymol* **399,** 717–728.

Zhang, J.-H., Chung, T. D. Y., and Oldenburg, K. R. (1999). A simple statistical parameter for use in evaluation and validation of high throughput screening assays. *J. Biomol. Screen.* **2,** 67–73.

[31] Systems Cell Biology Based on High-Content Screening

By KENNETH A. GIULIANO, PATRICIA A. JOHNSTON, ALBERT GOUGH, and D. LANSING TAYLOR

Abstract

A new discipline of biology has emerged since 2004, which we call "systems cell biology" (SCB). Systems cell biology is the study of the living cell, the basic unit of life, an integrated and interacting network of genes, proteins, and myriad metabolic reactions that give rise to function. SCB takes advantage of high-content screening platforms, but delivers more detailed profiles of cellular systemic function, including the application of advanced reagents and informatics tools to sophisticated cellular models. Therefore, an SCB profile is a cellular systemic response as measured by a panel of reagents that quantify a specific set of biomarkers.

Background

Until recently, the focus in drug discovery and basic biomedical research has been on simplifying the complexity of the living human organism to individual genes, single metabolic pathways, single proteins, and one potential modulating molecule, such as a small chemical compound or bioproducts to regulate complex functions. This one gene, one protein, one external modulating treatment concept dominated the drug discovery process and much of basic biomedical research since the early 1990s. This paradigm grew out of the promise of the human genome project and the theory that identifying all protein-coding genes would lead to much more rapid discovery of cures for human disease (Collins *et al.*, 2003a,b; Phillips and Van Bebber, 2005).

Unfortunately, this reductionist approach to drug discovery has not delivered the promised efficiencies. Instead, the number of hypotheses validated in the clinic, as measured by novel product launches, has stagnated. In part, this failure is due to an overreliance on the continuity of the stepwise one gene, one target, one molecule reverse genetics approach described earlier. Multiple genes, including both protein coding and noncoding genes, regulate most cellular processes (Costa, 2005; Cummins *et al.*, 2006; Huttenhofer *et al.*, 2005; Volinia *et al.*, 2006), and proteins are part of complex, interacting pathways with extensive compensatory capacities. Therefore, even when a single small molecule or bioproduct has a specificity for binding to a single protein, the

METHODS IN ENZYMOLOGY, VOL. 414   0076-6879/06 $35.00
DOI: 10.1016/S0076-6879(06)14031-8

impact on cellular and, therefore, tissue and organ function is much more complex than expected (Melnick *et al.*, 2006). In addition, absolute specificity of small molecules and biologics is rarely demonstrated and "off-target" effects must be understood for both efficacy and potential toxicity (Hopkins and Groom, 2002; Yang *et al.*, 2004). Finally, most diseases are multifactorial where the disease phenotype arises from the dysregulation of multiple genes, pathways, and proteins (Glocker *et al.*, 2006; Jain *et al.*, 2005; Nadeau *et al.*, 2003; Tuomisto *et al.*, 2005).

Nevertheless, drug discovery has come to rely on high-throughput screening (HTS) technologies that focus on isolated targets as a means to boost productivity. Fortunately, technologies such as high-content screening (HCS) have been introduced that combine the efficiencies of HTS with superior biological context of the intact cell. HCS makes it possible to gain deeper knowledge about the effect of experimental compounds on target proteins within the context of the living cell, and thus HCS has been an important step in changing the paradigm for drug discovery. The founders and early employees of Cellomics, Inc. introduced the first HCS platform for drug discovery in 1997 and enabled the field of cellomics (see Giuliano *et al.*, 1997; Taylor, 2006; Taylor *et al.*, 2006). The discipline of cellomics can be described as the first step toward understanding how the output of the foundational "omics" technologies, i.e., genomics, proteomics, and metabolomics, function in a living environment (Giuliano *et al.*, 1997, 2003; Taylor, 2006). However, cellomics fall short of elucidating cellular activity as the function of a system. Taylor *et al.* (2006) summarize the application of HCS to drug discovery and systems biology.

Application of the principles of systems biology to cellomics approaches using HCS generates the discipline of systems cell biology (SCB). Systems biology is the study of a whole organism viewed as an integrated and interacting network of genes, proteins, and metabolic reactions that give rise to life (Aloy and Russell, 2005; Bugrim *et al.*, 2004; Butcher *et al.*, 2004; Hood and Perlmutter, 2004; Westerhoff and Palsson, 2004). Thus, the "omics" focuses on the parts, whereas systems biology deals with the functional complexity of the whole. The "omics" approach is an oversimplification, whereas systems biology at the organismal level is a relatively low-throughput and expensive process. SCB harnesses the higher throughput capacity of HCS technology while avoiding the expense and potential confounding species-related problems (worms, flies, fish) associated with traditional organism-based systems biology.

Systems cell biology is defined as the study of the living cell, the basic "unit of life," an integrated and interacting network of genes, proteins, and a myriad of metabolic reactions that give rise to function (Giuliano *et al.*, 2005; Taylor, 2006; Taylor and Giuliano, 2005). One can think of the cell as

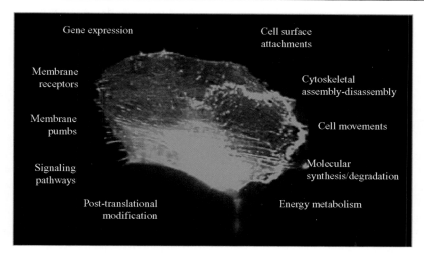

FIG. 1. Cells integrate many cellular processes, such as gene expression and energy metabolism, to yield normal functions. Diseases are due to the dysregulation of one or more of these cellular processes. Many of these processes share pathways, signals, and proteins and should be investigated as part of the cell system.

the simplest living "system" (Fig. 1): much less complex than a complete organism, but possessing the systemic functional complexity that facilitates a detailed understanding of the response of the integrated system to a perturbant such as a drug. Thus, the cell serves as a model in which to explore drug efficacy and potential toxicity, inexpensively and rapidly, across a broad range of doses and response times. An SCB profile is a cellular, systemic response as measured by a panel of reagents that quantify a specific set of biomarkers. The Federal Drug Administration (FDA) defines a biomarker as "a characteristic that is objectively measured and evaluated as an indicator of normal biologic or pathogenic processes." Selection of the optimal cell types, whether human primary cells or cell lines, coupled with focused SCB profiling assays, reagents, and informatics analysis, is the key to making systems cell biology a valuable approach to drug discovery and development, as well as biomedical research (Fig. 2).

SCB models elucidate the impact of small molecules and bioproducts on cell functions by characterizing the cellular responses with a large number (>4) of specific cellular parameters that yield a cell system profile (Giuliano *et al.*, 2005; Taylor and Giuliano, 2005). Multiplexing facilitates simultaneous capture of many parameters of cellular function, and informatics tools enable identification of the *interaction* of these parameters. These cellular responses and complex interactions can be related to a set of

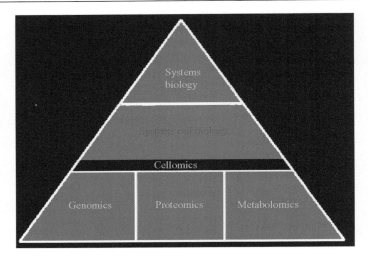

Fig. 2. Systems cell biology offers enough complexity, while allowing high-throughput and cost-effective assays. SCB and systems biology are based on the fundamental components of living systems represented by the "omics."

cellular function biomarkers: a unique cellular systemic profile. The cellular function biomarkers selected to represent the systemic response of particular cells are measured with a variety of fluorescence-based reagents. The response profile of a variety of human cells to perturbants is used to build a database that can be mined to relate the effect of "new" perturbants to those of known cellular mechanism (Fig. 3).

The Systems Cell Biology Toolbox

The toolbox for applying SCB profiling uses HCS platforms to yield data on the response of individual cells, as well as their subcellular responses. SCB leverages the power of HCS technology, which consists of readers, first-generation reagents, and basic informatics software, with proprietary panels of cellular function reagents, including more advanced reagents that can dynamically measure and manipulate cellular constituents. In addition, advanced SCB profiling informatics are used to generate new systems knowledge of disease, as well as profiles of the cellular systemic response to emerging drugs (Fig. 4).

HCS Instrumentation and Assay Design

There are now several commercially available HCS instruments and application software packages compatible with SCB so no platform

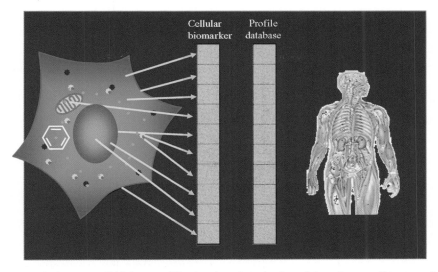

FIG. 3. Systems cell biology profiling involves the selection of the relevant cell types and cellular biomarkers of activity representing the cell as a system and not just a collection of targets and independent pathways. A database of responses can be used as a predictive tool in future profiles.

developments are required (Gough and Johnston, 2006). Although instrumentation and application software for HCS continues to evolve, there is adequate capability today for SCB.

HCS assays, the forerunner of SCB, begin with living cells that are treated with small molecules, bioproducts, and/or physical disruption. Cells are fixed at several time points and are subsequently labeled and read on an HCS reader. These assays reflect cellular activity at a moment in time and are known as "fixed end point assays." Sample preparation methods are available to automate all of these steps, making the assays fast and reproducible. Thus, the fixed end point approach can be a relatively high-throughput screening method, even for SCB. However, the time domain of the biology is limited to one time point. Therefore, the investigator must either create a time series by preparing multiple plates processed over time or initially define the half-time of some cellular process of interest and set the time of fixation accordingly.

Live cell HCS is possible with the integration of an on-board fluidics environmental chamber (Abraham *et al.*, 2004a,b; O'Brien and Haskins, 2006; O'Brien *et al.*, 2006). This can be accomplished with an integrated platform (for a review of HCS screening platforms, see Gough and

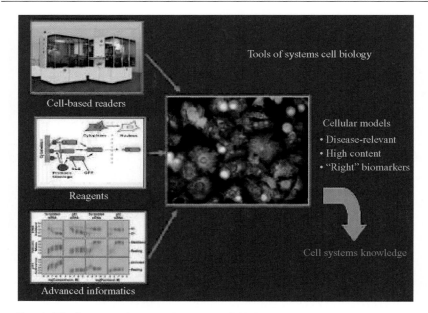

Fig. 4. The key new elements of systems cell biology profiling are the cellular models, advanced reagents, and informatics tools. High-content screening instrumentation and application software are available commercially and are useful now, although they will continue to evolve.

Johnston, 2006) or by applying add ons to a fixed end point reader. Live cell HCS assays or SCB profiles can be based on a single time point measurement using fluorescence-based reagents that are functional in living cells or the assays can be composed of multiple kinetic measurements that are made starting before the experimental treatment and continuing over time. Kinetic assays are useful in defining the half-times of specific biological processes and for critically defining the complex temporal–spatial dynamics of cells and their processes. These kinetics data are extremely valuable in the assay development phase.

Advanced Reagents to Measure and Manipulate Cellular Constituents

Tools that enable the selective and dynamic measurement and manipulation of cellular constituents and responses are important elements of the SCB approach to elucidating disease phenotypes (Table I). Multiple reagents exist for the measurement of up to six distinct parameters based on fluorescence in each well and more parameters can be analyzed by using more than one well to profile the same treatment; this is a key element of the SCB approach (Giuliano *et al.*, 2006).

TABLE I

SOME OF THE MAJOR REAGENTS USED TO MEASURE AND MANIPULATE CELLULAR CONSTITUENTS[a]

Selected reagents to measure	Selected reagents to manipulate
Antibodies	Directed siRNA to coding RNA
Fluorescent protein tagging	Random siRNA
Diffusible fluorescent probes	Gene switches
Fluorescent protein biosensors	Caged compounds
Positional biosensors	

[a] For a more extensive review, see Giuliano et al. (2006). These reagents are an integral part of the cellular models of disease analyzed by SCB profiling.

There are many choices of reagents to "measure" cellular responses (Giuliano et al., 2006). For example, antibodies are readily available for a large number of targets and have been multiplexed with up to four multi-colored fluorescent dyes for HCS assays. Panels of antibodies used to quantify SCB biomarkers in multiwell assays enable drug discovery applications from cellular models of disease to cytotoxicity profiling. Fluorescent proteins such as the green fluorescent protein from the jellyfish, as well as alternative versions, can be used to express fluorescently tagged target proteins in cells. Diffusible fluorescent probes are available that can label specific organelles and monitor a variety of cellular metabolites and ions. Some of the diffusible probes can be fixed for end point measurements. Fluorescent protein bio-sensors can be created to detect numerous protein activities through changes in the fluorescence properties (e.g., intensity, spectral shifts, aniso-tropy). Positional biosensors have been developed to detect some protein activities through changes in the distribution to different intracellular com-partments. It is anticipated that advances in both spectral selection technol-ogies and fluorescence-based reagents will extend the number of cellular parameters that can be selectively measured in the near future.

The number of reagents used to manipulate cell constituents is also growing rapidly (Giuliano et al., 2006). For example, directed siRNAs have become important tools in "knocking down" the expression of specific proteins to demonstrate the role of the protein in particular pathways and disease states (Carpenter and Sabatini, 2004; Huppi et al., 2005; Mattick, 2004; Sioud, 2004). The emerging importance of noncoding RNAs such as micro-RNAs in regulating cellular processes has created the oppor-tunity to correlate any RNA, coding or noncoding, with a particular cellu-lar process. Gene switches can be used to regulate the expression of either proteins or siRNAs (Gupta et al., 2004; Karzenowski et al., 2005; Kumar et al., 2004; Ventura et al., 2004). Multiple gene switches in the same cells permit more sophisticated manipulations of cellular processes

(Kumar *et al.*, 2004). Caged compounds are molecules whose normal activity is under control of illumination at a specific wavelength of light to release the activity within cells in time and space. Several classes of this type of manipulation reagent have been reviewed (Giuliano *et al.*, 2006). SCB applied to cellular models of disease integrates the "measure and manipulate" paradigm to better define mechanisms of disease.

Informatics for Systems Cell Biology

Informatics is an essential component of the automated work flow of HCS (Dunlay *et al.*, 2006; Taylor *et al.*, 2001). Commercial platforms for HCS typically include informatics software for extracting usually up to four channels of specific fluorescence measurements, but also dozens of cellular, morphometric features from each image. In addition, these basic informatics tools also include a database for managing the hundreds of images and millions of data points generated from each plate and client tools for reviewing the images and data resulting from a screen. These tools are adequate for HCS, which usually involves measuring and comparing a few features associated with a particular target (Giuliano *et al.*, 1997).

However, although SCB employs the existing toolbox of HCS informatics as a starting point, the requirements for analyzing and mining data extend well beyond the present generation of HCS informatics tools. SCB makes use of panels of molecularly specific features based on fluorescence with broad coverage of cell functions and a combination of multiplexed single cell features, as well as cell population features. In addition to standard HCS informatics, the power of SCB is enhanced by multiplexing of cellular measurements; correlating measurements within single cells; analysis of subpopulations of cells; clustering and pattern matching of multifactorial data sets; and modeling cellular systems. Recent developments in automated cell analysis and systems biology are providing some additional tools to fill this gap, including some efforts to develop more comprehensive data management systems for HCS and SCB (Goldberg *et al.*, 2005; Parvin and Callahan, 2002; Parvin *et al.*, 2003).

SCB starts with rich, highly multiplexed biomarkers that characterize a broad range of cellular functions (Giuliano *et al.*, 2005; Taylor and Giuliano, 2005). Algorithms in the portfolio of bioapplications available from HCS vendors provide some of this information but are often limited in scope to analysis of a particular cell target. There is also an unmet need for combining measurements from different algorithms in order to correlate a diversity of features in single cells. Improved methods of image analysis, such as pattern analysis, can be applied more generally to identify target localization to specific subcellular compartments, such as the nucleus, endoplasmic reticulum, golgi, and others, while requiring only a single fluorescent channel (Boland and Murphy, 2001; Murphy, 2005). In HCS, it is often sufficient to reduce the cell population response to a single value, such as the mean or median of the measurement. However, it is well known in cell cycle analysis that additional information is available by analyzing the population and segmenting it into subpopulations, which may respond differentially (Bocker et al., 1996; Giuliano et al., 2004). It has also been shown that direct comparison of population distributions, for example, by making use of the Kolomogorov–Smirnov (KS) test, provides a much more sensitive measure of the shift in a population response than the use of standard statistical parameters, such as the mean or median of the distribution (Giuliano et al., 2005; Peacock, 1983).

In addition to the tools required to generate the measurement data needed for systems cell analysis, tools are required for the analysis of these rich data sets. Network modeling methods that have been applied successfully to proteomic analysis (Janes and Lauffenburger, 2006), and cell signaling networks (Araujo and Liotta, 2006) are also of use in extracting more information from SCB data. Eventually, these models will take SCB analysis beyond the value of multiparameter profiles to a tool for diagnostics and use as surrogate end points for therapies (Danna and Nolan, 2006).

One representation of biomarker data resulting from SCB is a feature vector or "fingerprint" of the cellular responses across many assays for a specific compound or treatment. These fingerprints can then be mined to identify correlations between response profiles, which provide a means to predict functional effects, including a detailed mode of action. For example, even using basic HCS assays to classify a library of known toxic compounds has shown that compound clusters correlate with known modes of action (O'Brien and Haskins, 2006; O'Brien et al., 2006; Tencza and Sipe, 2004). In another example, about 70,000 compounds have been screened in vitro against 60 human cancer cell lines from different organs (Shi et al., 2000). Each compound is represented by a vector (e.g., "fingerprint") of 60 anti-cancer activity values, characteristic of the compound activity (Rabow et al., 2002). Although not an SCB application, since each assay is only a single measurement, clearly the data model and informatics approaches are similar, and the study illustrates the importance of measurements that span cell lines and cell types as well as cell functions. As a final example of systems cell analysis, compound effects on inflammation have been measured in multiple assays and shown to cluster by mode of action (Berg et al., 2006). Clearly some of the basic informatics tools required for SCB are already available and are being applied to related challenges, but it is equally clear that there is a need for more sophisticated tools, especially for the integration of these tools into a platform that will make SCB as accessible and useful as HCS.

Example Systems Cell Biology Profile

Cancer Model Using the Human Lung Carcinoma Cell Line (A549)
Expressing Wild-Type p53

In anticancer drug discovery, the myriad of cellular events regulated by the p53 tumor suppressor protein present an invaluable set of potential HCS targets (Giuliano *et al.*, 2004). That the mutation or deletion of p53 protein in many cancer cell types is often a determining factor in the chemotherapeutic outcome (Blagosklonny and Pardee, 2001; Borbe *et al.*, 1999; Osaki *et al.*, 2000) emphasizes the need for information on the cellular and molecular activities regulated by the p53 protein and the effect drugs have on these interrelationships. Thus, there is a need for new approaches to rapidly and precisely modulate components of the p53 signaling pathway, as well as complementary SCB methods to dissect the network of cellular and molecular activities regulated by this important tumor suppressor protein.

Therefore, the goal is to create a cell model where different components of the p53 pathway are manipulated so that the expression levels can be regulated to monitor the impact on the downstream cellular events and any off-target and/or compensatory effects. Giuliano *et al.* (2004) reported on the effects that p53 "knockdown" with RNAi has on the response of A549 cells to multiple anticancer agents. In this new example, p53 pathway regulation (Fig. 5) was "manipulated" by controlling the expression level of the p53-modulating protein HDM2 with a gene switch. A profile of the cellular response was built by "measuring" a panel of SCB biomarker features within the p53 pathway or other possible events either downstream (e.g., apoptosis, cell growth and division, cytoskeletal reorganization, and organelle function) or upstream (e.g., DNA damage) (Fig. 5). A single vector RheoSwitch (RheoGene; Norristown, PA) was used to regulate expression of a fluorescently labeled HDM2 protein in A549 cells using multiple concentrations of inducer molecule. Figure 6 shows that the expression level of fluorescent HDM2 was dependent on the concentration of the inducer molecule used to treat the cells as well as the total time of induction.

A fixed end point SCB assay was designed to generate a profile that elucidates the cell system response to upregulating HDM2 in the absence of any other manipulation. In addition to the quantification of HDM2 expression in each cell, Table II shows the additional cellular function biomarkers measured to build the profile. The measurements were chosen to look both at specific downstream activities and more general cellular processes. Example images of cells in which SCB reagents were used to measure and manipulate activity are shown in Fig. 7.

The considerable number of SCB measurements made after treatment of cells with multiple inducer molecule concentrations produced a large

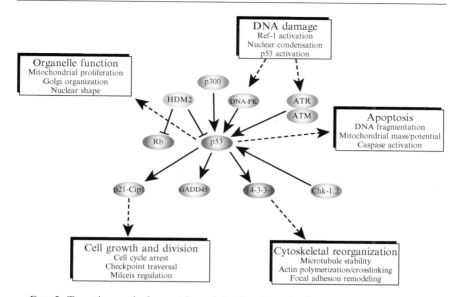

FIG. 5. Targeting a single protein activity for drug development, such as the tumor suppressor protein p53 in cancer (center), is challenging. Within the cell, this target protein and signaling pathway is part of an integrated and interacting network. Therefore, an SCB profiling approach is required to define efficacy, as well as potential off-target toxicity and compensatory activities. Thus, a cell model can be constructed where specific components of the p53 pathway can have the expression levels regulated by gene switches, short interfering RNAs, or combinations of both, and specific protein–protein interactions, molecular processes, and phenotypic responses can be measured as part of the SCB profile.

and complex SCB profile. As shown before, automated hierarchical clustering analysis based on KS population dissimilarity analysis provided an ideal first step in deciphering profiling data (Giuliano *et al.*, 2004, 2005). Figure 8 shows the clustered profiling results in the form of a heat map. Cell population responses that were not significantly different from control populations (e.g., KS <0.2) were encoded in the heat map using blue and black shades. Significant cell population responses induced by HDM2 (e.g., KS >0.2) were encoded in the heat map using yellow and red shades. In the overall response, there were three relatively distinct clusters: high responders (left), low responders (center), and intermediate responders (right).

It was expected (Fig. 8; right arrow) that the elevation of HDM2 would decrease the cellular level of p53, which did occur. However, simply raising the cellular concentration of HDM2 about five times had unexpected effects, such as stabilizing microtubules, increasing PI3 kinase activation, and increasing the level of DNA repair and redox of DNA-binding

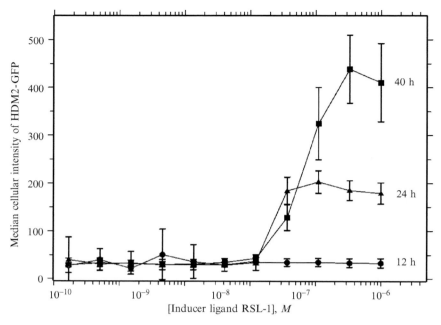

FIG. 6. Regulation of the expression level of a fluorescent-hdm2 under the control of a gene switch showing the expression levels after 12, 24, and 40 h of stimulation with the inducer.

proteins. To enable the building of connections between the cellular processes measured in the profile, we relied on the use of cell maps to visualize how two HCS feature measurements were related in the same cells under different conditions (Giuliano *et al.*, 2004, 2005). In this example, heat map data in Fig. 8 showed that the changes in phosphohistone H3 levels and the intracellular distribution of APE/Ref-1 as a function of HDM2 expression level were similar enough for them to be clustered together. Using the cell maps shown in Fig. 9, we could demonstrate the relationship between these two markers of cell cycle regulation and the DNA damage–oxidative stress response. The cell maps show that as the expression level of HDM2 increased in response to RSL-1, the phosphohistone H3 level decreased homogeneously to a lower level (vertical shift down) while the predominant nuclear localization of APE/Ref-1 increased (horizontal shift right) consistent with some common regulatory unit that coordinates the two processes. Thus, this brief example demonstrates the potential of an SCB approach to build new knowledge on the regulatory connections between cellular processes through the selective manipulation and measurement of cellular constituents.

TABLE II

FEATURES MEASURED IN THE SCB PROFILE AND THEIR ROLES IN THE REGULATION OF CELLULAR SYSTEMS

Measurement	Abbreviation for heat map	Cellular system regulation
Total nuclear intensity and variation of Hoechst 33342 label	dna, nu4	DNA content: cell cycle arrest DNA degradation
Projected nuclear area and shape defined by Hoechst 33342 label	nu1, nu2, nu3	Nuclear condensation or fragmentation
Nuclear intensity of phosphohistone H3 label	ph3	G_2/M cell cycle transition
Cytoplasmic intensity of α-tubulin label after detergent extraction	tub	Microtubule stability
Cytolasmic vinculin label at substrate	vin	Focal adhesion remodeling
Nuclear/cytoplasm ratio of p53 label	p53	Tumor suppressor involved in DNA damage response. HDM2 inhibits p53 upon complex formation
Nuclear/cytoplasm ratio of p21 label	p21	Tumor suppressor inhibitor of cell cycle progression. Active p53 upregulates p21 expression
Cytoplasmic intensity of phospho-Akt label	akt	Activated Akt promotes cell survival by causing the inhibition of apoptosis targets
Nuclear intensity of APE/Ref-1 label	ape	APE/Ref-1 is activated upon DNA damage or oxidative stress
Cytoplasmic level and organization of 14-3-3 protein	x43	14-3-3 proteins regulate cell survival, apoptosis, proliferation, and checkpoint control through phosphorylation-dependent protein–protein interactions
Organization and intensity of golgi and mitochondrial organelles	gol, mit	Organelles such as mitochondria play specific roles in many cellular processes, including apoptosis

Summary and Conclusions

The development of HCS was a major step in improving the drug discovery and development process, as well as creating a complementary high-throughput method for research imaging microscopy. The integrated use of HCS, as well as semiautomated microscopy and high-performance research imaging techniques, will continue to be a powerful set of tools used to define cellular functions of cell constituents and whole cells. SCB takes advantage of basic HCS platforms, while creating a cell "systems" response instead of focusing on one target or a few related cell factors

HDM2-GFP

Microtubule
cytoskeleton

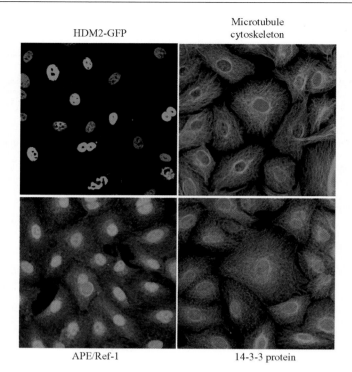

APE/Ref-1

14-3-3 protein

FIG. 7. Example images of cellular feature labeling for SCB analysis. HDM2-GFP is shown with expressing cells (green) overlaid with Hoechst 33342-labeled nuclei (blue). An antibody to α-tubulin was used to measure changes in the stability of the microtubule cytoskeleton. Antibodies against APE/Ref-1 labeled the dual function enzyme that takes part in DNA repair as well as regulating the redox state of DNA-binding proteins. Changes in the nuclear distribution of APE/Ref-1 were used to measure the extent of DNA-repair pathway activation. The 14-3-3 protein, through its association with the cytoskeleton, regulates the localization and activity of other proteins (e.g., Cdc25). The phosphorylation level of the cytoplasmically localized 14-3-3 protein was used as a measure of its protein–protein interaction activity.

in simple assays. More advanced classes of reagents that measure and manipulate cellular constituents, as well as more integrated and powerful informatics tools, will be the driving forces of SCB.

Prospectus

Systems cell biology will be applied to the whole continuum of the drug discovery and development process. Cellular models of disease are being created to better understand the selected targets, along with the systems

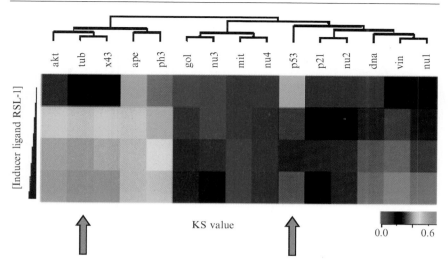

Pi3 kinase activation
MT stabilization
Regulation of distribution of key proteins (e.g., cdc25)
DNA repair activation and redox of DNA binding Proteins

FIG. 8. An SCB profile of a cellular model of cancer where a human lung cancer cell line was transduced with a fluorescently tagged protein, HDM2, a modulator of p53 that was regulated with the RheoSwitch gene switch. The level of expression of this tagged HDM2 was regulated by increasing the concentration of the inducer ligand (increased concentration of expression on the y axis). This allowed the "manipulation" of HDM2 levels in cells. The cellular systems response was monitored by measuring the KS statistics of 15 distinct cellular parameters (see abbreviations of parameters on the top with clustering) in a fixed end-point assay using high-content screening. The KS statistics were color coded to show those cellular parameters that did not change (blue), changed moderately (black), and changed extensively (yellow-red).

response of the cell models to optimize selection of lead compounds. In addition, cytotoxicity profiling, using the SCB approach, will provide a critical "filter" for prioritizing lead compounds and could evolve into a predictive tool. Patient sample profiling at the cell and tissue levels will become an important approach in defining patient subpopulations for more focused clinical trials and could also become a valuable cell-based diagnostic method. SCB will also have a major impact on basic biomedical research where complex processes such as the differentiation of stem cells are clearly a systems challenge. A major effort will be focused on creating powerful reagents that measure and manipulate specific biochemical/molecular events in a reversible manner, such as protein–protein

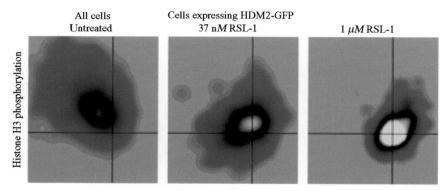

FIG. 9. Cell maps demonstrate the relationships between cellular processes within the same cells. Cell maps were prepared from SCB profiling data as described (Giuliano *et al.*, 2004, 2005). HCS measurements of histone H3 phosphorylation and APE/Ref-1 intracellular distribution made in the same cells were plotted against each other at various HDM2 expression levels that were induced with RSL-1. The fiduciary lines, which are constant in each map, show that cells exhibiting decreased phosphohistone H3 levels were, in large part, the same cells that exhibited predominantly nuclear-localized APE/Ref-1.

interactions in relation to other dynamic cellular processes. In addition, a new generation of informatics tools will allow understanding of the complex systems responses of cells to various manipulations.

Acknowledgments

The authors thank Drs. Dean Cress and Prasanna Kumar from RheoGene, Inc. for their collaborative efforts in gene switch technology development.

References

Abraham, V. C., Samson, B., Lapets, O., and Haskins, J. R. (2004a). Automated classfication of individual cellular responses across multiple targets. *Preclinica* **2**, 349–355.

Abraham, V. C., Taylor, D. L., and Haskins, J. R. (2004b). High content screening applied to large-scale cell biology. *Trends Biotechnol.* **22**, 15–22.

Aloy, P., and Russell, R. B. (2005). Structure-based systems biology: A zoom lens for the cell. *FEBS Lett.* **579**, 1854–1858.

Araujo, R. P., and Liotta, L. A. (2006). A control theoretic paradigm for cell signaling networks: A simple complexity for a sensitive robustness. *Curr. Opin. Chem. Biol.* **10**, 81–87.

Berg, E. L., Kunkel, E. J., Hytopoulos, E., and Plavec, I. (2006). Characterization of compound mechanisms and secondary activities by BioMAP analysis. *J. Pharmacol. Toxicol. Methods* **53**, 67–74.

Blagosklonny, M. V., and Pardee, A. B. (2001). Exploiting cancer cell cycling for selective protection of normal cells. *Cancer Res.* **61,** 4301–4305.

Bocker, W., Gantenberg, H. W., Muller, W. U., and Streffer, C. (1996). Automated cell cycle analysis with fluorescence microscopy and image analysis. *Phys. Med. Biol.* **41,** 523–537.

Boland, M. V., and Murphy, R. F. (2001). A neural network classifier capable of recognizing the patterns of all major subcellular structures in fluorescence microscope images of HeLa cells. *Bioinformatics* **17,** 1213–1223.

Borbe, R., Rieger, J., and Weller, M. (1999). Failure of taxol-based combination chemotherapy for malignant glioma cannot be overcome by G2/M checkpoint abrogators or altering the p53 status. *Cancer Chemother. Pharmacol.* **44,** 217–227.

Bugrim, A., Nikolskaya, T., and Nikolsky, Y. (2004). Early prediction of drug metabolism and toxicity: Systems biology approach and modeling. *Drug Discov. Today* **9,** 127–135.

Butcher, E. C., Berg, E. L., and Kunkel, E. J. (2004). Systems biology in drug discovery. *Nature Biotechnol.* **22,** 1253–1259.

Carpenter, A. E., and Sabatini, D. M. (2004). Systematic genome-wide screens of gene function. *Nature Rev. Genet.* **5,** 11–22.

Collins, F. S., Green, E. D., Guttmacher, A. E., and Guyer, M. S. (2003a). A vision for the future of genomics research. *Nature* **422,** 835–847.

Collins, F. S., Morgan, M., and Patrinos, A. (2003b). The human genome project: Lessons from large-scale biology. *Science* **300,** 286–290.

Costa, F. F. (2005). Non-coding RNAs: New players in eukaryotic biology. *Gene* **357,** 83–94.

Cummins, J. M., He, Y., Leary, R. J., Pagliarini, R., Diaz, L. A., Jr., Sjoblom, T., Barad, O., Bentwich, Z., Szafranska, A. E., Labourier, E., Raymond, C. K., Roberts, B. S., Juhl, H., Kinzler, K. W., Vogelstein, B., and Velculescu, V. E. (2006). The colorectal microRNAome. *Proc. Natl. Acad. Sci. USA* **103,** 3687–3692.

Danna, E. A., and Nolan, G. P. (2006). Transcending the biomarker mindset: Deciphering disease mechanisms at the single cell level. *Curr. Opin. Chem. Biol.* **10,** 20–27.

Dunlay, R. T., Czekalski, W. J., and Collins, M. A. (2006). Overview of informatics for HCS. *In* "High Content Screening: A Powerful Approach to Systems Cell Biology and Drug Discovery" (D. L. Taylor, J. R. Haskins, and K. A. Giuliano, eds.). pp. 269–280. Humana Press, Totowa, NJ.

Giuliano, K. A., Chen, Y. T., and Taylor, D. L. (2004). High-content screening with siRNA optimizes a cell biological approach to drug discovery: Defining the role of p53 activation in the cellular response to anticancer drugs. *J. Biomol. Screen* **9,** 557–568.

Giuliano, K. A., Cheung, W. S., Curran, D. P., Day, B. W., Kassick, A. J., Lazo, J. S., Nelson, S. G., Shin, Y., and Taylor, D. L. (2005). Systems cell biology knowledge created from high content screening. *Assay Drug Dev. Technol.* **3,** 501–514.

Giuliano, K. A., DeBiasio, R. L., Dunlay, R. T., Gough, A., Volosky, J. M., Zock, J., Pavlakis, G. N., and Taylor, D. L. (1997). High-content screening: A new approach to easing key bottlenecks in the drug discovery process. *J. Biomol. Screen* **2,** 249–259.

Giuliano, K. A., Haskins, J. R., and Taylor, D. L. (2003). Advances in high content screening for drug discovery. *ASSAY Drug Dev. Technol.* **1,** 565–577.

Giuliano, K. A., Taylor, D. L., and Waggoner, A. S. (2006). Reagents to measure and manipulate cell functions. *In* "High Content Screening: A Powerful Approach to Systems Cell Biology and Drug Discovery" (D. L. Taylor, J. R. Haskins, and K. A. Giuliano, eds.). pp. 141–164. Humana Press, Totowa, NJ.

Glocker, M. O., Guthke, R., Kekow, J., and Thiesen, H. J. (2006). Rheumatoid arthritis, a complex multifactorial disease: On the way toward individualized medicine. *Med. Res. Rev.* **26,** 63–87.

Goldberg, I. G., Allan, C., Burel, J. M., Creager, D., Falconi, A., Hochheiser, H., Johnston, J., Mellen, J., Sorger, P. K., and Swedlow, J. R. (2005). The Open Microscopy Environment (OME) data model and XML file: Open tools for informatics and quantitative analysis in biological imaging. *Genome Biol.* **6,** R47.

Gough, A. H., and Johnston, P. A. (2006). Requirements, features and performance of high content screening platforms. *In* "High Content Screening: A Powerful Approach to Systems Cell Biology and Drug Discovery" (D. L. Taylor, J. R. Haskins, and K. A. Giuliano, eds.). pp. 41–62. Humana Press, Totowa, NY.

Gupta, S., Schoer, R. A., Egan, J. E., Hannon, G. J., and Mittal, V. (2004). Inducible, reversible, and stable RNA interference in mammalian cells. *Proc. Natl. Acad. Sci. USA* **101,** 1927–1932.

Hood, L., and Perlmutter, R. M. (2004). The impact of systems approaches on biological problems in drug discovery. *Nature Biotechnol.* **22,** 1215–1217.

Hopkins, A. L., and Groom, C. R. (2002). Opinion: The druggable genome. *Nature Rev. Drug Discov.* **1,** 727–730.

Huppi, K., Martin, S. E., and Caplen, N. J. (2005). Defining and assaying RNAi in mammalian cells. *Mol. Cell* **17,** 1–10.

Huttenhofer, A., Schattner, P., and Polacek, N. (2005). Non-coding RNAs: Hope or hype? *Trends Genet.* **21,** 289–297.

Jain, S., Wood, N. W., and Healy, D. G. (2005). Molecular genetic pathways in Parkinson's disease: A review. *Clin. Sci. (Lond.)* **109,** 355–364.

Janes, K. A., and Lauffenburger, D. A. (2006). A biological approach to computational models of proteomic networks. *Curr. Opin. Chem. Biol.* **10,** 73–80.

Karzenowski, D., Potter, D. W., and Padidam, M. (2005). Inducible control of transgene expression with ecdysone receptor: Gene switches with high sensitivity, robust expression, and reduced size. *Biotechniques* **39,** 191–192, 194, 196 passim.

Kumar, M. B., Potter, D. W., Horman, R. E., Edwards, A., Tice, C. M., Smith, H. C., Dipietro, M. A., Polley, M., Lawless, M., Wolohan, P. R., Kethidi, D. R., and Palli, S. R. (2004). Highly flexible ligand-binding pocket of ecdysone receptor: A single amino acid change leads to discrimination between two groups of nonsteroidal ecdysone agonists. *J. Biol. Chem.* **279,** 27211–27218.

Mattick, J. S. (2004). RNA regulation: A new genetics? *Nature Rev. Genet.* **5,** 316–323.

Melnick, J. S., Janes, J., Kim, S., Chang, J. Y., Sipes, D. G., Gunderson, D., Jarnes, L., Matzen, J. T., Garcia, M. E., Hood, T. L., Beigi, R., Xia, G., Harig, R. A., Asatryan, H., Yan, S. F., Zhou, Y., Gu, X. J., Saadat, A., Zhou, V., King, F. J., Shaw, C. M., Su, A. I., Downs, R., Gray, N. S., Schultz, P. G., Warmuth, M., and Caldwell, J. S. (2006). An efficient rapid system for profiling the cellular activities of molecular libraries. *Proc. Natl. Acad. Sci. USA* **103,** 3153–3158.

Murphy, R. F. (2005). Location proteomics: A systems approach to subcellular location. *Biochem. Soc. Trans.* **33,** 535–538.

Nadeau, J. H., Burrage, L. C., Restivo, J., Pao, Y. H., Churchill, G., and Hoit, B. D. (2003). Pleiotropy, homeostasis, and functional networks based on assays of cardiovascular traits in genetically randomized populations. *Genome Res.* **13,** 2082–2091.

O'Brien, P. J., and Haskins, J. R. (2006). *In vitro* cytotoxicity assessement. *In* "High Content Screening: A Powerful Approach to Systems Cell Biology and Drug Discovery" (D. L. Taylor, J. R. Haskins, and K. A. Giuliano, eds.). pp. 415–426. Humana Press, Totowa, NJ.

O'Brien, P. J., Irwin, W., Diaz, D., Howard-Cofield, E., Krejsa, C. M., Slaughter, M. R., Gao, B., Kaludercic, N., Angeline, A., Bernardi, P., Brain, P., and Hougham, C. (2006). High concordance of drug-induced human hepatotoxicity with *in vitro* cytotoxicity measured in a novel cell-based model using high content screening. *Arch. Toxicol.* In press.

Osaki, S., Nakanishi, Y., Takayama, K., Pei, X. H., Ueno, H., and Hara, N. (2000). Alteration of drug chemosensitivity caused by the adenovirus-mediated transfer of the wild-type p53 gene in human lung cancer cells. *Cancer Gene Ther.* **7**, 300–307.

Parvin, B., and Callahan, D. E. (2002). BioSig: An informatics framework for representing the physiological responses of living cells. *BioSilico* **1**.

Parvin, B., Yang, Q., Fontenay, G., and Barcellos-Hoff, M. H. (2003). BioSig: An imaging bioinformatics system for phenotypic analysis. *IEEE Trans. Systems Man Cybernet. B* **33**, 814–824.

Peacock, J. A. (1983). Two-dimensional goodness-of-fit testing in astronomy. *Month. Notices R. Astronom. Soc.* **202**, 615–627.

Phillips, K. A., and Van Bebber, S. L. (2005). Measuring the value of pharmacogenomics. *Nature Rev. Drug Discov.* **4**, 500–509.

Rabow, A. A., Shoemaker, R. H., Sausville, E. A., and Covell, D. G. (2002). Mining the National Cancer Institute's tumor-screening database: Identification of compounds with similar cellular activities. *J. Med. Chem.* **45**, 818–840.

Shi, L. M., Fan, Y., Lee, J. K., Waltham, M., Andrews, D. T., Scherf, U., Paull, K. D., and Weinstein, J. N. (2000). Mining and visualizing large anticancer drug discovery databases. *J. Chem. Inf. Comput. Sci.* **40**, 367–379.

Sioud, M. (2004). Therapeutic siRNAs. *Trends Pharmacol. Sci.* **25**, 22–28.

Taylor, D. L. (2006). Insights into high content screening: Past, present and future. *In* "High Content Screening: A Powerful Approach to Systems Cell Biology and Drug Discovery" (D. L. Taylor, J. R. Haskins, and K. A. Giuliano, eds.). pp. 3–18. Humana Press, Totowa, NJ.

Taylor, D. L., and Giuliano, K. A. (2005). Multiplexed high content screening assays create a systems cell biology approach to drug discovery. *Drug Disc. Today Technol.* **2**, 149–154.

Taylor, D. L., Haskins, J. R., and Giuliano, K. A. (2006). High content screening: A powerful approach to systems cell biology and drug discovery. *In* "Methods in Molecular Biology" (J. Walker, Series, ed.). pp. 1–434. Humana Press, Totowa, NJ.

Taylor, D. L., Woo, E. S., and Giuliano, K. A. (2001). Real-time molecular and cellular analysis: The new frontier of drug discovery. *Curr. Opin. Biotechnol.* **12**, 75–81.

Tencza, S. B., and Sipe, M. A. (2004). Detection and classification of threat agents via high-content assays of mammalian cells. *J. Appl. Toxicol.* **24**, 371–377.

Tuomisto, T. T., Binder, B. R., and Yla-Herttuala, S. (2005). Genetics, genomics and proteomics in atherosclerosis research. *Ann. Med.* **37**, 323–332.

Ventura, A., Meissner, A., Dillon, C. P., McManus, M., Sharp, P. A., Van Parijs, L., Jaenisch, R., and Jacks, T. (2004). Cre-lox-regulated conditional RNA interference from transgenes. *Proc. Natl. Acad. Sci. USA* **101**, 10380–10385.

Volinia, S., Calin, G. A., Liu, C. G., Ambs, S., Cimmino, A., Petrocca, F., Visone, R., Iorio, M., Roldo, C., Ferracin, M., Prueitt, R. L., Yanaihara, N., Lanza, G., Scarpa, A., Vecchione, A., Negrini, M., Harris, C. C., and Croce, C. M. (2006). A microRNA expression signature of human solid tumors defines cancer gene targets. *Proc. Natl. Acad. Sci. USA* **103**, 2257–2261.

Westerhoff, H. V., and Palsson, B. O. (2004). The evolution of molecular biology into systems biology. *Nature Biotechnol.* **22**, 1249–1252.

Yang, Y., Blomme, E. A., and Waring, J. F. (2004). Toxicogenomics in drug discovery: From preclinical studies to clinical trials. *Chem. Biol. Interac.* **150**, 71–85.

[32] Digital Autofocus Methods for Automated Microscopy

By Feimo Shen, Louis Hodgson, and Klaus Hahn

Abstract

Automatic focusing of microscope images is an essential part of modern high-throughput microscopy. This chapter describes implementation of a robust autofocus system appropriate for using either air or oil immersion objectives in robotic imaging. Both hardware and software algorithms are described, and caveats of using viscous immersion media with multifield scanning are detailed.

Introduction

In traditional microscopy, it can be difficult to maintain focus when using objectives with high numerical aperture (NA) and low working distances. This problem is compounded during extended time-lapse studies when temperature fluctuations affect focus. Focus is especially critical in time-lapse high-content screening because this requires scanning of many fields of view (FOV) over large areas (Bajaj et al., 2000; Shen et al., 2006). High NA objectives characteristically have a very low depth of field (DOF); only a very thin section of the specimen falls in focus in the acquired image. In order to solve the problem of focal drift, a fast and robust autofocus method is required.

Potential sources of focus fluctuations are (1) the substrate supporting the specimen is not flat; (2) thermal fluctuation causes the distance between the sample and the optical elements to drift over time; and (3) fluid flow and exchange of medium in cell perfusion experiments cause vibration. Specimen supports are typically not flat. For example, the height of a 9 × 15-mm^2 glass coverslip can vary 6 μm across the diagonal (Bravo-Zanoguera et al., 1998). The specimen support used in larger area scans can vary even more, especially when level mounting of the specimen is difficult or there is a misalignment of the scanning stage. For microscopes that are aligned carefully, thermal fluctuations are the major culprit. The focal drift associated with temperature fluctuation can be as much as 1 μm per degree Celsius (data not shown). In live-cell perfusion experiments, microfluctuations in temperature additionally affect the focus position. Autofocus is essential in time-lapse studies to overcome these problems.

METHODS IN ENZYMOLOGY, VOL. 414
Copyright 2006, Elsevier Inc. All rights reserved.

0076-6879/06 $35.00
DOI: 10.1016/S0076-6879(06)14032-X

Commercially available options for autofocus can be categorized into three groups: (1) hardware-based solutions, (2) software-based solutions, and (3) fully integrated high-throughput microscopy (HTM) platforms. Hardware-based solutions that use RS-170 video signals include the MAC series of video autofocus processors from Ludl Electronic Products Ltd. (Hawthorne, NY) and ASI's video autofocus (Applied Scientific Instrumentation, Inc., Eugene, OR). Most makes of microscopy integration software include software-based autofocus in their packages. Example systems that allow a certain level of user customization include MetaMorph 6.3 (Molecular Devices Corp, Sunnyvale, CA) and SlideBook 4.1 (Intelligent Imaging Innovations, Inc., Denver, CO). Integrated HTM platforms such as the KineticScan HCS Reader (Cellomics, Inc., Pittsburgh, PA), Pathway (BD Biosciences, Rockwille, MD), and CellLab IC 100 (Beckman Coulter, Inc., Fullerton, CA) are easy-to-use, black-box systems that have autofocus capability.

All these premanufactured solutions are not readily adapted to many individual user requirements. Perhaps most importantly, they are ill suited to high-resolution imaging of dim fluorescence because they do not provide for using higher NA oil immersion lenses. Most can accommodate only dry objective lenses up to $40\times$, 0.95 NA. This chapter details the construction of an autofocus system that overcomes this difficulty using off-the-shelf materials and methods that can be engineered into any microscope for automation.

In the system described here, the hardware acquires a z stack of images for each focus attempt. Digital image processing software determines the high-frequency content of each image in the stack and uses the results to determine a "focus index" for each image. The sharpest image is acquired at the best focus position calculated from the focus indices.

Hardware

The components needed to carry out digital autofocus in a microscopy scan are as follows:

1. A microscope capable of differential interference contrast (DIC) imaging.
2. Hardware to precisely control the spacing between the objective and the specimen.
3. A motorized x–y stage with linear encoding feedback accuracy.
4. An equivalent of a computer with a Pentium 4 2.4-GHz central processing unit (CPU), 2-GB random access memory (RAM), and minimum of two peripheral component interconnect (PCI) expansion slots.
5. A scientific grade digital camera that can interface with a computer.

Figure 1 shows an example of an assembled system in our laboratory. We have an Olympus Model IX71 inverted microscope (Olympus America, Inc., Melville, NY) on which we have fitted the autofocus module. This is an infinity-corrected microscope equipped to perform DIC imaging. DIC imaging is preferred over other means of generating contrast because it does not attenuate fluorescence, is easy to set up, and has greater high-frequency content (Inoue and Spring, 1997). The latter is very important in our resolution-based autofocus. In contrast to DIC, bright-field images do

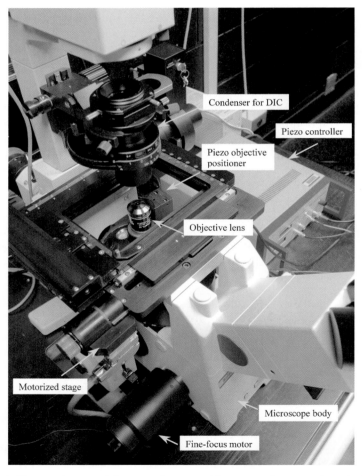

Fig. 1. The autofocus module on a microscope. The camera is mounted on the bottom port and is hidden from view.

not have sufficient contrast to allow autofocus because most live specimens are transparent and cannot be stained. Although phase-contrast imaging produces good contrast, phase objective lenses have a phase plate in the back focal plane that attenuates light when acquiring fluorescence images. A DIC-enabled microscope is ideal for our method because it both provides sufficient contrast and brings more high-frequency content in the spatial domain of the image.

For micropositioning the objective lens in its axial, or z position, various solutions are possible. We have used a piezo-electric transducer that attaches to the objective turret of the microscope below the objective lens. This Physik Instrumente P/N 7-721.10 PIFOC (Auburn, MA) had a limited travel range of 100 μm but an excellent theoretical precision of 24.4 nm. A Model P-725 PIFOC can also be used for a longer range of 400 μm with the same resolution. The PIFOC are controlled by an E-662 LVPZT amplifier with built-in servo feedback control. A better alternative is a piezo-stage insert that can be used to directly move the specimen for focusing. This insert is preferable because the conventional PIFOC is screwed onto the objective turret of the microscope. There it elevates the objective lens 13 mm from the original position, significantly degrading DIC images because of the additional space between the objective lens and the Nomarski prism.

An alternative to the PIFOC is an electric stepper motor attachment with linear-encoded feedback (Ludl Electronic Products Ltd., Hawthorne, NY: P/N 99A420 and 99A404 for Olympus IX71 microscope), installed on the fine focus knob of the microscope. Using this motor preserves DIC image quality. However, the linear encoder feedback precision is only 50 to 100 nm. The advantage of this focus z motor is the much longer range of motion (the entire range of the original focus travel of the microscope). One can use both the PIFOC and the z-axis motor on the same system for a microscope that has both the ideal range of the motor and the higher resolution and good repeatability of the PIFOC.

The motorized stage is used to translate the specimen in the x–y plane for scanning. We use a stepper motor stage from Ludl Electronic Products Inc. (P/N 99S108-02) for the inverted microscope. The stage has a linear encoder of 100-nm resolution for each axis and is capable of attaching a piezo-electric transducer add-on for vertical travel of the specimen. The stage motors are driven by a National Instruments MID-7604 power drive (National Instruments Corp., Austin, TX) and are interfaced with the computer.

Autofocus calculations are carried out by software on the computer. Therefore, focus speed increases as computer processing speed increases. Advances in microcomputers have made them powerful enough to carry out the calculation without resorting to the dedicated image processor hardware used in the past (Price and Gough, 1994). We use a Pentium Xeon dual 2.4-GHz CPU computer (model Precision 650, Dell, Round Rock, TX). During a scan, many processes are carried out by the computer, such as motion control of the stage and piezo device, timing, and computation of focus variables. Therefore, a high CPU clock speed is preferred. Parallelism such as multiple CPUs, dual-core CPUs, and Hyperthreading (HT) technology can provide an additional edge in multithreading these operations.

As other cameras and modules of the microscope require expansion slots, the host computer should be selected to have as many built-in PCI or PCI-X slots as possible. Our computer uses a PCI expansion slot on its main board for the image acquisition board. It uses another PCI expansion slot for an interface board (National Instruments Corp., Model PCI-7358) for the motor power drive. The same board also performs 16-bit digital-to-analog (D/A) conversion to control the piezo device by outputting a 0- to 10-V direct current (DC) voltage. The computer also has two IEEE 1394a ports for additional digital cameras.

The digital camera is interfaced with the computer to capture images for autofocus. The camera should have high bit depth, for example, 12-bit or greater, and be used for both autofocus and subsequent fluorescence image acquisition. In our system we have a Photometrics CoolSNAP ES monochrome camera (Roper Scientific, Inc., Tucson, AZ). This is a scientific grade camera that has 1392×1040 pixels in the CCD array. It has a fast 12-bit digitization rate of 20 MHz. Interfacing with the computer, we have a PVCAM PCI image grabber board (Roper Scientific, Inc). This camera coupled with the image grabber board can perform fast region-of-interest (ROI) acquisition for focusing on a specified region in the FOV. It also has an electronic shutter instead of a mechanical shutter for faster repeated digital autofocus and integrated fluorescent image acquisition.

Software

The software carries out the autofocus calculation instructions. It also controls the hardware, that is, stage movement, piezo movement, light path shuttering, and image acquisition via the camera. These controls utilize library functions from the motion control and image acquisition software development kits (SDKs) provided by the hardware manufacturers (Roper Scientific and National Instruments Corp.).

The autofocus method is based on image spatial resolution. When an in-focus image becomes out of focus, its two-dimensional spatial resolution decreases. The loss of resolution is manifested in decreased high-frequency content of the out-of-focus image. Our autofocus routine measures the high-frequency content at each plane of the image to determine optimal focus. High-frequency information is first separated from the full frequency spectrum of the image by applying a high-pass or band-pass filter. The filters can be implemented either with analog electronic devices or digitally with a computer. Digital filters are easier to implement than manufacturing electronic hardware boards (Bravo-Zanoguera *et al.*, 1998). C/C++ libraries such as Intel Integrated Performance Primitive 5.0 (Intel Inc., Santa Clara, CA) and IPL 98 (University of Denmark, Denmark) can be used to perform the filtering and calculations. The downloadable libraries are available on the Internet. High-level programming languages such as LabView 7.0 (National Instruments Corp.) and MATLAB 7.0 (Mathworks Inc., Natick, MA) are easier to use, but slower.

For each field, the micropositioner incrementally changes the distance between the specimen and the objective lens to search for the best focus position. An image is acquired into memory for each axial z position. Digital image processing is performed on each image to obtain its "focus index." A set of focus indices is then computed for each focus attempt, and these are fit to a curve to find the optimum focus. The focus measurement function used to compute the focus index for an image is

$$F(z) = \frac{\sum_x \sum_y [f(x, y) \otimes i_z(x, y)]^2}{\left[\sum_x \sum_y i_z(x, y)\right]^2}, \tag{1}$$

where (x, y) is the discrete coordinate in the image and $i_z(x,y)$ is the image pixel intensity acquired at each focus position, z, and $f(x,y)$ is a high-pass or band-pass filter. $\otimes$ stands for the convolution operator.

There are different choices for the digital filter, $f(x, y)$. Integer filter kernels are faster to compute than floating point. Custom-designed kernels usually do not use integers. Simple integer filters such as the one-dimensional derivative filter

$$[-1 \quad 0 \quad 1]$$

and the basic high-pass filter

$$\begin{bmatrix} -1 & -2 & -1 \\ -2 & 12 & -2 \\ -1 & -2 & -1 \end{bmatrix}$$

can be sufficient (Oliva *et al.*, 1999; Price and Gough, 1994). Specialized, custom-designed filters such as the 31-tap floating-point kernel in the Appendix can also be used to fight contrast reversal (Oliva *et al.*, 1999).

By going through each axial position in the specimen, a set of focus indices is obtained to calculate the best focus position by fitting a curve over the indices, as shown in Fig. 2A. Thin specimens produce an ideal curve that can be fit readily to a Gaussian or polynomial function. However, when thicker specimens have discernible objects in different planes, multiple peaks can be produced in the curve, as shown in Fig. 2B. This is more frequently a problem in images taken using high NA lenses because the DOF is much less than the cell thickness. In multimodal curves, the

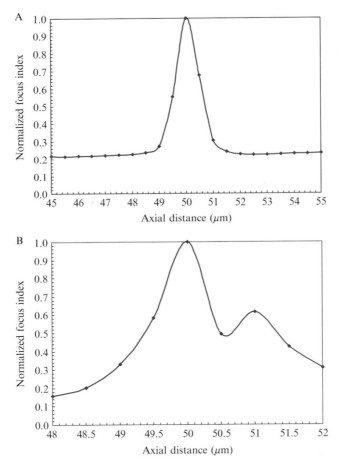

FIG. 2. Focus function curves from fixed fibroblasts. A curve from a thin specimen (A) is contrasted with that from a thicker specimen (B) in which more than one plane of focus provides discernible contrast. (Image a: 21 axial positions in the 10-μm search range, objective = Olympus PlanApo N TIRFM 60×/1.45 NA. Image b: 9 axial positions in the 4-μm search range, objective = Olympus UPlanFL N 40×/1.30 NA. Focus indices are normalized to one and connected using a smoothed line.)

highest peak represents the intracellular layer that contains the most detail. Curve fitting such data using a predefined function will fail. A weighted-average approach must be applied to counter this problem (Price and Gough, 1994).

The best focus position with power-weighted average is calculated as

$$B = \frac{\sum zF(z)^m}{\sum F(z)}, \tag{2}$$

where $F(z)$ is shown in Eq. (1) and m is an integer. The flowchart in Fig. 3 summarizes the process of finding the best focus z position. C++ code for the class member function "incremFocus" is listed in the Appendix.

When focusing a FOV, the in-focus position of the previous field is taken to be the middle axial position, on the assumption that the substrate supporting the cells does not change height abruptly. The very first field must be focused manually to set the starting in-focus point within the set focus range. The focus region of interest (ROI) is set and memory is allocated. The specimen is moved to the bottom of the focus range. From bottom up, a "for" loop is carried out to obtain a sequence of focus indices. In each step through the axial sampling, an image with pixel values of data type "int" is acquired. The pixel values must be scaled down to an 8-bit integer in order to prevent data overflow in calculations of the summations in Eq. (1). [Depending on the operating system, an "int" data type has different bit depths. For Windows 2000 and Windows XP, it is 32 bits (4 bytes) and has a range of 2^{32}. If 10- or 12-bit integers are used, the numerator and denominator in Eq. (1) can be greater than 2^{32} and cause computation errors.] The total illumination of the image is computed by summing all pixel grayscale values. Then the image is convolved with a filter kernel representing the high-pass or band-pass filter. The sum of the squared values of each element is then calculated for the convolution output array. Finally, the resultant sum is divided by the square of the total illumination to give the focus index. After all the planes are covered in the axial direction, memory for the image and its convolved outputs can be deallocated and a power-weighted average [Eq. (2)] is used to interpolate for the best focus position.

Viscoelasticity

When an oil or water immersion lens is used, the immersion medium between the specimen and the objective lens can hinder autofocus. This is because the specimen, the immersion medium, and the objective lens form an assembly that has prominent viscoelastic properties. When the objective lens is suddenly moved toward the specimen, the immersion medium is

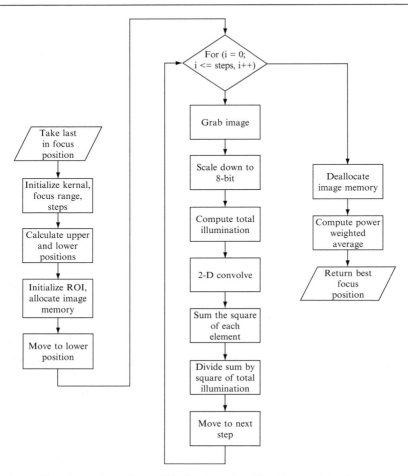

Fig. 3. Flowchart of autofocus. This flowchart describes the algorithm for carrying out digital autofocus of a single field. The function in the Appendix shows an example of the implementation. It can be called repeatedly to perform autofocus of all fields in a scan.

pushed toward the specimen. Because the medium is incompressible, the specimen is pushed away from the objective and the spring mechanism in the objective is pushed away from the specimen. The specimen and the objective spring are elastic and yield as the medium flows to the sides, and the system reaches a new state of equilibrium. In the case of an air objective lens, the viscosity of the medium (air) is so little that the speed of reaching equilibration is limited only by the hardware mechanics. When oil is present, its viscosity dictates the focus speed. Therefore, image acquisition must be delayed after each step during the focus attempt for reliable focusing. Different immersion oils can be used so the user can opt to maximize speed (using low viscosity) or resolution (higher viscosity). The fluorescence of the oil should be checked at the wavelengths of interest so it does not interfere with the experiments. We use Olympus immersion oil (Olympus America, Inc., Melville, NY), which has low viscosity (135 cStoke) at 23° and an index of refraction of 1.516 that matches normal glass coverslips. Using this oil in our system, we set a delay of 25 to 50 ms to achieve fast autofocus (<0.8 s per FOV) with minimal autofluorescence.

Multiple-Field Scans

The automated microscope can perform a time-lapse scan of either a predesignated area encompassing multiple adjacent fields (area scan) or a set of predesignated fields of interest that may be separated from one another (point scan). Different applications may also call for obtaining several images at each FOV before moving to the next FOV (time priority) or obtaining a single image at each FOV while cycling between fields (space priority). The autofocus search range will be different for each of these imaging modes. In an area scan, the specimen support can be assumed to be flat enough so that heights of consecutive fields vary little. Therefore, the best focus position calculated from the previous FOV can be used as the midpoint z position in the autofocus search range of the current field. Steps 1 to 3 in Fig. 3 describe this method. For a $40\times$ 1.30 NA lens, the autofocus search range can be as little as 1 μm. In a point scan, in-focus heights of different x–y locations may vary widely depending on how far a field is away from the previous field. Therefore, either the search range should be increased to accommodate all the fields or the user should log the in-focus positions for all the fields before the scan starts. If oil-immersion medium is used, enough oil should be applied to the scanned area so that the surface does not run dry, despite the sloughing of the oil caused by scanning.

Conclusion

This chapter summarizes the components and algorithm required to assemble a fully software-based autofocus system for automated microscopy screening. All components are off the shelf and are readily available. Software that carries out the autofocus can be written in any language that supports computer hardware data input/output and can resemble that listed in the Appendix. As modern microcomputers have become faster, as well as less costly, digital autofocus has become a solution that can be implemented easily. The autofocus module setup outlined here will repeatably produce sharp focus, despite the perturbations that impact scanning during microscopy-based high-content screening.

Appendix

What follows is the core of the autofocus, a member function "increm-Focus" of the "CAutofocus" class written in C++. It is an implementation of the flowchart shown in Fig. 3. National Instruments' "flexMotion" and Intel's Image Processing Library (IPL) v2.5 C libraries are used.

```
double CAutofocus::incremFocus(CImageGrabber
&grabOne, double cur_pos, double range, short steps)
    {
    float H[31]={0.00701,
    -0.00120, 0.00185, -0.01265, -0.01211, 0.08346,
    -0.04688, -0.18633, 0.27488, 0.13864, -0.58840,
    0.22454, 0.66996, -0.74667, -0.30163, 1.00000,
    -0.30163, -0.74667, 0.66996, 0.22454, -0.58840,
    0.13864, 0.27488, -0.18633, -0.04688, 0.08346,
    -0.01211, -0.01265, 0.00185, -0.00120, 0.00701};
    //int H[9]={-1, -2, -1, -2, 12, -2, -1, -2, -1};
    //int H[3]={1, 0, -1};
    IplConvKernelFP* pIplConvKern;
    double *FI;
    FI = new double [steps+1];
    double normConv, normIllum, powered, upper_pos,
lower_pos, pos;
    double maxFI=0, numerator=0.0, denominator=0.0;
    IPLStatus sta;
    int status;
    /* setup focus step locations */
    upper_pos = cur_pos + range/2;
    lower_pos = upper_pos - range;
    pIplConvKern = iplCreateConvKernelFP(31,1,15,0,
H);
    /* init ********************/
    grabOne.SetAFParams600,600,8;// center region
of field of view with 1x1 binning, low intg, 1x gain.
    IplImage* pInImg = iplCreateImageHeader(1,0,
IPL_DEPTH_16U,''GRAY'',''G'',IPL_DATA_
ORDER_PIXEL,
```

```
  IPL_ORIGIN_TL,IPL_ALIGN_QWORD,grabOne.
imgWidth,grabOne.imgHeight,NULL,NULL,NULL,NULL);
  iplAllocateImage(pInImg,0,0);
  IplImage* pSImg = iplCreateImageHeader(1,0,
IPL_DEPTH_32F,''
GRAY'',''G'',IPL_DATA_ORDER_PIXEL,
  IPL_ORIGIN_TL,IPL_ALIGN_QWORD,grabOne.
imgWidth,grabOne.imgHeight,NULL,NULL,NULL,NULL);
  iplAllocateImageFP(pSImg,0,0);
  IplImage* pMImg = iplCreateImageHeader(1,0,
IPL_DEPTH_32F,
''GRAY'',''G'',IPL_DATA_ORDER_PIXEL,
  IPL_ORIGIN_TL,IPL_ALIGN_QWORD,grabOne.
imgWidth,grabOne.imgHeight,NULL,NULL,NULL,NULL);
  iplAllocateImageFP(pMImg,0,0);
  /* end init ******************/
  pos = lower_pos;
  status=flex_load_dac(BOARD_ID,DAC4,posToDstep
(pos),0xFF); // output to analog out
  Sleep(20);
  /* the incremental process to calculate focus
index for each z plane (step) */
  for (int i=0; i<=steps; i++)
  {
  pInImg->imageData = (char*)grabOne.Acquire;
  sta = iplScaleFP(pInImg, pMImg, 0, 4095); // scale
down from 16 bit to 8 bit
  normIllum = iplNorm(pMImg, NULL, IPL_L1);
  iplConvolve2DFP(pMImg, pSImg, &pIplConvKern,1,
IPL_SUM);
  iplSquare(pSImg, pMImg);
  normConv = iplNorm(pMImg, NULL, IPL_L1);
  FI[i]=normConv/(normIllum*normIllum);
  if (FI[i]>maxFI)
  maxFI=FI[i];
  pos = pos + range/(double)steps;
  status=flex_load_dac(BOARD_ID,DAC4,posToDstep
(pos),0xFF); // output to analog out
  Sleep(50);
  }
  iplDeallocate( pInImg, IPL_IMAGE_ALL );
  iplDeallocate( pMImg, IPL_IMAGE_ALL );
```

```
iplDeallocate( pSImg, IPL_IMAGE_ALL );
iplDeleteConvKernelFP( pIplConvKern );
grabOne.Complete;
/* power-weighted average for best focus position
interpolation */
ofstream out(''D:\\work\\focusin.txt'');
for ( int i=0; i<=steps; i++)
{
FI[i]=FI[i]/maxFI;
powered = pow(FI[i], 12);
denominator += powered;
numerator += i * powered;
out << FI[i] << ''\n'';
}
out.close;
Sleep(50);
pos = lower_pos + (range/(double)steps) *
(numerator/denominator);
status=flex_load_dac(BOARD_ID,DAC4,posToDstep
(pos),0xFF); // output to analog out
Sleep(50);
delete [] FI;
FI=NULL;
return pos;
}
```

References

Bajaj, S., Welsh, J. B., Leif, R. C., and Price, J. H. (2000). Ultra-rare-event detection performance of a custom scanning cytometer on a model preparation of fetal nRBCs. *Cytometry* **39,** 285–294.

Bravo-Zanoguera, M., Massenbach, B., Kellner, A., and Price, J. (1998). High-performance autofocus circuit for biological microscopy. *Rev. Sci. Instr.* **69,** 3966–3977.

Inoue, S., and Spring, K. R. (1997). "Video Microscopy, the Fundamentals." Plenum, New York.

Oliva, M. A., Bravo-Zanoguera, M., and Price, J. (1999). Filtering out contrast reversals for microscopy autofocus. *Appl. Opt.* **38,** 638–646.

Price, J. H., and Gough, D. A. (1994). Comparison of phase-contrast and fluorescence digital autofocus for scanning microscopy. *Cytometry* **16,** 283–297.

Shen, F., Hodgson, L., Rabinovich, A., Hahn, K., and Price, J. (2006). Functional proteometrics for cell migration. *Cytometry* **69,** 555–572.

[33]   Fluorescence Lifetime Imaging Microscopy: Two-Dimensional Distribution Measurement of Fluorescence Lifetime

By MASANOBU FUJIWARA and WILLIAM CIESLIK

Abstract

A newly developed fluorescence lifetime imaging microscopy (FLIM) system has combined the high-temporal resolution of a streak camera with the high-spatial resolution of a microscope to obtain a two-dimensional distribution of fluorescence lifetimes within living cells. The temporal resolution is as short as 20 ps. The effective field of view is 48×45 μm with 0.2 μm resolution using a $60\times$ water immersion objective. Image acquisition time is as short as 3 s per image. Measured and published values of lifetime for standard fluophores are shown with good agreement. Examples of FLIM and fluorescence resonance energy transfer images are presented.

Introduction

Fluorescence is the emission of a longer wavelength photon from a molecule as it returns to its ground state energy level after being excited by a shorter wavelength photon. If a short-pulsed light source is used for excitation, an exponential decay of the fluorescence intensity will be observed, as in Fig. 1. The decay of the fluorescence intensity over time is characteristic of each fluorescent molecule and is known as the fluorescence lifetime (t_F).

Unlike intensity-based measurements of fluorescence, the lifetime (t_F) is constant, irrespective of concentration, path length, and illumination variations within the sample. However, very subtle changes in the local environment of the molecule, such as pH, ion concentration, polarity, and viscosity, can produce changes in the lifetime. Molecular nonradiative energy transfers between fluorescent molecules will produce lifetime changes that are detected easily.

Fluorescence lifetime imaging microscopy (FLIM) extends this powerful technique into the microscopic domain. High-resolution imaging of intracellular structure can now include changes in the fluorescent lifetime of fluorescently labeled components. Structural, biochemical, and photodynamic processes can be observed in living cells, as shown in Fig. 2. Fluorescence resonance energy transfer (FRET) benefits from the high temporal resolution and speed of the system. Observation of

METHODS IN ENZYMOLOGY, VOL. 414   0076-6879/06 $35.00
          DOI: 10.1016/S0076-6879(06)14033-1

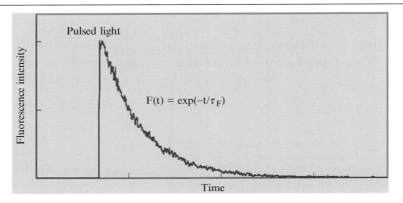

FIG. 1. Typical fluorescence decay curve.

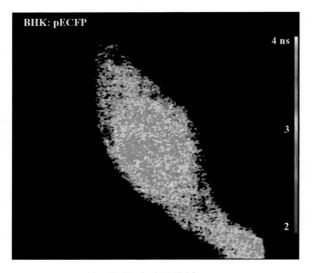

FIG. 2. Typical FLIM image.

protein–protein interactions in living cells, among other biological imaging applications, offers new possibilities in research.

Operating Principle of a Streak Camera

The streak camera is key to the FLIM system, which detects ultrafast light phenomena while converting it to spatial information in the form of a streak image. The operating principle is shown in Fig. 3.

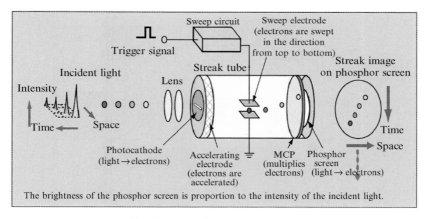

Fig. 3. Schematic of streak camera.

For illustration purposes, four pulses of light with varying intensity are separated in time and space and input to the streak camera. The pulses are projected onto the slit-shaped photocathode through a relay lens and are converted to electrons sequentially by a photocathode. The electrons are accelerated and deflected by a pair of high-voltage sweep electrodes. During the sweep, the electrons are deflected at slightly different angles depending on the arrival time at the electrodes. The electrons are then multiplied by a microchannel plate (MCP) and converted back to photons by a phosphor screen to form a streak image. The streak image displays four optical pulses across the vertical direction according to the arrival time to the streak camera. The earliest pulse is located at the uppermost position, and the latest pulse is located at the bottommost position. The spatial position of each light pulse is maintained in horizontal direction, as well as the light intensity, or number of photons within each pulse. A high-speed CCD camera is used to digitize and transfer the streak image to a PC computer for lifetime processing.

Conventional techniques for time-domain imaging such as multigate detection can provide a temporal resolution of a few hundred picoseconds. The time-correlated single photon counting technique can have a temporal resolution as short as tens of picoseconds; however, long acquisition times can prohibit applications, such as live cell imaging. The streak camera in the FLIM system provides 20 ps temporal resolution at multiple positions simultaneously, providing both high-temporal resolution and high-speed image acquisition.

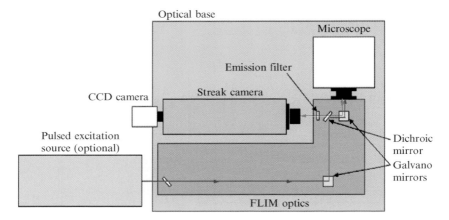

FIG. 4. Configuration of FLIM system.

Configuration of the FLIM System

The configuration of the FLIM system is shown in Fig. 4, including a pulsed excitation source, microscope, streak camera, and FLIM optics. A pulsed laser such as a mode-locked Ti:Sapphire or gain-switched diode laser is used for the excitation source. The FLIM optics and streak camera are attached to the side port of the microscope. The FLIM optics include two galvano mirrors that deflect the excitation light over the sample in two directions, both X and Y, while the streak camera captures the fluorescence emission that pass through the dichroic and emission filter. The CCD camera digitizes and transfers the streak image to a PC computer for processing.

Streak Image

A fluorescence intensity image (left side) and a typical streak image (right side) are shown in Fig. 5. For illustration purposes, a single line is extracted from the intensity image and displayed as a streak image. The vertical axis in the streak image represents the temporal content, or fluorescence decay data within this single line.

Measurement Principle of FLIM System

The measurement principle of the FLIM system is illustrated in Fig. 6. One galvano mirror scans the excitation light in the horizontal X direction and aligns with the horizontal input slit of the streak camera. The streak

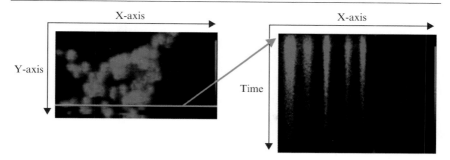

FIG. 5. Fluorescence intensity image and streak image.

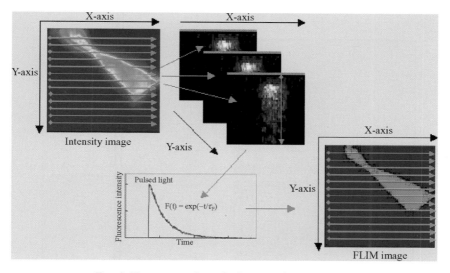

FIG. 6. Fluorescence intensity image and FLIM image.

camera captures the fluorescence light associated with each spatial position in this horizontal X axis. The second galvano mirror scans the excitation light in vertical Y direction and is synchronized with CCD readout. The scan rate per line is limited by the readout speed of the CCD camera, which in this case is 5 ms. A streak image stack is collected and stored in a PC memory buffer. The fluorescence lifetime is calculated from the exponential decay profile at each location along the X axis. The FLIM image is constructed by plotting the lifetime at each location along the horizontal X axis and vertical Y axis.

The effective field of view is 48×45 μm with 0.2 μm resolution, using a $60\times$ water immersion objective. The shortest image acquisition time is 3 s for full area with 600 horizontal lines. The temporal resolution is typically less than 20 ps for a 1-ns full timescale.

System Calibration

Standard fluophore solutions of various fluorescence dyes are used for the system calibration. Lifetime images and histograms are shown in Fig. 7: rhodamine 6G in ethanol (a and b), rose bengal in acetone (C and D), and rose bengal in ethanol (E and F). Solid lines in the histograms are Gaussian fits to data. Measured and published (in parentheses) values of lifetime are shown in the histograms, and there is a very good agreement among them.

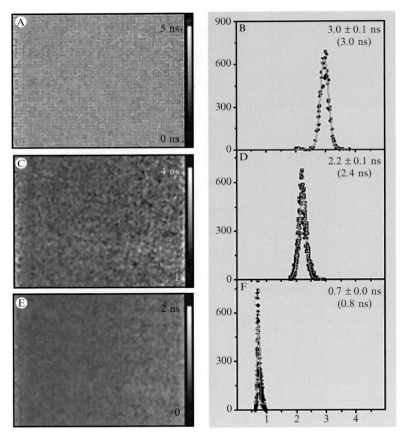

FIG. 7. FLIM image and histogram (Krishnan et al., 2003a).

Lifetime Imaging in Cells

Lifetime images of BHK cells expressing cytosololic ECFP (Cyto-ECFP) and fusion protein Bax-ECFP (Cyto-BaxECFP) are shown in Figs. 8 and 9 (Krishnan *et al.*, 2003b). The mean lifetime values of Cyto-ECFP and Cyto-Bax-ECFP are 3.2 ± 0.3 and 2.8 ± 0.3 ns, respectively.

Lifetime images of cells expressing mitochondrial targeted ECFP (Mito-ECFP) and membrane-targeted ECFP (Mem-ECFP) are shown in

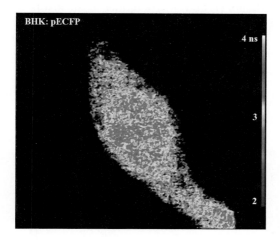

FIG. 8. Cyto-ECFP.

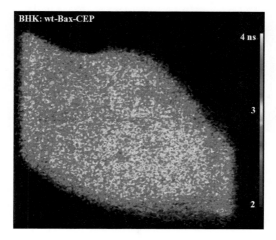

FIG. 9. Cyto-BaxECFP (Krishnan *et al.*, 2003b).

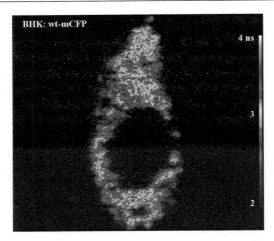

FIG. 10. Mito-ECFP (Krishnan *et al.*, 2003b).

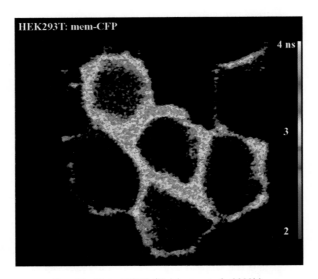

FIG. 11. Mem-ECFP (Krishnan *et al.*, 2003b).

Figs. 10 and 11 (Krishnan *et al.*, 2003b). Both mean lifetime values of Mito-ECFP and Mem-ECFP are 2.9 ± 0.1 ns.

FRET Imaging in Cells

Fluorescence resonance energy transfer is a distance-dependent non-radiative energy transfer interaction between two electronic excited states of fluorescent substances. Excitation energy is transferred from donor to acceptor fluorescent substances. Using this phenomenon, real-time imaging of protein structural dynamics, protein–protein interactions, enzyme activity, and structural change can be observed and quantified.

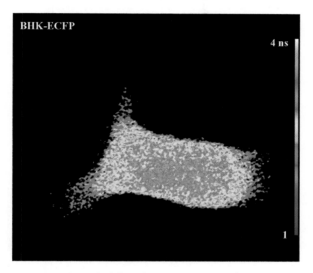

FIG. 12. ECFP (Herman *et al.*, 2003).

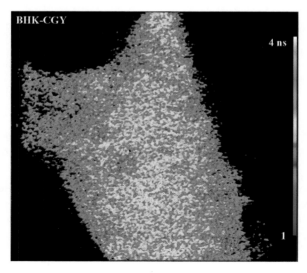

FIG. 13. CGY (Herman *et al.*, 2003).

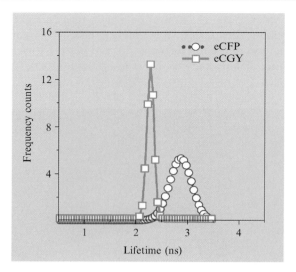

FIG. 14. Histogram (Herman *et al.*, 2003).

Lifetime images of BHK cells expressing ECFP, CFP-polyglycine-YFP, and the histogram of their lifetimes are shown in Figs. 12, 13, and 14, respectively (Herman *et al.*, 2003). The mean lifetime of ECFP is 2.9 ns, whereas that of CGY is 2.3 ns. The decrease in mean lifetime of CGY compared to that of ECFP is due to the FRET process (energy transfer between CFP site and YFP site).

Acknowledgments

I thank Mr. Takashi Ito and Ms. Ayumi Ishima from Hamamatsu Photonics K. K. for providing useful information. This work is supported by Dr. R. V. Krishnan and Dr. Brian Herman, Department of Cellular and Structural Biology, University of Texas Health Science Center.

References

Herman, B., Krishnan, R. V., Zhang, J., Masuda, A., Lopez-Cruzan, M., Zhang, Y., and Centonze, V. E. (2003). FRET and FLIM microscopy in the study of the role of oxidative stress and apoptosis in aging. *Microsc. Microanal.* **9**(Suppl. 2), 226–227.

Krishnan, R. V., Biener, E., Zhang, J.-H., Heckel, R., and Herman, B. (2003b). Probing subtle fluorescence dynamics in cellular proteins by streak camera based fluorescence lifetime imaging. *Appl. Phys. Lett.* **83**, 4558–4660.

Krishnan, R. V., Saitoh, H., Terada, H., Centonze, V. E., and Herman, B. (2003a). Development of multiphoton fluorescence lifetime imaging microscopy (FLIM) system using a streak camera. *Rev. Sci. Instrum.* **74**, 2714–2721.

Author Index

A

Abe, M., 101
Abe, T., 22
AbelMalik, P., 160
Abney, J. R., 163
Abraham, V. C., 229, 360, 464, 484, 605
Abramson, R. D., 534
Abruzzo, L. V., 318, 319
Acharya, M., 22, 23
Adachi, M., 38
Adair, R., 141
Adam, L., 168
Adams, C. L., 230, 440, 441, 461, 485, 497, 504
Adams, D. R., 516, 518
Adams, J. A., 302
Addison, C. L., 249
Adesnik, H., 160, 165
Adie, E. J., 263
Adjei, A. A., 153
Adkins, W. N., 141
Afendis, S. J., 156
Agami, R., 522, 535, 536
Agard, D., 234
Agarwal, N., 318
Agbaria, R., 580
Aggarwal, B. B., 267
Agoulnik, I. U., 200, 203
Aguilera-Moreno, A., 151, 155, 157
Ahn, S., 118
Ahnert-Hilger, G., 157
Ai, Y., 364, 367, 368, 385, 418, 420
Aikawa, Y., 22
Akashi, K., 22
Akong, M., 164, 178, 189, 201, 301
Alanine, A. I., 248
Albagli, O., 22
Albrecht-Buehler, G., 243
Albrizio, S., 101
Aleem, E., 1
Alexandrov, Y., 365, 420
Alivisatos, A. P., 212
Allal, C., 130

Allan, C., 179, 273, 608
Alldred, M., 159
Allen, K. M., 141
Allgrove, J., 141
Allman, D. R., 252
Almholt, D. L., 121, 122, 365, 386, 390, 420, 436, 437
Alouani, S., 125
Aloy, P., 602
Altenberg, G., 319
Altschul, S. F., 534
Altschuler, S. J., 189, 230, 442, 465
Alvarez, R., 517
Alvarez-Barrientos, A., 153
Amazit, L., 188
Ambrose, C. M., 153, 154, 158
Ambs, S., 601
Ames, R., 113
Amos, W. B., 2
An, S., 157, 243
Anbazhagan, R., 22, 23
Anderson, L., 485
Anderson, N. L., 465
Andexinger, S., 126
Andrade-Gordon, P., 252
Andrew, D., 141
Andrews, C., 349
Andrews, D. T., 447, 609
Angeline, A., 605, 609
Anisowicz, A., 141
Ansanay, H., 168
Antony, S., 2
Aoki, C., 151
Aota, S., 235, 243
Apolloni, A., 154
Apte, S., 141
Araujo, R. P., 609
Archer, T. K., 43
Arendt, T., 2
Arguin, C., 122
Arkhammar, P. O., 3, 248, 365, 420
Arlotta, P., 343
Armknecht, S., 229

643

E

W

Subject Index

A

Acumen Explorer, *see* Laser-scanning microplate cytometry

Adenoviral sensors
  comparison with other viral vectors, 249–250
  design and construction, 249, 251–252
  glucocorticoid receptor–green fluorescent protein translocation sensor
    image acquisition and analysis, 258
    incubation conditions, 257–258
    materials, 257
    overview, 255–257
  high-content analysis
    applications, 254–255
    requirements, 248
  nuclear factor of activated T cells–nitroreductase reporter sensor
    image acquisition and analysis, 260–261
    incubation conditions, 260
    materials, 260
  prospects, 262–263
  stable transfectant limitations, 248–249
  STAT studies, 263
  transduction, 249, 251–252
  validation, 254

Adhesion, *see* Cell adhesion

β_2-Adrenergic receptor, *see* G protein-coupled receptors

Akt, Forkhead Redistribution assay
  Akt inhibition and nuclear translocation, 522
  isoform-specific knockdown, 522, 525
  small-interfering RNA transfection, 525–526

Androgen receptor
  function, 198, 200
  high-throughput microscopy
    overview, 198, 200
    translocation and nuclear variance assays, 200–201, 203
  pathology, 200

Apoptosis, multiplexed automated fluorescence microscopy assays
  caspase-3 cleavage detection, 293
  cell cycle/apoptosis assay multiplexing, 296, 298–299
  morphology change detection, 295
  overview, 292–293
  poly(ADP)-ribose polymerase cleavage detection, 293–294
  TUNEL assay, 294–295, 298–299

AR, *see* Androgen receptor

ArrayScan, *see* Extracellular signal-regulated kinases; Jun N-terminal kinase; Nuclear factor-κB; p38 mitogen-activated protein kinase

Arrestin, *see* G protein-coupled receptors

ATPase, plate::explorer assays, 595–597

Autofocus, *see* Digital autofocus

Automated fluorescence microscopy
  adenoviral sensors, *see* Adenoviral sensors
  calcium imaging, *see* Calcium flux
  cell cycle sensors, *see* Cell cycle
  cellular phenotypes for compound classification, *see* Cytometrix technologies
  confocal microscopy limitations in high-content screening, 435
  epigenetic modulator screening, *see* Epigenetic modulators
  functional genomic libraries, *see* Genomic library high-content screening
  G protein-coupled receptor screening, *see* G protein-coupled receptors
  high-content screening versus high-throughput screening, 122
  high-resolution light microscopy, *see* Automatic high-resolution light microscopy
  high-throughput microscopy overview, 188–189
  inflammation studies, *see* Inflammatory response